COLLEGE ALGEBRA AND TRIGONOMETRY

COLLEGE ALGEBRA AND TRIGONOMETRY

Robert D. Bechtel
Arthur A. Finco
Robert B. Kane
PURDUE UNIVERSITY

D. Van Nostrand Company
NEW YORK CINCINNATI TORONTO LONDON MELBOURNE

D. Van Nostrand Company Regional Offices:
New York Cincinnati Millbrae

D. Van Nostrand Company International Offices:
London Toronto Melbourne

Library of Congress Catalog Card Number 73-15297
ISBN: 0-442-24264-6

Published by D. Van Nostrand Company
450 West 33rd Street, New York, N.Y. 10001

Published simultaneously in Canada by
Van Nostrand Reinhold Ltd.

10 9 8 7 6 5 4 3 2 1

PREFACE

College Algebra and Trigonometry covers the topics usually treated in precalculus freshman mathematics courses. The text is written both for students who wish to enter occupational fields which apply algebra and trigonometry and for students who wish to continue in one of the scientific disciplines.

The book presents the basic elements of algebraic systems—real numbers, polynomials, and complex numbers—as early as possible, and then develops their properties, largely in the framework of the theory of equations. No attempt is made to prove or have the students prove all of the results. Rather, an understanding and a facility to use the principles and results are emphasized.

The first chapter is a brief treatment of real number properties and establishes much of the notation for the text. Polynomial and rational expressions introduced in Chapter 2 are viewed as elements in mathematical systems in which one performs operations within a structure. Next follow four chapters which develop results concerning real numbers and polynomials: Chapter 3 on linear equations and inequalities; Chapter 4 on exponents and radicals; Chapter 5 on quadratic equations and inequalities; and Chapter 6 on polynomial and rational functions. Complex numbers are presented in the next two chapters, as well as in Chapter 15 after the trigonometric topics have been established. Chapter 9 treats exponential and logarithmic functions, and the optional (but useful) Chapter 10 treats elementary linear algebraic topics. Chapters 11 through 14 contain the main body of the trigonometry portion of the book, treating successively the trigonometric functions and their graphs, trigonometric equations, and the trigonometry of triangles. Finally, Chapter 16 is an introductory treatment of the useful concepts of sequences and series. The book uses over 200 illustrations in the development of appropriate concepts, thereby incorporating elementary concepts of analytic geometry throughout the text.

This text was written under the assumption that (1) an instructor would be available to the student on some basis and (2) the student should read the material as well as solve the exercises. The self-test paragraphs (which appear as "boxes") are intended to aid the student in periodically assessing his reading comprehension and understanding of the material being studied; the answers appear in the margin so as not to interrupt the student's reading. The number and variety of exercises make it possible to tailor assignments to the needs of the class; in general, exercises for a section are ordered according to difficulty, and selected answers are found at the back of the book. The development in the text is mathematically sound, but the emphasis is *not* on theory, special topics, or theorem proving. These considerations are left for the more advanced junior-senior courses.

The book is suitable for a one-semester course of five credit hours, or a two-course sequence of two or three semester hours per course. Similar suitability applies to quarter-hour analogs of these courses. The text can be adapted to many shorter courses, in which elementary topics, especially in algebra, can be reviewed either quickly or as needed. The comprehensive treatment provides the review and reference material often needed in these shorter courses. A semiprogrammed student study guide is available as a supplementary publication

Robert D. Bechtel
Arthur A. Finco
Robert B. Kane

CONTENTS

1. THE REAL NUMBER SYSTEM

1-1 SUBSETS OF THE REAL NUMBERS

The set of real numbers, which we denote by R, underlies elementary mathematics. There are precisely enough real numbers to match each one with a point on a line and precisely enough points on a line to match each point with a real number. This one-to-one correspondence of real numbers and the points on a line yields what is referred to as the real number line. A few points of the real number line are labeled in Figure 1.

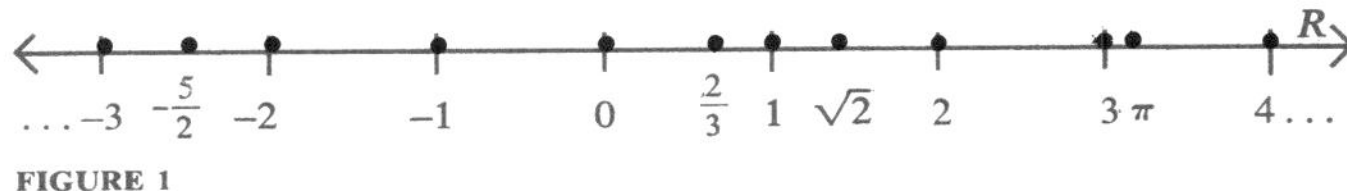

FIGURE 1

The notation $a \in R$ means that a is an element or member of the set R. Consider $P = \{1, \sqrt{2}, 4\}$. Since every member of P is also a real number (a member of R) we say that P is a subset of R and write $P \subset R$.

Definition A set A is a **subset** of a set B (denoted $A \subset B$) if and only if each element of A is an element of B.

You have studied certain subsets of the real numbers before, such as the natural (counting) numbers, $N = \{1, 2, 3, \ldots\}$; the whole numbers, $W = \{0, 1, 2, 3, \ldots\}$; the integers, $I = \{\ldots, -2, -1, 0, 1, 2, \ldots\}$; and the rational numbers, Q, composed of those members of R which can be represented by $\frac{a}{b}$, where a and b are integers and where $b \neq 0$. There are points on the real number line that do not correspond to rational numbers. The positive number whose square is 2, $\sqrt{2}$, is not a rational number. The circumference of a circle with unit diameter, denoted by π, is not a rational number. There are many such numbers, which we call the irrational numbers—those real numbers which are not rational numbers. These sets are diagramed in Figure 2. The shaded region represents the set of irra-

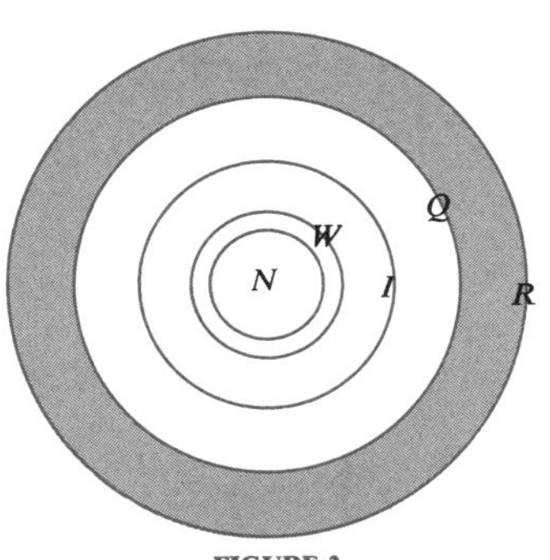

FIGURE 2

tional numbers. Note that N is a subset of W. In fact:

$$N \subset W \subset I \subset Q \subset R$$

The subset of integers, $\{-2, -1, 0, +1, +2, +3, \ldots\}$, may be written $\{x \in I\colon x > -3\}$. The colon is read "such that" and the expression is read, "the set of all integers, x, such that x is greater than negative three." Since we will be referring to real numbers so often we replace "$x \in R$" by "x." Thus $\{x \in R\colon x > -3\}$ and $\{x\colon x > -3\}$ both name the set of all real numbers, x, such that $x > -3$.

The real numbers together with addition, multiplication and the basic properties of these operations form the *real number system*. You have probably studied the basic properties before. We list them here for your review.

Closure Properties: For all a and b, there exists a unique corresponding element $a + b$ in R called the sum of a and b and a unique corresponding element $a \cdot b$ in R called the product of a and b.

Commutative Properties: For all a and b, $a + b = b + a$ and $a \cdot b = b \cdot a$.

Associative Properties: For all a, b, and c, $(a + b) + c = a + (b + c)$ and $(a \cdot b) \cdot c = a \cdot (b \cdot c)$.

Identity Properties: There exists $0 \in R$ such that $a + 0 = 0 + a = a$ for every a. There exists $1 \in R$ such that $1 \neq 0$, and $a \cdot 1 = 1 \cdot a = a$ for every a. The element 0 is called the **identity for addition** or the **additive identity element**; 1 is called the **identity for multiplication** or the **multiplicative identity element.**

Inverse Properties: For each a, there exists a unique real number, $-a$, such that $a + (-a) = 0$ and if $a \neq 0$ there exists a unique real number a^{-1}, such that $a \cdot a^{-1} = 1$. The element $(-a)$ is called the **additive inverse** of a and the element a^{-1} is called the **multiplicative inverse** of a.

Distributive Property for Multiplication over Addition: For all a, b, and c, $a(b + c) = ab + ac$.

There is a **completeness property**, which, when added to this list would insure that the system we are describing is precisely the real number system. However, a careful statement of a completeness property is beyond the scope of this book. We will only mention that this additional property insures 1-to-1 correspondence between the set of real numbers and the points on a line.

In discussing the real number system we have used the equality relation, $=$, for numbers in R. If a and $b \in R$, then $a = b$ means a is the same number as b. Hence, $\frac{1}{2} = 0.5$, $4 = 4$, and $1 + 1 = 2$. Note that the equality relation is

reflexive: $a = a$
symmetric: If $a = b$, then $b = a$
and transitive: If $a = b$ and $b = c$, then $a = c$.

Exercises

1. Classify each number according to set membership in N, W, I, Q, the set of irrational real numbers, and R.

   a. $\frac{1}{2}$ b. $\frac{\pi}{2}$ c. $3 - 3$ d. -1

   e. $2 - \sqrt{2}$ f. 9 g. $\frac{-4}{2}$ h. $0.66666\ldots$

   i. $\sqrt{2} \cdot \sqrt{3}$ j. $3\pi \cdot 0$ k. $0.123\overline{123}$ l. $\frac{1}{3}\pi(\sqrt{2})^2$

Which of the real numbers in Exercises 2–7 are irrational?

2. The length of the diagonal of a square which has a side of length 1.
3. The area of a circular region, if the radius of the circle has length 1.
4. The perimeter of a rectangle, if the lengths of two adjacent sides are π and $9 - \pi$.
5. The volume of a spherical region, if the radius of the sphere has length 2.
6. The area of a square, if the length of a side is $\sqrt{2}$.
7. The circumference of a circle, if the length of the radius is $1 \div \pi$.

Exhibit a 1-to-1 correspondence between the sets in Exercises 8–11.

8. N and $\{x \in N$: x is an odd number$\}$
9. N and $\{x \in I\colon\ x \notin W\}$
   ($x \notin W$ means that x is *not* a member of W.)
10. N and W
11. N and I
12. Which basic property is illustrated by each statement?

    a. $17 + 94 = 94 + 17$ b. $\pi \cdot 1 = \pi$

    c. $\sqrt{3} \cdot \frac{1}{\sqrt{3}} = 1$ d. $2(9 + 1) = 18 + 2$

    e. $\sqrt{2} + (-\sqrt{2}) = 0$ f. $7 \cdot 50 = 35 \cdot 10$

    g. $\sqrt{2} + 0 = \sqrt{2}$ h. $70 + 5 = 60 + 15$

    i. $\pi \cdot 2 = 2 \cdot \pi$ j. $1 \cdot \frac{\pi}{2} = \frac{\pi}{2}$

Find the sum for each pair in Exercises 13–17.

13. 43, −79; −482, −713; −82, 91.
14. $-\frac{3}{7}, \frac{2}{5}$; $-\frac{7}{3}, -\frac{5}{2}$; $-2.304, 0.0278$.
15. π, π; $\sqrt{2}, \sqrt{2}$; $\frac{\pi}{2}, \frac{\pi}{2}$; 0, 9; 1, 9.

16. 49, −72; −72, 49; $3\frac{1}{9}$, $-4\frac{3}{5}$; $-\left(4\frac{3}{5}\right)$, $3\frac{1}{9}$.

17. 5, 7 + 3; 5 + 7, 3; $-\pi$, π; $\sqrt{2}$, $-\sqrt{2}$.

Find the product for each pair in Exercises 18-20.

18. Exercise 13.

19. Exercise 14.

20. $(5,\ 7 \cdot 3)$, $(5 \cdot 7,\ 3)$, $\left(\frac{2}{3}, \frac{3}{2}\right)$, $\left(\frac{\pi}{2}, \frac{2}{\pi}\right)$.

1-2 ELEMENTARY THEOREMS

Using the basic properties allows us to prove theorems about the real number system. Sometimes we will include proofs, sometimes we will leave the proof as an exercise and sometimes theorems with which you are familiar from earlier work or whose proofs are tedious or difficult will be presented without reference to their proofs.

First we list some theorems which you may recall from earlier work in algebra.

Theorem 1 For all real numbers a, b and c:

(i) If $a = b$, then $a + c = b + c$

(ii) If $a = b$, then $ac = bc$.

Theorem 2 The product of 0 and any real number is 0, and if the product of two numbers is 0, then at least one of the numbers is 0. Thus, $ab = 0$ if and only if $a = 0$ or $b = 0$.

Subtraction and division are based on the operations of addition and multiplication and are called the **inverse operations** for addition and multiplication respectively. Thus $a - b = c$ and c uniquely satisfies $c + b = a$. Similarly $a \div b = c$ provided $c \in R$ and c uniquely satisfies $c \cdot b = a$. The method of dividing and multiplying by the reciprocal of the divisor is justified by Theorem 3.

Theorem 3 For $b \neq 0$, $a \div b = ab^{-1}$.

In particular if $a = 1$, then $\frac{1}{b} = b^{-1}$. Note the use of the alternative notation $\frac{a}{b}$ for $a \div b$. Since division by zero leads to incongruities, we stipulate that

$b \neq 0$. For example, suppose $\frac{5}{0} = q$. Then $5 = q \cdot 0$. But $q \cdot 0 = 0$ and $5 \neq 0$. Now suppose $\frac{0}{0} = q$. Then $0 = q \cdot 0$. Since every real number q satisfies this equation, $\frac{0}{0}$ does not denote a unique real number.

We use the symbol "-" to represent different ideas. For example, in the phrase $\pi - (-\pi)$, the first "-" sign indicates *subtraction* whereas the second "-" indicates that $-\pi$ is the *additive inverse* of π or that $-\pi$ is the real number, *negative pi*. If a represents any nonzero real number, it is impossible to tell if its additive inverse, $-a$, is positive or negative. However, if we specify that a is positive, then its additive inverse, $-a$, is negative. If a is negative, its additive inverse, $-a$, is the positive real number $-a$.

What number is the additive inverse of each element in $\{3, 7, \pi, 200, 0\}$? In $\{-5, -\sqrt{2}, -43, -400\}$?

$\{-3, -7, -\pi, -200, 0\}$ • $\{5, \sqrt{2}, 43, 400\}$

The real number -4 has the property that $-4 + 4 = 0$. Similarly, $-(-4) + (-4) = 0$. Thus both 4 and $-(-4)$ are inverses of -4. Since additive inverses are unique $-(-4) = 4$. Theorem 4 generalizes this result.

Theorem 4 For every a, $-(-a) = a$.

Theorem 5 For every a and b, $-(a + b) = (-a) + (-b)$.

Theorem 5 states that the additive inverse of a sum is the sum of the additive inverses of each addend. This provides a method for adding two negative real numbers such as -5 and -9:

$$(-5) + (-9) = -(5 + 9) = -14.$$

Identify $(-b)$, $a + (-b)$, and $a - b$ if $a = -4$, $b = 3$; if $a = 4$, $b = 7$; if $a = \pi$, $b = -2\pi$.

a	b	$-b$	$a + (-b)$	$a - b$
-4	3	-3	-7	-7
4	7	-7	-3	-3
π	-2π	2π	3π	3π

Examine the results listed for $a + (-b)$ and $a - b$ in the self test exercise above. Theorem 6 generalizes these results.

Theorem 6 For every a and b, $a - b = a + (-b)$.

The next theorem summarizes several computational patterns used in multiplying real numbers.

Theorem 7 For every a and b:

(i) $(-a)b = -(ab)$

(ii) $a(-b) = -(ab)$

(iii) $(-a)(-b) = ab$.

Notice that if $a = 1$, then $(-a)b = (-1)b = -b$. This fact is used extensively in computing.

We often use fractions to denote real numbers. Patterns used to compute with rational numbers are valid for real numbers. They are collected here for your reference.

Theorem 8 For $b \neq 0$, $\frac{-a}{b} = \frac{a}{-b} = -\left(\frac{a}{b}\right)$ and $\frac{-a}{-b} = \frac{a}{b}$. For $a \neq 0, b \neq 0$, $\frac{1}{\frac{a}{b}} = \frac{b}{a}$.

$\frac{1}{2} = -\frac{-1}{2} = -\frac{1}{-2} = \frac{-1}{-2}$,
$\frac{-4}{5} = -\frac{4}{5} = \frac{4}{-5} = -\frac{-4}{-5}$;

• $-\frac{1}{2} = -\frac{-1}{-2} = \frac{-1}{2} = \frac{1}{-2}$,
$-\frac{-4}{5} = \frac{4}{5} = \frac{-4}{-5} = -\frac{4}{-5}$

Express $\frac{1}{2}$ and $\frac{-4}{5}$ in three alternate forms; express the additive inverse of $\frac{1}{2}$ and $\frac{-4}{5}$ in three alternate forms.

Theorem 9 For a, b, c, d, where $b \neq 0$ and $d \neq 0$, $\frac{a}{b} = \frac{c}{d}$ if and only if $ad = bc$.

If a, b, c, and d are integers, Theorem 9 is the definition for equal rational numbers. For example, $\frac{14}{42} = \frac{1}{3}$ because $14 \cdot 3 = 42 \cdot 1$.

Theorem 10 For every a, b, c, where $b \neq 0$ and $c \neq 0$, $\frac{ac}{bc} = \frac{a}{b}$.

The patterns in Theorem 11 are the usual methods of rational number arithmetic.

Theorem 11 For $b \neq 0$ and $d \neq 0$:

(i) $\frac{a}{b} + \frac{c}{b} = \frac{a + c}{b}$

(ii) $\frac{a}{b} + \frac{c}{d} = \frac{ad + bc}{bd}$

(iii) $\frac{a}{b} - \frac{c}{d} = \frac{ad - bc}{bd}$

(iv) $\frac{\frac{a}{b}}{\frac{c}{d}} = \frac{ad}{bc}$, for $c \neq 0$.

Exercises

1. Which theorem or property is illustrated by each sentence?
   a. If $3 \cdot c = 0$ then $c = 0$.
   b. If $0 = c$ then $5 + 0 = 5 + c$.
   c. If $b = 3$ then $a \cdot b = a \cdot 3$.

2. Translate into inverse element notation:
   a. $\frac{a}{b}, \frac{0}{b}, \frac{b}{b}$ where $b \neq 0$
   b. $\frac{1}{bd}, \frac{1}{b}, \frac{1}{d}$ where $b \neq 0, d \neq 0$
   c. $\frac{ac}{bd}, \frac{a}{b}, \frac{c}{d}$ where $b \neq 0, d \neq 0$

3. Translate into the form $\frac{a}{b}$ (meaning $a \div b$):
   a. $(-a)(b^{-1})$; $-[a(-b)^{-1}]$; $(-a)(-b)^{-1}$; $(a)(b^{-1})$
   b. $(1)[a(b^{-1})]^{-1}$; $1 \div (a \div b)$; $(1 \div a) \div b$

4. In general, is it true that $a - b = b - a$? If $a - b = b - a$, what can you say about a and b?

5. Perform Exercise 4 with the operation of subtraction replaced by the operation of division.

6. If $a \neq 0$, what number is $0 \div a$?

7. Use one of the symbols . . . , $-3, -2, -1, 0, 1, 2, 3,$ . . . in identifying each real number
   a. $43 + (-79)$, $(-482) + (-713)$, $91 + (-82)$, $[(-43) + 75] + (-89)$
   b. $43 - 79$, $(-482) - 713$, $91 - 82$, $(75 - 43) - 89$

8. Find the additive inverse of each.
   a. $(-\pi)$
   b. $-(-\pi)$
   c. $(3 + 4)$
   d. $[(-3) + (-4)]$
   e. $[3 + (-4)]$
   f. $[(-3) + 4]$
   g. $(7 - 10)$
   h. $-[7 + (-10)]$
   i. $(6 - \sqrt{3})$
   j. $(-2)(7)$
   k. $(-2)(-7)$
   l. $(2a)$

9. Find $(-4)(7 - 10)$. Find $(-4)(7) - (-4)(10)$. Is $(-4)(7 - 10) = (-4)(7) - [(-4)(10)]$ a true statement?

10. Denote the multiplicative inverse of $\frac{-3}{4}$ by at least six different numerals; likewise, the additive inverse of $\frac{-3}{4}$.

11. Find simple, familiar notation for each.
   a. $\frac{-4}{7} + \frac{2}{9}$
   b. $\left(\frac{6}{-7}\right)\left(\frac{-4}{13}\right)^{-1}$
   c. $\left(\frac{\pi}{\sqrt{2}}\right)\left(\frac{\pi}{2}\right)^{-1}$
   d. $\left(-\frac{43}{\sqrt{17}}\right)\left(\frac{82}{17}\right)^{-1}$
   e. $\frac{2}{3} - \frac{4}{5}$
   f. $\dfrac{\frac{11}{5}}{\frac{4}{3}}$

1-3 ORDER RELATIONS AND ABSOLUTE VALUE

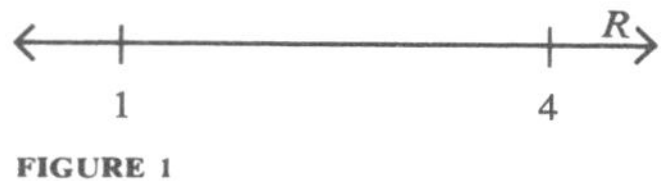

FIGURE 1

For any real numbers r and s, the relation $r < s$ can be interpreted on the number line as "s is to the right of r." For example, we want 4 to be to the right of 1 [Figure 1]. The positive number 3 indicates how far 4 is to the right of 1. Note that $4 - 1$ is this positive number 3.

Definition For any a and $b, a < b$ (read "a is less than b") if and only if $b - a$ is a positive real number; $a > b$ (read "a is greater than b") if and only if $b < a$; $a \leqslant b$ means $a < b$ or $a = b$; $a \geqslant b$ means $a > b$ or $a = b$. Note that if $a < 0$, then a is negative; if $a > 0$, then a is positive. For any real number a, exactly one of the following is true: $a < 0, a = 0$, or $a > 0$. Next we list a few results which can be used as computational aids in working with the relations $<, \leqslant, >$, and $\geqslant$.

Theorem 1 For any a, b, c, if $a < b$, then $a + c < b + c$.

Theorem 2 For any a, b, c, (i) if $a < b$ and $c > 0$ then $ac < bc$, and (ii) if $a < b$ and $c < 0$ then $ac > bc$.

Theorem 2 states that multiplying both members of an equality by a positive number preserves the sense of the inequality and multiplying by a negative number reverses the sense of the inequality.

The order relation $<$ is a **dense order**; that is, for every two distinct real numbers there exists another real number which is between the given two. An easy way to show this is to consider the average of two distinct numbers. The average will always be between the two given numbers.

By considering points on a number line, we can refer to "a set of points between the point corresponding to 2 and the point corresponding to -4" or make a statement such as "the distance between 2 and -4 is 6."

The concept of absolute value plays an important role in the geometric notion of distance between points. Associated with each point r on the number line is a nonnegative real number that describes the distance between r and 0 (called the origin). What nonnegative number would you use to denote the distance between each of the points A, B, C, D, and E and the origin in Figure 2?

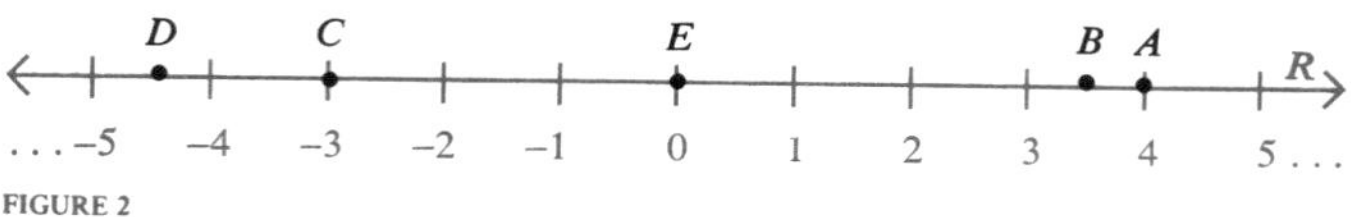

FIGURE 2

For each real number x, we introduce the notation $|x|$ (called the **absolute value** of x) to denote the nonnegative real number that is the distance between the point x and the origin. Hence, for the points A, B, C, D, and E

we have the distance between

$$A \text{ and the origin: } |4| = |4 - 0| = 4,$$

$$B \text{ and the origin: } \left|\frac{7}{2}\right| = \left|\frac{7}{2} - 0\right| = \frac{7}{2},$$

$$C \text{ and the origin: } |-3| = |-3 - 0| = 3,$$

$$D \text{ and the origin: } \left|-\frac{9}{2}\right| = \left|-\frac{9}{2} - 0\right| = \frac{9}{2},$$

$$E \text{ and the origin: } |0| = |0 - 0| = 0.$$

What is a geometric interpretation for $|-2|$? For $|(-2) - 0|$? For $\left|\frac{7}{2}\right|$? For $\left|\frac{7}{2} - 0\right|$? For $|z|$? For $|z - 0|$?

Distance between -2 and the origin; Same interpretation as for $|-2|$; Distance between $\frac{7}{2}$ and the origin; Same as the preceeding; Distance between z and the origin; Same as the preceeding

In considering distance, must one of the points be the origin? What is the distance between A and C? Between A and D? The distance between

$$A \text{ and } C \text{ is: } |4 - (-3)| = 7,$$

$$A \text{ and } D \text{ is: } \left|4 - \left(-\frac{9}{2}\right)\right| = \frac{17}{2},$$

$$C \text{ and } A \text{ is: } |(-3) - 4| = 7.$$

The basic principle is that we may interpret $|x - y|$ as the distance between x and y. How does the number $|x - y|$ relate to $|y - x|$?

We can use this geometric interpretation to find elements of a set, such as

$$S = \{x\colon\ |x - 5| = 2\}.$$

This set notation is read "the set of all real numbers x such that $|x - 5| = 2$" or "the set of all x such that the distance between x and 5 is 2." The distance between 3 and 5 is 2 ($|3 - 5| = 2$), and the distance between 7 and 5 is 2 ($|7 - 5| = 2$). Thus, $S = \{3, 7\}$ [Figure 3].

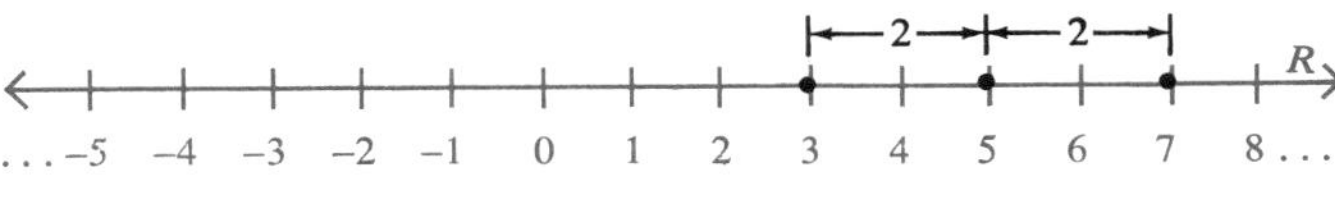

FIGURE 3

$T = \{-1, 5\}$

$U = \{-6, 2\}$

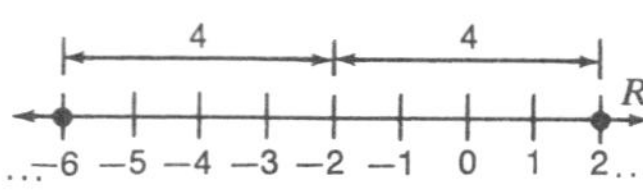

What elements are in the following sets? Illustrate your conclusions with diagrams.

$$T = \{x: |x - 2| = 3\}$$

$$U = \{x: |x + 2| = 4\}$$

For set U note that $|x + 2| = |x - (-2)|$; hence, -2 is an important reference point.

What is the geometric interpretation for this set?

$$V = \{x: |x - 1| \leqslant 4\}$$

The condition of membership in V states that the distance between x and 1 is less than or equal to 4. Which points of V are farthest from 1? The points for the set V form what type of geometric figure [Figure 4]?

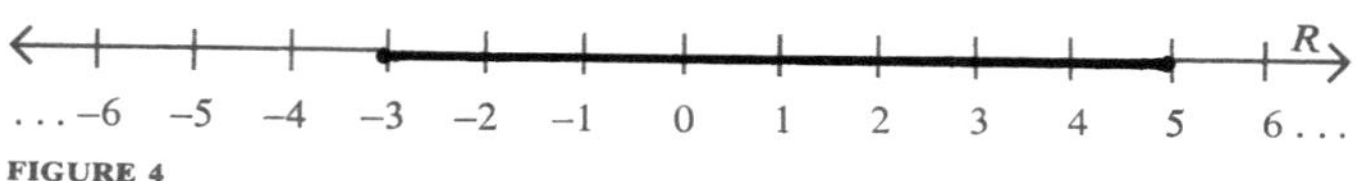

FIGURE 4

$W = \{x: x \geqslant 1 \text{ or } x \leqslant 5\}$

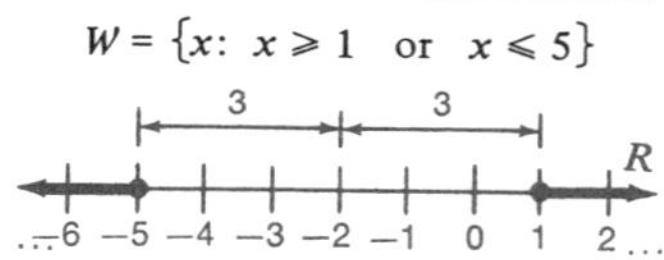

- If $a > 0$, $|a| = |a - 0| = a$;
- $|-\pi| = |-(\pi) - 0| = -(-\pi)$;
- If $b < 0$, $|b| = |b - 0| = -b$

Draw a figure that corresponds to the set:

$$W = \{x: |x - (-2)| \geqslant 3\}.$$

If $a > 0$ what number is $|a| = |a - 0|$? What number is $|-\pi| = |(-\pi) - 0|$? Remember that distance is expressed by a nonnegative real number. If $b < 0$ what number is $|b| = |b - 0|$?

In introducing absolute value we emphasized its geometric interpretation as a distance concept. However, absolute value may be defined by using only properties of R.

Definition For any a:

$$|a| = \begin{cases} a, & \text{if } a \geqslant 0 \\ -a, & \text{if } a < 0. \end{cases}$$

This definition agrees with the preceeding discussion for if $x < 0$ then $-x$ is positive.

Exercises

1. Write true statements of the form $a < b, a \leqslant b, a > b, a \geqslant b$ for each pair.
   a. 1.9, 1.99 b. $-1.9, -1.99$ c. 1.9, 1.09
   d. $-1.09, -1.9$ e. -1.9, 1.09 f. 1.9, -1.09
2. Is it true that $1 < 1$? That $1 \leqslant 1$? That $1 > 1$? That $1 \geqslant 1$?
3. For every a, is it true that $a < a$? That $a \leqslant a$? That $a > a$? That $a \geqslant a$?
4. Find c and d such that $-\frac{1}{100} < c < d < 0$. Now find e such that e is between c and d.
5. Find a, b, c such that $a < b$ and $ac < bc$. Now find a, b, c such that $a < b$ and $bc < ac$. Find a, b, c, such that $ac = bc$ and $a \neq b$.
6. What is the largest real number? The largest negative integer? The smallest real number? The smallest positive integer?
7. Give a geometric interpretation on the number line for adding c to the numbers a, b where $a < b$, (a) if $c > 0$ and (b) if $c < 0$.
8. As in Exercise 7, give a geometric interpretation of multiplying a by c (a) if $c > 0$ and (b) if $c < 0$.
9. Use a number line as an aid in identifying each.
   a. $|7|$ b. $|-2|$
   c. $|8 - 0|$ d. $|0 - (-3)|$
   e. $|2 - 5|$ f. $|12 - (-2)|$
   g. $|(-4) + 3| = |(-4) - (-3)|$ h. $|2 \cdot (-3)|$
   i. $|2| + |3|$ j. $|2 + 3|$
   k. $|2 + (-3)|$ l. $|2| + |(-3)|$
10. Use a geometric interpretation to list the elements of each set.
    a. $\{x\colon |x| = 4\}$ b. $\{x \in I\colon |x| = 0 \text{ and } |x| = 5\}$
    c. $\{x\colon |x| = 0\}$ d. $\{x\colon |x - 4| = 2\}$
    e. $\{x\colon |x + 3| = |x - (-3)| = 4\}$ f. $\{x \in N\colon |x - 5| = 3\}$
    g. $\{x \in N\colon |5 - x| = 3\}$ h. $\{x \in I\colon |x + 1| = 5\}$
11. Use a geometric interpretation to list the elements of the sets
    a. $\{x \in N\colon |x - 5| < 3\}$
    b. $\{x \in N\colon |5 - x| < 3\}$
    c. $\{x \in I\colon |x + 1| > 5\}$
12. For each set, indicate by means of a drawing the corresponding points on a number line.
    a. $\{x\colon x \geqslant 2\}$ b. $\{x\colon |x - (-5)| < 2\}$
    c. $\{x\colon x < -1\}$ d. $\{x\colon |x - 4| \geqslant 3\}$
13. Use the absolute value concept to describe the sets.

    $$\{x\colon x > 4\} \cup \{x\colon x < -4\}$$

    $$\{x\colon x > 2\} \cup \{x\colon x < -6\}$$
14. If $|a| = |b|$, how are a and b related?

Review Exercises

1. Illustrate each basic property using the numbers $3, \sqrt{5}, \frac{2}{15}$ **1-1**
   a. Associative property for multiplication
   b. Distributive property for multiplication over addition.
2. What is the multiplicative inverse of -2 in I? In Q? In R?
3. What theorem justifies each? **1-2**
   a. $-6 + -9 = -(6 + 9)$.
   b. To subtract a real number a, add the additive inverse of a.
   c. To divide by a non-zero real number b, multiply by the multiplicative inverse of b.
4. List the elements of the set, $\{x \in I\colon\ |x - 3| \leqslant 4\}$ **1-3**
5. Graph this set on the number line, $\{x\colon\ |x + 1| > 3\}$

2. POLYNOMIALS AND RATIONAL EXPRESSIONS

2-1 ADDITION AND SUBTRACTION

Powers of a symbol x:

$$x, x^2, x^3, \ldots$$

are used to form polynomials in algebra such as

$$4x + 9, \quad 2x^2 + 1x + 3, \quad \text{and} \quad 3x^3 + 0x^2 + 5x + 6.$$

The numbers preceeding the powers of x are called **coefficients.** Unless otherwise noted, a polynomial in this text will have coefficients which are real numbers. The nonnegative integers used as superscripts for x are called **exponents.** We agree that $x^1 = x$ and $x^0 = 1$ for the symbol x. Thus $4x + 9$ and $4x^1 + 9x^0$ name the same polynomial.

Definition A **polynomial in x** is an element of the form

$$a_n x^n + a_{n-1} x^{n-1} + \cdots + a_2 x^2 + a_1 x^1 + a_0 x^0$$

where $a_n, a_{n-1}, \ldots, a_2, a_1$, and a_0 are real numbers and n is a nonnegative integer.

For the polynomial $4x^3 + 9x^2 + 5x^1 + 7x^0$, $n = 3, a_3 = 4, a_2 = 9, a_1 = 5$ and $a_0 = 7$.

What are $n, a_n, a_{n-1}, \ldots, a_2, a_1$, and a_0 for the polynomial $5x^0$? For $1x^1 + 3x^0$? For $6x^2 + 0x^1 + 1x^0$? For $\frac{3}{2}x^3 + \sqrt{2}x^2 + (-1)x^1 + 4\pi x^0$?

$n = 0, a_0 = 5$; • $n = 1, a_1 = 1, a_0 = 3$; • $n = 2, a_2 = 6, a_1 = 0, a_0 = 1$; • $n = 3, a_3 = \frac{3}{2}, a_2 = \sqrt{2}, a_1 = -1, a_0 = 4\pi$

If we use the notation

$$a_n x^n + a_{n-1} x^{n-1} + \cdots + a_2 x^2 + a_1 x + a_0$$

we usually omit a coefficient of 1 and the parts of the notation that have coefficients of 0. Thus, $2x^2 + x + 3$ is the same polynomial as $2x^2 + 1x^1 + 3x^0$; $6x^2 + 1$ is the same polynomial as $6x^2 + 0x^1 + 1x^0$; $2x$ is the same polynomial as $0x^2 + 2x^1 + 0x^0$; and $3x^3 + 5x + 6$ is the same polynomial as $3x^3 + 0x^2 + 5x^1 + 6x^0$. Notice that $3x^0 = 3$, $-\pi x^0 = -\pi$, and $0x^0 = 0$ are polynomials because $x^0 = 1$; that is, each real number is a polynomial. Finally, if a coefficient is a negative real number, it is often convenient to write $2x^2 - x - 3$ instead of $2x^2 + (-1)x + (-3)$.

Answers may vary:

a. $2x$

b. $1x^5 + 0x^4 + 0x^3 + 0x^2 + 1x + 0x^0$

c. $-2x^4 - 5\sqrt{2}x^3 + 14x^2 - 2\pi x - \frac{7}{2}$

d. $13x^3 + (-3\sqrt{7})x^2 + \frac{14}{9}x + \left(-\frac{9}{2\pi}\right)$

e. $4x + 1$

f. $3x^2 + 12x^1 + 9x^0$

g. $\sqrt{3}x^4 - 4x^3 + \frac{9}{20}x^2 + \frac{2}{3}x - \sqrt{2}$

Rename each of these polynomials:

a. $0x^3 + 0x^2 + 2x + 0$ b. $x^5 + x$

c. $(-2)x^4 + (-5\sqrt{2})x^3 + 14x^2 + (-2\pi)x + \left(-\frac{7}{2}\right)$

d. $13x^3 - 3\sqrt{7}x^2 + \frac{14}{9}x - \left(\frac{9}{2\pi}\right)$

e. $4x^1 + 1x^0$ f. $3x^2 + 12x + 9$

g. $\frac{3}{\sqrt{3}}x^4 + (8 - 12)x^3 + \left(\frac{1}{4} + \frac{1}{5}\right)x^2 + \left(-\frac{3}{7} \cdot \frac{14}{-9}\right)x + (-\sqrt{2})$

In a polynomial each part of the form $a_i x^i$ is called a **term**. Four terms are indicated for the polynomial

$$\frac{3}{2}x^3 + 0x^2 + (-1)x^1 + (-4\pi)x^0$$

but only three of these terms have nonzero coefficients.

Since

$$\frac{6}{4} = \frac{3}{2}, \quad \frac{2}{-2} = -1, \quad \text{and} \quad \frac{4\pi}{-1} = (-4\pi):$$

$$\frac{6}{4}x^3 + \left(\frac{2}{-2}\right)x^1 + \left(\frac{4\pi}{-1}\right)x^0$$

and

$$\frac{3}{2}x^3 + (-1)x^1 + (-4\pi)x^0$$

name the same polynomial. We define equality for polynomials in terms of their coefficients.

Definition Polynomials

$$a_n x^n + a_{n-1}x^{n-1} + \cdots + a_2 x^2 + a_1 x^1 + a_0 x^0$$

and

$$b_n x^n + b_{n-1} x^{n-1} + \cdots + b_2 x^2 + b_1 x^1 + b_0 x^0$$

are equal if and only if

$$a_0 = b_0, \quad a_1 = b_1, \quad a_2 = b_2, \ldots, \quad a_{n-1} = b_{n-1}, \quad a_n = b_n.$$

A polynomial that has at most one nonzero term is called a **monomial**; a polynomial that contains exactly two nonzero terms is called a **binomial.** Terms of polynomials having the same exponent of x are called **similar terms.**

Definition The largest exponent for a nonzero term of a polynomial p is called the **degree** of the polynomial (denoted deg p).

The degree of each of these polynomials: $4x^3 + 2x + 1$, $0x^5 + 3x^4 + 2$, $2x^1$, $3x^0$, $(-4)x^0$, $0x^2 + 0x + 0$, and 0, is 3, 4, 1, 0, and 0, respectively for the first five polynomials. In the last two cases, there is no nonzero term, and hence the polynomial $0x^2 + 0x + 0 = 0$ has no degree. Each polynomial with the exception of the zero polynomial has a degree.

Polynomials are added by summing coefficients of similar terms. For example:

$$\begin{array}{rl} & 6x^2 + 4x + 1 \\ (+) & 2x^2 + 5x + 3 \\ \hline \end{array}$$

$$(6 + 2)x^2 + (4 + 5)x + (1 + 3) = 8x^2 + 9x + 4.$$

The addition of polynomials is commutative.

Example 1

$$(2x + 3) + (3x + 1) = (2 + 3)x + (3 + 1) = 5x + 4$$

$$(3x + 1) + (2x + 3) = (3 + 2)x + (1 + 3) = 5x + 4$$

The addition of polynomials is associative.

Example 2

$$[(x + 1) + (x + 2)] + (x + 3) = (2x + 3) + (x + 3) = 3x + 6$$

$$(x + 1) + [(x + 2) + (x + 3)] = (x + 1) + (2x + 5) = 3x + 6$$

The zero polynomial is the additive identity element for polynomial addition.

Can you find a polynomial which when added to

$$3x^4 + 4x^3 + 2x^2 + 6x + 5$$

gives the zero polynomial? This polynomial is the additive inverse of the given polynomial and is

$$(-3)x^4 + (-4)x^3 + (-2)x^2 + (-6)x + (-5).$$

The preceding examples illustrate that polynomial addition has properties similar to real number addition; the operation is closed, commutative, associative, 0 is the identity polynomial, and each polynomial has an additive inverse.

The difference of two polynomials is found by subtracting coefficients of similar terms. For example:

$$\begin{array}{r} 7x^2 - 4x + 3 \\ (-)\ 4x^2 + 5x + 4 \\ \hline (7-4)x^2 + (-4-5)x + (3-4) = 3x^2 - 9x - 1. \end{array}$$

A second method of computing the difference is given in the following theorem.

Theorem For any two polynomials p and q, $p - q = p + (-q)$, where $-q$ denotes the additive inverse of q.

Exercises

1. For each polynomial identify the nonnegative integer n and the coefficients $a_n, a_{n-1}, \ldots, a_2, a_1$, and a_0 in the definition of a polynomial in x.
   a. $\frac{1}{2}x^3 + \pi x^2 + 3x + \sqrt{2}$ b. -7
   c. $\frac{4}{9}x^3 + 3x - 2$ d. $-x + \pi$
   e. $-x^2 + 1$ f. $0x + 0$
   g. $0x^4 + 4x^3 + 0x^2 + 0x - 3$ h. $\sqrt{\pi}x^{100}$
   i. 0 j. $4x^3 - 0x^2 + 0x - 3$
2. Which polynomials in Exercise 1 are monomials? Binomials?
3. Which polynomials in Exercise 1 have a degree? Find the degree for each of these.
4. In Exercise 1, find the examples that name the same polynomial.
5. Find the sum and difference of each pair of polynomials.
   a. $5x^3 + 3x^2 + (-2)x + 9$, $4x^3 + (-3)x^2 + 5x + 7$
   b. $6x^4 - 3x^2 + 2x - 7$, $9x^3 - 14x - 12$
   c. $3x^5 + 2x^2 - 7x + 8$, $-3x^5 - 2x^2 + 7x - 7$
   d. $2x^7 - 3x^4 + 7$, $0x^7 + 0x^6 + 0x^5 + 0x^4 + 0x^3 + 0x^2 + 0x + 0$
   e. 0, $24x^8 - 1$
   f. $x^4 - 1$, $1 - x^4$
6. State the degree (if it exists) for each polynomial given in Exercise 5; for each sum and difference found for Exercise 5.

7. Express in polynomial form:
   a. $(2x^3 - 4x^2 + 7x - 5) + (-4x^3 - 4x^2 - 7x + 5)$
   b. $(7x^5 + 2x^3 - 3x) + (-2x^4 + 4x - \pi)$
   c. $(2x^4 - 3x^3 + 7x^2 - 4x + 3) + (3x^4 + 5x^3 - 7x^2 - 4x + 10)$
   d. $\left(4x^5 + \frac{1}{2}x^4 - 3x^3 - x + 2\right) + \left(\frac{7}{2}x^5 + 4x^3 - 3x^2 - \frac{1}{7}x + \sqrt{2}\right)$
   e. $(4x^3 - 3x^2 + 2x - 1) - (2x^3 + 3x^2 + 2x - 1)$
   f. $(4x^3 - 3x^2 + 2x - 1) - (-2x^3 - 3x^2 - 2x + 1)$
   g. $(14x^7 - 4x^3 - 3x) - (6x^4 - 4x^2 + 5)$
   h. $(0x^3 + 0x^2 + 0x + 0) - (2x^2 + 4x + 5)$
   i. $(-2x^3 - 4x^2 + 3x - 1) - (-2x^3 - 4x^2 + 3x - 1)$

8. Find the additive inverse of:
   a. $7x^5 + 3x^4 + 2x^2 + 5$ b. $-7x^4 + 3x^3 - 2x + 4$

9. Carefully describe the polynomials that have degree 2; that have degree 1; that have degree 0; that do not have a degree.

10. Observe the three different uses of the "+" symbol in the example below.

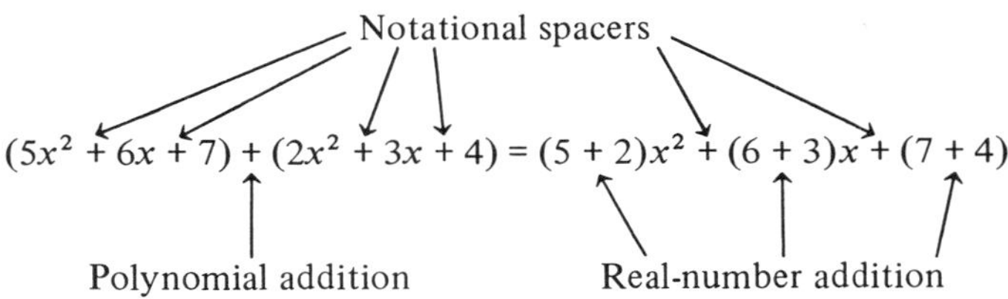

Add the polynomials $4x^2$, $7x$, and 3 to convince yourself that the '+' symbols which separate terms in a polynomial are symbols for polynomial addition as well as spacing devices. Explain why this is so.

11. Since the "+" separator in a polynomial can be interpreted as polynomial addition,

$$\begin{aligned}(3x) + (4x^2) + (1) &= (0x^2 + 3x + 0) + (4x^2 + 0x + 0) + (0x^2 + 0x + 1)\\ &= 4x^2 + 3x + 1\end{aligned}$$

and

$$\begin{aligned}(1) + (3x) + (4x^2) &= (?) + (?) + (?)\\ &= 4x^2 + 3x + 1.\end{aligned}$$

The order in which the terms of a polynomial are written is arbitrary. Arrange the terms of:

$$2 + 4x^2 + 7x^3 + 2x^5 - x + x^4$$

$$-4x^3 + 7x - 2x^2 + 5$$

$$8x^2 - 4 + 2x^3 - 4x^2 + 7x - 2 + 5x^3$$

a. according to the exponents of x such that the exponents decrease **(descending order);**
b. according to the exponent of x such that the exponents increase **(ascending order);**

12. Assume that p, q, and r are the polynomials

$$p = a_n x^n + a_{n-1} x^{n-1} + \cdots + a_2 x^2 + a_1 x + a_0,$$
$$q = b_n x^n + b_{n-1} x^{n-1} + \cdots + b_2 x^2 + b_1 x + b_0,$$

and

$$r = c_n x^n + c_{n-1} x^{n-1} + \cdots + c_2 x^2 + c_1 x + c_0.$$

Prove that:

a. $p = p$
b. If $p = q$ then $q = p$.
c. If $p = q$ and $q = r$ then $p = r$.
d. $p + q = q + p$
e. $(p + q) + r = p + (q + r)$
f. $p + 0 = 0 + p = p$

13. Subtraction can be defined for polynomials p, q, and r by $p - q = r$ if and only if $r + q = p$. Use the example in the text to illustrate this definition.

2-2 MULTIPLICATION AND DIVISION

Multiplication of polynomials is based on the assumptions that (1) multiplication of polynomials is commutative, (2) multiplication distributes over addition, (3) $x^m \cdot x^n = x^{m+n}$ for every nonnegative integer m and n, and (4) the product of the polynomials a ($a \in R$) and x^n is ax^n. Thus:

$$(5x^3)(7x^5) = 5[x^3(7x^5)] = 5[(x^3 \cdot 7)x^5] = 5[(7 \cdot x^3)x^5]$$
$$= 5[7(x^3 \cdot x^5)] = (5 \cdot 7)(x^3 \cdot x^5) = 35 \cdot x^8 = 35x^8.$$

In general, for ax^n and bx^m, $(ax^n)(bx^m) = (ab)x^{n+m}$.

$2x^4$; • $-7x^3$; • 15; • $49x^2$; • $168x^7$; • $-6x^6$; • x^7; • x^6; • x^8

Find the products for the following pairs of polynomials: $2, x^4$; $-7, x^3$; $3x^0, 5x^0$; $7x, 7x$; $14x^3, 12x^4$; $3x^4, (-2)x^2$; x^3, x^4; x^2, x^4; x^4, x^4.

The distributive property plays a key role in multiplication of polynomials.

Examples In horizontal form:

$$4x^2(2x + 3) = (4x^2 \cdot 2x) + (4x^2 \cdot 3) = 8x^3 + 12x^2$$

$$\begin{aligned}
&(2x^2 + x + 1)(3x + 2) \\
&\quad = (2x^2 + x + 1) \cdot 3x + (2x^2 + x + 1) \cdot 2 \\
&\quad = (2x^2 \cdot 3x + x \cdot 3x + 1 \cdot 3x) + (2x^2 \cdot 2 + x \cdot 2 + 1 \cdot 2) \\
&\quad = (6x^3 + 3x^2 + 3x) + (4x^2 + 2x + 2) \\
&\quad = 6x^3 + 7x^2 + 5x + 2
\end{aligned}$$

Using a vertical form:

$$\begin{array}{rl}
2x^2 + x + 1 & \\
(\times) \qquad \qquad 3x + 2 & \\
\hline
4x^2 + 2x + 2 & \longleftrightarrow (2x^2 + x + 1)2 \\
6x^3 + 3x^2 + 3x \qquad \qquad & \longleftrightarrow (2x^2 + x + 1)3x \\
\hline
6x^3 + 7x^2 + 5x + 2 & \longleftarrow (+)
\end{array}$$

The general multiplication algorithm is: For polynomials p and $q, p \cdot q$ is the sum of the products of each term in p with every term in q.

Polynomial products of the type $(2x + 5)[4x + (-3)]$—the product of two binomials—occur so frequently that considering the general case is quite useful.

$$\begin{aligned}
(ax + b)(cx + d) &= (ax + b)(cx) + (ax + b)d \\
&= [(ax)(cx) + b(cx)] + [(ax)d + bd] \\
&= (ac)x^2 + [(bc)x + (ad)x] + bd \\
&= (ac)x^2 + (bc + ad)x + bd
\end{aligned}$$

You should develop facility in writing polynomial products after inspecting the two binomials:

$$(4x - 3)(2x + 5) \quad \text{or} \quad [4x + (-3)](2x + 5) = 8x^2 + 14x - 15$$

By choosing binomials with special coefficients we have the following cases:

$$(ax - b)(ax + b) = a^2x^2 - b^2$$

$$(x - b)(x + b) = x^2 - b^2$$

$$(ax + b)(ax + b) = a^2x^2 + (2ab)x + b^2$$

$$(x + b)(x + b) = x^2 + (2b)x + b^2$$

$$(ax - b)(ax - b) = a^2x^2 - (2ab)x + b^2$$

$$(x - b)(x - b) = x^2 - (2b)x + b^2$$

There is a division procedure for polynomials whereby, given a dividend and nonzero divisor, there exists a corresponding unique pair of polynomials called the quotient and remainder. This is similar to the division algorithm for whole numbers where 5 is the quotient and 2 is the remainder if 17 is divided by 3. As a polynomial example, if the dividend is

$$x^3 + 2x + 3$$

and the divisor is $2x + 4$, then there exists a unique quotient

$$\frac{1}{2}x^2 - x + 3$$

and remainder (-9) such that

$$(x^3 + 2x + 3) = \left(\frac{1}{2}x^2 - x + 3\right)(2x + 4) + (-9).$$

The computational procedure for finding the polynomial quotient and remainder is similar to that for the system of integers.

$$\begin{array}{r l}
 & \frac{1}{2}x^2 - x + 3 \\
2x + 4 \,\big) & x^3 \qquad\quad + 2x + 3 \\
 & \underline{x^3 + 2x^2} \\
 & -2x^2 + 2x \\
 & \underline{-2x^2 - 4x} \\
 & \qquad 6x + 3 \\
 & \qquad \underline{6x - 12} \\
 & \qquad\quad -9
\end{array}$$

The quotient is $\frac{1}{2}x^2 - x + 3$ and the remainder is $-9 = -9x^0$. Note that the remainder has degree 0 which is less than the degree of the divisor.

Theorem If D and $d \neq 0$ are polynomials, there exists a unique pair of polynomials q and r such that $D = d \cdot q + r$ and $\deg r < \deg d$ or r has no degree ($r = 0$).

If $D = 4x^4 - 2x^2 + x + 3$ and $d = x^2 - x + 1$:

$$
\begin{array}{r|l}
 & 4x^2 + 4x - 2 = q \\
\hline
x^2 - x + 1 & 4x^4 \qquad\qquad - 2x^2 + x + 3 \\
 & 4x^4 - 4x^3 + 4x^2 \\
\cline{2-2}
 & \qquad 4x^3 - 6x^2 + x \\
 & \qquad 4x^3 - 4x^2 + 4x \\
\cline{2-2}
 & \qquad\qquad - 2x^2 - 3x + 3 \\
 & \qquad\qquad - 2x^2 + 2x - 2 \\
\cline{2-2}
 & \qquad\qquad\qquad - 5x + 5 = r
\end{array}
$$

Exercises

Express each of the products in Exercises 1–22 as a polynomial in the form $a_nx^n + a_{n-1}x^{n-1} + \cdots + a_2x^2 + a_1x + a_0$.

1. $4x^3(7x^4 + 8)$
2. $-3x^2(8x^3 - 7x + 3)$
3. $17(3x^3 - 4x^2 + 2x - 7)$
4. $2x(3x^2 + 2)$
5. $-2x^3(7x^3 - 3x^2 + 2x - 4)$
6. $(5x - 7)(5x + 7)$
7. $(3x + 7)(2x - 4)$
8. $(9x - 11)(7x - 8)$
9. $(8x + 13)(7x + 4)$
10. $(2 - x)(3 - x)$
11. $(3x - 5)(2x + 7)$
12. $(x + 7)(x + 7)$
13. $(17x - 2)(3x + 14)$
14. $5(6x^3 + 2x^2 + 3)$
15. $(x + 2)(x - 3)$
16. $(2 - x)(3 + x)$
17. $(3x - 4)(3x + 4)$
18. $(100x - 9)(100x + 9)$
19. $x^7(4x^2 + 2x - 1)$
20. $(x + 9)(x + 5)$
21. $(x - 5)(x + 5)$
22. $(2x + 11)(3x - 5)$

Use the vertical technique to express the products in Exercises 23-31.

23. $(x^3 + 3x^2 + 2x - 1)(2x + 3)$
24. $(x^5 + 3x^3 + 7)(4x^2 - 2x + 5)$
25. $(5x^4 + 3x^2 + 7x - 4)(6x^2 - x + 3)$
26. $(x^2 - x + 1)(x + 1)$
27. $(x^4 - 3x^2 + 2)(x^3 + 4x)$
28. $(x^4 + x^3 + x^2 + x + 1)(x - 1)$
29. $[(x + 2)(x + 2)](x + 2)$
30. $[(x + 3)(x + 3)](x + 3)$
31. $(x^6 - 5x^2 + 9x + 3)(3x^4 + x^2 - 4)$
32. If $r \in R$, prove that $(x + r)^3 = x^3 + 3rx^2 + 3r^2x + r^3$.
33. If r is a real number, what polynomial is $(x - r)^3$?

If D is the dividend and d is the divisor, find the quotient and the remainder for Exercises 32–42.

34. $D = 4x^3 + 3x^2 - x + 1, \quad d = 2x + 5$
35. $D = x^4 - x^3 + x^2 - x + 1, \quad d = x + 1$
36. $D = 3x^4 + 7x^3 - 2x^2 + 4x - 7, \quad d = x + 3$

37. $D = x^7 - 3x^4 + 2x + 1, \quad d = 3x^2 + 5$

38. $D = 8x^4 + 8x^3 + 12x^2 + 6x - 1, \quad d = 2x^2 + x + 3$

39. $D = x^6 - 4x^3 - 2x^2 - 4, \quad d = x^3 + 1$

40. $D = x^{12} + 3x^4, \quad d = x^{25} + 2x^4 - 1$

41. $D = x^6 - 1, \quad d = x^2 + 4x + 1$

42. $D = x^6 + 6x^4 + 12x^2 + 8, \quad d = x^2 + 2$

2-3 SYNTHETIC DIVISION

Many times the division algorithm for polynomials is applied where the divisor is a first degree polynomial, $x - a$. For this special situation there is a procedure called **synthetic division.**

In the following example, the first division algorithm is applied, then a few modifications are introduced, and then the concise synthetic division procedure is derived.

$$\begin{array}{r|llll}
 & 2x^2 & +\,6x & +\,17 & \\ \hline
x - 3 & 2x^3 & & -\;x & -\,14 \\
 & 2x^3 & -\,6x^2 & & \\ \hline
 & & 6x^2 & -\;x & \\
 & & 6x^2 & -\,18x & \\ \cline{3-4}
 & & & 17x & -\,14 \\
 & & & 17x & -\,51 \\ \cline{4-5}
 & & & & 37
\end{array}$$

If descending order is used the powers of x can be deleted. The order of the numerals suffices for recovering the terms of the polynomial.

$$\begin{array}{r|rrrr}
 & 2 & 6 & 17 & \\ \hline
-3 & 2 & 0 & -1 & -14 \\
 & 2 & -6 & & \\ \hline
 & & 6 & -1 & \\
 & & 6 & -18 & \\ \cline{3-4}
 & & & 17 & -14 \\
 & & & 17 & -51 \\ \cline{4-5}
 & & & & 37
\end{array}$$

There is needless duplication as indicated by the "circled" pairs. We omit the bottom coefficient from each pair.

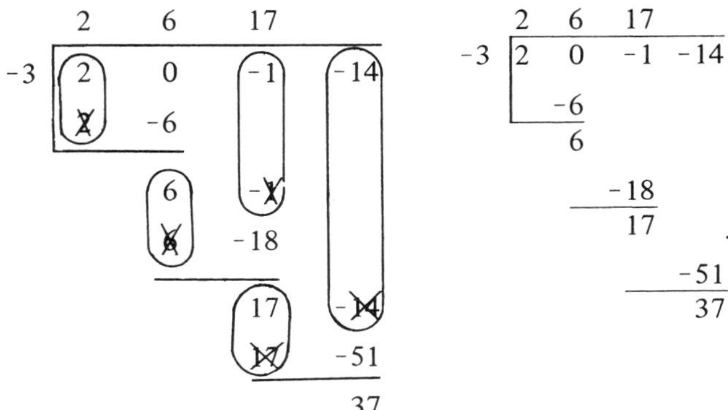

We can conserve space by pushing the symbols up from below and forming fewer rows.

	(2)	~~6~~	~~17~~	
−3	2	0	−1	−14
		−6	−18	−51
	—	6	17	37

We place 2 in the first position of the last row. Then the last number in this line is the remainder (37), and the other numbers (2, 6, 17) are the coefficients of the quotient, in order.

−3	2	0	−1	−14
		−6	−18	−51
	2	6	17	37

The third line is derived by subtracting multiplies of −3 which can also be accomplished by adding multiples of 3. Hence, the final form is:

3	2	0	−1	−14
		6	18	51
	2	6	17	37

It is easy to operate with only the final version by performing the steps below.

1. Arrange the coefficients of the dividend in descending order. Isolate the number a of the divisor $x - a$ on the left (3 in this case).

3	2	0	−1	−14

2. Use the first number of row 1 as the first number below the line. Add the product of the

3	2	0	−1	−14
		$3 \cdot 2 = 6$		
	2	6		

number below the line and a to the next number of row 1.

3. Multiply the last number below the line by a and add this product to the next number in row 1. Continue in this manner until the numbers in row 1 are exhausted.

$$\begin{array}{r|rrrr} 3 & 2 & 0 & -1 & -14 \\ & & 3\cdot 2=6 & 3\cdot 6=18 & 3\cdot 17=51 \\ \hline & 2 & 6 & 17 & 37 \end{array}$$

4. The last number below the line is the remainder. The other numbers in order are the coefficients of the quotient.

$$\begin{array}{rrr|r} \hline 2 & 6 & 17 & \boxed{37} \end{array}$$

Quotient: $2x^2 + 6x + 17$
Remainder: 37

1, 3, −2, 1; 5, 4, 0, −7; 1, −1, 1, 0;
• a: 2, −2, −5, −3

List the coefficients of each of the following polynomials in descending order: $x^3 + 3x^2 - 2x + 1$, $5x^3 + 4x^2 - 7$, and $x^3 - x^2 + x$. Find a if $x - a = x - 2; x + 2; x - (-5); 3 + x$.

By synthetic division, we find the quotient polynomial and remainder if $2x^3 - x - 14$ is divided by $x - 2$.

$$\begin{array}{r|rrrr} 2 & 2 & 0 & -1 & -14 \\ & & 4 & 8 & 14 \\ \hline & 2 & 4 & 7 & \boxed{0} \end{array}$$

The remainder is 0 and the quotient is $2x^2 + 4x + 7$. Similarly, let us divide $2x^3 - x - 14$ by $x + 2 = x - (-2)$:

$$\begin{array}{r|rrrr} (-2) & 2 & 0 & -1 & -14 \\ & & -4 & 8 & -14 \\ \hline & 2 & -4 & 7 & \boxed{-28} \end{array}$$

the quotient is $2x^2 - 4x + 7$ and the remainder is -28.

Exercises

The division problems in Exercises 1–4 are performed by synthetic division. Fill in the missing parts and identify the dividend, divisor, quotient, and remainder.

1. $\begin{array}{r|rrrr} 2 & 1 & 0 & 3 & 2 \\ & & & & 14 \\ \hline & 1 & & 7 & \end{array}$

2. $\begin{array}{r|rrrrr} 1 & 2 & 1 & -3 & 4 & -2 \\ & & & & & 4 \\ \hline & 2 & & 0 & & \end{array}$

3. $\begin{array}{r|rrrr} (-2) & 4 & 1 & 0 & 3 \\ & & & & \\ \hline & & & -7 & \boxed{-25} \end{array}$

4. $\begin{array}{r|rrrrr} (-1) & 4 & 0 & 0 & 0 & 3 \\ & & & & & \\ \hline & 4 & & & & \boxed{7} \end{array}$

Use synthetic division to compute the quotient and remainder for each pair of polynomials in Exercises 5–12 where the first polynomial is the dividend and the second is the divisor.

5. $2x^3 - x^4 + 3x - 1,\ x - 1$
6. $x^3 + 4x - 3,\ x + 2$
7. $x^4 - 3,\ x - 2$
8. $x^4 + 3x^3,\ x + 3$
9. $3x^3 + 2x^2 - 4x + 5,\ x - 3$
10. $3x^3 + x^2 - 1,\ x + 1$
11. $x^4 - x^3 - 4x + 3,\ x - 1$
12. $x^5 + x - 1,\ x + 2$
13. Find the quotient and remainder when $x^3 - 7x + 6$ is divided by each polynomial.
    a. $x - 1$ b. $x - 2$ c. $x - 3$
    d. $x + 1$ e. $x + 2$ f. $x + 3$
14. Find the quotient and remainder when $x^3 - 3x^2 - x - 3$ is divided by each polynomial.
    a. $x - 1$ b. $x - 2$ c. $x - 3$
    d. $x + 1$ e. $x + 2$ f. $x + 3$
15. Find the quotient and remainder when $x^3 - \frac{3}{2}x^2 + \frac{3}{2}x - \frac{1}{2}$ is divided by:
    a. $x - 1$ b. $x - \frac{1}{2}$ c. $x - \frac{1}{4}$
    d. $x + 1$ e. $x + \frac{1}{2}$ f. $x + \frac{1}{4}$
16. Using synthetic division find which of the following divisors of $x^4 + 2x^3 - 7x^2 - 8x + 12$ yield the zero polynomial as the remainder.
    a. $x - 1$ b. $x + 1$ c. $x - 2$
    d. $x + 2$ e. $x - 3$ f. $x + 3$
17. Which of the following divisors of $6x^4 + x^3 - 8x^2 - x + 2$ yield the zero polynomial as the remainder?
    a. $x - 1$ b. $x + 1$ c. $x - \frac{1}{3}$
    d. $x + \frac{1}{3}$ e. $x - \frac{1}{2}$ f. $x + \frac{1}{2}$
    g. $x - \frac{2}{3}$ h. $x + \frac{2}{3}$ i. $x - 0$

2-4 FACTORING POLYNOMIALS

If a, b, and p are polynomials and if $p = a \cdot b$, then a and b are **factors** of p, and p is a **multiple** of both a and b.

Definition If p is a polynomial with $\deg p \geq 1$, then p is a **prime** if and only if for every product $a \cdot b = p$, either a or b is a real number.

Any polynomial of degree greater than zero which is not prime is **composite**; that is, it is factorable.

To classify polynomials as prime or composite, we recall some elementary techniques for factoring polynomials.

Monomial factors can be found using a direct application of the distributive property.

Examples
a. $5x^7 - 10x^3 + 15x^2 = 5x^2(x^5 - 2x + 3)$
b. $12x^4 - 21x^3 = 3x^3(4x - 7)$

You have learned how to find the product $(ax + b)(cx + d)$. Now we reverse this process. That is, given a polynomial such as $12x^2 + 64x + 45$, we want to find first degree factors such that

$$(ax + b)(cx + d) = acx^2 + (ad + bc)x + bd = 12x^2 + 64x + 45.$$

If a, b, c, and d are integers the condition $ac = 12$ suggests that we examine the following pairs for $\{a, c\}$:

$$\{1, 12\}, \quad \{2, 6\}, \quad \{3, 4\}, \quad \{-1, -12\}, \quad \{-2, -6\}, \quad \{-3, -4\}.$$

Since $bd = 45$, we might examine the following pairs for $\{b, d\}$:

$$\{1, 45\}, \quad \{3, 15\}, \quad \{5, 9\}, \quad \{-1, -45\}, \quad \{-3, -15\}, \quad \{-5, -9\}.$$

We select a, b, c, and d so that $ad + bc = 64$: $a = 2, b = 9, c = 6, d = 5$, and

$$(2x + 9)(6x + 5) = 12x^2 + 64x + 45.$$

Factor each of the following composite polynomials as a product of prime polynomials:

a. $2x^2 + x$	b. $x^2 - x - 2$	c. $x^3 + x^2 + x$
d. $3x^2 + 16x + 5$	e. $x^3 + 2x^2 + x$	f. $3x^2 - 5x - 2$
g. $16x^2 - 9$	h. $25x^2 + 10x + 1$	i. $x^2 - 1$

a. $x(2x + 1)$
b. $(x - 2)(x + 1)$
c. $x(x^2 + x + 1)$
d. $(3x + 1)(x + 5)$
e. $x(x + 1)(x + 1)$
f. $(3x + 1)(x - 2)$
g. $(4x - 3)(4x + 3)$
h. $(5x + 1)(5x + 1)$
i. $(x - 1)(x + 1)$

If you notice that

$$4x^2 - 1 = 2^2x^2 - 1^2$$

$$x^2 - 5 = x^2 - (\sqrt{5})^2$$

$$3x^2 + 2\sqrt{3}x + 1 = (\sqrt{3})^2x^2 + (2 \cdot \sqrt{3} \cdot 1)x + 1^2$$

then you can factor these polynomials:

$$4x^2 - 1 = (2x + 1)(2x - 1)$$

$$x^2 - 5 = (x + \sqrt{5})(x - \sqrt{5})$$

$$3x^2 + 2\sqrt{3}x + 1 = (\sqrt{3}x + 1)(\sqrt{3}x + 1).$$

(See the equations on page 19.) Since

$$(2x - 3)(4x^2 + 6x + 9) = 8x^3 - 27,$$

and

$$(x + 4)(x^2 - 4x + 16) = x^3 + 64$$

$8x^3 - 27$ and $x^3 + 64$ are composite polynomials. The general statement of the patterns found in these examples is:

$$a^3x^3 + b^3 = (ax + b)(a^2x^2 - abx + b^2)$$

and

$$a^3x^3 - b^3 = (ax - b)(a^2x^2 + abx + b^2).$$

Useful special cases of this pattern are:

$$x^3 + b^3 = (x + b)(x^2 - bx + b^2)$$

and

$$x^3 - b^3 = (x - b)(x^2 + bx + b^2).$$

In factoring polynomials having integral coefficients, sometimes one only uses factors which also have integral coefficients. Consider the polynomial $x^2 - 2$. We know that

$$x^2 - 2 = (x - \sqrt{2})(x + \sqrt{2}).$$

But $(x - \sqrt{2})$ and $(x + \sqrt{2})$ are not polynomials with integral coefficients. Thus the polynomial $x^2 - 2$ cannot be expressed as a product of polynomials with integral coefficients.

Prime polynomials with integral coefficients are defined analogously to the previous definition for polynomials over the real numbers. A polynomial p with integral coefficients and with $\deg p \geqslant 1$ is prime, if and only if for every product $a \cdot b = p$, where a and b have integral coefficients either a or b is equal to 1 or -1. Thus factoring a polynomial is relative to the particular set of coefficients being discussed.

When attempting to factor $x^2 + 5x + 2$ using integral coefficients, we are interested in finding two first degree factors. If

$$x^2 + 5x + 2 = (ax + b)(cx + d)$$

what can you say about a, b, c, and d if each represents an integer? What are the products of ac and bd? The following list contains all of the cases that yield these products:

$$(x + 1)(x + 2)$$

$$(-x + 1)(-x + 2) = (-1)^2(x - 1)(x - 2)$$

$$(x - 1)(x - 2)$$

$$(-x - 1)(-x - 2) = (-1)^2(x + 1)(x + 2).$$

In no case is the product $x^2 + 5x + 2$.

Testing $x^2 + 5x + 2$ or any polynomial with integral coefficients for factors of degree greater than or equal to one as a member of the larger system of polynomials with rational coefficients will not aid in the search for such factors. Although there are more polynomials available they are not the right type.

Exercises

Classify each of the polynomials in Exercises 1-11 as prime, composite, or neither, using real coefficients.

1. $\pi x - \sqrt{3}$ 2. πx^2 3. $x^4 - 625$ 4. $x^3 - 1$
5. $400x - 200$ 6. $0x^0$ 7. $-13x^0$ 8. $14x^3 - 8x^2$
9. $4x^2 - 10x - 6$ 10. $x^3 - 9x$ 11. $125x^3 + 216$

12. Factor each composite polynomial in Exercises 1-11 as a product of prime polynomials.

13. Factor each of these polynomials as a product of prime polynomials using real coefficients.
 a. $8x^2 + 2x - 15$ b. $8x^2 - 2x - 15$ c. $8x^2 - 14x - 15$ d. $x^2 - 7$

Express each of the polynomials in Exercises 14-16 as a product of a real number and a polynomial with only integers as coefficients.

14. $\frac{1}{2}x + \frac{1}{3}$ 15. $\frac{2}{3}x^2 + \frac{4}{7}x + \frac{3}{5}$ 16. $\frac{1}{2}x^3 + \frac{1}{3}x^2 + \frac{1}{4}x + \frac{1}{5}$

In Exercises 17-19 determine which of the polynomials are prime, using only integral coefficients for factors:

17. a. $x^2 - 3$ b. $125x^3 - 1$
 c. $x^2 - 16$ d. $x^3 + 3$
 e. $x^5 - x$ f. $x^2 - \frac{9}{4}$
 g. $x^4 - 2$ h. $8x^2 - 9$
 i. $x^2 - 9$ j. $4x^2 - 25$

18. a. $x^2 + 1$ b. $3x^3 + 4x^2$
 c. $2x^2 - 1$ d. $x^2 + x + 1$
 e. $16x^2 - 9$ f. $4x^2 - 3$

19. a. $x^2 + 10x + 13$ b. $x^2 + 9x + 12$
 c. $4x^2 + 4x + 15$ d. $3x^2 + 9x + 4$
 e. $x^2 - 3x + 15$ f. $4x^2 + 24x + 18$
 g. $2x^2 + 4x - 3$ h. $2x^2 + 8x - 15$

2-5 POLYNOMIALS IN SEVERAL VARIABLES

The idea of a polynomial can be extended to include more than one symbol. For example,

$$4x^3 + 3x^2y + 5y^3 = 4x^3y^0 + 3x^2y^1 + 5x^0y^3$$

is a polynomial in x and y and

$$4x^3y^2z + 3x^2yz^2 + 5z^4 = 4x^3y^2z^1 + 3x^2y^1z^2 + 5x^0y^0z^4$$

is a polynomial in x, y, and z.

The polynomial $4x^3y^2z + 3x^2yz^2 + 5x^4$ has three nonzero terms: $4x^3y^2z$, $3x^2yz^2$, and $5x^4$. The degree of a nonzero term is the sum of the exponents for the symbols of the polynomial set. For the term

$$4x^3y^2z = 4x^3y^2z^1$$

the degree is $3 + 2 + 1 = 6$. The term $5z^4 = 5x^0y^0z^4$ has degree $0 + 0 + 4 = 4$. What is the degree of the term $3x^2yz^2$? The maximum degree of the nonzero terms of a polynomial is the degree of the polynomial. Thus the degree of the polynomial $4x^3y^2z + 3x^2yz^2 + 5x^4$ is six. Recall that polynomial addition is performed by adding coefficients of similar terms (terms whose symbols have identical exponents). For example

$$(3x^2 + 2xy + 4y^3) + (2x^2 - 5xy + 7x^3)$$

$$= (3x^2 + 2x^2) + (2xy - 5xy) + (4y^3 + 0y^3) + (0x^3 + 7x^3)$$

$$= (3 + 2)x^2 + (2 - 5)xy + (4 + 0)y^3 + (0 + 7)x^3$$

$$= 5x^2 - 3xy + 4y^3 + 7x^3.$$

The product

$$(3x^2 + 2xy + 4y^3)(2x^2 - 5xy)$$

$$= (3x^2 + 2xy + 4y^3)(2x^2) + (3x^2 + 2xy + 4y^3)(-5xy)$$

$$= 6x^4 + 4x^3y + 8x^2y^3 - 15x^3y - 10x^2y^2 - 20xy^4$$

$$= 6x^4 - 11x^3y + 8x^2y^3 - 10x^2y^2 - 20xy^4.$$

In this and the following section you will study standard types of polynomial factoring. The examples and exercises will be restricted to polynomials with integral coefficients; however, the principles involved extend to polynomials with real number coefficients.

The distributive property is illustrated by

$$3(4+5) = 3 \cdot 4 + 3 \cdot 5$$

$$4x^2(x+y) = 4x^3 + 4x^2y$$

$$(2x+y)(x-y) = 2x(x-y) + y(x-y).$$

The left-hand members give factorizations for the polynomials in the right-hand members. In these cases, we say that "a common factor has been removed."

The factors of $3x^2y^2 + 6x^3y + 9xy^3$ common to each term are 3, x, $3x$, y, $3y$, xy, and $3xy$. So $3xy$ is the factor which is the greatest common divisor of the coefficients times a polynomial of largest degree. For the polynomial $12x^2y - 3x^2y^3 + 9x^3y^2$, this common factor of the terms is $3x^2y$; for the polynomial $15xy^2 + 6xy^3 - 10xy^4$, this factor is xy^2.

Exercises

1. Find the degree of these polynomials.
   a. $4x^2yz + 17xyz$ b. $7x^3 - x^2 + 5$ c. $x^3 + y^3 + z^3$
   d. $4x^2y - 3z^3$ e. $\frac{1}{2}x^3y - \pi x^2z$ f. $3x^2 - y$
   g. $\frac{1}{4}$ h. $4y^2 - 2y$ i. $x^2 + 2xy + y^2 - z^2$
   j. $\sqrt{2}$ k. $4x^2 + 3y^2 + 2z^2 + xyz$
2. Find the sum and the degree of the sum for
$$(4x^2yz + 17xyz) + (4x^2 + 3y^2 + 2z^2 + xyz) + (x^2 + y^2 + z^2).$$

Find the sum of each pair of polynomials in Exercises 3-5.

3. $2x^3y^2 + 5x^2y^3 + 7x + 8y,\ 3x^2y^2 + 5x^2y^3 - 2y$
4. $4x^2 - 2xy + y^2 - x - y + 3,\ 7x^2 - y^2 + 5$
5. $3x^2y + 4xy^2 + 7y^2,\ 5xy + x - y$

Express each product in Exercises 6-27 as a sum of terms.

6. $x(3xy + z)$
7. $xz(x^2y^2 - 2)$
8. $(x + y + z)(x - z)$
9. $x(2x - y)(2x + y)$
10. $(y - \pi x)(y - \pi x)$
11. $(2x^2 - y^2)(2x^2 - y^2)$
12. $(x + \pi y)(x - \pi y)$
13. $(\sqrt{2}x + \sqrt{2}y)(\sqrt{2}x - \sqrt{2}y)$
14. $(3x + 4y)(5x - 6y)$
15. $(x - y)(x + y)$
16. $(x + y)(x + y)$
17. $(7x + 5y)(7x + 5y)$
18. $\left(x - \frac{2}{3}y\right)\left(9x + \frac{1}{3}y\right)$
19. $4xy(x^2 - 2xy^2 + y^3)$
20. $xyz(x^2 + x + y^3 + y + z^4 + z)$
21. $3x(x + 5y - 2)$
22. $(-4y)(xz + 5yz - 2xyz)$
23. $(3x - 4y)(x + 5y - 2)$

24. $(x + y)(x^2 - xy + y^2)$

25. $(x - y)(x^2 + xy + y^2)$

26. $(x - y)(x^4 + x^3y + x^2y^2 + xy^3 + y^4)$

27. $(4x - 3y + 2z)(x^2 + 2xy + y^2 + xyz + z^2)$
(Use a vertical form.)

In Exercises 28–49 use integral coefficients.

28. Factor the following polynomials.
a. $x^2y - xy^2$
b. $4x^4y^3 - 9x^2y^3$
c. $4x^2 - 7xy + 3y^2$
d. $49x^2 - 25y^2$

29. What is the factor with largest degree and greatest coefficient for the terms of each of these polynomials?
a. $6xy - 9y$
b. $xy - x$
c. $24xy + 16x$
d. $3x^2y + 9x^2y^2$
e. $35x^3y^2 - 84x^2y^3$
f. $99x^2 + 41y^2$

Use the fact that multiplication is distributive over addition to factor each polynomial in Exercises 30–49.

30. $4x^2 + 8x$

31. $6xy^2 + 18x^3y^3 - 6xy^3$

32. $16x^3y^3 - 8x^2y + 12x^3y$

33. $8x^2 + 8x$

34. $10x^2y^2 - 15xy^2 + 5xy$

35. $6xy - 4y^2$

36. $4x^2y^3 - 5x^2y - 7x^2$

37. $6x^2y^3 - 8xy^2 + 4xy^3$

38. $9xy + 18xy^2$

39. $36x^2y - 18xy^2 + 9xy$

40. $4y(x - 2) - 5(x - 2)$

41. $2x(x - y) + (x - y)$

42. $x^2(y - 1) - 2x(y - 1) + (y - 1)$

43. $x^2(y^2 - 5y + 1) - 2(y^2 - 5y + 1)$

44. $4xy(y + 1) - 6x^2(y + 1)$

45. $4y(x + y) - x(x + y)$

46. $\frac{1}{2}x^3y^3 + \frac{1}{4}y^2 - \frac{1}{8}xy = \frac{1}{8}y(\qquad)$

47. $\frac{1}{3}x^4y^3 - \frac{1}{6}x^2y^2 - \frac{2}{3}x^3y = \frac{1}{6}x^2y(\qquad)$

48. $x^2(y^2 - 2y + 1) - 16(y^2 - 2y + 1)$

49. $x^2(y^2 + 2y + 1) - (y^2 + 2y + 1)$

2-6 FACTORING SPECIAL POLYNOMIALS

The polynomials $x^2 - y^2$, $x^3 - y^3$, $x^3 + y^3$, and similar polynomials are easily factored and follow the same patterns that we saw for the one variable cases on pages 19, 20, and 27. Thus,

$$x^2 - y^2 = (x + y)(x - y)$$

$$x^3 - y^3 = (x - y)(x^2 + xy + y^2)$$

$$x^3 + y^3 = (x + y)(x^2 - xy + y^2).$$

Examples a. $25x^2 - 9y^2 = (5x)^2 - (3y)^2$

$$= (5x + 3y)(5x - 3y)$$

b. $64x^3 - 27y^3 = (4x)^3 - (3y)^3$

$$= (4x - 3y)[(4x)^2 + (4x)(3y) + (3y)^2]$$

$$= (4x - 3y)(16x^2 + 12xy + 9y^2)$$

What are the factors of $64x^3 + 27y^3 = (4x)^3 + (3y)^3$?

A product such as

$$(2x + 3y)(x - 4y)$$

is easily found to be the polynomial $2x^2 - 5xy - 12y^2$. Now we are concerned with reversing the process. Given the polynomial

$$2x^2 - 5xy - 12y^2$$

we want to find its factors. The factors of $2x^2$ and $-12y^2$ to be considered are $(2x)(x)$, and $(-y)(12y)$, $(y)(-12y)$, $(2y)(-6y)$, $(-2y)(6y)$, $(3y)(-4y)$, $(-3y)(4y)$. In testing the "xy term" of each product, we find $(2x + y)[x + (-4y)]$ is a factorization of $2x^2 - 5xy - 12y^2$.

Other basic patterns of the form

$$(ax + by)(cx + dy) = acx^2 + (ad + bc)xy + bdy^2$$

are represented by the polynomials

$$x^2 + 2xy + y^2 = (x + y)^2$$

and

$$x^2 - 2xy + y^2 = (x - y)^2.$$

Then to find the factors of the polynomial $25x^2 - 20xy + 4y^2$, use the form $(5x)^2 - 2(5x)(2y) + (2y)^2$. Thus

$$25x^2 - 20xy + 4y^2 = (5x - 2y)^2.$$

It is sometimes helpful to arrange or group certain terms in your initial steps. For example

$$\begin{aligned} 2x - 2y + x^2 - y^2 &= (2x - 2y) + (x^2 - y^2) \\ &= 2(x - y) + (x + y)(x - y) \\ &= [2 + (x + y)](x - y) \\ &= (2 + x + y)(x - y). \end{aligned}$$

The strategy is to group the terms in such a way that factoring relative to a few terms of the polynomial will lead eventually to factoring the polynomial. In factoring the polynomial $x^3 - 2x - 3x^2 + 6$, we find that

$$(x^3 - 2x) + (-3x^2 + 6) = x(x^2 - 2) - 3(x^2 - 2) = (x - 3)(x^2 - 2).$$

Another approach for factoring the same polynomial would be

$$\begin{aligned} x^3 - 2x - 3x^2 + 6 &= (x^3 - 3x^2) - (2x - 6) \\ &= x^2(x - 3) - 2(x - 3) \\ &= (x^2 - 2)(x - 3). \end{aligned}$$

Exercises

Factor each of the polynomials using integral coefficients.

1. $y^2 - x^2$
2. $8x^2 - 50y^2$
3. $49 - y^2$
4. $x^2y^2 - 100$
5. $9(x + y)^2 - 16$
6. $16y^4 - x^4$
7. $4x^2 - y^2$
8. $1 - x^2$
9. $4x^2 - 9y^2$
10. $(x - 1)^2 - 9$
11. $3x^3 - 3xy^2$
12. $(x + 1)^2 - 25y^2$
13. $8x^3 - 125y^3$
14. $225 - x^2$
15. $8x^3 + 125y^3$
16. $x^4 - y^4$
17. $9x^2(y^2 - x^2) - 25y^2(y^2 - x^2)$
18. $x^3y^3 + 1$
19. $0.008x^3 - 0.027y^3$
20. $1 + (x - y)^3$
21. $(2x - y)^3 - 1$
22. $x^3 + 1000$
23. $y^3 - x^3$
24. $16y^3 - 2x^3y^3$
25. $x^6y^6 + 216x^3$
26. $8(2x - y)^3 - 1$
27. $x^2 - 8x + 7$
28. $x^2 + 8x + 15$
29. $y^2 - 4y + 4$
30. $x^2y^2 - xy - 20$
31. $2x^2 - 8x + 6$
32. $y^2 + 2y + 1$
33. $y^2 + 4y + 4$
34. $x^2 + 2xy - 3y^2$
35. $8x^2 - 26xy + 15y^2$
36. $4x^2 + 4xy + y^2$

37. $x^2 - 4xy + 4y^2$
38. $x^3y^2 + 6x^2y^2 - 7xy^2$
39. $x^2 - 10xy + 25y^2$
40. $4x^2 + 12xy + 9y^2$
41. $25x^2 + 10xy + y^2$
42. $10x^2y^2 + 10xy - 200$
43. $45x - 18xy - 27xy^2$
44. $8x^2 + 30xy - 27y^2$
45. $8x^2 - 73xy + 9y^2$
46. $24x^2y^2 + 38xy^3 + 15y^4$
47. $25x^2y + 70xy^2 + 49y^3$
48. $x^4 + 2x^2 + 1$
49. $16x^2 - 56xy + 49y^2$
50. $15x^2 - 37xy + 20y^2$
51. $(x + y)^2 + (x + y) - 12$
52. $24x^4 - 46x^3y + 15x^2y^2$
53. $81x^2 - 198xy + 121y^2$
54. $x^4 - 2x^2 + 1$
55. $x^2 + 2x + 1 - y^2$
56. $4x^3 + 4x^2 - x - 1$
57. $x^3 + 2x^2 + x + 2$
58. $y^2 - x^2 + 2x - 1$
59. $x^3 + x - 2x^2 - 2$
60. $x^2y + 3x^2 - 2y - 6$
61. $9 - 4x^2 - 12xy - 9y^2$
62. $x^3 + 2xy - x^2 - 2y$
63. $x^3 - y^3 - x^2 + xy$
64. $y^3 - x^3 - y + x$
65. $2x - 2y + x^2 - y^2$
66. $x^3 - x^2 + y - xy$
67. $x^2y - y^3 + 2y^2 - 2x^2$
68. $8x^3 + 4x^2 - 6x - 3$
69. $x^3 + y^3 + x + y$
70. $2x^3 - 2xy^2 - 3y^2 + 3x^2$
71. $x^2 + 6x - y^2 + 9$
72. $x^3 + x^2 + 2xy + y^2 + y^3$

2-7 RATIONAL EXPRESSIONS

In arithmetic we use whole numbers as numerators and nonzero whole numbers as denominators to denote rational numbers. In earlier work in algebra you used polynomials as numerators and denominators. The resulting objects were called **rational expressions.** Several examples of rational expressions are:

$$\frac{x+1}{x^2 - 3x + 2}, \quad \frac{-3x^2 + 4x + 7}{x^5 + 2x + 9}, \quad \frac{x-1}{1}.$$

The zero polynomial is never a denominator.

Computing with rational expressions is analogous to computing with rational numbers. For rational numbers it is true that $\frac{a}{b} = \frac{c}{d}$ if and only if $a \cdot d = b \cdot c$. Equality for rational expressions is defined in a similar fashion.

Definition For polynomials $p, q \neq 0, r$, and $t \neq 0, \frac{p}{q} = \frac{r}{t}$ if and only if $p \cdot t = q \cdot r$.

For example: $(x - 3)(2x + 8) = 2x^2 + 2x - 24$ and $2(x+4)(x - 3) = 2x^2 + 2x - 24$ so $\frac{x - 3}{2(x + 4)} = \frac{x - 3}{2x + 8}$.

Use this definition to justify that the following rational expression is correctly named in several ways:

$$\frac{6x^2 - 15x - 9}{12x^2 + 54x + 24} = \frac{3(2x + 1)(x - 3)}{6(2x + 1)(x + 4)} = \frac{x - 3}{2(x + 4)} = \frac{x - 3}{2x + 8}.$$

We classify the last two forms in the self test section above as simplest forms since the numerator and denominator have no common factor except 1 or -1. If the denominator is 1, we identify the polynomial in the numerator with this rational expression. Thus $\frac{x - 1}{1}$ and $x - 1$ denote the same rational expression.

$3x^2 - 2x + 1$: polynomial, rational expression; ● $\frac{5}{4}$: all three categories; ● $x^2 + 12x + 11$: polynomial, rational expression; ● $(x^2 + 2x + 1) \div (-x^2 + 2x + 1)$: rational expression; ● $x^7 + \frac{3}{0}$: none of these; ● $\frac{0}{x^7} + 3$: all three categories

Classify each of the following as a rational number, a polynomial, a rational expression, none of these: $3x^2 - 2x + 1, \frac{5}{4}, x^2 + 12x + 11, (x^2 + 2x + 1) \div (-x^2 + 2x + 1), \frac{x^7 + 3}{0}$, and $\frac{0}{x^7 + 3}$.

Example 1 Find a simplest form for the rational expression

$$\frac{x^2 - 1}{x^2 + 3x + 2}.$$

Solution: We begin by factoring the numerator and denominator using integral coefficients:

$$\frac{x^2 - 1}{x^2 + 3x + 2} = \frac{(x - 1)(x + 1)}{(x + 2)(x + 1)} = \frac{x - 1}{x + 2}.$$

Example 2 Find a simplest form for the rational expression

$$\frac{x^3 + x^2 - xy^2 - y^2}{-x^2 - x^2y + y^3 + y^2}.$$

Solution: By factoring the numerators and denominators:

$$\frac{x^3 + x^2 - xy^2 - y^2}{-x^2 - x^2y + y^3 + y^2}$$

$$= \frac{x^3 - xy^2 + x^2 - y^2}{-x^2(1 + y) + y^2(y + 1)} = \frac{x(x^2 - y^2) + (x^2 - y^2)}{y^2(y + 1) - x^2(y + 1)}$$

$$= \frac{(x + 1)(x^2 - y^2)}{(y^2 - x^2)(y + 1)} = \frac{(x^2 - y^2)(x + 1)}{(-1)(x^2 - y^2)(y + 1)}$$

$$= \frac{x + 1}{-y - 1}.$$

To find the product or the sum of two rational expressions, we take our cue from the rational number system.

Definition For polynomials $p, q \neq 0, r$, and $t \neq 0$

$$\frac{p}{q} \cdot \frac{r}{t} = \frac{p \cdot r}{q \cdot t}.$$

Example 3 Find the product of $\frac{2x + 3}{3x - 4}$ and $\frac{x^2 + x - 2}{x^2 - x - 2}$.

Solution:

$$\frac{2x + 3}{3x - 4} \cdot \frac{x^2 + x - 2}{x^2 - x - 2} = \frac{(2x + 3)(x^2 + x - 2)}{(3x - 4)(x^2 - x - 2)}$$

$$= \frac{2x^3 + 5x^2 - x - 6}{3x^3 - 7x^2 - 2x + 8}$$

Example 4 Multiply $\frac{x - 1}{x^2 - x - 6}$ and $\frac{x + 2}{x^2 + 2x - 3}$.

Solution:

$$\frac{x - 1}{x^2 - x - 6} \cdot \frac{x + 2}{x^2 + 2x - 3} = \frac{(x - 1)(x + 2)}{(x^2 - x - 6)(x^2 + 2x - 3)}$$

$$= \frac{(x - 1)(x + 2)}{(x - 3)(x + 2)(x + 3)(x - 1)}$$

$$= \frac{1}{x^2 - 9}$$

To add rational expressions rename the addends so that the denominators are the same, then add the numerators and use the common denominator.

Example 5 Add $\frac{x + 1}{2x^2 - 5x + 3}$ and $\frac{x - 2}{3x^3 - 8x + 5}$.

Solution:

$$\frac{x+1}{2x^2-5x+3}+\frac{x-2}{3x^2-8x+5}$$

$$=\frac{x+1}{(2x-3)(x-1)}+\frac{x-2}{(3x-5)(x-1)}$$

$$=\frac{(x+1)(3x-5)}{(2x-3)(x-1)(3x-5)}+\frac{(x-2)(2x-3)}{(3x-5)(9x-1)(2x-3)}$$

$$=\frac{(x+1)(3x-5)+(x-2)(2x-3)}{(2x-3)(x-1)(3x-5)}$$

$$=\frac{3x^2-2x-5+2x^2-7x+6}{(2x-3)(x-1)(3x-5)}$$

$$=\frac{5x^2-9x+1}{6x^3-25x^2+34x-15}$$

$\frac{2}{3}: \frac{10}{15} = \frac{2}{3} \cdot \frac{5}{3}$ ● $4x - 3: \frac{4x^2-7x+3}{x-1}$ ● $\frac{3}{x-y}: \frac{-3}{y-x}$ ● $\frac{x+1}{x^2-x-6}: \frac{2x^2-3x-5}{(x-3)(x+2)(2x-5)}$ ● $\frac{1}{x-y}: \frac{xy}{x^2y-xy^2}$

Rename each of the following rational expressions so that the denominator is the polynomial shown in brackets: $\frac{2}{3}$, [15]; $4x - 3$, $[x - 1]$; $\frac{3}{x-y}$, $[(y - x)]$; $\frac{x+1}{x^2-x-6}$, $[(x - 3)(x + 2)(2x - 5)]$; and $\frac{1}{x-y}$, $[x^2y - xy^2]$.

The following definition indicates how to find the sum of any two rational expressions r_1, and r_2.

Definition Let r_1 and r_2 be two rational expressions. Let r_1 and r_2 be named so they have a common denominator: $r_1 = \frac{p_1}{q}$ and $r_2 = \frac{p_2}{q}$, then

$$r_1 + r_2 = \frac{p_1}{q} + \frac{p_2}{q} = \frac{p_1 + p_2}{q}.$$

$x + 1$, $\frac{y+x}{xy}$, $\frac{x^2+5}{x}$, $\frac{5+x^2}{x}$

Find a simplest form for each of the following elements

$$\frac{x^2+2x}{x+1}+\frac{1}{x+1};\ \frac{1}{x}+\frac{1}{y};\ x+\frac{5}{x};\ \frac{5}{x}+x$$

To subtract rational expressions rename the rational expressions so that the denominators are the same, then subtract the numerators and use the common denominator.

Example 6 Subtract $\frac{x-2}{x^2-x-6}$ from $\frac{x+1}{x^2+x-2}$.

Solution:

$$\frac{x+1}{x^2+x-2} - \frac{x-2}{x^2-x-6}$$

$$= \frac{x+1}{(x+2)(x-1)} - \frac{x-2}{(x-3)(x+2)}$$

$$= \frac{(x+1)(x-3)}{(x+2)(x-1)(x-3)} - \frac{(x-2)(x-1)}{(x-3)(x+2)(x-1)}$$

$$= \frac{(x^2-2x-3)-(x^2-3x+2)}{(x+2)(x-1)(x-3)}$$

$$= \frac{x^2-2x-3-x^2+3x-2}{(x+2)(x-1)(x-3)}$$

$$= \frac{x-5}{x^3-2x^2-5x+6}$$

Example 7 Find a simplest form for $\frac{1}{x+y} - \frac{x-y}{x^2+xy} - \frac{x}{xy+y^2}$.

Solution:

$$\frac{1}{x+y} - \frac{x-y}{x^2+xy} - \frac{x}{xy+y^2}$$

$$= \frac{xy}{(x+y)xy} - \frac{(x-y)y}{(x+y)xy} - \frac{x^2}{(x+y)xy}$$

$$= \frac{xy-[(x-y)y]-x^2}{(x+y)xy}$$

$$= \frac{xy-xy+y^2-x^2}{(x+y)xy}$$

$$= \frac{(y-x)(y+x)}{(x+y)xy}$$

$$= \frac{y-x}{xy}$$

Exercises 1. Which pair of symbols denote the same rational expression? Use the definition of equality of two rational expressions to justify your answer in each case.

a. $\frac{8}{12}, \frac{2}{3}$ b. $\frac{63}{20}, \frac{21}{6}$ c. $\frac{x-3}{2(x+4)}, \frac{x-3}{2x+8}$

d. $\frac{-(x+1)}{x-1}, \frac{x+1}{-(x-1)}$ e. $\frac{x+1}{x^2-1}, \frac{1}{x-1}$ f. $\frac{x^2-5}{x}, \frac{x+5}{1}$

2. Find a simplest form for rational expression.

a. $\frac{35}{79}$ b. $\frac{4x^2}{6x^3-8x^2}$

c. $\frac{10x^3}{4x^4-12x^3}$ d. $\frac{2x^2-5x-3}{2x^2-7x+3}$

e. $\frac{2x^2-x-15}{2x^2-3x-20}$ f. $\frac{123}{111}$

g. $\frac{34x^4y^6}{51x^3y^7}$ h. $\frac{14x^7y^4}{77x^8y^3}$

i. $\frac{6x^2+8x-8}{6x^2+2x-4}$ j. $\frac{x^2-3xy-4y^2}{x^2-5xy+4y^2}$

k. $\frac{x^3-3x^2-x+3}{x^3-4x^2-x+4}$ l. $\frac{x^3-5x^2-2x+24}{x^3-x^2-14x+24}$

m. $\frac{x^4-3x^2-4}{x^4+5x^2+4}$ n. $\frac{xy-4y-x+4}{xy-y-x+1}$

o. $\frac{x^4-2x^2y^2+y^4}{(x+y)^2(x-y)^2}$ p. $\frac{x^4-2x^2y^2+y^4}{y^3-x^2y+xy^2-x^3}$

q. $\frac{x^3+2x^2-13x+10}{x^4+4x^3-9x^2-16x+20}$

r. $\frac{x^4-x^2-x^2y^2+y^2}{-x^4+2x^3y-x^2y^2+x^2-2xy+y^2}$

3. Find a simplest form for each product.

a. $\frac{2x-4}{x+7}\cdot\frac{x-7}{2x+3}$ b. $\frac{6x^3}{2x-1}\cdot\frac{4x-2}{3x^5}$

c. $\frac{4x-2}{3x^5}\cdot\frac{6x^3}{2x-1}$ d. $\frac{3x}{4x^2-9}\cdot\frac{2x-3}{7x^3}$

e. $\frac{6x^2-5x-6}{8x^2+10x-3}\cdot\frac{4x-1}{x^2+x}$ f. $\frac{x-y}{x+y}\cdot\frac{x^2-y^2}{x^2+xy}$

g. $\frac{3y^4}{7x^3}\cdot\frac{21x^4}{22y^5}$ h. $\frac{x-2}{x^2-x-6}\cdot\frac{x^2+2x-3}{x+1}$

i. $\frac{x-y}{x+2y}\cdot\frac{x^2-4y^2}{x^2-xy}$ j. $\frac{3x^2y^2+9xy+7}{x^3y^2}\cdot\frac{144x^9+x}{144x^9+x}$

k. $\frac{y^2-9x^2}{y^2-xy}\cdot\frac{x^2-xy}{6x^2-xy-y^2}$ l. $\frac{x^2-1}{6x^2+5x-6}\cdot\frac{3x^2-8x+4}{x^2-4}$

m. $\dfrac{x^4 - y^4}{x^4 + 2x^2y^2 + y^4} \cdot \dfrac{x^2 + y^2}{x^2 - y^2}$

n. $\dfrac{15x^3y^2}{19x^2y^4} \cdot \dfrac{57x^4y}{70xy^3}$

o. $\left(\dfrac{x^2}{x + 1} \cdot \dfrac{x - 1}{x^3}\right) \cdot \dfrac{x + 2}{x^2 - 1}$

p. $\dfrac{x^3 - y^3}{x - y} \cdot \dfrac{y^3 - x^3}{y^3 - x^3}$

q. $\dfrac{2x^3 - 7x^2 - 7x + 12}{2x^3 - 15x^2 + x + 42} \cdot \dfrac{x^2 - 9x + 14}{x^2 - 5x + 4}$

r. $\dfrac{144x^2 + 12x + 6}{42xy} \cdot \dfrac{42xy}{144x^2 + 12x + 6}$

s. $\dfrac{7x^2 - 22x + 3}{2x^2 - x - 15} \cdot \dfrac{6x^2 - x - 40}{21x^2 - 59x + 8}$

t. $\dfrac{(x^3 - y^3)(x^2 - y^2)}{(x - y)^2} \cdot \dfrac{(x - y)^2}{(x^3 - y^3)(x^2 - y^2)}$

u. $\dfrac{x^3 - 3x^2 + 2x - 6}{6x^3 + 19x^2 - 52x + 15} \cdot \dfrac{2x^3 + 5x^2 - 22x + 15}{x^3 + x^2 + 2x + 2}$

v. $\dfrac{x^2 + 7x + 7}{2x^2 + 7x - 4} \cdot \dfrac{3x^2 + 19x + 28}{2x^2 - 3x + 3} \cdot \dfrac{2x^2 - 7x + 3}{3x^2 - 8x - 35}$

4. Find a simplest form for each rational expression.

a. $\dfrac{4}{9} + \dfrac{5}{12}$

b. $\dfrac{7}{10} - \dfrac{3}{15}$

c. $\dfrac{4}{3x - 5} + \dfrac{6}{3x - 1}$

d. $\dfrac{1}{5x - 2} - \dfrac{3}{x - 3}$

e. $\dfrac{2}{3x - 4} + \dfrac{1}{2x + 1}$

f. $\dfrac{3}{x - y} + \dfrac{2}{y - x}$

g. $\dfrac{2}{x^2 + 2x - 3} + \dfrac{1}{x^2 + 5x + 6}$

h. $\dfrac{1}{x + y} + \dfrac{1}{x - y}$

i. $\dfrac{1}{x} + \dfrac{1}{y} - \dfrac{1}{x - y}$

j. $\dfrac{x + 1}{x^2 - x - 6} - \dfrac{2x - 1}{2x^2 - 11x + 15}$

k. $4x - 3 + \dfrac{1}{x - 1}$

l. $3x - 1 + \dfrac{1}{2x - 7}$

m. $\dfrac{1}{x} - \dfrac{1}{y}$

n. $\dfrac{x}{y} - \dfrac{y}{x}$

o. $\dfrac{1}{x + y} - \dfrac{1}{x - y}$

p. $x^2 + 2 - \dfrac{x - 1}{2x^2 - 3x + 8}$

q. $\dfrac{2x - 1}{x^2 + 2x - 8} + \dfrac{x + 1}{x^2 + x - 12}$

r. $\dfrac{3x - 1}{x^2 + 2x - 3} - \dfrac{x - 2}{x^2 + 5x + 6}$

s. $\dfrac{2x + 3}{x^2 - 9} - \dfrac{3}{x + 3}$

t. $\dfrac{3x - 1}{x^2 - 2x} + \dfrac{1}{2 - x}$

u. $2 - \dfrac{3}{x} + \dfrac{1}{x^2 - 2x}$

v. $\dfrac{2x - y}{x^2 - y^2} + \dfrac{x - 2y}{y^2 - x^2}$

w. $\dfrac{2x+1}{x^2+2x-3}+\dfrac{x-4}{2x^2+3x-5}-\dfrac{3x-2}{2x^2+11x+15}$

x. $\dfrac{3x-4y}{x^2-y^2}+\dfrac{2}{x+y}-\dfrac{3}{y-x}$

y. $\dfrac{x-3}{2x^2+x-6}-\dfrac{2x+1}{2x^2-7x+6}+\dfrac{x+5}{4-x^2}$

z. $\dfrac{x+4}{x^2+x-6}+\dfrac{x-4}{x^2-x-2}-\dfrac{x^2+x+2}{x^3+2x^2-5x-6}$

2-8 DIVISION OF RATIONAL EXPRESSIONS

As with the other operations, division of rational expressions is very similar to division with rational numbers. Division is defined in terms of the primary operation of multiplication.

Definition For rational expressions r_1 and r_2, $r_1 \div r_2 = r_3$ provided r_3 is a rational expression, and r_3 uniquely satisfies $r_3 \cdot r_2 = r_1$.

As you know, the basic computational procedure for finding quotients of rational numbers involves changing to the multiplication operation. For example

$$\frac{2}{5} \div \frac{3}{7} = \frac{2}{5} \cdot \frac{7}{3}.$$

This is justified by showing that $\left(\dfrac{2}{5} \cdot \dfrac{7}{3}\right) \cdot \dfrac{3}{7}$ is, indeed, $\dfrac{2}{5}$. A similar working principle is used for computation with rational expressions.

Theorem For rational expressions $\dfrac{p}{q}$ and $\dfrac{r}{t}$ and $\dfrac{r}{t} \neq 0$, $\dfrac{p}{q} \div \dfrac{r}{t} = \dfrac{p}{q} \cdot \dfrac{t}{r}$.

Proof: Since $\left(\dfrac{p}{q} \cdot \dfrac{t}{r}\right) \cdot \dfrac{r}{t} = \dfrac{p}{q}$, then $\dfrac{p}{q} \div \dfrac{r}{t} = \dfrac{p}{q} \cdot \dfrac{t}{r}$. (Why?)

Example Divide $\dfrac{2x^2-15x+18}{2x^2+5x+3}$ by $\dfrac{2x^2+x-6}{2x^2+x-3}$.

Solution:

$$\frac{2x^2-15x+18}{2x^2+5x+3} \div \frac{2x^2+x-6}{2x^2+x-3}$$

$$= \frac{2x^2 - 15x + 18}{2x^2 + 5x + 3} \cdot \frac{2x^2 + x - 3}{2x^2 + x - 6}$$

$$= \frac{(2x - 3)(x - 6)}{(x + 1)(2x + 3)} \cdot \frac{(x - 1)(2x + 3)}{(2x - 3)(x + 2)}$$

$$= \frac{x^2 - 7x + 6}{x^2 + 3x + 2}$$

Divide $\dfrac{6x^2 + x - 1}{3x^2 - 5x - 2}$ by $\dfrac{2x^2 + 7x + 3}{3x^2 - 7x + 2}$; divide $\dfrac{x^3 - y^3}{x^3 + x^2y + xy^2 + y^3}$ by $\dfrac{x^3 + 2x^2y + 2xy^2 + y^3}{x^3 - x^2y + xy^2 - y^3}$.

$\dfrac{9x^2 - 6x + 1}{3x^2 + 10x + 3}$; • $\dfrac{x^2 - 2xy + y^2}{x^2 + 2xy + y^2}$

Another notation is commonly used for division. If $\frac{p}{q}$ and $\frac{r}{t}$ are rational expressions, then

$$\frac{\frac{p}{q}}{\frac{r}{t}} \text{ means } \frac{p}{q} \div \frac{r}{t}.$$

This notation is also used with rational numbers:

$$\frac{\frac{2}{5}}{\frac{3}{7}} = \frac{2}{5} \div \frac{3}{7} = \frac{2}{5} \cdot \frac{7}{3} = \frac{14}{15}.$$

If this alternative notation is used, a simplest form of the rational number is obtained by multiplying by $1 = \frac{m}{m}$, where m is the least common multiple of the denominators. The least common multiple of 5 and 7 is 35. In the above example, we have

$$\frac{\frac{2}{5}}{\frac{3}{7}} \cdot \frac{35}{35} = \frac{\frac{2}{5} \cdot 35}{\frac{3}{7} \cdot 35} = \frac{14}{15}.$$

This procedure is applicable to quotients of rational expressions.

$$\frac{\frac{1}{x-2}+\frac{1}{x}}{\frac{1}{x-3}+\frac{2}{x}} = \frac{\frac{1}{x-2}+\frac{1}{x}}{\frac{1}{x-3}+\frac{2}{x}} \cdot \frac{(x-2)\cdot x\cdot(x-3)}{(x-2)\cdot x\cdot(x-3)}$$

$$= \frac{x(x-3)+(x-2)(x-3)}{(x-2)x+2(x-2)(x-3)} = \frac{2x^2-8x+6}{3x^2-12}.$$

Exercises Find a simplest form for each rational expression in Exercises 1-22.

1. $\dfrac{1-\frac{3}{5}}{2+\frac{4}{5}}$

2. $\dfrac{3+\frac{\frac{2}{3}}{5}}{\frac{\frac{4}{2}}{5}-\frac{\frac{4}{5}}{2}}$

3. $\dfrac{2x^2-13x-70}{3x^2-16x-12} \div \dfrac{3x+2}{x-6}$

4. $\dfrac{2x^2-7x+3}{x^2-2x-8} \div \dfrac{4x^2-4x+1}{2x^2+5x+2}$

5. $\left(\dfrac{1}{x}+\dfrac{1}{y}\right) \div \left(\dfrac{1}{x}-\dfrac{1}{y}\right)$

6. $2 \div \dfrac{2x}{3x+4}$

7. $\dfrac{2x^2-5xy+2y^2}{x^2-3xy+2y^2} \div \dfrac{x+y}{x-2y}$

8. $\dfrac{x-3}{2x+5} \div \dfrac{x^3-3x^2+x-3}{2x^3+5x^2-2x-5}$

9. $\dfrac{x+\frac{3}{2-x}}{x-2+\frac{1}{2+x}}$

10. $\dfrac{1-\frac{x}{y}}{1+\frac{x}{y}}$

11. $1+\dfrac{\frac{1}{1+x}}{\frac{1}{1-x}}$

12. $\dfrac{x+\frac{3y}{4}}{\frac{y}{x}-\frac{x}{2y}}$

13. $\dfrac{y+\frac{1}{x}}{x-\frac{1}{y}}$

14. $\dfrac{\frac{1}{x}-\frac{1}{x-y}}{\frac{1}{x+y}+\frac{1}{y}}$

15. $\dfrac{\frac{x}{x+y}-\frac{x}{x-y}}{\frac{y}{x+y}+\frac{y}{x-y}}$

16. $x-\dfrac{1}{x-\frac{1}{1-x}}$

17. $\dfrac{6x^2 + x - 12}{8x^2 - 14x + 5} \cdot \dfrac{10x^2 - x - 2}{12x^2 - 19x + 4} \div \dfrac{10x^2 + 19x + 6}{16x^2 - 24x + 5}$

18. $\dfrac{\dfrac{1}{(x+y)^2} - \dfrac{1}{x^2}}{y}$

19. $\dfrac{\dfrac{1}{x+y} - \dfrac{1}{x}}{y}$

20. $\dfrac{x^3 - y^3}{x^2 + y^2} \cdot \dfrac{x + y}{x^2 - y^2} \div \dfrac{x^2 + xy + y^2}{x^4 - y^4}$

21. $\dfrac{\dfrac{x - 4}{x^2 + 9x - 10}}{\dfrac{x - 1}{x^2 + 6x - 40}}$

22. $\dfrac{2x + 1 - \dfrac{1}{x + 1}}{\dfrac{1}{x - 1} + \dfrac{2x - 1}{x + 1}}$

23. Complete the proof of the Theorem on page 41.

Review Exercises

1. What is the degree of each polynomial? 2-1
   a. $3x^2 + 2x + 4$ b. $2x^2 + 7$ c. 0 d. $2x + 4$ e. 3

2. a. Find the sum of the five polynomials in Exercise 1.
   b. Find the additive inverse of this sum.

3. What polynomial is $(6x - 7)^3$? 2-2

4. Find the quotient and remainder if the dividend is $x^7 + 4x^5 + 2x^3 - 3x^2 + 4$ and the divisor is $x^4 - 2x^2 + 5x - 1$.

5. Use synthetic division to compute the quotient and remainder if $x^5 + 3x^3 + 6x^2 - 5$ is divided by 2-3
   a. $x - 2$ b. $x + 3$

6. Find the prime factors of $x^4 - 6x^2 + 5$ 2-4
   a. using real coefficients.
   b. using integral coefficients.

7. Find the sum and product for the polynomials $4x^3y - 3xy^3 + 7xy$ and $xy - 5y$. 2-5

Factor each polynomial in Exercises 8–11 using integral coefficients.

8. $xy^2 - x^2y$ 2-6

9. $216y^3 + x^3$

10. $24x^2y - 6xy^2 - 3y^3$

11. $3x^2y - 4x^2 - 3y^3 + 4y^2$

Use integral coefficients in 12–16.

12. Find a simplest form for $\dfrac{x^4 - y^4}{2x^4 - x^2y^2 - 3y^4}$. 2-7

13. Express the product $\frac{x^3 - y^3}{x^3 + y^3} \cdot \frac{x^2y - xy^2 + y^3}{x^3 + x^2y + xy^2}$ in a simplest form.

14. Express the sum $\frac{x - y}{x^2 - xy} + \frac{x + y}{xy - y^2}$ in a simplest form.

15. Find a simplest form for $\frac{x - 2y}{(x - 2y)(2x + 3y)} - \frac{x - y}{(x + 2y)(2x + 3y)}$.

16. Find a simplest form for each rational expression. 2-8

a. $\frac{14x^2 - 37xy + 5y^2}{y^2 - 6yx - 7x^2} \div \frac{5y^2 - 7xy + 2x^2}{x^2 - y^2}$

b. $\frac{(y + 1) + \frac{1}{x + 1}}{(x + 1) - \frac{1}{y + 1}}$

3. LINEAR EQUATIONS AND INEQUALITIES

3-1 SENTENCE AND SOLUTION SETS

The real number line is a one-to-one correspondence between the real numbers and points on a line. Thus each real number is represented by a point on the line, and each point represents a real number. With it we are able to analyze problems both numerically and geometrically.

Ordered pairs of real numbers identify points in a plane. We denote the set of all ordered pairs of real numbers by the symbol $R \times R$ (read "R cross R"), where

$$R \times R = \{(x, y):\ x \in R \text{ and } y \in R\}.$$

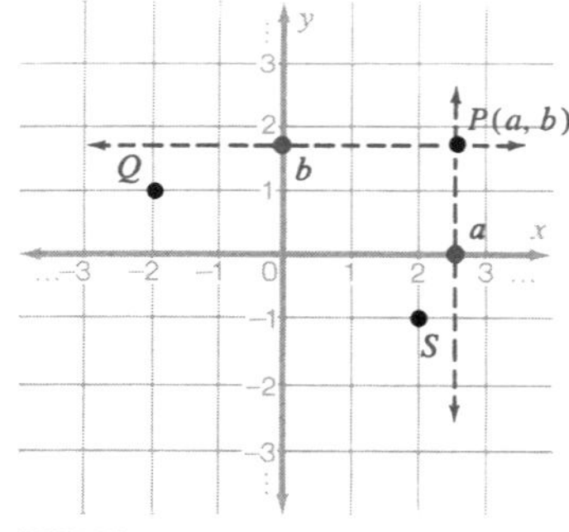

FIGURE 1

In a plane, perpendicular real number lines which intersect at their origins are used as reference axes. The horizontal line is called the ***x*-axis**; the vertical line, the ***y*-axis**. Points on these lines correspond to real numbers as indicated in Figure 1.

Let P be a point in the plane. If the line containing P and perpendicular to the x-axis intersects this horizontal axis at $a \in R$ and if the line containing P and perpendicular to the y-axis intersects this vertical axis at $b \in R$, then we identify P with the ordered pair (a, b) [Figure 1]. If the ordered pair is given, we can determine the corresponding point in the plane. This arrangement of identifying points of a plane and the elements of $R \times R$ is called a **coordinate plane**. This identification of points with ordered pairs establishes a one-to-one correspondence between the points on a plane and the set of ordered pairs, $R \times R$. Thus, we also have a correspondence between geometric figures in the plane and subsets of $R \times R$. For example, if

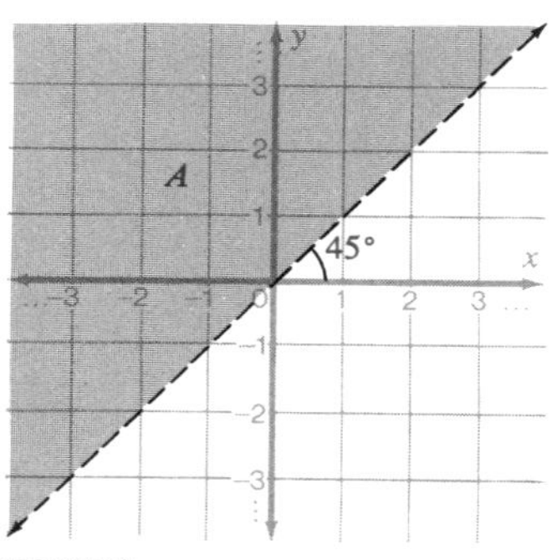

FIGURE 2

$$A = \{(x, y) \in R \times R:\ x < y\}$$

then A is a subset of $R \times R$ and the graph of A is the half-plane consisting of those points above the line in Figure 2. Henceforth, the reference set for (x, y) is understood to be $R \times R$ unless specified otherwise and we will

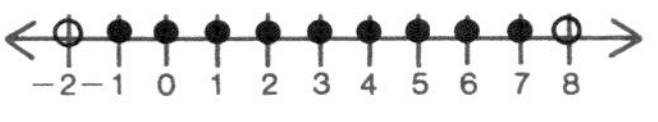

FIGURE 3

eliminate "$\in R \times R$" from our notation and write

$$\{(x, y)\colon\ x < y\} \text{ instead of } \{(x, y) \in R \times R\colon\ x < y\}.$$

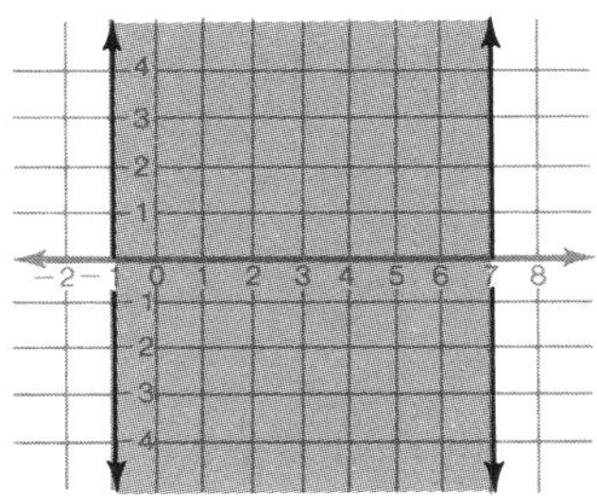
FIGURE 4

Consider the equation

$$|x - 3| = 4.$$

Theoretically we test each real number; that is, we say that the replacement set for the variable x in the equation $|x - 3| = 4$ is R. The real numbers that yield true statements, 7 and -1, are called **solutions** of the equation. The set $\{7, -1\}$ is called the **solution set** of the equation.

Until the replacement set has been clearly identified, we cannot meaningfully find a solution for a sentence. We will use the term sentence to mean any declarative sentence of grammar that is true or false, or becomes true or false when the variables are replaced by appropriate elements. By a statement we will mean a sentence that is true or false. Consider different replacement sets for the variable in the inequality

$$|x - 3| \leqslant 4.$$

1. If I is the replacement set for x, then $\{-1, 0, 1, 2, 3, 4, 5, 6, 7\}$ is the solution set [Figure 3].

2. If the replacement set for x is R, find the solution set.

The solution set is the set of all points between -1 and 7, and the points -1 and 7; that is,

$$\{x\colon\ -1 \leqslant x \leqslant 7\}$$

3. If the replacement set for x is R and the replacement set for y is R (the replacement set for (x, y) is $R \times R$), then the sentence $|x - 3| \leqslant 4$ places no restriction on y. Thus, the solution set is $\{(x, y)\colon\ -1 \leqslant x \leqslant 7\}$ [Figure 4].

Constructing graphs of the replacement and solution sets for these three cases gives another method of representing the solution sets of equations and inequalities.

Exercises

Describe the geometric figure that corresponds to each set.

1. $\{x\colon\ 0 \leqslant x \leqslant 4\}$
2. $I \times I$
3. $\{(x, y)\colon\ x = y\}$
4. $\{(x, y) \in I \times I\colon\ x = 4 \text{ and } y = 2\}$
5. $\{(x, y)\colon\ xy = 0\}$

Graph each subset of $R \times R$ and describe the geometric figure involved.

6. $\{(x, y)\colon\ x + y = 8\}$
7. $\{(x, y)\colon\ x \geqslant 2\}$
8. $\{(x, y)\colon\ 2 \leqslant x \leqslant 4 \text{ and } y \geqslant 0\}$
9. $\{(x, y)\colon\ x \geqslant 2 \text{ and } y \leqslant -1\}$

Graph the replacement and solution sets for each sentence in Exercises 10–13.

	Sentence	*Replacement set(s) for x or (x, y)*	
10.	$\lvert x + 2 \rvert \leqslant 1$	a. I	b. R
11.	$\lvert 2x - 3 \rvert > 4$	a. I	b. R
12.	$\lvert x \rvert + \lvert y \rvert \leqslant 4$	a. $I \times I$	b. $R \times R$
13.	$xy = 0$ and $\lvert x \rvert \leqslant 4$	a. $I \times I$	b. $R \times R$

14. Draw the graph of the solution set of the sentence $\lvert x \rvert < 4$ for each replacement set.
a. I b. $I \times I$ c. R d. $R \times R$

3-2 LINEAR EQUATIONS AND INEQUALITIES IN ONE VARIABLE

If the replacement set for x is R in the equation $7x + 3 = 4x - 5$, we can test various real numbers to find those that yield true statements. Or we can transform $7x + 3 = 4x - 5$ into one whose solution set is obvious. Since x in the equation is to be replaced by real numbers to find the solution set, we can treat members of equations using properties of real numbers. Thus the following equations are equivalent.

$$7x + 3 = 4x - 5$$

$$7x + (-4x) + 3 = 4x + (-4x) - 5 \qquad \text{(Add } -4x \text{ to both members.)}$$

$$3x + 3 = -5$$

$$3x + 3 + (-3) = -5 + (-3) \qquad \text{(Add } -3 \text{ to both members.)}$$

$$3x = -8$$

$$\frac{1}{3}(3x) = \frac{1}{3}(-8) \qquad \left(\text{Multiply both members by } \frac{1}{3}.\right)$$

$$x = -\frac{8}{3}$$

The last equation is taken as the final stage, and we have solved the equation.

Each of the equations in the preceding sequence is equivalent to the other equations. By **equivalent** equations, we mean that they have the same solution set. Notice that the equation $x = -\frac{8}{3}$ is in such a simple form that the solution set is obvious. An equation whose solution set is readily apparent is precisely what we are seeking.

To solve an equation, we usually write only a brief sequence of equivalent equations.

Definition **A linear equation in one variable**, x, is an equation that is equivalent to $ax + b = 0$, where $a \neq 0$.

Thus, the sentence $7x + 3 = 4x - 5$ is an example of a linear equation in one variable since it is equivalent to $3x + (-8) = 0$. **Solving** a linear equation in one variable means finding an equivalent equation of the form $x = r$.

Solve the equation $4x - 9 = 0$.

$4x - 9 = 0$
$4x - 9 + 9 = 0 + 9$
$4x = 9$
$\frac{1}{4}(4x) = \frac{1}{4}(9)$
$x = \frac{9}{4}$

Theorem 1 The linear equation $ax + b = 0$ is equivalent to the equation $x = -\frac{b}{a}$.

Definition **A linear inequality in one variable** x is an inequality equivalent to one of the following:

$$ax + b < 0$$

$$ax + b \leqslant 0$$

$$ax + b > 0$$

$$ax + b \geqslant 0.$$

Solving the inequality $x + 4 < 7$ is very similar to solving the linear equation $x + 4 = 7$. We find an equivalent inequality that has a solution set which is immediately apparent. The x for inequalities is replaced by real numbers, hence the members of inequalities can be subjected to changes based on number properties. If R is the replacement set, the following inequalities are equivalent:

$$x + 4 < 7$$

$$x + 4 + (-4) < 7 + (-4) \qquad \text{(Add } -4 \text{ to both members.)}$$

$$x < 3.$$

In solving linear inequalities, we search for an inequality from one of the forms, $x < r$ or $x > r$.

Theorem 2 The linear inequality $ax + b > 0$ is equivalent to $ax > -b$ for $a \neq 0$.

There are versions of Theorem 2 for the inequalities $ax + b \geqslant 0$, $ax + b \leqslant 0$, and $ax + b < 0$.

Another basic pair of equivalent inequalities is illustrated by: $-3x > 5$,

$x < -\frac{5}{3}$. The second inequality is derived by multiplying the members of the first by the *negative* real number, $-\frac{1}{3}$. Multiplication by a negative number changes the sense of the inequality. Multiplication by a positive number preserves the sense of the inequality. Theorem 3 summarizes these statements.

Theorem 3 The inequality $ax > d$ is equivalent to $x > \frac{d}{a}$ if $a > 0$, or to $x < \frac{d}{a}$ if $a < 0$.

Other versions of this theorem involve the relations $\geqslant, <$, and $\leqslant$.

We use these theorems to solve linear inequalities. To solve $4 - 3x > 0$, we have

$$-3x + 4 > 0$$

$$-3x > -4$$

$$x < \frac{4}{3}.$$

Since $4 - 3x > 0$ is equivalent to $x < \frac{4}{3}$, the graph of the solution set is a half-line:

Describe the graph of the solution set for $4 - 3x > 0$.

We conclude with a somewhat more complicated example:

$$2(x - 3) + 4 > x - 3(5 - 2x)$$

$$2x - 2 > 7x - 15$$

$$-5x > -13$$

$$x < \frac{13}{5}.$$

The graph of the solution set on the real number line can be described as the set of all points extending to the left of $\frac{13}{5}$.

Exercises

Solve each equation in Exercises 1-16.

1. $14x - 3 = 6x + 13$
2. $4x + 9 = 9x - 4$
3. $4(3x - 5) = 7x - 5$
4. $2x - \sqrt{2} = x + 1$
5. $2(3 - x) - 4(x + 5) = 16$
6. $-3(2 - 4x) = x - 2(7 - x)$
7. $1 - \pi x = \pi^2 + x$
8. $6x + 17 = 10x - 3$
9. $11 - 3x = 2x - 7$
10. $x - 10 = -2(x - 7)$

11. $\pi x - 7 = 2x + 3$

12. $4(x - 7) - (2x + 3) = 2 - x$

13. $1 - 7(3 - 4x) = 2x - 5(2 + x)$

14. $2(x - \sqrt{3}) + 1 = 1 - \sqrt{2}x$

15. $4 - [3x - 2(1 - x)] = 4 - 2(x + 7)$

16. $13 - 8x = 1 - [1 - (x - 1) + 3(2 - x)]$

Show that each pair of equations in Exercises 17-20 is equivalent.

17. $7x + 3 = 8,\ x = \frac{5}{7}$

18. $4x = 7x - 10,\ x = \frac{10}{3}$

19. $3x - 7 = 6x + 12,\ x = -\frac{19}{3}$

20. $3(5x - 4) = 3 - 2x,\ x = \frac{15}{17}$

Consider each underlined symbol in Exercises 21-26 a variable and the remaining symbols as specific real numbers. Solve the equations.

21. In geometry the volume V of a prism with base of area b and height h is $V = b\underline{h}$.

22. A basic law of physics relating force F, mass m, and acceleration a is $F = \underline{m}a$.

23. The volume V of a cone having height h and radius of the base r is $V = \frac{1}{3}r^2\underline{h}$.

24. Temperature measured on the Fahrenheit scale F and the Celsius (centigrade) scale C are related by the equation $C = \frac{5}{9}(\underline{F} - 32)$.

25. The distance s an object falls in a vacuum is given by $s = \frac{1}{2}\underline{g}t^2$ where g is the acceleration due to gravity and t is the time.

26. The amount A for principal P invested at rate r after a time period of length t is given by $A = P(1 + \underline{r}t)$.

Solve each inequality in Exercises 27-42. What geometric figure is the graph of each solution set?

27. $2x - 5 > 0$

28. $11 - 3x \leqslant 7$

29. $7x + 4 > 9$

30. $\frac{1}{7}x - \frac{1}{3} \geqslant 14$

31. $4x + 9 > -3x + 7$

32. $\frac{1}{3}x + 4 \geqslant 0$

33. $-3x - 7 < -3$

34. $7 - 4x \leqslant 9$

35. $\frac{2}{3}x + \frac{1}{4} < -\frac{5}{6}$

36. $9 - 5x \geqslant 2x + 11$

37. $4(x - 3) \leqslant -3(2 - x)$

38. $x - (3 - 2x) > 7 - 5x$

39. $-2(3 - x) > x - (7 - x)$

40. $2 - 3x < 1 - 10(x + 1)$

41. $12(3 - 2x) \geqslant 17 + 23(1 - x)$ 42. $3x + 7 \leqslant x - 2(3 - x)$

43. In solving the linear inequality $-3x + 5 \geqslant 12$, we know that the graph of its solution set is a closed ray. To find the endpoint of this ray, solve the equation $-3x + 5 = 12$. $\left(x = -\frac{7}{3}.\right)$ The endpoint of the ray is $-\frac{7}{3}$ but what is its direction? Test any one point other than $-\frac{7}{3}$, say 0. Is it true that $-3(0) + 5 \geqslant 12$? (No!) Hence, the ray extends to the left. The equivalent inequality is $x \leqslant -\frac{7}{3}$. Use this example to generate a technique of solving inequalities by first solving equations. Apply your techniques to Exercises 27-36.

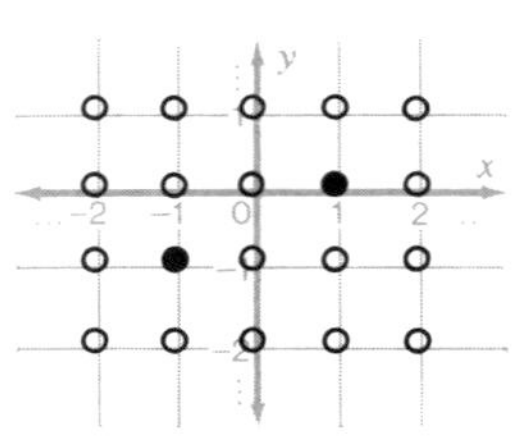

FIGURE 1

Prove that each pair of equations in Exercises 44-47 is equivalent, where $a \neq 0$ and $a \neq c$.

44. $ax + b = cx + d, \ (a - c)x + (b - d) = 0$

45. $x + b = d, \ x = d - b$

46. $ax = b, \ x = \frac{a}{b}$

47. $ax + b = cx + d, \ x = \frac{d - b}{a - c}$

$\{(x, y): x - 2y = 1\}$

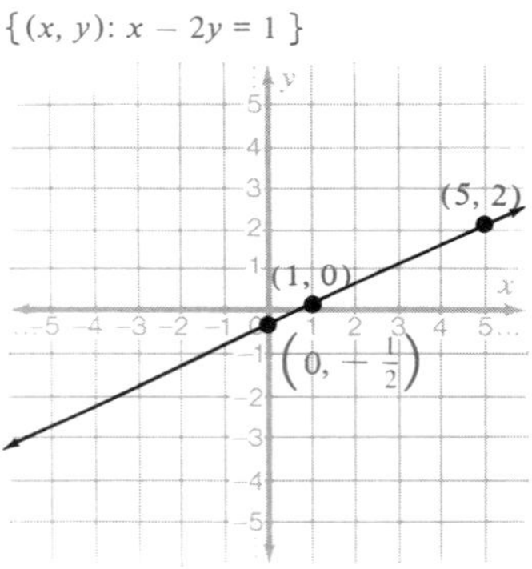

FIGURE 2

3-3 LINEAR EQUATIONS AND INEQUALITIES IN TWO VARIABLES

$\{(x, y): x - 2y = 4\}$

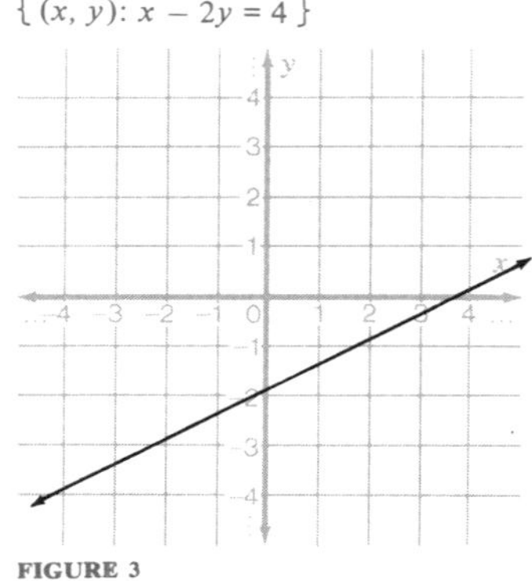

FIGURE 3

If x is replaced by one of the numbers $-2, -1, 0, 1$, or 2 and y is replaced by one of the numbers $-2, -1, 0$, or 1 in the equation $x - 2y = 1$, the graph of the replacement set for (x, y) and the solution set for this equation is found in Figure 1.

If $R \times R$ is the replacement set for (x, y) in the same equation, the graph of the solution set is a *line*. We can sketch the graph by finding two points in the solution set. It is easy to see that $(1) - 2(0) = 1$ is a true statement; hence, $(1, 0)$ belongs to the solution set. The element $\left(0, -\frac{1}{2}\right)$ is also in the solution set, and $(5, 2)$ is included as a check. If we have made no errors, the three points are collinear [Figure 2].

Definition Equations which are equivalent to an equation of the form $ax + by = c$, where either $a \neq 0$ or $b \neq 0$ are called **linear equations in two variables**,

If $b = 0$ the equation is equivalent to $ax = c$. Its graph is a vertical line through $\left(\frac{c}{a}, 0\right)$; $2x = -5$ is equivalent to $2x + 0 \cdot y = -5$ and also to $x = -\frac{5}{2}$. If $a = 0$ the graph is a horizontal line through $\left(0, \frac{c}{b}\right)$.

$\{(x, y)\colon 2x = -5\}$

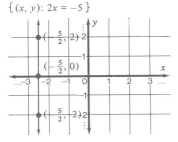

x and y. Also, inequalities equivalent to $ax + by > c$ or $ax + by \geqslant c$ are called **linear inequalities in two variables.**

The term *linear* is appropriate, since the graphs of the solution sets of these equations are lines.

What if $b = 0$? Graph the solution set for $2x = -5$ if $R \times R$ is the replacement set. Describe the graph of the solution set for a linear equation in two variables if $a = 0$.

Associated with each of the linear inequalities is the linear equation $ax + by = c$, which plays a key role in determining the graph of the solution set for the corresponding inequality. For example, if we are trying to identify the graph for the solution set of $x - 2y > 4$, we consider the corresponding equation $x - 2y = 4$. The graph for the solution set of $x - 2y = 4$ is a line [Figure 3]. From geometry we know that a line separates the plane into two half-planes. The points of the line are in neither half-plane. The graph for the solution set of $x - 2y > 4$ is one of the half-planes determined by the line for $x - 2y = 4$. Which side of the line is it? All we need to find is one point in this half-plane. Let us try the point (1, 1): $(1) - 2(1) \not> 4$. Hence, we are on the wrong side. The graph is the half-plane determined by the equation $x - 2y = 4$ and containing the point, say $(3, -2)$ [Figure 4].

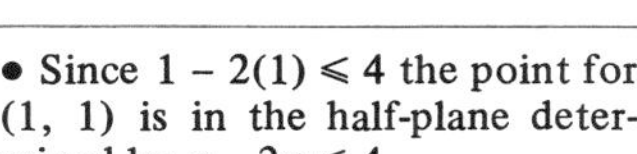

• Since $1 - 2(1) \leqslant 4$ the point for (1, 1) is in the half-plane determined by $x - 2y \leqslant 4$.

$\{(x, y)\colon x - 2y \leqslant 4\}$

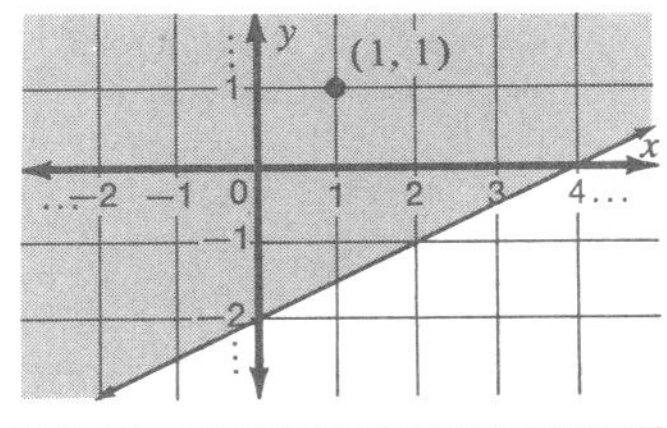

$\{(x, y)\colon x - 2y > 4\}$

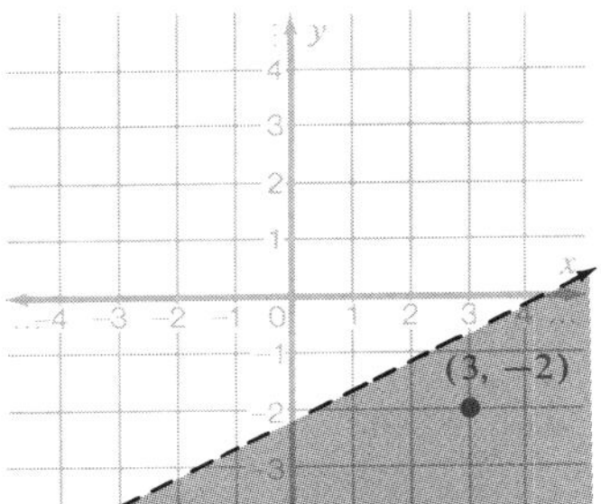

FIGURE 4

$\{(x, y)\colon 5x + 3y \geqslant 15\}$

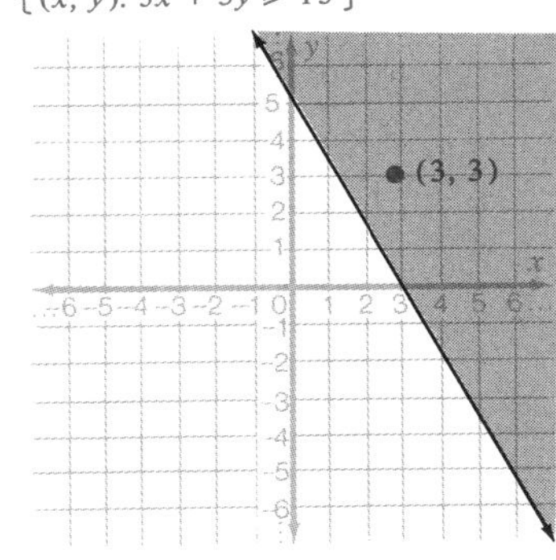

FIGURE 5

Find the graph for the solution set of $x - 2y \leqslant 4$. Is the point (1, 1) in the half-plane determined by this equation?

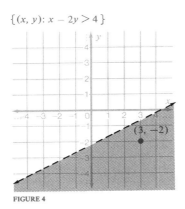

Similarly we find that the graph for the solution set of $5x + 3y \geqslant 15$ is composed of the points in the half-plane determined by the equation $5x + 3y = 15$ together with the points on the line [Figure 5].

Exercises Sketch the graph of the solution sets for each equation in Exercises 1-10.

1. $x + 2y = 2$
2. $x = 1$
3. $3x - 2y = 2$
4. $y = 1$
5. $x + 2y = 5$
6. $7x + 4y = 20$
7. $2x + y = 2$
8. $4x - 5y = 10$
9. $x + y = 3$
10. $4x - y = 2$
11. Prove that the equations $ax + by = c$ and $y = \left(-\frac{a}{b}\right)x + \frac{c}{b}$ are equivalent, where $b \neq 0$.
12. On the same set of coordinate axes, sketch the graphs of the solution sets for the equations $y = ax + (1 - a)$ and a has the different values 0, 1, -1, $\frac{1}{2}$, $-\frac{1}{2}$, 2, -2, $\frac{1}{4}$, $-\frac{1}{4}$, 4, and -4.

Graph the solution sets for each inequality if $(x, y) \in R \times R$.

13. $x + 2y \geqslant 2$
14. $x \leqslant 1$
15. $3x - 2y - 2 > 0$
16. $y > 1$
17. $x + 2y < 5$
18. $7x + 4y < 20$
19. $2x + y > 2$
20. $4x - 5y - 10 \leqslant 10$
21. $x + y - 3 \geqslant 0$
22. $4x - y - 2 > 0$

In Exercises 23-25 sketch the graph of the solution set for the inequality of each part and sketch the intersection of these solution sets.

23. $x + 2y \leqslant 5,\ x \geqslant 0,\ y \geqslant 0$
24. $x + y < 0,\ x - y > 0$
25. $3x - 7y + 27 \geqslant 0,\ 2x - y - 4 \leqslant 0,\ 5x + 3y + 1 \geqslant 0$

3-4 LINES IN THE PLANE

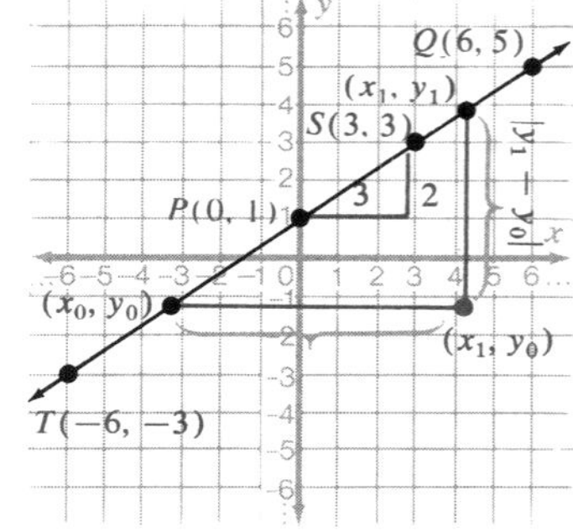

FIGURE 1

Consider the line [Figure 1] through the points $P(0, 1)$ and $Q(6, 5)$. The points $S(3, 3)$ and $T(-6, -3)$ are also on this line. A number is assigned to this line in the following way.

Definition For every two points (x_0, y_0) and (x_1, y_1) on a line, the number

$$\frac{y_1 - y_0}{x_1 - x_0}$$

is the same and is called the **slope** of the line.

Using points P and Q, P and S, T and S, and S and T, in the example, we see that in each case the slope is $\frac{2}{3}$:

$$\frac{5-1}{6-0} = \frac{3-1}{3-0} = \frac{3-(-3)}{3-(-6)} = \frac{(-3)-3}{(-6)-3}.$$

The definition of slope is based on the general result stated in Theorem 1.

Theorem 1 Let L be a nonvertical line [Figure 2] through the two points $P(a, b)$ and $Q(c, d)$. Then for any two points $P_0(x_0, y_0)$ and $P_1(x_1, y_1)$ on L

$$\frac{y_1 - y_0}{x_1 - x_0} = \frac{d-b}{c-a}.$$

Points	*Slope*
(0, 0) and (9, 4)	$\frac{4}{9}$
(0, 0) and (4, 9)	$\frac{9}{4}$
(−7, 6) and (2, −3)	−1
(−1, −3) and (5, −17)	$-\frac{7}{3}$
(−2, 1) and (−2, 43)	no slope

What is the slope of the line through the points (0, 0) and (9, 4)? (0, 0) and (4, 9)? (−7, 6) and (2, −3)? (−1, −3) and (5, −17)? (−2, 1) and (−2, 43)?

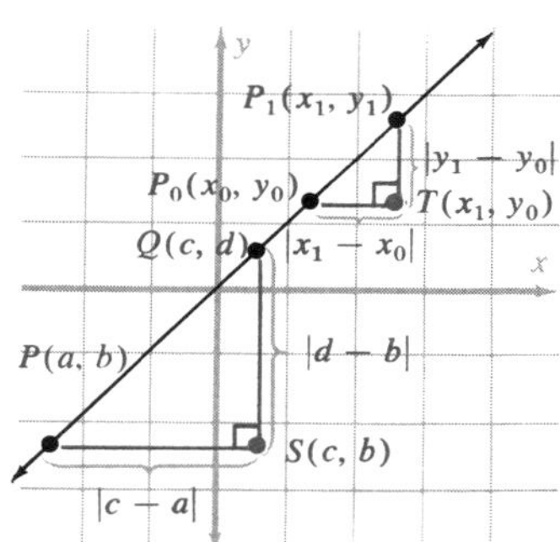

FIGURE 2

Note that vertical lines do not have a slope. Why?

Now let us turn our attention to a characterization of the lines in a coordinate plane. The set of graphs for solution sets in $R \times R$ of the linear equations $ax + by = c$, $a \neq 0$ or $b \neq 0$ is equal to the set of lines in the coordinate plane. All the points on a vertical line [Figure 3] have the same first coordinate, call it d. Also, the graph of the solution set for the linear equation $x = d$ (or $1 \cdot x + 0 \cdot y = d$) is this same vertical line. Similarly, a horizontal line [Figure 4] is the graph of the solution set of linear equation $y = f$ (or $0 \cdot x + 1 \cdot y = f$), and conversely. The solution set for the equation $ax + by = c$ is a vertical line if $a \neq 0$ and $b = 0$. If $b \neq 0$ then it can be shown that the solution set has a graph which is a nonvertical line with slope $\frac{-a}{b}$.

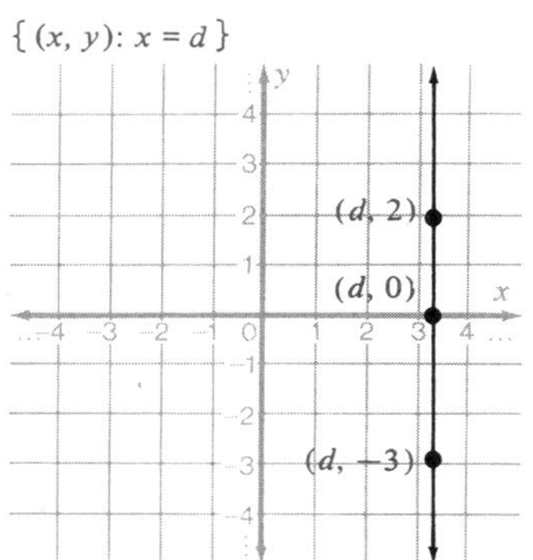

FIGURE 3

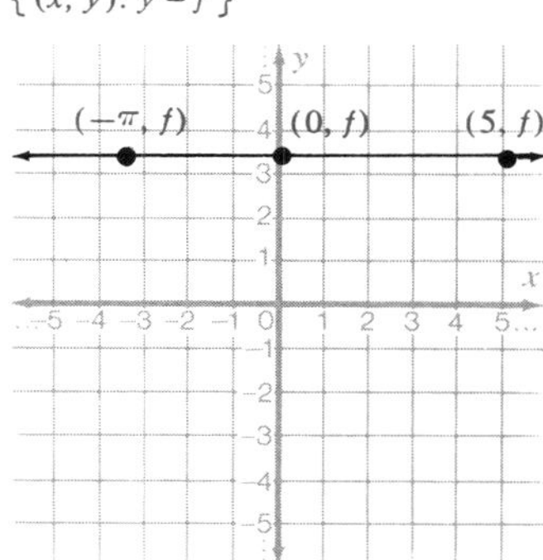

FIGURE 4

The linear equation $(-3)x + 2y = (-4)$ is equivalent to the equation

$$y = \frac{3}{2}x + (-2).$$

The element in the solution set with first coordinate 0 is $(0, -2)$, and is on the y-axis. Also, if (x_0, y_0) and (x_1, y_1) are two points of the line which is the graph for the equation, then the slope of the line is

$$\frac{y_1 - y_0}{x_1 - x_0} = \frac{\left[\frac{3}{2}x_1 + (-2)\right] - \left[\frac{3}{2}x_0 + (-2)\right]}{x_1 - x_0} = \frac{\frac{3}{2}(x_1 - x_0)}{x_1 - x_0} = \frac{3}{2}.$$

The equation $y = \frac{3}{2}x + (-2)$, and in general, any equation whose form is $y = mx + b$ is said to be in **slope-y-intercept form.**

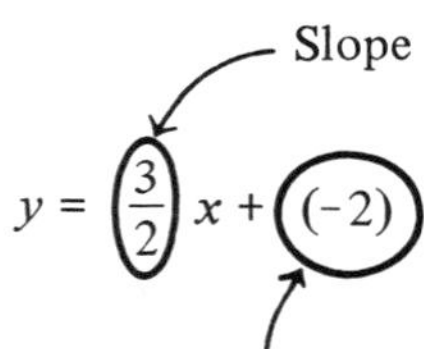

The second coordinate of the point on the y-axis

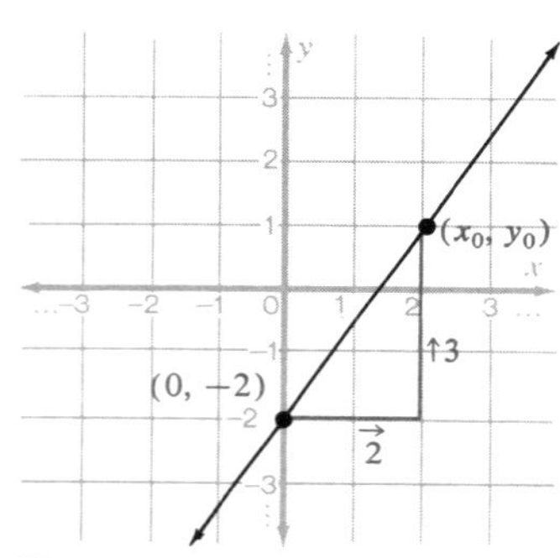

FIGURE 5

To sketch the graph of the solution set for this linear equation, locate the point $(0, -2)$ and draw the line through this point with slope $\frac{3}{2}$ [Figure 5]. The graph of the solution set of $y = mx + b$ is a line through $(0, b)$ [**y-intercept**] with slope m.

Find the slope-y-intercept form for each of the following linear equations: $3x + 2y = 4$, $3x - 2y = 4$, $x - y = 0$, $x + y = 4$.

Linear Equation	*Slope-y-intercept form*
$3x + 2y = 4$	$y = -\frac{3}{2}x + 2$
$3x - 2y = 4$	$y = \frac{3}{2}x + (-2)$
$x - y = 0$	$y = x$ $[y = (1)x + 0]$
$x + y = 4$	$y = -x + 4$ $[y = (-1)x + 4]$

Given the point $(-1, 2)$ and slope (-3), there is a unique line through this point with this slope. Thus, from the definition of slope the set of points of this line is precisely the graph for the equation

$$\frac{y - 2}{x - (-1)} = -3.$$

Find an equivalent form for the above equation of the form $ax + by = c$; of the slope-y-intercept form.

Equivalent equations:

$y - 2 = -3[x - (-1)]$
$y - 2 = -3x - 3$
$3x + 1 \cdot y = -1$ (type $ax + by = c$)
$y = -3x + (-1)$ [slope-y-intercept type]

In general, if (x_1, y_1) is a specific point on a line and m is the slope of the line, then

$$\frac{y - y_1}{x - x_1} = m$$

is a linear equation for this line. This form for a linear equation is called the **point-slope form.**

From elementary geometry we know that two points determine a line. If the points are $(-1, -2)$ and $(2, 7)$ the slope of the line is

$$\frac{7 - (-2)}{2 - (-1)} = \frac{9}{3} = 3.$$

If we use the point $(-1, -2)$ and the point-slope form, we have

$$\frac{y - (-2)}{x - (-1)} = 3$$

which is equivalent to $3x - y = -1$. If we use the point $(2, 7)$ and the point-slope form an equivalent equation is derived.

If (x_1, y_1) and (x_2, y_2) are two points then a linear equation for the line through these points is

$$\frac{y - y_1}{x - x_1} = \frac{y_2 - y_1}{x_2 - x_1}$$

provided $x_2 \neq x_1$.

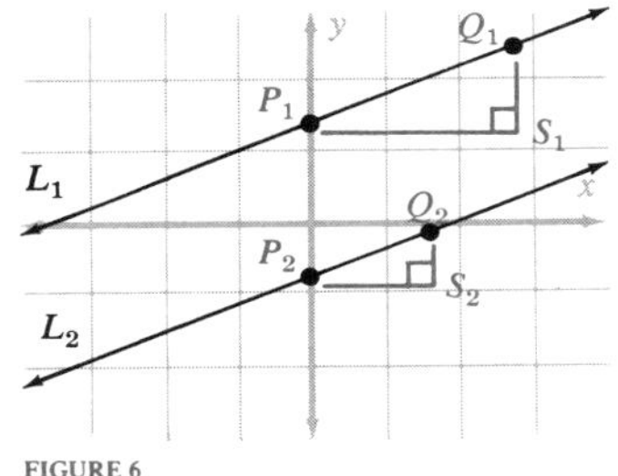

FIGURE 6

In discussing linear equations and lines in the plane, we introduced a number for each nonvertical line, called its slope. We conclude this section with two theorems which show how to use the slopes of two lines to see if the lines are parallel or if they are perpendicular to each other. Proofs of these results are based on concepts from elementary geometry.

Theorem 2 Two lines L_1 and L_2 with slopes m_1 and m_2 are parallel if and only if $m_1 = m_2$.

Proof: If the lines are parallel and have slopes, then each line crosses the y-axis at a different point, P_1 and P_2 [Figure 6]. Let Q_1 and Q_2 be other points on L_1 and L_2, respectively. If L_1 and L_2 are horizontal, what are their slopes? If the lines are not horizontal then the vertical lines through Q_1 and Q_2, and the horizontal lines through P_1 and P_2 form similar right triangles, $\Delta P_1Q_1S_1$ and $\Delta P_2Q_2S_2$. Using these similar triangles, it follows that $m_1 = m_2$. The remainder of the proof is omitted.

Theorem 3 Two lines L_1 and L_2 with slopes m_1 and m_2, respectively, are perpendicular if and only if $m_1 = -\dfrac{1}{m_2}$; that is, if and only if $m_1m_2 = -1$.

Exercises

Find two points of the solution set for each equation in Exercises 1-8. Use these points to find the slope of the line (if it has one).

1. $4x + y = 5$ 2. $4x - y = 5$ 3. $x + 4y = 3$ 4. $x - 4y = 3$

5. $x + 4y = 0$ 6. $x - 4y = 0$ 7. $3x + 0 \cdot y = 5$ 8. $0 \cdot x + 3y = -5$

9. Find the linear equation for each line in the form $ax + by = c$ for the line containing:
   a. (0, 0) and (0, 5) b. (7, 2) and (-2.5, 2)
   c. (0, 0) with slope -1 d. (0, 0) with slope 2
   e. (0, 0) with slope $\frac{1}{2}$ f. (0, 2) with slope 0
   g. (0, 0) with slope 0 h. (0, 0) with no slope

For Exercises 10–21 find the slope-y-intercept form for each linear equation. Sketch the graph of the line. State the slope and coordinates of the point on the y-axis in each case.

10. $6x - 3y = -5$ 11. $6x + 3y = 5$ 12. $6x - 3y = 5$

13. $6x + 3y = -5$ 14. $x - 2y = 0$ 15. $x - 2y = -8$

16. $x + 2y = 0$ 17. $x + 2y = -8$ 18. $y = 0$

19. $5x - 2y = 10$ 20. $2x + 5y = 10$ 21. $2x = 3$

22. Find a linear equation of the form $ax + by = c$ for a line through
   a. (-1, -1) and (5, 5) b. (2, 6) and (-1, 9)
   c. (3, 7) and (3, -2) d. $\left(\frac{1}{2}, \frac{2}{3}\right)$ and $\left(\frac{1}{10}, \frac{1}{10}\right)$
   e. (-1, -1) with slope -1 f. (2, 6) with slope 0
   g. (3, -2) with slope $-\frac{1}{3}$ h. (3, -2) with slope -3.

23. If lines L_1, L_2, L_3, L_4 are graphs for the solution set of the linear equations $3x - y = -2$, $-x + 3y = 3$, $x + 3y = 3$, $3x - y = 2$, find the pairs of parallel lines; the pairs of perpendicular lines.

24. Find 5 linear equations for lines parallel to the line for each equation.
   a. $7x - y = 0$ b. $2x + 3y = 5$.

25. Find a linear equation for the line through point (2, 3) which is parallel to the line for the equation $3x + y = -2$.

26. Find a linear equation for the line through the point (2, 3) which is perpendicular to the line for the equation $3x + y = 2$.

27. If the line for the equation $ax + by = c$ intersects both axes, but not the origin, find the coordinates of these points.

28. Why is the equation

$$\frac{x}{a} + \frac{y}{b} = 1$$

referred to as the **intercept form** for a linear equation? Find the intercept form for $2x + 3y = 6$; for $2x - 5y = 10$. Sketch the graphs of these lines with the aid of the intercept forms.

29. If $b \neq 0$, the linear equation $ax + by = c$ is equivalent to

$$y = \left(-\frac{a}{b}\right)x + \frac{c}{b}.$$

Use the second form of this equation to find the y-intercept.

30. In Exercise 29, let (x_0, y_0) and (x_1, y_1) be elements in the solution set. Then $y_0 = \left(-\frac{a}{b}\right)x_0 + \frac{c}{b}$, and $y_1 = \left(-\frac{a}{b}\right)x_1 + \frac{c}{b}$. What is the slope of the line through these points?

3-5 SYSTEMS OF LINEAR EQUATIONS IN TWO VARIABLES

$\{A = (x, y): 2x + 3y = 11\}$
$\{B = (x, y): x - 2y = 5\}$

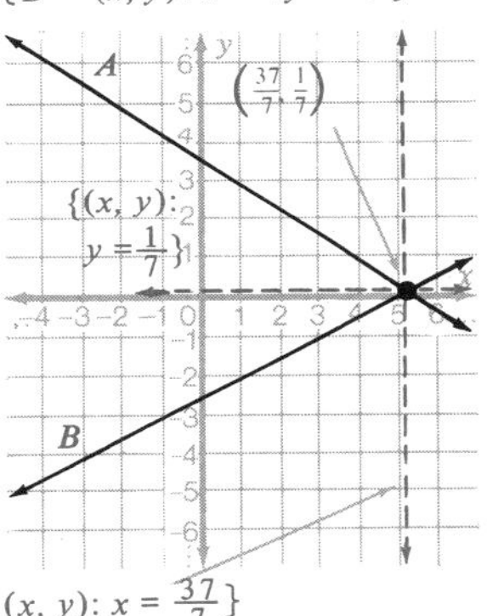

FIGURE 1

The solution set for a system of linear equations is the intersection of the solution sets of the linear equations. For example, the solution set of

$$(1) \quad \begin{cases} 2x + 3y = 11 \\ x - 2y = 5 \end{cases}$$

contains one point which is the intersection of the graphs for $2x + 3y = 11$ and $x - 2y = 5$ [Figure 1]. Notice that the system

$$(2) \quad \begin{cases} x = \dfrac{37}{7} \\ y = \dfrac{1}{7} \end{cases}$$

is a very simple system in that its solution set is obviously the set $\left\{\left(\frac{37}{7}, \frac{1}{7}\right)\right\}$. This system has the same solution set as system (1). Such systems are called **equivalent**. Finding system (2) is called **solving** system (1). There are two conventional methods for solving this type of system, the "addition-subtraction" method and the "substitution" method.

Let us illustrate the "addition-subtraction" method by solving the system

$$\begin{cases} 2x + 3y = 11 \\ x - 2y = 5. \end{cases}$$

Using an equivalent system and "subtracting"

$$\begin{array}{r} 2x + 3y = 11 \\ (-)\ 2x - 4y = 10 \\ \hline 7y = 1 \end{array} \qquad [2(x - 2y) = 2(5)].$$

Retaining the original first equation with the derived equation $7y = 1$ we have the equivalent system

$$\begin{cases} 2x + 3y = 11 \\ y = \dfrac{1}{7} & \left[\dfrac{1}{7}(7y) = \dfrac{1}{7}(1)\right] \end{cases}$$

And finally we have the system

$$\begin{cases} x = \dfrac{37}{7} & \left[2x + 3\left(\dfrac{1}{7}\right) = 11\right] \\ y = \dfrac{1}{7}. \end{cases}$$

Usually only a brief computational listing is written, with the underlying systems being understood.

We supply the computations for a similar system. Can you supply the equivalent systems for each of the five systems below?

$$\begin{cases} 9x - 8y = 1 \\ 6x + 12y = 5 \end{cases}$$

① $2(9x) - 2(8y) = 2(1)$

$3(6x) + 3(12y) = 3(5)$

② $(-16 - 36)y = 2 - 15$

③ $y = \dfrac{-13}{-52} = \dfrac{1}{4}$

④ $9x - 8\left(\dfrac{1}{4}\right) = 1$

⑤ $9x = 3$

⑥ $x = \dfrac{1}{3}$

$$\begin{cases} x = \dfrac{1}{3} \\ y = \dfrac{1}{4} \end{cases}$$

Equivalent systems:

① $\begin{cases} 18x - 16y = 2 \\ 18x + 36y = 15 \end{cases}$

② $\begin{cases} 9x - 8y = 1 \\ -52y = -13 \end{cases}$

③ $\begin{cases} 9x - 8y = 1 \\ y = \dfrac{1}{4} \end{cases}$

④ $\begin{cases} 9x - 8\left(\dfrac{1}{4}\right) = 1 \\ y = \dfrac{1}{4} \end{cases}$

⑤ $\begin{cases} 9x = 3 \\ y = \dfrac{1}{4} \end{cases}$

⑥ $\begin{cases} x = \dfrac{1}{3} \\ y = \dfrac{1}{4} \end{cases}$

By inspection, $\left\{\left(\dfrac{1}{3}, \dfrac{1}{4}\right)\right\}$ is the solution set for the last system. Since all of these systems are equivalent, this is also the solution set of the original system.

Verify that $\left\{\left(\dfrac{1}{3}, \dfrac{1}{4}\right)\right\}$ is the solution set of the original system.

We now consider the system

$$\begin{cases} 2x + 3y = 11 \\ x - 2y = 5 \end{cases}$$

in terms of the "substitution" method, indicating only the basic computational steps:

$$x - 2y = 5$$

$$x = 5 + 2y$$

$$2x + 3y = 11$$

$$2(5 + 2y) + 3y = 11$$

$$7y = 1$$

$$y = \frac{1}{7}$$

$$x = 2\left(\frac{1}{7}\right) + 5 = \frac{37}{7}.$$

Note that a "substitution" step is performed in the "addition-subtraction" method. See the previous examples.

If the lines for a system of two linear equations intersect in a point, the system is called **independent**.

Two special cases of systems of linear equations are considered in some detail in Exercises 9–14. In one case, the linear equations in the system are equivalent and the graph of each solution set is the same line. For example

$$\begin{cases} 3x - 4y = 1 \\ 8y = 6x - 2 \end{cases}$$

is such a system. The graph of the solution set for this system is the line in Figure 2. These systems are called **dependent** systems.

In the other case, the linear equations in the system have parallel lines. There are no common points in this case. Such a system is called **inconsistent**. Let us try to apply the "addition-subtraction" method to the system

$$\begin{cases} y = -2x + 5 \\ 3 + y + 2x = 0. \end{cases}$$

If we make the assumption that there are real numbers r_1 and r_2 such that (r_1, r_2) is a member of the solution set of the system, then

$$r_2 = -2r_1 + 5 \qquad\qquad 2r_1 + r_2 = 5$$
$$3 + r_2 + 2r_1 = 0 \quad \text{and} \quad \underline{2r_1 + r_2 = -3}$$
$$0 = 5 - (-3) = 8$$

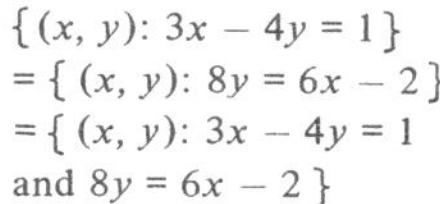

$\{(x, y)\colon 3x - 4y = 1\}$
$= \{(x, y)\colon 8y = 6x - 2\}$
$= \{(x, y)\colon 3x - 4y = 1$
and $8y = 6x - 2\}$

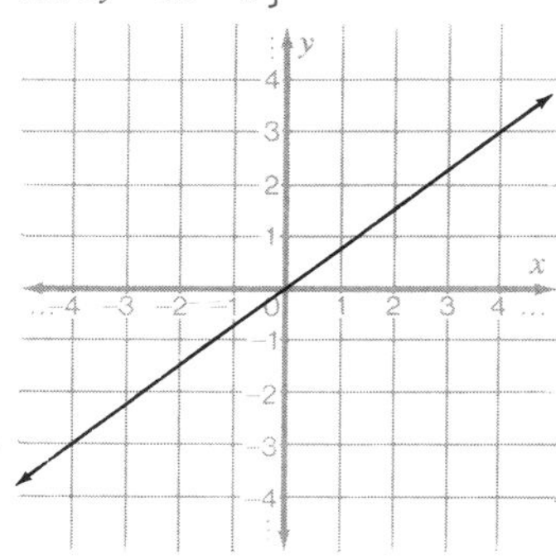

FIGURE 2

which is absurd since $0 \neq 8$. Our assumption is false, and the system has no solution. The graph for the equations of the system is a pair of parallel lines.

Systems of equations are useful as problem solving tools. The following examples and exercises should help you develop facility in formulating equations for solving verbal problems.

Example 1 The difference of four times a particular natural number and three times a second natural number is 11. Twice the first number is 475 decreased by 5 times the second number. What are these two natural numbers?

Solution: Let n_1 and n_2 be the first and second of these numbers, respectively. From the first sentence, we know that (n_1, n_2) is in the solution set for the sentence

$$4x - 3y = 11.$$

After reading the second sentence, we know that (n_1, n_2) must also be a member of the solution set of the equation

$$2x = 475 - 5y.$$

Thus, we are interested in the system comprised of these two sentences, which you can show is equivalent to the system

$$\begin{cases} x = 35 \\ y = 43. \end{cases}$$

The solution set for this system is $\{(35, 43)\}$; hence, $n_1 = 35$ and $n_2 = 43$.

Example 2 The perimeter of a rectangle is 108 feet. If the length is doubled and the width is halved, then the perimeter is 174 feet. Find the length and width of the original rectangle.

Solution: Let l feet and w feet be the dimensions of the original rectangle. Hence (l, w) belongs to the solution set of the system

$$\begin{cases} 2x + 2y = 108 \\ 2(2x) + 2\left(\frac{1}{2}y\right) = 174. \end{cases}$$

We solve this system by deriving the equivalent system:

$$\begin{cases} x = 40 \\ y = 14. \end{cases}$$

The solution set is $\{(40, 14)\}$; $l = 40$, and $w = 14$.

Example 3 In mixing punch for a party, two batches were mixed. The first batch consisted of six quarts of a 25% ginger ale and 75% fruit juice mixture. The second batch was a 50%–50% mixture of these two ingredients. How many quarts of punch would there be if just enough of the second mixture were poured into the six-quart batch so that the result was a 40% ginger ale–60% fruit juice drink.

Solution: Let p be the number of quarts in the final batch, and a be the number of quarts poured into the six-quart batch. Then (a, p) is a member of the solution set for the sentence

$$6 + x = y.$$

How many quarts of ginger ale are in the 6-quart batch? How many quarts of ginger ale are added? Try to express the number of quarts of ginger ale in the final solution by two methods, and hence, derive the sentence

$$(0.25)(6) + 0.5x = 0.4y.$$

You should be able to show that

$$\begin{cases} 6 + x = y \\ (0.25)(6) + 0.5x = 0.4y \end{cases} \quad \text{and} \quad \begin{cases} x = 9 \\ y = 15 \end{cases}$$

are equivalent systems. Hence, the solution set is $\{(9, 15)\}$ and $a = 9$. Then there would be $p = 15$ quarts in the final batch of punch.

Exercises

Find equivalent systems of the form $\begin{cases} x = a \\ y = b \end{cases}$ for each system of linear equations. Graph the solution set of each equation in the original system and in the final system.

1. $\begin{cases} 7x + 5y = 4 \\ 2x - y + 11 = 0 \end{cases}$

2. $\begin{cases} 2x + 5y + 11 = 0 \\ 3x + 7y + 13 = 0 \end{cases}$

3. $\begin{cases} 5x + 12y = 15 \\ 5x - 8y + 30 = 0 \end{cases}$

4. $\begin{cases} x + 6y = 6 \\ 6x + 22y - 15 = 0 \end{cases}$

5. $\begin{cases} 4x - 11y + 2 = 0 \\ 5x + 2y + 34 = 0 \end{cases}$

6. $\begin{cases} 7x + 4y + 5 = 0 \\ 2x + y + 2 = 0 \end{cases}$

7. $\begin{cases} 15x + 10y = 7 \\ 3x - 15y = 49 \end{cases}$

8. $\begin{cases} 7x + 14y = 13 \\ 5x + y + 1 = 0 \end{cases}$

Sketch the graph for each dependent system in Exercises 9–11.

9. $\begin{cases} 2x - 4y = 3 \\ 6x - 9 = 12y \end{cases}$

10. $\begin{cases} 2x - 3 = 5 \\ 8 - x = 4 \end{cases}$

11. $\begin{cases} x + 7y = 3 \\ 3.5y = 1.5 - 0.5x \end{cases}$

12. If $b_1 \neq 0$ and $b_2 \neq 0$ in an inconsistent system

$$\begin{cases} a_1x + b_1y = c_1 \\ a_2x + b_2y = c_2 \end{cases}$$

what is the relationship among the numbers a_1, a_2, b_1, b_2, c_1, and c_2?

13. In an inconsistent system what is the relationship among the numbers a_1, a_2, b_1, b_2, c_1, and c_2 if $b_1 = 0$?

14. Use two methods to show that each system is inconsistent.

    a. $\begin{cases} 4x + 3y = 1 \\ 6y + 8x + 5 = 0 \end{cases}$ b. $\begin{cases} 4x = 10y + 3 \\ 2x = 5y \end{cases}$

Derive systems of equations to aid you in solving these verbal problems.

15. The difference of two natural numbers is 151. If the larger number is divided by one-half the smaller, then the quotient is 4 and the remainder is 29. Find these two natural numbers.

16. Mr. V. I. Person has a total annual income of $106.75 from two particular investments. The rates for these investments are $5\frac{1}{2}\%$ and $4\frac{1}{4}\%$. If the total amount invested in these two ventures is $2100, how much is invested at each rate?

17. The length of a rectangle is twice the width. If the length is increased by 10 feet and the width by 5 feet, the perimeter becomes 300 feet. Find the length and width of the original rectangle.

18. A board 5 feet 2 inches in length is cut into two pieces such that one is 5 inches longer than the other. Find the length of each piece.

19. A rectangle is twice as long as it is wide. The perimeter is 40 feet. What are the dimensions of this rectangle?

20. Jim leaves a motel and travels down the highway at 60 miles per hour. Thirty minutes later Bill leaves the same motel and travels along the highway at a rate of 65 miles per hour in the same direction as Jim. In how many hours will Bill pass Jim? How far from the motel will they be when Bill passes Jim?

21. A particular integer is 5 more than twice another integer. The second integer increased by 37 is the additive inverse of the first integer. What are these two integers?

22. A 12-gallon container is partly filled with a 50% acid solution. If the container is filled with water, a 12.5% solution is obtained. How many gallons of 50% solution were originally in the tank?

23. An airplane is flown 480 miles in 3 hours 12 minutes with the aid of a tail wind. If the plane has been flown the same distance at the same air speed, but into the wind, the time would have been 4 hours. What was the air speed of the plane and the wind current?

24. How many pounds of candy worth $1.20 a pound should be mixed

with a certain number of pounds of candy worth \$.90 a pound to yield 15 pounds worth \$1.00 a pound?

25. Two planes were flown at the same air speed. One plane was flown for 2 hours into the wind and traveled 750 miles. The other plane was flown for 3 hours with the tail wind and traveled 1,575 miles. What was the air speed of each? What was the wind current?

26. One hundred cubic centimeters of a 70% acid solution is diluted with water. How many cubic centimeters of water must be added in order to obtain a 25% acid solution?

27. A two-mile race is conducted on a quarter-mile track. Two runners pace themselves so that they run at a constant speed. One runner's lap-time is 70 seconds, and the other man's time is 82 seconds. During the race will the faster runner "lap" the slower runner? If so, how far will the slower man have run when this takes place?

Review Exercises

1. Indicate the set of points in a coordinate plane that corresponds to $\{(x, y):\ |x| \leqslant 1 \text{ and } y \geqslant 0\}$. **3-1**

2. If R is the replacement set for x solve the equation **3-2**

$$3(x - 4) - 5(2 - x) = 1 - 4x.$$

3. If R is the replacement set for x, graph the solution set for the inequality $3 - 2x \leqslant 4(2x - 7)$.

4. Sketch the graph of the solution set for $2x - 3y = 8$ if the replacement set for (x, y) is $R \times R$. **3-3**

5. Sketch the graph of the solution set for $x - 3y \geqslant 5$ if $R \times R$ is the replacement set for (x, y).

6. Sketch the line through the point $(2, -3)$ with each slope. **3-4**

   a. 1 b. -1 c. $\frac{1}{2}$ d. $-\frac{1}{2}$ e. 2 f. -2

7. Find a linear equation $ax + by = c$ such that the graph of its solution set is the line through $(-2, 3)$ and $(1, -4)$.

8. Find a linear equation such that the graph of its solution set is a line through $(2, 3)$ and perpendicular to the line for the equation $x + 2y = 9$.

Find the solution set for each system. **3-5**

9. $\begin{cases} 7x - 3y = 5 \\ 4x + 5y = 2 \end{cases}$ 10. $\begin{cases} y + 3 = \frac{2}{3}x \\ 2x - 3y = 9 \end{cases}$ 11. $\begin{cases} 7x - 3y = 5 \\ 1.5y + 5 = 3.5x \end{cases}$

12. One integer is 100 more than twice a second integer. Ten times the first integer is 1057 more than the second integer. Find these two integers.

4. EXPONENTS AND RADICALS

4-1 RATIONAL EXPONENTS $\frac{1}{n}$

In this chapter we will develop properties of exponents that are rational numbers and indicate how notation involving radicals may be used interchangeably with exponential notation.

In deciding how to define rational exponents, we recall that all integers are rational numbers. Hence, the definition of rational exponents should preserve the definition and the properties of integral exponents. For this reason, we first review exponents that are integers.

Definition For any real number b

$$b^n = \underbrace{b \cdot b \cdot b \cdots b}_{n \text{ factors}}, \quad \text{if } n \in I \text{ and } n > 1$$

$$b^1 = 1.$$

For any nonzero real number b

$$b^0 = 1$$

$$b^{-n} = \frac{1}{b^n}, \quad \text{if } n \in I \text{ and } n \geqslant 0.$$

The real number b^m is called the **mth power** of b; b is called the **base**; and m is called the **exponent.** The symbol 0^0 is not assigned a meaning.

Why is $b = 0$ excluded from the definition of negative integral exponents? Evaluate each of the following: 3^5, 4^{-3}, 7^2, 5^3, 8^{-2}, $3^3 \div 5^4$, and $3(10^6) + 5(10^4) + 2(10^2) + 3(10^0) + 9(10^{-2}) + 7(10^{-4})$.

Division by zero is undefined; ●

$3^5 = 243$; $4^{-3} = \frac{1}{64}$; $7^2 = 49$;

$5^3 = 125$; $8^{-2} = \frac{1}{64}$;

$3^3 \div 5^4 = \frac{27}{625}$; ●

$3(10^6) + 5(10^4) + 2(10^2) + 3(10^0) + 9(10^{-2}) + 7(10^{-4}) = 3{,}050{,}203.0907$

The usual properties of integral exponents are listed in the following theorem.

Theorem For any nonzero real numbers a and b and any integers m and n:

(i) $b^m \cdot b^n = b^{m+n}$

(ii) $(ab)^m = a^m b^m$

(iii) $(b^m)^n = b^{mn}$

(iv) $\frac{b^n}{b^m} = b^{n-m}$.

Rational exponents should be defined so that the definition and properties for integral exponents are preserved. For example, if rational exponents are to satisfy (iii) above, then

$$(b^{1/n})^n = b^{1/n \cdot n} = b^1 = b.$$

We make the following definition.

Definition Any number that is a solution of the equation

$$x^n = b \qquad (n \text{ a positive integer})$$

is called an ***n*th root** of b.

If $b > 0$ and if n is even, there are exactly two real nth roots of b, one positive and one negative. For example, the real 4th roots of 16 are 2 and -2. If $b > 0$ and if n is odd, there is exactly one real nth root of b. For example, the 5th root of 32 is 2. If $b < 0$ and if n is even, there is no real nth root of b since the even power of any real number is nonnegative. Finally, if $b < 0$ and if n is odd, there is exactly one real nth root of b. For example, the 5th root of -32 is -2.

In order to avoid ambiguity in the case where two real nth roots of b exist, mathematicians have defined what is called the principal nth root of b.

Definition The **principal *n*-th root** of b, denoted $b^{1/n}$, is the real number if one exists and the positive number if two exist, such that

$$(b^{1/n})^n = b.$$

What is the principal square root of 81? What is the principal cube root of 27? Of -27? Of 125? What is the principal fourth root of 256? Of -256?

Principal square root of 81 is 9; • Principal cube root of 27 is 3; • of -27 is -3; • of 125 is 5; • Principal fourth root of 256 is 4; • of -256 is undefined.

Exercises

1. Write a numeral without exponents that names each number.
   a. $(1^0)(2^0)(3^0)$
   b. $(4^3)(-2)^2$
   c. $\left(\frac{3}{4}\right)^{-2}\left(\frac{4}{3}\right)^{-2}$
   d. $(5^{-2})(-3)^{-3}$
   e. $(-3)^{-3}$
   f. $(3^2 - 5^2)^{-1}$
   g. $(-3)^3 \div (-3)^5$
   h. $\left(\frac{2}{3}\right)^{-1} \div \left(\frac{3}{2}\right)^{-1}$
   i. $\left(\frac{1}{3}\right)^{-2}(-3)^{-2}$
   j. $\left(\frac{2}{3}\right)^{-8}\left(\frac{2}{3}\right)^{6}$
   k. $\left(\frac{2}{3}\right)^{-2} \div \left(\frac{2}{5}\right)^{-5}$
   l. $(2^2)^2$
   m. $(2^{-2} + 3^{-3})^{-1}$
   n. $2^{-2} + 3^{-1}$
   o. $(2^0)(2^1)(2^4)$
   p. $(6 + 4)^{-2}(10^2)$
   q. $10^6 + 10^4 + 10^2 + 10^0 + 10^{-2}$
   r. $10^0 + 10^{-1} + 3 \cdot 10^{-2} + 7 \cdot 10^{-5}$
2. Write each expression without negative exponents and indicate any necessary restrictions on the variables.
   a. $x^2 \cdot x^{-3}$
   b. $y^{-5} \cdot y^5$
   c. $u^2 \div u^{-2}$
   d. $t^1 \div t^0$
   e. $(ab)(a^{-2}b^2)^{-2}$
   f. $1^{-3} \div x^0$
   g. $(st)^0 \cdot s^{-2}$
   h. $(x^{-1} + y^{-1})^{-2}$
   i. $z^{-2} + z^{-3}$
   j. $(u^{-3})^2$
   k. $\left(\frac{x}{y}\right)^{-1}\left(\frac{y}{x}\right)^{-2}$
   l. $a^2 + a^{-1} + a^3 - a$
   m. $(xy^0)^{-5}$
   n. $x^{-2}y^{-3} \div xy^{-1}$
   o. $(u^2v^{-3}w^4)^{-2}$
   p. $(1 \div u^3)^{-2}$
   q. $(x^2y^3) \div (xy^2)^{-1}$
   r. $(x^2y^{-3}z^0)^{-2}$
3. Evaluate each of the following expressions if $m = 3$ and $n = 2$.
   a. $\left(\frac{1}{2}\right)^{m-n}$
   b. $(2 - 3m)^{-n}$
   c. $2^n - \frac{1}{2} + m^n$
   d. $10^{-m} + 10^n - 10^{m-n}$
   e. $(m^2)^n \div (n^2)^m$
   f. $\left(\frac{m}{n}\right)^{n-m}$
   g. $(m^{-1} + n^{-1})^n$
   h. $(2 - mn) \div (mn - 2)^{-1}$
   i. $(mn)^{3n-2m}$
   j. $(2^{-n} + 3^{-m})^{-n}$
4. Write each expression without negative exponents if $m = n + 2$ and $n = -1$. Indicate any necessary restrictions on the variables.
   a. $(2x)^{m-n}$
   b. $\left(\frac{1}{a} - \frac{1}{b}\right)^{mn}$
   c. $(x^{-n}y^{-m})^{2m}$
   d. $(x^m + y^n)^{(m+n)}$
   e. $[(xy)^m]^n$
   f. $a^m - a^{-n} + a^{2n}$
   g. $(x^2 - y^{m^2})^{-2m}$
   h. $(s^m - t^n) \div (st)^{mn}$

5. In words, describe the meaning of each power in terms of the principal nth root of the base. Write an equation expressing this relationship.

   a. $19^{1/7}$ b. $90^{1/2}$ c. $(-216)^{1/3}$

   d. $(-57)^{1/7}$ e. $32^{1/8}$ f. $12^{1/9}$

6. What is the principal nth root of b for each indicated value of b and n?

   a. $b = 32,\ n = 5$ b. $b = -243,\ n = 5$

   c. $b = -625,\ n = 4$ d. $b = 36,\ n = 2$

   e. $b = 729,\ n = 6$ f. $b = -64,\ n = 6$

7. What real number does each power name?

   a. $25^{1/2}$ b. $16^{1/4}$ c. $121^{1/2}$

   d. $125^{1/3}$ e. $-9^{1/2}$ f. $64^{1/3}$

   g. $(-125)^{1/3}$ h. $(-625)^{1/4}$ i. $-36^{1/2}$

   j. $343^{1/3}$ k. $(0.01)^{1/2}$ l. $(-0.008)^{1/3}$

8. Show that -2 and 2 are solutions of the equation $x^2 = 4$.

9. Show that -2 is the only real solution of the equation $x^3 = -8$.

4-2 RATIONAL EXPONENTS $\frac{m}{n}$

Recall that a nonzero rational number can always be expressed in the form $\frac{m}{n}$ where m is an integer and n is a positive integer. If property iii of integral exponents is to hold for rational exponents

$$(b^{1/n})^m = b^{(1/n)\cdot m} = b^{m/n}$$

which suggests the following definition for rational exponents.

Definition For $b \geqslant 0$ and rational number $\frac{m}{n}$, where n is any positive integer, and for $b < 0$ and rational number $\frac{m}{n}$, where n is an odd positive integer:

$$b^{m/n} = (b^{1/n})^m.$$

For numbers b and n for which $b^{1/n}$ is defined in R, $b^{m/n}$ is the mth power of the principal nth root of b.

$x \in A$ and $x \in R$	$x \in A$ and $x \in R$
$16^{2/4} = (16^{1/4})^2 = 4$	$27^{-2/3} = \frac{1}{9}$
$16^{1/2} = 4$	$16^{-3/4} = \frac{1}{8}$
$27^{2/3} = 9$	$243^{-3/5} = \frac{1}{27}$
$16^{3/4} = 8$	$9^{-4/4} = \frac{1}{9}$
$243^{3/5} = 27$	$(-27)^{2/3} = 9$
$9^{4/4} = 9$	$(-243)^{3/5} = -27$
$16^{-2/4} = \frac{1}{4}$	$(-27)^{-2/3} = \frac{1}{9}$
$16^{-1/2} = \frac{1}{4}$	$(-243)^{-3/5} = -\frac{1}{27}$

Which elements of $A = \{16^{2/4}, 16^{1/2}, 27^{2/3}, 16^{3/4}, 243^{3/5}, 9^{4/4}, 16^{-(2/4)}, 16^{-(1/2)}, 27^{-(2/3)}, 16^{-(3/4)}, 243^{-(3/5)}, 9^{-(4/4)}, (-16)^{2/4}, (-16)^{1/2}, (-27)^{2/3}, (-16)^{3/4}, (-243)^{3/5}, (-9)^{4/4}, (-16)^{-(2/4)}, (-16)^{-(1/2)}, (-27)^{-(2/3)}, (-243)^{-(3/5)}, (-9)^{-(4/4)}\}$ are real numbers? Write a numeral without exponents that names each real number.

Since $\frac{m}{n} = \left(\frac{1}{n}\right) m = m\left(\frac{1}{n}\right)$, it seems reasonable to suspect that the order in which we perform the operations of raising a power and taking roots does not affect the result. In the following theorem, presented without proof, we see that this generalization is true to an extent.

Theorem 1 For $b \geqslant 0$ and rational number $\frac{m}{n}$, where n is any positive integer, and for $b < 0$ and rational number $\frac{m}{n}$, where n is an odd positive integer:

$$(b^{1/n})^m = (b^m)^{1/n}.$$

Example 1 What real number is $(-32)^{3/5}$? $[(-32)^{1/5}]^3$? $[(-32)^3]^{1/5}$?

Solution: Using the definition of rational exponents and then Theorem 1, we have

$$(-32)^{3/5} = [(-32)^{1/5}]^3 = [(-32)^3]^{1/5} = -8$$

and the order in which we cube and extract principal fifth roots does not affect the value of $(-32)^{3/5}$.

Example 2 What real number is $(-32)^{6/10}$? $[(-32)^{1/10}]^6$? $[(-32)^6]^{1/10}$?

Solution: Since the base of $(-32)^{6/10}$ is negative and the denominator of the exponent is even, $(-32)^{6/10}$ is not defined in R. Similarly, $(-32)^{1/10}$ does not denote a real number; therefore, $[(-32)^{1/10}]^6$ is undefined in R. However

$$[(-32)^6]^{1/10} = [(32)^6]^{1/10} = [(32)^{1/10}]^6 = (2^{1/2})^6 = 8.$$

Therefore, the order in which we raise to the 6th power and extract principal tenth roots is an important consideration.

Elements of set A affected by order of raising to power and taking roots: $(-16)^{2/4}$, $(-9)^{4/4}$, $(-16)^{-2/4}$, $(-9)^{-4/4}$

For which elements of set A on this page does the order in which we perform the operations of raising to a power and taking roots affect the result?

When confronted with the symbol $(-27)^{4/6}$, there is a great temptation to write $(-27)^{4/6} = (-27)^{2/3}$ because $\frac{2}{3} = \frac{4}{6}$. Since $(-27)^{4/6}$ is not defined in R and $(-27)^{2/3} = [(-27)^{1/3}]^2 = 9$, obviously

$$(-27)^{2/3} \neq (-27)^{4/6}.$$

In general when $b < 0$, m and n are even, and n is positive $b^{m/n}$ is not a real number and $b^{m/n} \neq b^{m'/n'}$ where $m = 2m'$ and $n = 2n'$. Thus, when $b < 0$, $b^{2/2} \neq b$. In particular, if $b = -2$, $(-2)^{2/2} \neq -2$.

The difficulties encountered in the preceding paragraph vanish when b is a nonnegative number.

A symbol such as $(b^2)^{1/2}$ presents a tempting situation. It seems natural to write $(b^2)^{1/2} = b^{2/2}$ which is true if $b \geqslant 0$ but false if $b < 0$. Yet, $(b^2)^{1/2}$ is defined whether $b \geqslant 0$ or $b < 0$. For example, when $b = 2$, $(2^2)^{1/2} = 2$; when $b = -2$, $[(-2)^2]^{1/2} = 2$. Since b^2 is nonnegative and the square root of a nonnegative number is nonnegative:

$$(b^2)^{1/2} = |b|.$$

More generally, when n is an even positive integer

$$(b^n)^{1/n} = |b|.$$

All of the properties of integral exponents hold for rational exponents as well. These properties are summarized below.

Theorem 2 For any nonzero a and b and rational numbers $\frac{m}{n}$ and $\frac{p}{q}$ for which the symbols $a^{m/n}$, $b^{m/n}$, $b^{p/q}$ represent real numbers:

(i) $b^{m/n}b^{p/q} = b^{(m/n)+(p/q)}$

(ii) $a^{m/n}b^{m/n} = (ab)^{m/n}$

(iii) $(b^{m/n})^{p/q} = b^{[(m/n)\cdot(p/q)]}$

(iv) $\dfrac{b^{m/n}}{b^{p/q}} = b^{[(m/n)-(p/q)]}$

Note that b may be negative only if n and q are odd integers. Thus, $16^{3/4} \cdot 16^{1/2} = 16^{3/4+1/2} = 16^{5/4} = (16^{1/4})^5 = 2^5 = 32$ and $(-3)^{2/3} \cdot (-3)^{4/3} = (-3)^{(2/3)+(4/3)} = (-3)^2 = 9$.

Is it true that $(-27)^{2/3} \cdot 8^{1/3} = (-3)^2 \cdot 2 = 18$? Does $[(-8) \cdot 16]^{1/2} = (-8)^{1/2} \cdot 16^{1/2}$ exist as a real number? Why or why not?

Yes: $(-27)^{2/3} \cdot 8^{1/3} = [(-27)^{1/3}]^2 \cdot 8^{1/3} = (-3)^2 \cdot 2 = 18$; • No;

$$[(-8) \cdot 16]^{1/2} = (-128)^{1/2}$$

which is undefined.

Exercises

1. What real number does each power in parts a–p name?

a. $\left(\frac{125}{64}\right)^{2/3}$ b. $\left(-\frac{27}{64}\right)^{2/3}$ c. $16^{-(3/4)}$

d. $81^{-(3/4)}$ e. $\frac{4}{9}^{3/2}$ f. $-256^{2/4}$

g. $(0.0004)^{1/2}$ h. $(0.0016)^{1/2}$ i. $-81^{2/4}$

j. $(-2)^{4/4}$ k. $8^{-(2/3)}$ l. $32^{-(4/5)}$

m. $(-243)^{-(2/5)}$ n. $\left(-\frac{27}{125}\right)^{-(2/3)}$ o. $\left(\frac{1}{27}\right)^{1/3} \cdot \left(-\frac{1}{8}\right)^{-(1/3)}$

2. Which represent real numbers:

a. $(-2)(-2)^{1/2}$ b. $9^{1/2}$ c. $(-3)^{2/3}$

d. $-3^{2/3}$ e. $(-2)^{1/2}(-3)^{1/3}$ f. $(-3)^{1/3}(-5)^{1/5}$

g. $(-8)^{1/4}$ h. $(-8)^{8/6}$ i. $(-9)^{4/6}$

3. Write each expression using a single exponent. Indicate any restrictions on the variables.

a. $t^{1/3} \cdot t^{1/2}$ b. $(x^{1/2})^{1/3}$ c. $y^{1/2} \div y^{1/3}$

d. $a^{1/2} \div a^{2/3}$ e. $(3x)^{2/3} \cdot 9x^2$ f. $(xy)^{4/3} \cdot (xy)^{-(1/2)}$

g. $(x^{1/2})^2(x^2)^{1/2}$ h. $(3y)^{4/5} \div (3y)^{2/3}$ i. $(xy)^{2/5} \cdot (tz)^{2/5}$

j. $(2t)^{-(1/2)} \cdot (2t)^{1/2}$ k. $x^{-(4/5)} \div x^{-(1/2)}$ l. $(z^{1/2})^3$

m. $(x^{2/3})^{5/6} \div x^{1/2}$ n. $a^{1/2} \div 3a^{1/4}$ o. $x^{1/2} \cdot x^{1/2}$

p. $s^{2/3} \cdot t^{2/3}$ q. $(x^{3/4})(y^{3/4})$ r. $(x^{1/2})^{-3} \div x^{2/3}$

s. $x^{1/4} \cdot y^{1/2} \cdot x^{1/4}$ t. $(x^4)^{1/6}$ u. $2(x^2)^{1/2}$

4. Under what restrictions on b, q, and n is part i of Theorem 2 true?

5. Under what conditions does property ii hold?

6. State the conditions under which property iii is true.

7. Write a statement of the conditions under which property iv is true.

4-3 RADICALS

When $4^{1/2}$ and $(-8)^{1/3}$ were defined, you may have observed that they name the same numbers as $\sqrt{4}$ and $\sqrt[3]{-8}$, symbols with which you are already familiar. The equivalence of these two kinds of notation for real numbers is generalized in the following definition.

Definition For $b \geqslant 0$, any integer m, and any positive integer n; and for $b < 0$, any integer m, and any odd positive integer n:

$$b^{m/n} = (\sqrt[n]{b})^m.$$

The symbol $\sqrt[n]{b}$ is called a **radical**. The positive integer n is called the **index** of the radical, and b is called the **radicand**. Recall that if the index is 2 we omit the index and write $\sqrt{b}$ instead of $\sqrt[2]{b}$. Thus, $\sqrt{4} = 4^{1/2} = 2$, $\sqrt[3]{-27} = (-27)^{1/3} = -3$, and $\sqrt[3]{8} = 8^{1/3} = 2$. Also, $\sqrt{-9} = (-9)^{1/2}$ is undefined in R.

Since radical notation is equivalent to rational-exponential notation, the properties developed for rational exponents (Section 4-2) are valid when expressed in terms of radicals. Statements of these properties are presented below in terms of radicals.

Theorem For $b \geqslant 0$, any integer m, and any positive integer n; and for $b < 0$, any integer m, and any odd positive integer n:

(i) $(\sqrt[n]{b})^m = \sqrt[n]{b^m}$.

For any a and b and positive integers m and n for which the symbols below represent real numbers:

(ii) $(\sqrt[n]{b})^n = b$

(iii) $\sqrt[n]{a} \cdot \sqrt[n]{b} = \sqrt[n]{ab}$

(iv) $\sqrt[m]{\sqrt[n]{b}} = \sqrt[mn]{b}$

(v) $\dfrac{\sqrt[n]{a}}{\sqrt[n]{b}} = \sqrt[n]{\dfrac{a}{b}}$.

Use rational exponents to denote each of the following real numbers: $\sqrt[6]{7}$, $(\sqrt[3]{9})^5$, $(\sqrt{10})^3$, and $\sqrt[6]{(-27)^4}$.

$7^{1/6}$, $9^{5/3}$, $10^{3/2}$, $[(-27)^4]^{1/6}$

Notice that if $a < 0$ or $b < 0$, then n must be an odd positive integer. Both m and n must be odd if $b < 0$ in (iv) and $b \neq 0$ in (v).

Write another radical for $\sqrt[3]{5} \cdot \sqrt[3]{6}$, $\sqrt{8} \div \sqrt{2}$, $\sqrt{\dfrac{4}{9}}$, $\sqrt{\sqrt{16}}$, $\sqrt{9(16)}$, and $\sqrt[6]{4}$.

$\sqrt[3]{5} \cdot \sqrt[3]{6} = \sqrt[3]{30}$,

$\sqrt{8} \div \sqrt{2} = \sqrt{\dfrac{8}{2}} = \sqrt{4}$,

$\sqrt{\dfrac{4}{9}} = \dfrac{\sqrt{4}}{\sqrt{9}}$, $\sqrt{\sqrt{16}} = \sqrt[4]{16}$,

$\sqrt{9(16)} = \sqrt{9} \cdot \sqrt{16}$,

$\sqrt[6]{4} = \sqrt{\sqrt[3]{4}}$ or $\sqrt[6]{4} = \sqrt[3]{\sqrt{4}} = \sqrt[3]{2}$

Recall from Section 4-2 that

$$(b^n)^{1/n} = |b|$$

for any b and even positive integer n. Using radical notation, we have the equivalent sentence

$$\sqrt[n]{b^n} = |b|$$

for $b \in R$ and n a positive even integer. In particular:

$$\sqrt{b^2} = \sqrt[4]{b^4} = \sqrt[6]{b^6} = |b|.$$

These results can be used to simplify certain radicals. Patterns for simplifying $\sqrt[n]{x}$ where x can be factored as the nth power of some real number, $\sqrt[n]{x} = \sqrt[m]{y}$ for $m < n$, and where x is denoted by a fraction are presented in the examples below. The numerals in parentheses refer to parts of the Theorem, this section.

Example 1 a. Simplify $\sqrt{98}$.

Solution:

$\sqrt{98} = \sqrt{49 \cdot 2}$	factoring
$= \sqrt{49} \cdot \sqrt{2}$	(iii)
$= 7 \cdot \sqrt{2}$	(ii)

b. Simplify $\sqrt[3]{-16}$.

Solution:

$\sqrt[3]{-16} = \sqrt[3]{-8 \cdot 2}$	factoring
$= \sqrt[3]{-8} \cdot \sqrt[3]{2}$	(iii)
$= -2\sqrt[3]{2}$	(ii)

c. Simplify $\sqrt[3]{(-27)^2}$.

Solution:

$\sqrt[3]{(-27)^2} = (\sqrt[3]{-27})^2$	(i)
$= (-3)^2$	(ii)
$= 9$	

d. Simplify $\sqrt[6]{64}$.

Solution:

$\sqrt[6]{64} = \sqrt[3]{\sqrt{64}}$	(iv)
$= \sqrt[3]{8}$	(ii)
$= 2$	(ii)

e. Simplify $\sqrt{\dfrac{81}{100}}$.

Solution:

$$\sqrt{\frac{81}{100}} = \frac{\sqrt{81}}{\sqrt{100}} \qquad \text{(v)}$$

$$= \frac{9}{10} \qquad \text{(ii)}$$

Example 2 Show that $\sqrt{\sqrt{x^4}} = |x|$.

Solution: If x is any real number, then $x^4 \geqslant 0$; hence, $\sqrt{x^4}$ and $\sqrt{\sqrt{x^4}}$ represent real numbers. We have:

$$\sqrt{\sqrt{x^4}} = \sqrt[4]{x^4} \qquad \text{(iv)}$$

$$= |x|$$

by the property of equal even indices and exponents (page 71).

Example 3 Show that $\sqrt{8x^2y^3} = 2|x||y|\sqrt{2y}$, provided the expression denotes a real number.

Solution: The radical represents a real number when x is any real number and y is nonnegative. Thus:

$$\sqrt{8x^2y^3} = \sqrt{4x^2y^2 \cdot 2y} \qquad \text{factoring}$$

$$= \sqrt{4x^2y^2} \cdot \sqrt{2y} \qquad \text{(iii)}$$

$$= 2|x||y|\sqrt{2y} \qquad \text{Example 2}$$

Example 4 a. Show that $\sqrt[9]{x^6} = \sqrt[3]{x^2}$.

Solution: The radical denotes a real number.

$$\sqrt[9]{x^6} = \sqrt[3]{\sqrt[3]{x^6}} \qquad \text{(iv)}$$

$$= \sqrt[3]{\sqrt[3]{x^3 \cdot x^3}} \qquad \text{factoring}$$

$$= \sqrt[3]{\sqrt[3]{x^3} \cdot \sqrt[3]{x^3}} \qquad \text{(iii)}$$

$$= \sqrt[3]{x \cdot x} \qquad \text{(ii) and (i)}$$

$$= \sqrt[3]{x^2}$$

b. Show that $\sqrt[5]{\dfrac{x^5}{y^{10}}} = \dfrac{x}{y^2}$.

Solution: The radical denotes a real number, provided $y \neq 0$.

$$\sqrt[5]{\frac{x^5}{y^{10}}} = \frac{\sqrt[5]{x^5}}{\sqrt[5]{y^{10}}} \qquad \text{(v)}$$

$$= \frac{x}{\sqrt[5]{y^5 \cdot y^5}} \qquad \text{(i), (ii), and factoring}$$

$$= \frac{x}{y^2} \qquad \text{(i), (ii), and (iii)}$$

Exercises

1. Write the equivalent rational-exponent form of each radical. Simplify the exponent if it is permissible.

a. $\sqrt[5]{7}$ b. $\sqrt[3]{7}$ c. $(\sqrt{4})^3$
d. $\sqrt{x^{-2}}$ e. $\sqrt[5]{3^2}$ f. $(\sqrt[3]{2})^5$
g. $\sqrt[4]{x^4}$ h. $\sqrt{x^2}$ i. $\sqrt{3x^5y}$
j. $\sqrt{\sqrt{9^2}}$ k. $\sqrt{\sqrt{x}}$ l. $\sqrt{\sqrt{\sqrt{3^{-2}}}}$
m. $\sqrt{\sqrt[3]{(-5)^{12}}}$ n. $\sqrt[5]{x^2y^3}$ o. $\sqrt{(x-2)^2}$

2. Specify the values of the variables for which each radical is defined in R and simplify.

a. $\sqrt{24}$ b. $\sqrt[3]{z^5}$ c. $\sqrt[5]{-160}$
d. $\sqrt[3]{40}$ e. $\sqrt{128x^3y^4}$ f. $\sqrt{\dfrac{y^6}{y^3}}$
g. $\sqrt{\dfrac{12x^3}{49y^2}}$ h. $\sqrt[8]{x^8y^{16}}$ i. $\sqrt[6]{4^3}$
j. $\sqrt[5]{x^{10}}$ k. $\sqrt{16x^3y^4}$ l. $\sqrt[4]{(3x^2)(27x^2)}$
m. $\sqrt{\dfrac{49}{144}}$ n. $\sqrt{\dfrac{x^5}{x^3}}$ o. $\sqrt[3]{\dfrac{96x^3}{y^3}}$
p. $\sqrt[3]{\dfrac{16}{2}}$ q. $\sqrt[3]{-64x^7}$ r. $\sqrt{3 \cdot 27}$
s. $\sqrt[3]{(11xy^2)(121x^2y)}$ t. $\sqrt{\sqrt{64}}$ u. $\sqrt[3]{\sqrt{x^6}}$
v. $\sqrt{9x^2+12xy+4y^2}$ w. $\sqrt{\dfrac{(x+2)^3}{x^2+4x+4}}$ x. $\sqrt{\dfrac{\sqrt[3]{x^7}}{9\sqrt{3^3\sqrt{3^2}}}}$

3. Specify the values of the variables for which each radical is defined in R and simplify.

a. $\sqrt[5]{(3^{-4})(3^9)}$ b. $\sqrt{4 - x^2}$

c. $\sqrt{x^2 + 4x + 4}$ d. $\sqrt{4x^2 + 12xy + 9y^2}$

e. $\sqrt{(x - y)(x^2 - y^2)}$ f. $\sqrt[3]{(2x + 1)(4x^2 + 4x + 1)}$

g. $\sqrt{\dfrac{x^2}{9x^2 - 12x + 4}}$ h. $\sqrt{\dfrac{(x + 3)^5}{x^3 + 9x^2 + 27x + 27}}$

4-4 OPERATING WITH RADICAL EXPRESSIONS

If the same radical (same index and radicand) appears in two addends, the distributive property for real numbers may be applied to simplify the expression. When the same radical appears in addends, they are like or similar radicals.

Example 1 Add $5\sqrt[3]{x}$ and $2\sqrt[3]{x}$; $\sqrt{x^3}$ and $\sqrt{49x}$; $9\sqrt{5}$ and $8\sqrt[3]{5}$. If possible, simplify each result.

Solution: Using distributivity, we have

$$5\sqrt[3]{x} + 2\sqrt[3]{x} = (5 + 2)\sqrt[3]{x} = 7\sqrt[3]{x}.$$

In the sum $\sqrt{x^3} + \sqrt{49x}$ the radicals must be simplified before the existence of similar radicals is apparent. Hence, we have

$$\sqrt{x^3} + \sqrt{49x} = |x|\sqrt{x} + 7\sqrt{x} = (|x| + 7)\sqrt{x}$$

provided $x \geqslant 0$. Since $x \geqslant 0$, $|x| = x$ and $(|x| + 7)\sqrt{x}$ could be denoted by $(x + 7)\sqrt{x}$.

The sum $\sqrt[9]{5} + 8\sqrt[3]{5}$ cannot be indicated by one term because the addends do not possess similar radicals. Furthermore, we cannot obtain similar radicals since the indices are different.

Example 2 Find the sum of $7\sqrt[3]{9}$, $2\sqrt[4]{9}$, and $3\sqrt[4]{9}$

Solution:

$$7\sqrt[3]{9} + 2\sqrt[4]{9} + 3\sqrt[4]{9} = 7\sqrt[3]{9} + (2 + 3)\sqrt[4]{9} = 7\sqrt[3]{9} + 5\sqrt[4]{9}.$$

Since the radicals in this sum and the addend $7\sqrt[3]{9}$ have different indices, the sum cannot be indicated by one term.

Example 3 What real number is the difference

$$2\sqrt{3} - (\sqrt{27} + \sqrt{147})?$$

Solution:

$$2\sqrt{3} - (\sqrt{27} + \sqrt{147}) = 2\sqrt{3} - (3\sqrt{3} + 7\sqrt{3}) = 2\sqrt{3} - 10\sqrt{3}$$
$$= -8\sqrt{3}$$

Simplify: $3\sqrt[3]{x} + 5\sqrt[3]{x}$, $\sqrt[3]{27x^5} - \sqrt[3]{8y^3x^2}$, $3\sqrt{5} + \sqrt{125} - 6\sqrt{5}$, $3\sqrt[3]{24x^5} + 2\sqrt[3]{3y^3x^2} - 4\sqrt[3]{81x^8}$.

$8\sqrt[3]{x}$, $\sqrt[3]{x^2}(3x - 2y)$, $2\sqrt{5}$, $2\sqrt[3]{3x^2}(3x + y - 6x^2)$

Example 4 illustrates how part iii of the Theorem on p. 73 can be used to multiply certain expressions involving radicals. In Example 5, we use the fact that multiplication is distributive over addition.

Example 4 a. $\sqrt{2} \cdot \sqrt{8} = \sqrt{16} = 4$

b. $\sqrt[3]{4} \cdot \sqrt[3]{16} = \sqrt[3]{4 \cdot 16} = \sqrt[3]{64} = 4$

c. $\sqrt{3x} \cdot \sqrt{3x^3} = \sqrt{9x^4} = \sqrt{9} \cdot \sqrt{x^4} = 3x^2$, provided $x \geqslant 0$

Example 5 a. $\sqrt{2}(6 + 3\sqrt{3}) = 6\sqrt{2} + 3\sqrt{2}\sqrt{3} = 6\sqrt{2} + 3\sqrt{6}$

b. $\sqrt[5]{x^2}(\sqrt[5]{x^3} + 2) = \sqrt[5]{x^2}\sqrt[5]{x^3} + 2\sqrt[5]{x^2} = \sqrt[5]{x^5} + 2\sqrt[5]{x^2} = x + 2\sqrt[5]{x^2}$

Recall that the distributive property provides the rationale for multiplying expressions with more than one addend.

Example 6 a. $(\sqrt{2} + \sqrt{3})(3\sqrt{2} - 4\sqrt{3}) = 3(\sqrt{2})^2 - 4\sqrt{2}\sqrt{3} + 3\sqrt{2}\sqrt{3} - 4(\sqrt{3})^2$

$$= 6 - 7\sqrt{6} - 12 = -6 - 7\sqrt{6}$$

b. $(x + \sqrt{2})(2x - 3\sqrt{2}) = 2x^2 - 3x\sqrt{2} + 2x\sqrt{2} - 3(\sqrt{2})^2$

$$= 2x^2 - x\sqrt{2} - 6$$

c. $(\sqrt{2} + \sqrt{3})(\sqrt{2} - \sqrt{3}) = (\sqrt{2})^2 - (\sqrt{3})^2 = -1$

Note that the product of a sum and difference of radicals with index two does not involve radicals. Similarly, products such as

$$(2\sqrt[3]{x} - \sqrt[3]{a})(4\sqrt[3]{x^2} + 2\sqrt[3]{ax} + \sqrt[3]{a^2})$$
$$= 8\sqrt[3]{x^3} + 4\sqrt[3]{ax^2} + 2\sqrt[3]{a^2x} - 4\sqrt[3]{ax^2} - 2\sqrt[3]{a^2x} - \sqrt[3]{a^3}$$
$$= 8x - a$$

are free of radicals. Observe that the expressions being multiplied are the factors of the difference of two cubes and recall that for any real numbers u and v

$$(u - v)(u^2 + uv + v^2) = u^3 - v^3$$

and

$$(u + v)(u^2 - uv + v^2) = u^3 + v^3$$

The next example illustrates how part v of the Theorem on p. 73 can be used to divide certain expressions involving radicals. Note that the radicand in the denominator is a factor of the radicand in the numerator in each case.

Example 7 a. $\dfrac{\sqrt{x^5}}{\sqrt{x^3}} = \sqrt{\dfrac{x^5}{x^3}} = \sqrt{x^2} = |x|$

b. $\dfrac{\sqrt[3]{-16}}{\sqrt[3]{2}} = \sqrt[3]{\dfrac{-16}{2}} = \sqrt[3]{-8} = -2$

c. $\dfrac{\sqrt{(x+2)^3}}{\sqrt{x^2+4x+4}} = \sqrt{\dfrac{(x+2)^3}{x^2+4x+4}} = \sqrt{\dfrac{(x+2)^3}{(x+2)^2}}$

$= \sqrt{x+2}$, provided $x > -2$

Consider finding a rational approximation for $\dfrac{1}{\sqrt{2}} = \dfrac{\sqrt{2}}{2}$. Using 1.414 for $\sqrt{2}$, approximations for the two quotients can be obtained by dividing 1 by 1.414 or 1.414 by 2. Given a choice you would probably prefer to do the second division. The form $\dfrac{\sqrt{2}}{2}$ provides us with easy access to a rational approximation for the quotient. For this reason, it is often desirable to free denominators of radicals should they contain any. This process is called **rationalizing the denominator**. To rationalize the denominator of a fraction $\dfrac{a}{b}$, we multiply by $1 = \dfrac{c}{c}$ where c is chosen so that the product $b \cdot c$ contains no radicals.

Example 8 a. $\sqrt{\dfrac{3}{5}} = \dfrac{\sqrt{3}}{\sqrt{5}} \cdot \dfrac{\sqrt{5}}{\sqrt{5}} = \dfrac{\sqrt{15}}{5}$

b. $\sqrt[3]{\dfrac{aw^2}{z}} = \dfrac{\sqrt[3]{aw^2}}{\sqrt[3]{z}} = \dfrac{\sqrt[3]{aw^2}}{\sqrt[3]{z}} \cdot \dfrac{\sqrt[3]{z^2}}{\sqrt[3]{z^2}} = \dfrac{\sqrt[3]{aw^2z^2}}{\sqrt[3]{z^3}} = \dfrac{\sqrt[3]{aw^2z^2}}{z}$

$$\frac{1}{\sqrt{7}} = \frac{1}{\sqrt{7}} \cdot \frac{\sqrt{7}}{\sqrt{7}} = \frac{\sqrt{7}}{7},$$

$$\frac{1}{\sqrt{3}} = \frac{\sqrt{3}}{3},$$

$$\frac{2}{\sqrt{5}} - \frac{1}{\sqrt{2}} = \frac{(2\sqrt{2} - \sqrt{5})\sqrt{10}}{10}$$

$$= \frac{2\sqrt{5}}{5} - \frac{\sqrt{2}}{2},$$

$$\sqrt[3]{\frac{a}{w}} = \frac{\sqrt[3]{aw^2}}{w},$$

$$\sqrt{\frac{x}{y^5}} = \frac{\sqrt{xy}}{y^3}, \text{ provided } y > 0,$$

$$\sqrt[3]{\frac{b}{x^2y}} = \frac{\sqrt[3]{bxy^2}}{xy}$$

In each case the denominator can also be rationalized by multiplying the radicand by 1:

$$\sqrt{\frac{3}{5}} = \sqrt{\frac{3}{5} \cdot \frac{5}{5}} = \sqrt{\frac{15}{25}} = \frac{\sqrt{15}}{\sqrt{25}} = \frac{\sqrt{15}}{5}$$

$$\sqrt[3]{\frac{aw^2}{z}} = \sqrt[3]{\frac{aw^2}{z} \cdot \frac{z^2}{z^2}} = \sqrt[3]{\frac{aw^2z^2}{z^3}} = \frac{\sqrt[3]{aw^2z^2}}{\sqrt[3]{z^3}} = \frac{\sqrt[3]{aw^2z^2}}{z}$$

Rationalize the denominators of each of the following and simplify:

$$\frac{1}{\sqrt{7}}, \frac{1}{\sqrt{3}}, \frac{2}{\sqrt{5}} - \frac{1}{\sqrt{2}}, \sqrt[3]{\frac{a}{w}}, \sqrt{\frac{x}{y^5}}, \sqrt[3]{\frac{b}{x^2y}}.$$

To rationalize a denominator such as $(\sqrt{2} + \sqrt{3})$, we choose $\frac{\sqrt{2} - \sqrt{3}}{\sqrt{2} - \sqrt{3}}$ as the appropriate name for 1.

Example 9

a. $$\frac{2}{\sqrt{5} + \sqrt{7}} = \frac{2}{\sqrt{5} + \sqrt{7}} \cdot \frac{\sqrt{5} - \sqrt{7}}{\sqrt{5} - \sqrt{7}}$$

$$= \frac{2(\sqrt{5} - \sqrt{7})}{(\sqrt{5})^2 - (\sqrt{7})^2}$$

$$= -\sqrt{5} + \sqrt{7}$$

b. $$\frac{1}{\sqrt[3]{a} + \sqrt[3]{b}} = \frac{1}{\sqrt[3]{a} + \sqrt[3]{b}} \cdot \frac{\sqrt[3]{a^2} - \sqrt[3]{ab} + \sqrt[3]{b^2}}{\sqrt[3]{a^2} - \sqrt[3]{ab} + \sqrt[3]{b^2}}$$

$$= \frac{\sqrt[3]{a^2} - \sqrt[3]{ab} + \sqrt[3]{b^2}}{a + b}$$

Example 10 Divide $1 - \sqrt{2}$ by $\sqrt{2} - 5$ and simplify.

Solution:

$$\frac{1 - \sqrt{2}}{\sqrt{2} - 5} = \frac{1 - \sqrt{2}}{\sqrt{2} - 5} \cdot \frac{\sqrt{2} + 5}{\sqrt{2} + 5} = \frac{(1 - \sqrt{2})(\sqrt{2} + 5)}{(\sqrt{2})^2 - 5^2}$$

$$= \frac{3 - 4\sqrt{2}}{-23}$$

Divide $2 + 3\sqrt{3}$ by $-1 - \sqrt{3}$; 2 by $3\sqrt{2} + 4\sqrt{3}$; a by $c + d\sqrt{5}$; 1 by $\sqrt[3]{2} + 2\sqrt[3]{4}$. Simplify each result.

$$\frac{11 - \sqrt{3}}{-2}, \frac{3\sqrt{2} - 4\sqrt{3}}{-15}$$

$$\frac{ac - ad\sqrt{5}}{c^2 - 5d^2}, \frac{\sqrt[3]{4} + 8\sqrt{2} - 4}{34}$$

Exercises

1. Perform the indicated operations and simplify.

a. $8\sqrt{3} - 5\sqrt{3}$

b. $3\sqrt[3]{y} - \sqrt[3]{y}$

c. $4\sqrt{3} - \sqrt{12}$

d. $3x\sqrt[3]{y^4} - x\sqrt[3]{8y^7}$

e. $7\sqrt{2} + 2\sqrt{2} - \sqrt{32}$

f. $(3\sqrt{5} - 4\sqrt{2})^2$

g. $2\sqrt[3]{16} - 9\sqrt[3]{54} - 2\sqrt[3]{128}$

h. $\sqrt{x^2y^3} - 2\sqrt{x^4y^5} + 7\sqrt{y}$

i. $9\sqrt{xy^2} + 7\sqrt{4xy^2} - 3\sqrt{y^2}$

j. $x\sqrt[3]{x^4} - 3x^2\sqrt[3]{x} + 4\sqrt[3]{x^8}$

k. $\sqrt{3} \cdot \sqrt{6}$

l. $(\sqrt{x} - 3)(2\sqrt{x} + 3)$

m. $(\sqrt[3]{2} + \sqrt[3]{3})(\sqrt[3]{4} - \sqrt[3]{6} + \sqrt[3]{9})$

n. $x\sqrt[3]{(x+y)^4} - y\sqrt[3]{(x+y)^4}$

2. Multiply and simplify.

a. $\sqrt{6} \cdot \sqrt{8}$

b. $3\sqrt{2}(5\sqrt{2} - \sqrt{3})$

c. $\sqrt{x}(5\sqrt{x} + 7\sqrt{x})$

d. $a\sqrt{6}(a\sqrt{8} - b\sqrt{3})$

e. $\sqrt[3]{x^2}(3\sqrt[3]{x} - \sqrt[3]{7x})$

f. $(\sqrt{2} + \sqrt{3})^2$

g. $(2\sqrt{7} - 3\sqrt{6})^2$

h. $(7\sqrt{11} - 1)(6\sqrt{11} - 2)$

i. $(2\sqrt{3} + 7)(2\sqrt{3} - 7)$

j. $(\sqrt{5} - \sqrt{6})(\sqrt{5} + \sqrt{6})$

k. $(3\sqrt{7} + 2\sqrt{5})(3\sqrt{7} - 2\sqrt{5})$

l. $(a\sqrt{3} + b\sqrt{2})(a\sqrt{3} - b\sqrt{2})$

m. $(\sqrt{3} + \sqrt{5})(2\sqrt{3} - 7\sqrt{5})$

n. $(5\sqrt{7} - 3\sqrt{2})(3\sqrt{7} - 2\sqrt{2})$

3. Perform the indicated operations and simplify.

a. $\sqrt{0.50} + 3.4\sqrt{9x^2} - 3\sqrt{0.02x^2}$

b. $\sqrt[4]{81x^4} + x\sqrt[4]{16} - \sqrt[8]{x^8}$

c. $3.2\sqrt[3]{0.008z^4} - 0.7\sqrt[3]{-0.027z} + 0.1\sqrt[3]{0.064z^7}$

d. $x\sqrt[3]{7x^4} - \sqrt[3]{56x^7} + 4x^2\sqrt[3]{189}$

e. $x^2\sqrt[3]{-0.000008} - \sqrt[3]{0.064x^6} + x\sqrt[3]{0.001x^3}$

f. $\sqrt{\dfrac{x^2}{9}} + \dfrac{1}{x}\sqrt{x^5} + \dfrac{\sqrt{x^4}}{\sqrt{16x}} \quad (x > 0)$

g. $6z^2\sqrt[3]{z} + \dfrac{\sqrt[3]{81z^8}}{\sqrt[3]{3z}} - \sqrt[3]{z}\,\sqrt[3]{z^5} \quad (z \neq 0)$

h. $\dfrac{\sqrt{18z}}{\sqrt{2}} - \sqrt{\dfrac{4}{9z^{-3}}} + 9\sqrt{z} \quad (z > 0)$

i. $\sqrt{x^3 + 3x^2y + 3xy^2 + y^3} - \sqrt{x+y} + \sqrt{\dfrac{1}{(x+y)^{-3}}} \quad (x > 0, y > 0)$

j. $3x\sqrt{(x+y)^3} - xy\sqrt{x+y} - 5y\sqrt{(x+y)^3} \quad (x \geqslant 0, y \geqslant 0)$

4. Multiply and simplify.

a. $(\sqrt[4]{2} + 3)(\sqrt[4]{2} - 3)$

b. $(1 + \sqrt{2})\left(\dfrac{1}{2}\right)$

c. $\left(\dfrac{2\sqrt{3} - 4\sqrt{5}}{3}\right)^3$

d. $(\sqrt[3]{3} - \sqrt[3]{4})(\sqrt[3]{9} + \sqrt[3]{12} + \sqrt[3]{16})$

e. $(\sqrt[3]{5} + \sqrt[3]{3})(\sqrt[3]{25} - \sqrt[3]{15} - \sqrt[3]{9})$

f. $(\sqrt[3]{a} - \sqrt[3]{b})(\sqrt[3]{a^2} + \sqrt[3]{ab} + \sqrt[3]{b^2})$

5. Rationalize the denominators and simplify.

a. $\frac{3}{\sqrt{3}}$ b. $\frac{-5}{\sqrt{2}}$ c. $\frac{1}{\sqrt[3]{5}}$

d. $\frac{-10}{\sqrt[4]{49}}$ e. $1 - \frac{1}{\sqrt{2}}$ f. $2 + \frac{2}{\sqrt{3}}$

g. $3 - \frac{1}{\sqrt[3]{7}}$ h. $6 - \frac{1}{\sqrt[3]{4}}$ i. $\frac{5}{\sqrt{7}} - 4$

j. $\frac{9 - \sqrt{2}}{\sqrt{2}}$ k. $\frac{\sqrt{2} - 3}{\sqrt{3}}$ l. $\sqrt{\frac{5}{7}}$

m. $\sqrt[3]{\frac{1}{3}}$ n. $\sqrt{\frac{7}{8}}$ o. $\sqrt[3]{-\frac{1}{2}}$

p. $\sqrt[4]{\frac{20}{9}}$ q. $\sqrt{\frac{x}{2}}$ r. $\sqrt[5]{\frac{1}{2}}$

s. $\sqrt[3]{\frac{ax}{zb}}$ t. $\sqrt{\frac{x}{7y}}$ u. $\sqrt[5]{\frac{16y}{3x}}$

6. Rationalize the denominators and simplify.

a. $\frac{4}{1 - \sqrt{3}}$ b. $\frac{1 - \sqrt{2}}{1 + \sqrt{2}}$ c. $\frac{4}{\sqrt{5} - \sqrt{7}}$

d. $\frac{15}{2\sqrt{5} + \sqrt{3}}$ e. $\frac{\sqrt{2} - \sqrt{3}}{\sqrt{3} - \sqrt{2}}$ f. $\frac{\sqrt{5} - \sqrt{3}}{\sqrt{5} + \sqrt{3}}$

g. $\frac{5}{\sqrt{2}} + \frac{10}{\sqrt{2}}$ h. $\frac{1}{\sqrt{2}} - \frac{1}{\sqrt{3}}$ i. $\frac{5}{\sqrt{2}} - \frac{6}{\sqrt{3}}$

7. Rationalize the denominators and simplify.

a. $\frac{3}{\sqrt[3]{7} - \sqrt[3]{4}}$ b. $\frac{9}{\sqrt[3]{5} + 2\sqrt[3]{2}}$ c. $\frac{-3}{2 + \sqrt[3]{2}}$

d. $\frac{2}{1 - 3\sqrt[3]{5}}$ e. $\sqrt{\frac{1}{\sqrt{5} - 1}}$ f. $\sqrt{\frac{5}{\sqrt{21} - \sqrt{5}}}$

g. $\sqrt{\frac{1}{\sqrt{2}}}$ h. $\sqrt{\frac{7}{3 - \sqrt{2}}}$

8. Rationalize the denominators and simplify.

a. $\frac{1}{4\sqrt[3]{4} - 6\sqrt[3]{6} + 9\sqrt[3]{9}}$ b. $\frac{1}{\sqrt[3]{16} + 7\sqrt[3]{20} + 49\sqrt[3]{25}}$

c. $\frac{1}{a^2\sqrt[3]{b^2} + ac\sqrt[3]{bd} + c\sqrt[3]{d^2}}$ d. $\frac{1}{4\sqrt[3]{x^4} - 6\sqrt[3]{x^2 z} + 9\sqrt[3]{z^2}}$

Review Exercises

1. Evaluate $4(10^4) + 3(10^3) + 1(10^2) + 9(10^1) + 6(10^0) + 3(10^{-1}) +$ 4-1
$7(10^{-2}) + 8(10^{-3})$.

2. What is the principal cube root of -216? The principal fourth root of 81?

3. Describe the meaning of $7^{1/5}$ and $(-7)^{1/5}$ in terms of the principal fifth root of the base. Write an equation expressing each relationship.

4. Which elements of $A = \{(-7)^{2/3}, 13^{3/4}, (-16)^{5/4}, 9^{2/5}\}$ are real 4-2
numbers?

5. For what real values of x is it true that $x^{2/2} = x$?

6. Does the symbol $[(-3)^4]^{1/2}$ represent a real number?

7. Write a rational-exponent form of $(\sqrt[3]{5})^7$ and $\sqrt{\sqrt[3]{x^2}}$. 4-3

8. Specify the values of the variable for which each radical denotes a real number and simplify

   a. $\sqrt{4x^3y^2}$　　b. $\sqrt[3]{\dfrac{81u^7}{v^3}}$

9. Perform the indicated operations and simplify. 4-4

   a. $3\sqrt[3]{27} + \sqrt[3]{-81} + 4\sqrt[3]{3}$

   b. $7\sqrt{\dfrac{x}{y^2}} - 6\sqrt{\dfrac{9x}{y^2}} + 5\sqrt{\dfrac{1}{y^2}}$

10. Multiply and simplify.

    a. $(\sqrt{5} - 7\sqrt{3})(2\sqrt{5} + 6\sqrt{3})$　　b. $\sqrt[4]{x}(7\sqrt[4]{x} - \sqrt[4]{2x})$

11. Rationalize the denominators and simplify.

    a. $\dfrac{1 - \sqrt{3}}{6 - 3\sqrt{3}}$　　b. $\dfrac{1}{\sqrt[3]{5} - \sqrt[3]{3}}$

5. QUADRATIC EQUATIONS AND INEQUALITIES

5-1 SOLVING QUADRATIC EQUATIONS BY FACTORING

Definition An equation in one variable is a **quadratic equation** if and only if it is equivalent to $ax^2 + bx + c = 0$ where $a \neq 0$.

In many cases the mathematical model that we obtain for a situation is a quadratic equation. Try the following example.

Example 1 Find two consecutive even integers whose product is 528.

Solution: The smaller integer must be a solution of the equation

(1) $x(x + 2) = 528.$

Equation (1) is equivalent to

(2) $x^2 + 2x - 528 = 0.$

which is a quadratic equation in one variable. Factoring the left member of (2) yields

(3) $(x + 24)(x - 22) = 0.$

Since at least one of the factors must equal zero, equation (3) is equivalent to

(4) $x + 24 = 0 \quad \text{or} \quad x - 22 = 0.$

Thus, the solution set for any of these equivalent sentences is $\{-24, 22\}$. The pairs of even integers whose product is 528 are -24 and -22, or 22 and 24.

Equivalent equation(s)	*a*	*b*	*c*
a. $1x^2 + (-2)x + 1 = 0$	1	−2	1
b. $3x^2 + 47x + (-68) = 0$	3	47	−68
c. $4x^2 + (-3)x + 1 = 0$	4	−3	1
$-4x^2 + 3x + (-1) = 0$	−4	3	−1
$4kx^2 + (-3k)x + k = 0$ where $k \in R$ and $k \neq 0$	$4k$	$-3k$	k
d. $22x^2 + 17x + (-10) = 0$	22	17	−10

Verify that each of the following sentences is a quadratic equation by identifying a, b, and c.

a. $2x = x^2 + 1$ **b.** $(3x - 4)(x + 17) = 0$
c. $3x^2 + 4x + 9 = 7x^2 + x + 10$ **d.** $5x(2x + 1) = 2(5 - 6x - 6x^2)$

If the polynomial $ax^2 + bx + c$ can be factored as $(a_1x + b_1)(a_2x + b_2)$, then the quadratic equation $ax^2 + bx + c = 0$ is equivalent to

$$(5) \qquad (a_1x + b_1)(a_2x + b_2) = 0.$$

Furthermore, (5) is equivalent to the sentence

$$a_1x + b_1 = 0 \quad \text{or} \quad a_2x + b_2 = 0$$

which you have learned how to solve.

$\{-1\}$ since $(x + 1)(x + 1) = 0$ is equivalent to $x + 1 = 0$ or $x + 1 = 0$; • $\{-1, 1\}$ since $(x + 1)(x - 1) = 0$ is equivalent to:

$$x + 1 = 0 \quad \text{or} \quad x - 1 = 0;$$

• ϕ (If $r \in R$, then $r^2 \geqslant 0$ and $r^2 + 1 \geqslant 1$. Thus, $r^2 + 1 \neq 0$.)

Find the solution set of the following quadratic equations: $(x + 1)(x + 1) = 0$; $(x + 1)(x - 1) = 0$; and $x^2 + 1 = 0$.

Example 2 Find the solution set of $5x(2x + 1) = 2(5 - 6x - 5x^2)$.

Solution:

$$5x(2x + 1) = 2(5 - 6x - 5x^2)$$

$$10x^2 + 5x = 10 - 12x - 10x^2$$

$$20x^2 + 17x - 10 = 0$$

$$(4x + 5)(5x - 2) = 0$$

$$4x + 5 = 0 \quad \text{or} \quad 5x - 2 = 0$$

$$x = -\frac{5}{4} \quad \text{or} \quad x = \frac{2}{5}.$$

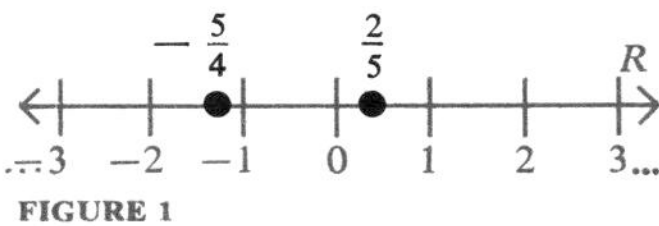

FIGURE 1

The graph of the solution set on the real number line is two points [Figure 1].

Exercises

Solve each quadratic equation in Exercises 1–22 by factoring.

1. $x^2 - 8x - 84 = 0$
2. $x^2 - 24x + 143 = 0$
3. $x^2 + 3x - 340 = 0$
4. $x^2 - 48x + 215 = 0$
5. $x^2 + 37x + 186 = 0$
6. $x^2 - 7x - 294 = 0$
7. $21x^2 + x - 2 = 0$
8. $6x^2 + 41x + 44 = 0$

9. $30x^2 - 83x + 13 = 0$
10. $60x^2 + 64x - 7 = 0$
11. $6x^2 + 23x + 7 = 0$
12. $27x^2 - 51x + 10 = 0$
13. $x(x - 8) = 5(8 - x)$
14. $x(x + 2) + 1 = 100$
15. $x(x + 2) + 3 = 2(40 - x)$
16. $15(x^2 + x - 1) = x - 7$
17. $x(x + 1) + 117 = -43(x + 1)$
18. $6(x^2 + 1) = 7 - x$
19. $(x - 3)(x + 3) = 280$
20. $(x + 7)(x + 2) = 500$
21. $(2x + 3)(3x + 1) = 588$
22. $(4x - 1)(2x - 3) = 297$
23. What pairs of consecutive integers have the property that their product is 342?
24. The length of a rectangular garden is four feet longer than the width. If the area of the garden is 285 square feet, find the dimensions of the garden.
25. Two-inch squares are cut from the corners of a rectangular piece of tin which is 10 inches longer than it is wide. The edges are turned up to form a pan having a volume of 400 cubic inches. Find the dimensions of the original piece of tin.
26. The graph for the solution set of the sentence $y = x^2 + 3x - 28$ has points on the x-axis. Explain how these points relate to the members of the solution set for the sentence $x^2 + 3x - 28 = 0$.
27. In Exercise 26 consider the sentence $y = ax^2 + bx + c$ where $a \neq 0$, and the corresponding sentence $ax^2 + bx + c = 0$. How many numbers can there be in the solution set for this second equation? How does this relate to the x-axis and the graph of the solution set of the first sentence?

5-2 THE QUADRATIC FORMULA

There are other standard methods for solving quadratic equations besides the factoring method which was discussed in Section 5-1. One of these is known as "completing the square," and a special version of this method is called the "quadratic formula."

To solve a given quadratic equation by completing the square we first find an equivalent equation of the form

(1) $$x^2 + dx = e.$$

Then, if the square of one-half the coefficient of x, $\left(\frac{d}{2}\right)^2$, is added to both members, we obtain

(2) $$x + dx + \left(\frac{d}{2}\right)^2 = \left(x + \frac{d}{2}\right)^2 = e + \left(\frac{d}{2}\right)^2.$$

Note that the left member is the square of the linear polynomial $x + \frac{d}{2}$. Now let $\frac{d}{2} = -t$ and let $e + \left(\frac{d}{2}\right)^2 = s^2$, provided $e + \left(\frac{d}{2}\right)^2 \geqslant 0$. Thus we find an equivalent equation of the form

(3) $$(x - t)^2 = s^2$$

where s and t are real numbers. This in turn is equivalent to

(4) $$x - t = s \quad \text{or} \quad x - t = -s$$

composed of linear equations which are easy to solve.

Example 1 Using form (1)

$$12x^2 + 5x - 2 = 0$$

is equivalent to

$$x^2 + \frac{5}{12}x = \frac{1}{6}.$$

We add the square of one-half the coefficient of x to get

$$x^2 + \frac{5}{12}x + \left(\frac{1}{2} \cdot \frac{5}{12}\right)^2 = \frac{1}{6} + \left(\frac{1}{2} \cdot \frac{5}{12}\right)^2$$

$$\left(x + \frac{5}{24}\right)^2 = \left(\frac{11}{24}\right)^2.$$

This final equation fits the form of equation (3) where $t = -\frac{5}{24}$ and $s = \frac{11}{24}$, and so from (4) the final equation is equivalent to

$$x + \frac{5}{24} = \frac{11}{24} \quad \text{or} \quad x + \frac{5}{24} = -\frac{11}{24}.$$

Example 2 We solve another equation similarly.

$$x^2 - 10x - 11 = 0$$

$$x^2 - 10x = 11$$

Solution set of $x^2 + 0 \cdot x - 1 = 0$.

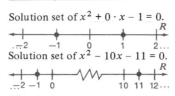

$\ldots -2 \quad -1 \quad 0 \quad 1 \quad 2 \ldots$ R

Solution set of $x^2 - 10x - 11 = 0$.

$\ldots -2 \quad -1 \quad 0 \quad 10 \quad 11 \quad 12 \ldots$ R

● Equivalent equations are derived below.

$x^2 - 3 = 0$
$(x - 0)^2 = (\sqrt{3})^2$

$x^2 - 4x = 0$
$x^2 - 4x + 4 = 4$
$(x - 2)^2 = 2^2$

$x^2 + 4x = 0$
$x^2 + 4x + 4 = 4$
$(x + 2)^2 = 2^2$
$[x - (-2)]^2 = 2^2$

$x^2 + 4x + 1 = 0$
$x^2 + 4x + 4 = 3$
$[x - (-2)]^2 = (\sqrt{3})^2$

$x^2 + dx = e$

$$x^2 + dx + \left(\frac{d}{2}\right)^2 = e + \left(\frac{d}{2}\right)^2$$

$$\left[x - \left(-\frac{d}{2}\right)\right]^2 = \left(\sqrt{e + \left(\frac{d}{2}\right)^2}\right)^2$$

$$x^2 - 10x + \left(\frac{1}{2}(-10)\right)^2 = 11 + \left(\frac{1}{2}(-10)\right)^2$$

$$x^2 - 10x + 25 = 11 + 25$$

$$(x - 5)^2 = 6^2.$$

Here $t = 5$ and $s = 6$ and the last line is equivalent to

$$x - 5 = 6 \quad \text{or} \quad x - 5 = -6.$$

On the real number line sketch the solution set of $x^2 + 0 \cdot x - 1 = 0$; of $x^2 - 10x - 11 = 0$. Find an equivalent equation of the form $(x - t)^2 = s^2$ for each of the following quadratic equations: $x^2 - 3 = 0$; $x^2 - 4x = 0$; $x^2 + 4x = 0$; $x^2 + 4x + 1 = 0$; and $x^2 + dx = e$.

Example 3 Find the solution set for the quadratic equation $x^2 - 12x + 32 = 0$ by completing the square.

Solution: The following sentences are equivalent:

$$x^2 - 12x + 32 = 0$$

$$x^2 - 12x = -32$$

$$x^2 - 12x + \left(\frac{1}{2}(-12)\right)^2 = -32 + \left(\frac{1}{2}(-12)\right)^2$$

$$x^2 - 12x + 36 = 4$$

$$(x - 6)^2 = 2^2$$

$$x - 6 = 2 \quad \text{or} \quad x - 6 = -2$$

$$x = 8 \quad \text{or} \quad x = 4.$$

Thus, the solution set is $\{8, 4\}$.

If the method of completing the square is applied to the quadratic equation $ax^2 + bx + c = 0$ where $a \neq 0$, we have:

$$ax^2 + bx + c = 0$$

$$x^2 + \frac{b}{a}x = -\frac{c}{a}$$

$$x^2 + \frac{b}{a}x + \left(\frac{b}{2a}\right)^2 = -\frac{c}{a} + \left(\frac{b}{2a}\right)^2$$

$$\left(x + \frac{b}{2a}\right)^2 = \frac{b^2 - 4ac}{4a^2}$$

$$x + \frac{b}{2a} = \sqrt{\frac{b^2 - 4ac}{4a^2}} \quad \text{or} \quad x + \frac{b}{2a} = -\sqrt{\frac{b^2 - 4ac}{4a^2}}$$

$$x = \frac{-b + \sqrt{b^2 - 4ac}}{2a} \quad \text{or} \quad x = \frac{-b - \sqrt{b^2 - 4ac}}{2a}.$$

This result is called the **quadratic formula.** The last sentence can be derived only if $b^2 - 4ac \geqslant 0$. The quadratic formula is often abbreviated by

$$x = \frac{-b \pm \sqrt{b^2 - 4ac}}{2a}.$$

Example 4 Find the solution set of $x^2 + 3x - 5 = 0$ using the quadratic formula.

Solution: In this equation $a = 1$, $b = 3$, and $c = -5$. The given equation is equivalent to

$$x = \frac{-3 \pm \sqrt{(3)^2 - 4(1)(-5)}}{2 \cdot 1}.$$

The solution set is $\left\{\frac{-3 + \sqrt{29}}{2}, \frac{-3 - \sqrt{29}}{2}\right\}$.

Example 5 Use the quadratic formula to solve $3x^2 - 5x + 1 = 0$.

Solution: In the formula $a = 3$, $b = -5$, and $c = 1$. This equation is equivalent to

$$x = \frac{-(-5) \pm \sqrt{(-5)^2 - 4 \cdot 3 \cdot 1}}{2 \cdot 3} \quad \text{or} \quad x = \frac{5 \pm \sqrt{13}}{6}.$$

a	b	c	*Equivalent equation*
1	-3	1	$x = \frac{3 \pm \sqrt{5}}{2}$
1	3	1	$x = \frac{-3 \pm \sqrt{5}}{2}$
3	-4	-1	$x = \frac{4 \pm \sqrt{28}}{6}$ or $x = \frac{2 \pm \sqrt{7}}{3}$

Use the quadratic formula to solve the equations:
$x^2 - 3x + 1 = 0$; $x^2 + 3x + 1 = 0$; $3x^2 - 4x - 1 = 0$.

In the system of real numbers, $\sqrt{r}$ denotes the nonnegative real number such that $(\sqrt{r})^2 = r$. The number r, itself, must be a nonnegative real number. Thus, $\sqrt{b^2 - 4ac}$ in the quadratic formula denotes a nonnegative real number if and only if $b^2 - 4ac \geqslant 0$. Using this fact together with the quadratic formula, we can determine if the solution set of a quadratic equation contains any real numbers.

1, 2, 0; $\left\{-\frac{b}{2a}\right\}$, $\left\{\frac{-b \pm \sqrt{b^2 - 4ac}}{2a}\right\}$, ϕ.

How many elements are in the solution set if $b^2 - 4ac = 0$? If $b^2 - 4ac > 0$? If $b^2 - 4ac < 0$? Write the solution set in each case.

The number $b^2 - 4ac$ is called the **discriminant** for the quadratic equation.

Exercises

Find the solution set for each of the following quadratic equations by the method of completing the square.

1. $x^2 + 2x - 35 = 0$
2. $x^2 + 10x + 24 = 0$
3. $x^2 - 10x + 9 = 0$
4. $x^2 - 5x - 24 = 0$
5. $x^2 + 13x + 36 = 0$
6. $x^2 - 19x + 84 = 0$
7. $x^2 + x - 12 = 0$
8. $x^2 - 8x + 15 = 0$
9. $x^2 + x - 20 = 0$
10. $x^2 - 19x + 78 = 0$
11. $x^2 + 8x + 12 = 0$
12. $x^2 - 12x + 27 = 0$
13. $x^2 + 11x + 18 = 0$
14. $x^2 + 5x - 66 = 0$
15. $15x^2 + x - 2 = 0$
16. $8x^2 - 2x - 3 = 0$
17. $9x^2 - 9x - 10 = 0$
18. $49x^2 - 84x + 32 = 0$
19. $25x^2 + 45x + 18 = 0$
20. $121x^2 + 242x - 23 = 0$
21. $36x^2 - 17x - 35 = 0$
22. $6x^2 - 11x - 35 = 0$
23. $24x^2 - 46x + 21 = 0$
24. $7x^2 + 16x - 15 = 0$

Find the discriminant for each quadratic equation in Exercises 25–46. Use the quadratic formula to solve each equation whose discriminant is nonnegative.

25. $x^2 - 4x + 3 = 0$
26. $x^2 + 6x + 9 = 0$
27. $x^2 + 2x + 3 = 0$
28. $3x^2 + 7x + 5 = 0$
29. $x^2 - 4x + 1 = 0$
30. $2x^2 + 4x - 1 = 0$
31. $x^2 - 4x + 5 = 0$
32. $x^2 + 5x + 7 = 0$
33. $3x^2 - 4x + 1 = 0$
34. $2x^2 + 7x + 5 = 0$
35. $x^2 + 2x = 5$
36. $7 = 2x^2 + 6x$
37. $121x^2 + 49 = 154x$
38. $2x^2 + 1 = 5x$
39. $2x^2 + 7x + 3 = 0$
40. $x^2 + 3x = 6$
41. $4x^2 - 12x + 9 = 0$
42. $3x - 1 = x^2$
43. $2x^2 - 6x = 3$
44. $4x^2 + 12x + 9 = 0$
45. $3x^2 + 5x = 1$
46. $4x^2 + 5 = 13x$
47. Find the lengths of the legs of a right triangle if one leg is 2 feet longer than the other leg and the length of the hypotenuse is 10 feet.

48. If r_1 and r_2 are the elements of the solution set for the quadratic equation $ax^2 + bx + c = 0$ and $b^2 - 4ac \geqslant 0$, prove that $r_1 + r_2 = -\dfrac{b}{a}$.

49. In Exercise 48, prove that $r_1 \cdot r_2 = \dfrac{c}{a}$.

50. Find a real number c such that -1 is in the solution set for the quadratic equation $x^2 - 5x + c = 0$.

51. Find a quadratic equation such that 3 and the multiplicative inverse of 3 are the elements of the solution set.

5-3 QUADRATIC INEQUALITIES IN ONE VARIABLE

In considering the general inequalities

$$ax^2 + bx + c < 0 \quad \text{and} \quad ax^2 + bx + c > 0$$

we can use the solution set for the quadratic equation $ax^2 + bx + c = 0$ to locate important reference points. Consider the quadratic inequality

$$6x^2 + x - 12 < 0.$$

The solution set for the corresponding equation

$$6x^2 + x - 12 = 0$$

which, using factoring, is equivalent to

$$(3x - 4)(2x + 3) = 0$$

is $\left\{\dfrac{4}{3}, -\dfrac{3}{2}\right\}$.

To solve the inequality, use the factored form

$$(3x - 4)(2x + 3) < 0$$

or

$$\left(x - \frac{4}{3}\right)\left(x + \frac{3}{2}\right) < 0$$

and consider the numbers on the line between, to the right, and to the left of the solutions of the corresponding equation.

Show that the following pairs are equivalent: $2x + 3 < 0$, $x < -\frac{3}{2}$; $2x + 3 > 0, x > -\frac{3}{2}$; $3x - 4 < 0, x < \frac{4}{3}$; $3x - 4 > 0, x > \frac{4}{3}$.

●
$$2x + 3 < 0$$
$$2x < -3$$
$$x < -\frac{3}{2}$$
are equivalent. ● Similarly,
$$3x - 4 < 0$$
$$3x < 0$$
$$x < \frac{4}{3}$$
are equivalent. The inequalities are reversed for the other two pairs.

To satisfy the inequality the replacement must make one factor negative and the other positive. Figure 1 shows the information used for checking the product of the factors.

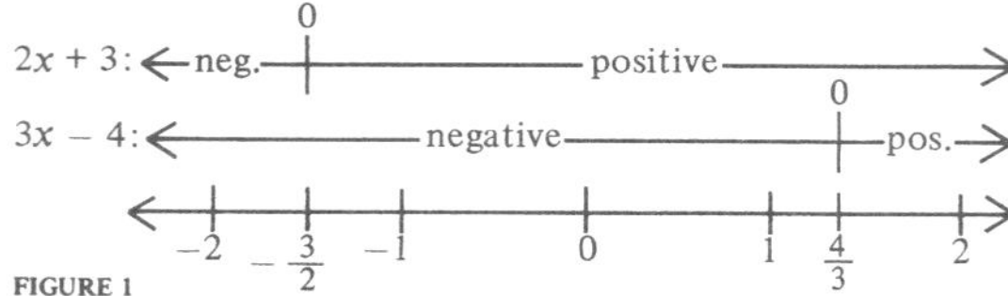

FIGURE 1

The numbers between $-\frac{3}{2}$ and $\frac{4}{3}$ satisfy the inequality; that is, the solution set is $\left\{r\colon -\frac{3}{2} < r < \frac{4}{3}\right\}$.

Example 1 Find the graph for the inequality $2x^2 - x - 15 > 0$.

Solution: The corresponding equation

$$2x^2 - x - 15 = 0$$

is equivalent to

$$(2x + 5)(x - 3) = 0.$$

Its solution set is $\left\{-\frac{5}{2}, 3\right\}$. The given inequality is equivalent to

$$(2x + 5)(x - 3) > 0$$

The product of two factors is positive if both factors are positive or if both factors are negative. From Figure 2

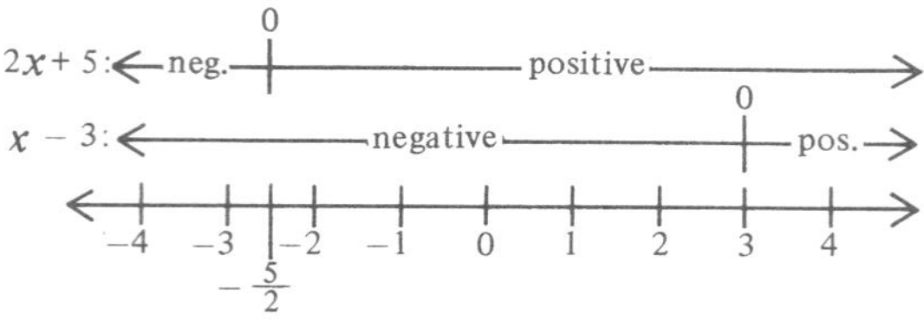

FIGURE 2

we see that the solution set is $\left\{r\colon r < -\frac{5}{2} \text{ or } r > 3\right\}$.

If the quadratic inequality contains the relation $\leqslant$ or $\geqslant$ then the solutions of the corresponding equation are also members of the solution set for the inequality. For example, the solution set for

$$2x^2 - x - 15 \geqslant 0$$

or

$$(2x + 5)(x - 3) \geqslant 0$$

is $\left\{r\colon\ r \leqslant -\frac{5}{2} \text{ or } r \geqslant 3\right\}$.

In solving the preceding quadratic inequalities we have relied on the solutions for the corresponding equation of the form $ax^2 + bx + c = 0$. However, we know that if the discriminant, $b^2 - 4ac$, is negative there are no real number solutions for this equation. If $a > 0$ and $b^2 - 4ac < 0$, we can use the method of completing the square to show that the following inequalities are equivalent:

$$ax^2 + bx + c < 0$$

$$\left(x + \frac{b}{2a}\right)^2 < \frac{b^2 - 4ac}{4a^2}.$$

Since $\frac{b^2 - 4ac}{4a^2}$ is a negative real number, no real number makes this last inequality a true statement. Thus the solution set for the inequality is the empty set.

Example 2 Solve $x^2 + 1 < 0$.

Solution: The quadratic inequality $x^2 + 1 < 0$ has the form $ax^2 + bx + c < 0$ where $a = 1$, $b = 0$, and $c = 1$. Note that $b^2 - 4ac = -4 < 0$. $x^2 + 1 < 0$ is equivalent to $x^2 < -1$. No real number satisfies this inequality; hence the solution set is ϕ.

If $a > 0$ and $b^2 - 4ac < 0$ the method of completing the square can be used to show that $ax^2 + bx + c > 0$ is equivalent to

$$\left(x + \frac{b}{2a}\right)^2 > \frac{b^2 - 4ac}{4a^2}.$$

Show that R is the solution set for the inequality $x^2 + 1 > 0$.

Since $x^2 + 1 > 0$ and $x^2 > -1$ are equivalent and the square of any real number is nonnegative, the solution set is R.

If $a > 0$ and $b^2 - 4ac < 0$ then the solution set for the inequality $ax^2 + bx + c > 0$ is the set of real numbers, R.

Exercises

Find the solution set of each quadratic inequality in Exercises 1-26 and sketch the graph.

1. $x^2 < 25$
2. $4x^2 - 9 > 0$
3. $x^2 - 4 \leqslant 0$
4. $49x^2 - 4 \geqslant 0$
5. $x^2 + 4x - 21 < 0$
6. $x^2 + 10x > 24$
7. $x^2 + 2x - 15 < 0$
8. $x^2 - 2x - 24 > 0$
9. $2x^2 + 5x - 12 \geqslant 0$
10. $25x^2 \leqslant 15x + 18$
11. $x^2 + 3x + 5 < 0$
12. $2x^2 - x + 3 > 0$
13. $3x + 4 - x^2 \geqslant 0$
14. $x^2 \geqslant 5x + 14$
15. $14x^2 - 29x - 15 < 0$
16. $100x^2 > 20x + 63$
17. $15x^2 - 41x + 14 < 0$
18. $45x^2 + 128x + 48 > 0$
19. $x^2 + 5x + 3 \leqslant 0$
20. $x^2 - 3x - 2 \geqslant 0$
21. $4x^2 - 12x + 9 < 0$
22. $x^2 + 2x + 1 > 0$
23. $x^2 + x + 1 < 0$
24. $3x^2 - x + 1 > 0$
25. $x^2 + 2x + 7 \leqslant 0$
26. $x^2 - 2x + 2 \geqslant 0$
27. If $a > 0$ find the solution sets for
    a. $x^2 < a$ **b.** $x^2 > a$ **c.** $x^2 \leqslant a$ **d.** $x^2 \geqslant a$
28. State results for the relations $\leqslant$ and $\geqslant$ analogous to the statements on page 93.
29. Prove that $ax^2 + bx + c < 0$ and $-ax^2 - bx - c > 0$ are equivalent sentences. How does this relate to the results of this section?

5-4 QUADRATIC SENTENCES AND ABSOLUTE VALUE

Recall that $|a - b|$ can be interpreted as the nonnegative distance between the points corresponding to a and b on the number line. We have seen that the quadratic inequality

$$x^2 - 2x + 15 < 0$$

is equivalent to

$$-3 < x < 5.$$

Thus, any solution must lie in the open segment determined by -3 and 5 [Figure 1]. Another way to specify the inequality is by using the distance from the midpoint of the segment. The midpoint of a segment divides it into two segments of equal length. In our example, the total length of the segment is $|5 - (-3)| = 8$. The midpoint of this segment is the point corresponding to 1, and the distance between any point on this

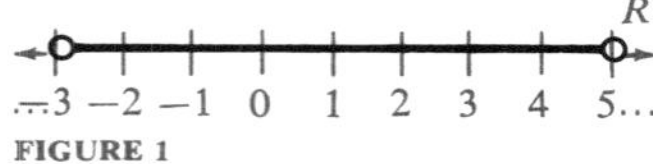

FIGURE 1

segment and the midpoint is less than 4. Using absolute value notation, we have the equivalent sentence

$$|x - 1| < 4.$$

Show that the pairs of sentences are equivalent: $-2 < x < 10$, $|x - 4| < 6$; $-3 < x < 1$, $|x + 1| < 2$; $-\frac{4}{3} \leqslant x \leqslant \frac{11}{3}$, $\left|x - \frac{7}{6}\right| \leqslant \frac{5}{2}$.

Interpret $|x - 4| < 6$ as those points which are less than a distance of 6 from the point for 4; hence, $4 - 6 < x < 4 + 6$ or $-2 < x < 10$; • $|x + 1| < 2$ is equivalent to $|x - (-1)| < 2$. We want those points which are less than a distance of 2 from the point for (-1); that is, $-3 < x < 1$;

• In the inequality, $\left|x - \frac{7}{6}\right| \leqslant \frac{5}{2}$, we want those points which are a distance less than or equal to $\frac{5}{2}$ from $\frac{7}{6}$; that is:

$$\frac{7}{6} - \frac{5}{2} \leqslant x \leqslant \frac{7}{6} + \frac{5}{2}$$

or

$$-\frac{4}{3} \leqslant x \leqslant \frac{11}{3}.$$

The graph of the solution set for the quadratic inequality

$$x^2 + 4x \geqslant 21$$

is the union of two disjoint rays with endpoints 3 and -7 [Figure 2]. This set contains all points whose distance from -2 is greater than or equal to 5. Hence, an equivalent sentence for this quadratic inequality is

$$|x - (-2)| \geqslant 5.$$

or

$$|x + 2| \geqslant 5.$$

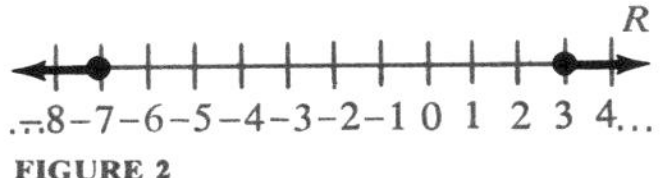

FIGURE 2

Examine these equivalent sentences:

$$|2x + 3| < 4$$

$$|2x - (-3)| < 4$$

$$\left|2\left[x - \left(-\frac{3}{2}\right)\right]\right| < 4$$

$$2\left|x - \left(-\frac{3}{2}\right)\right| < 4$$

$$\left|x - \left(-\frac{3}{2}\right)\right| < 2$$

The graph of the solution set is the set of all points whose distance from the point for $-\frac{3}{2}$ is less than 2 [Figure 3].

What quadratic inequality has the same graph? We know that Figure 3 is also the graph of the solution set for $-\frac{7}{2} < x < \frac{1}{2}$. Therefore

$$|2x + 3| < 4$$

FIGURE 3

and

$$-\frac{7}{2} < x < \frac{1}{2}$$

have the same solution set and are equivalent. Furthermore, each of the following sentences is equivalent to $-\frac{7}{2} < x < \frac{1}{2}$:

$$-\frac{7}{2} < x \quad \text{and} \quad x < \frac{1}{2}$$

$$x > -\frac{7}{2} \quad \text{and} \quad x < \frac{1}{2}$$

$$\left(x + \frac{7}{2}\right) > 0 \quad \text{and} \quad \left(x - \frac{1}{2}\right) < 0$$

$$\left(x + \frac{7}{2}\right)\left(x - \frac{1}{2}\right) < 0$$

$$x^2 + 3x - \frac{7}{4} < 0$$

$$4x^2 + 12x - 7 < 0.$$

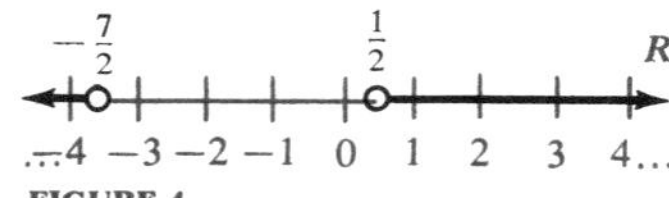

FIGURE 4

The quadratic inequality that we are seeking is $4x^2 + 12x - 7 < 0$.

Similarly, Figure 4 is the graph of the solution set of sentences:

$$|2x + 3| > 4$$

$$x < -\frac{7}{2} \quad \text{or} \quad x > \frac{1}{2}$$

and

$$4x^2 + 12x - 7 > 0.$$

Exercises

Use absolute value notation to write a sentence equivalent to each sentence in Exercises 1–8.

1. $-9 < x < 1$
2. $-10 \leqslant x \leqslant 10$
3. $x^2 - 11x + 18 < 0$
4. $x^2 - 4x + 4 \leqslant 0$
5. $x > 7$ or $x < -11$
6. $x \geqslant 7$ or $x \leqslant -7$
7. $x^2 + 14x + 33 > 0$
8. $15x^2 + 2x - 8 \geqslant 0$

Sketch the graph of the solution set for each inequality in Exercises 9–26.

9. $|x - 4| < 3$

10. $\left|x - \frac{1}{2}\right| \leqslant \frac{3}{4}$

11. $|x + 7| \leqslant 4$

12. $|x| \leqslant 3$

13. $|2x - 3| < 1$

14. $|x| - 3 > 2$

15. $\left|x + \frac{3}{7}\right| > \frac{4}{7}$

16. $|x + 3| > 4$

17. $|x| - 3 \leqslant 2$

18. $\left|x - \frac{3}{2}\right| < \frac{1}{2}$

19. $|x - 2| > 3$

20. $|x| > 6$

21. $|3x + 4| > 2$

22. $|3x + 2| \leqslant 5$

23. $|2x - 7| \geqslant 6$

24. $\left|x + \frac{4}{5}\right| \geqslant \frac{6}{5}$

25. $\left|\frac{1}{2}x - \frac{2}{3}\right| > \frac{1}{4}$

26. $\left|\frac{1}{3}x + \frac{1}{2}\right| > 1$

27. Write a quadratic inequality that is equivalent to each inequality in Exercises 11, 12, 18 and 22.

28. Write a quadratic inequality that is equivalent to each inequality in Exercises 15, 19, 20 and 25.

5-5 RADICAL EQUATIONS

If a variable in an equation appears in a radicand, the equation is called a **radical** or **irrational equation.** The equations

$$\sqrt{x - 5} = 9$$

$$3\sqrt{16 - x} = 5$$

$$\sqrt{3 - x} - \sqrt{x - 2} = 6$$

$$\frac{5}{\sqrt{x^2 - 4}} = 3$$

are examples of radical equations. The following theorem provides a method for solving radical equations.

Theorem For any real number r and nonzero real numbers s and t, if $s = t$ then $s^r = t^r$.

A statement which can be derived from the theorem is as follows: Any solution of the equation $u = v$ is a solution of the equation $u^x = v^x$, or the solution set of the equation $u = v$ is a subset of the solution set of the equation $u^x = v^x$. Therefore, to find the solution set of a radical equation (such as $\sqrt{3x + 4} = x$, where R is the replacement set for x), we perform the following steps.

1. Free the equation of radicals by raising both members of the equation to the power equal to the index of the radicals involved:

$$(\sqrt{3x+4})^2 = x^2$$

$$3x + 4 = x^2.$$

2. Find the solution set of the radical-free equation:

$$x^2 - 3x - 4 = 0$$

$$(x - 4)(x + 1) = 0$$

$$x - 4 = 0 \quad \text{or} \quad x + 1 = 0.$$

The solution set of $3x + 4 = x^2$ is $\{4, -1\}$.

3. Test solutions of the radical-free equation in the original equation to find which subset of the solution set for the radical-free equation is the solution set of the radical equation:

$$\sqrt{3(4) + 4} = \sqrt{16} = 4$$

$$\sqrt{3(-1) + 4} = \sqrt{1} = 1, \text{ but } 1 \neq -1.$$

Therefore, $\{4\}$ is the solution set for $\sqrt{3x + 4} = x$.

Perform (1), (2), and (3) on the following equations where R is the replacement set for x: $\sqrt{x - 6} = 7$; $\sqrt{x - 6} + 3 = \sqrt{x + 9}$; $\sqrt[3]{x} + 6 = 1$.

Equation	*Solution set*
$\sqrt{x-6} = 7$	$\{55\}$
$\sqrt{x-6} + 3 = \sqrt{x+9}$	$\{7\}$
$\sqrt[3]{x} + 6 = 1$	$\{-125\}$

A common error made when solving radical equations is to assume the converse of the theorem—that if $s^r = t^r$ then $s = t$. The following example shows that this converse is not true. Consider the statement:

If $z^2 = 3^2$, then $z = 3$.

Since -3 is a solution of $z^2 = 3^2$ but not of $z = 3$, not every solution of $z^2 = 3^2$ is a solution of $z = 3$. In general it is necessary to check each solution of the radical-free equation by substituting the number into the radical equation.

Example Find the solution set for $\sqrt{1-y}+\sqrt{1-5y}=2$ if R is the replacement set for y.

Solution: When there are two radical addends, we find an equivalent equation by adding the negative of one of the radicals to both members before squaring:

$$(1)\qquad \sqrt{1-y}=2-\sqrt{1-5y}.$$

Squaring each member of (1), we obtain

$$1-y=4-4\sqrt{1-5y}+1-5y$$

$$4y-4=-4\sqrt{1-5y}$$

$$(2)\qquad y-1=-\sqrt{1-5y}.$$

Squaring both members of (2), we obtain

$$y^2-2y+1=1-5y$$

which is radical free and equivalent to each of the following:

$$y^2+3y=0$$

$$y(y+3)=0$$

$$y=0 \quad \text{or} \quad y=-3.$$

It is now necessary to check each solution in the original equation:

$$\sqrt{1-0}+\sqrt{1-5(0)}=1+1=2.$$

Since the original equation is true when $y=0$, zero is a solution of the radical equation.

$$\sqrt{1-(-3)}+\sqrt{1-5(-3)}=2+4=6.$$

The original equation is not true when $y=-3$. The solution set of the radical equation is $\{0\}$.

A solution of the radical-free equation that is not a solution of the parent radical equation is sometimes called an **extraneous** solution of the radical equation.

Exercises

Find the solution set of each equation relative to R.

1. $\sqrt{x}=5$
2. $\sqrt{s-1}=s-3$
3. $\sqrt{x-5}=3$
4. $\sqrt{y}-5=3$
5. $\sqrt{y^2+27}-2y=0$
6. $\sqrt{2z-13}+3=0$

7. $\sqrt[3]{z+3} = -2$
8. $(v-3)^{1/3} = -1$
9. $\sqrt{t^2+2} = 11$
10. $\sqrt{2r+12} - r = 2$
11. $\sqrt{t+1} + \sqrt{t-1} = 3$
12. $\sqrt{2x-1} - \sqrt{x-1} = 1$
13. $\sqrt{r^2-4} = 1$
14. $\sqrt{r-3} - \sqrt{2-r} = 0$
15. $\sqrt[4]{3x+2} = 1$
16. $\sqrt{w^2} - 2 = 3$
17. $\sqrt[3]{u-3} = -1$
18. $\sqrt{\sqrt{x}+3} = 2$
19. $\sqrt{z^2+8} - 2 = z$
20. $\sqrt{2-\sqrt{u}} - \sqrt{4-u} = 0$
21. $u^{2/3} = (4u-3)^{1/3}$
22. $(x-5)^{1/2} = 3 - x^{1/2}$
23. $\sqrt{3y+7} = \sqrt{3y} - \sqrt{y-2}$
24. $\sqrt[3]{\sqrt{x}+1} = 2$
25. $(9y-4)^{1/3} - 3(y-1)^{1/3} = 0$
26. $\sqrt{r^2} - 4 = 0$
27. $2\sqrt{z} - \sqrt{2z+1} = \sqrt{z-3}$
28. $\dfrac{1}{\sqrt{x}} - \sqrt{x} = 0$
29. $(x-1)^{2/3} = (-2x)^{1/3}$
30. $\dfrac{3\sqrt{x}}{\sqrt{x-2}} = \sqrt{x+4}$
31. Find the solution set of the equation $\sqrt{x^2} - x = 0$ if $x \in R$.

5-6 GRAPHING $y = ax^2 + bx + c$

The general quadratic equation is

$$ax^2 + bxy + cy^2 + dx + ey + f = 0$$

$\{(x, y): y = x^2\}$

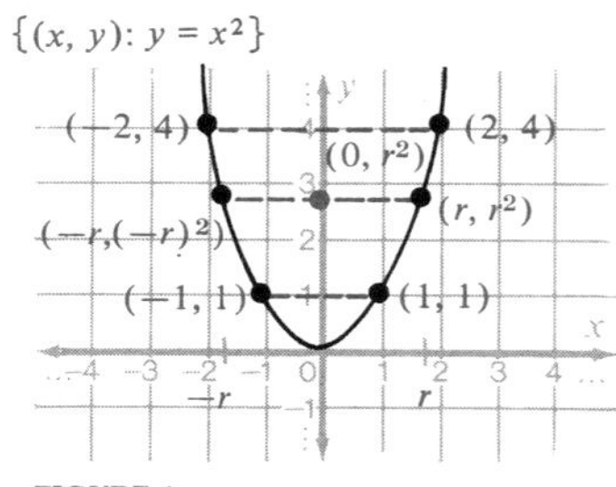

FIGURE 1

where $a, b, c, d, e, f \in R$ and a, b, and c are not all zero. We will study several special types of these equations.

Consider the simplest quadratic equation $y = x^2$. Since $r^2 \geqslant 0$ for every real number r and $r^2 = 0$ if and only if $r = 0$, (0, 0) is the lowest point of the graph of $y = x^2$. Also, since $(-r)^2 = r^2$ for every $r \in R$, the graph contains points (r, r^2) and $(-r, (-r)^2) = (-r, r^2)$. These are the endpoints of a horizontal segment with midpoint $(0, r^2)$ on the y-axis [Figure 1]. If (a, b) is a point of the graph, then $(-a, b)$ is a point of the graph and the midpoint of the segment determined by these points is $(0, b)$. A graph with this property is said to be symmetric with respect to the y-axis.

Definition If S is a subset of $R \times R$, then the graph of S is **symmetric with respect to the y-axis** if and only if for each (a, b) in S, $(-a, b)$ is also in S.

Sentence	Symmetric w.r.t. y-axis	
$y = 4x^2$	Yes	If $b = 4a^2$, then $4(-a)^2 = 4(a)^2 = b$
$y = -2x^2$	Yes	
$y^2 - x^2 = 0$	Yes	
$y \leqslant x^2$	Yes	
$x + y = 0$	No	$1 + (-1) = 0$ but $(-1) + (-1) \neq 0$

Which of the following sentences have graphs which are symmetric with respect to the y-axis: $y = 4x^2$; $y = -2x^2$; $y^2 - x^2 = 0$; $y \leqslant x^2$; $x + y = 0$?

The graph for the equation $y = x^2$ belongs to a collection of figures called **parabolas**. In the example above, the "low" point is called the vertex, and the y-axis is called the **axis of symmetry**.

The graph for the equation $y = -x^2$ is a reflection of the graph for $y = x^2$ about the x-axis [Figure 2]. The "high" point, (0, 0) is the vertex, and the y-axis is the axis of symmetry.

In considering the graphs for the equations $y = ax^2$, we can use the graphs for $y = x^2$ and $y = -x^2$ as references. For example, the parabola for $y = 4x^2$ has the same "low" point as the graph for $y = x^2$ but is "thinner" than this reference parabola [Figure 3]. However, the graph of $y = \frac{1}{4}x^2$ with the same "low" point is "wider" than the reference parabola [Figure 4].

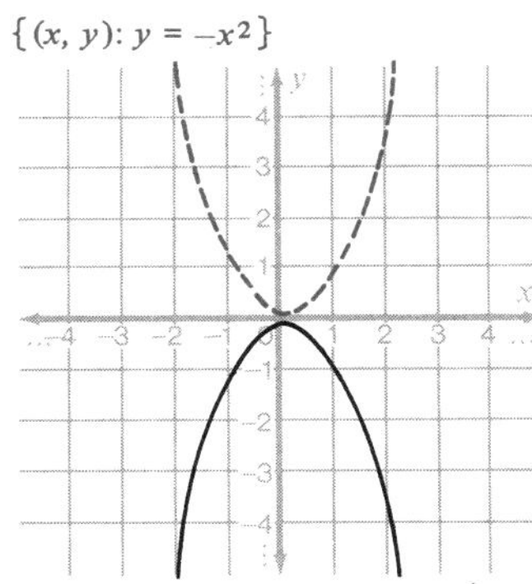

FIGURE 2

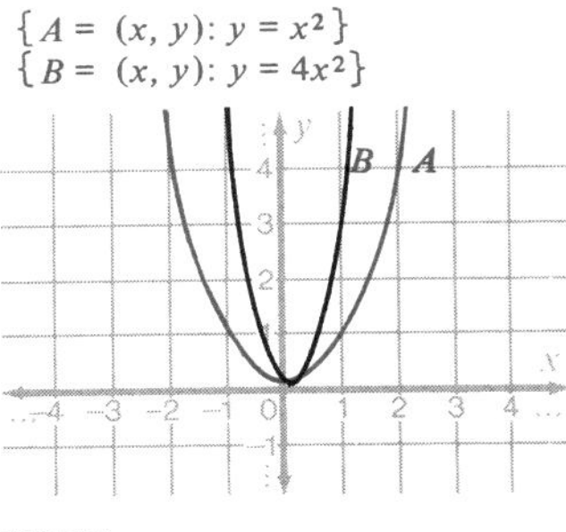

FIGURE 3

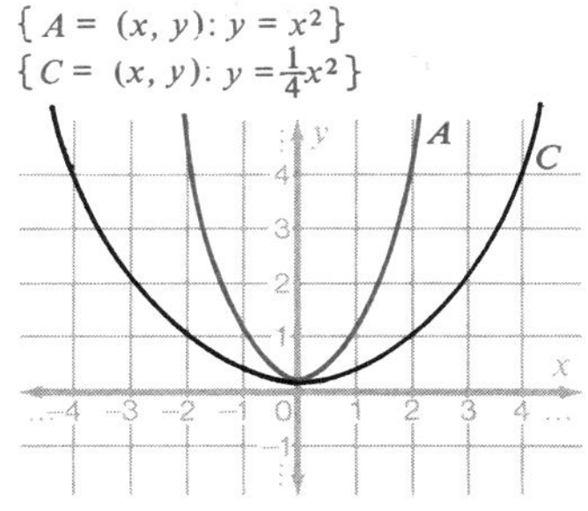

FIGURE 4

We can "slide" a parabola for $y = ax^2$ in a vertical direction to find the graph of the solution set for an equation of the type $y = ax^2 + c$. For example, the parabola for $y = \frac{1}{4}x^2$ furnishes an easy reference for locating

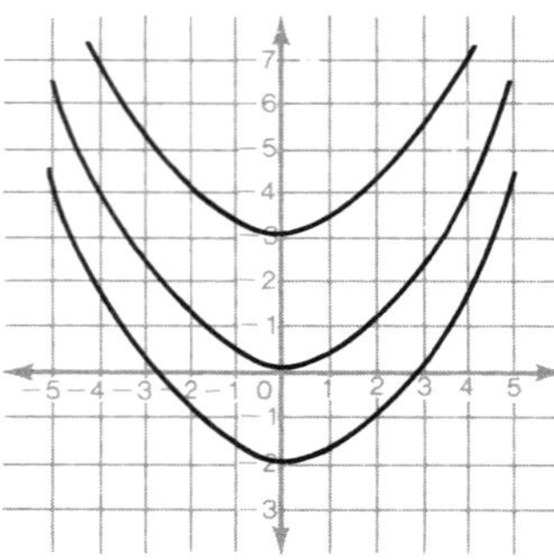
FIGURE 5

the parabolas for the equations $y = \frac{1}{4}x^2 + 3$ and $y = \frac{1}{4}x^2 + (-2)$. The vertices are $(0, 3)$ and $(0, -2)$, respectively [Figure 5]. The y-axis is the axis of symmetry in each case.

In using $y = ax^2$ as a reference for the equation $y = ax^2 + c$, if $a < 0$ or $a > 0$, the parabola has a vertex on the y-axis and the vertex is a "high" or "low" point, respectively, of the graph.

What if $b \neq 0$ in the equation $y = ax^2 + bx + c$? For example, the graph for $y = \frac{1}{4}x^2 - x - 3$ is a parabola with a vertical axis of symmetry and with vertex the "low" point as in the case for the equation $y = \frac{1}{4}x^2$. However, the axis of symmetry is not the y-axis.

Although the point for $(0, -3)$ is not the vertex, it is a point on the parabola since

$$-3 = \frac{1}{4}(0)^2 - (0) - 3.$$

We know that the horizontal line through the point $(0, -3)$ must intersect the parabola at another point [Figure 6]. That is, a point $(r, -3)$ must

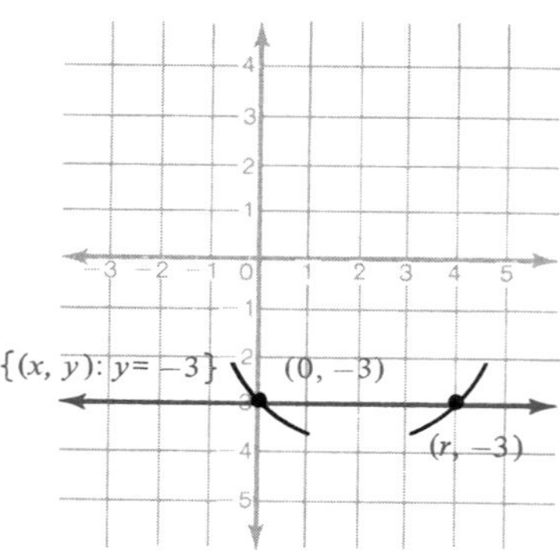

FIGURE 6

belong to the parabola for some nonzero real number r. If we find this point, then we can use symmetry of the parabola to locate the axis of symmetry and also the vertex.

Since the point $(r, -3)$ is on the parabola $y = \frac{1}{4}x^2 - x - 3$, we know that

$$-3 = \frac{1}{4}r^2 - r - 3.$$

Therefore

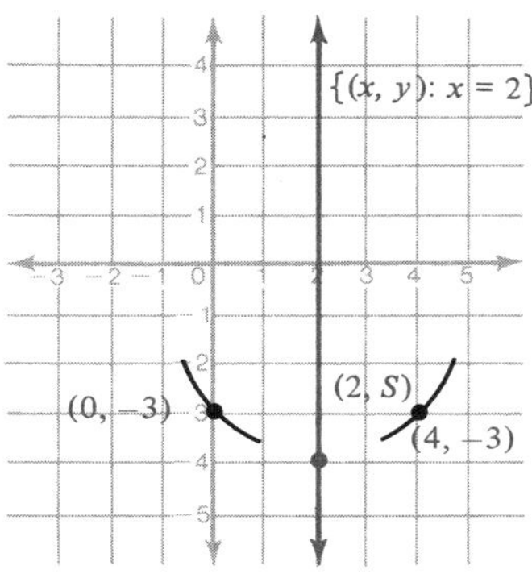

FIGURE 7

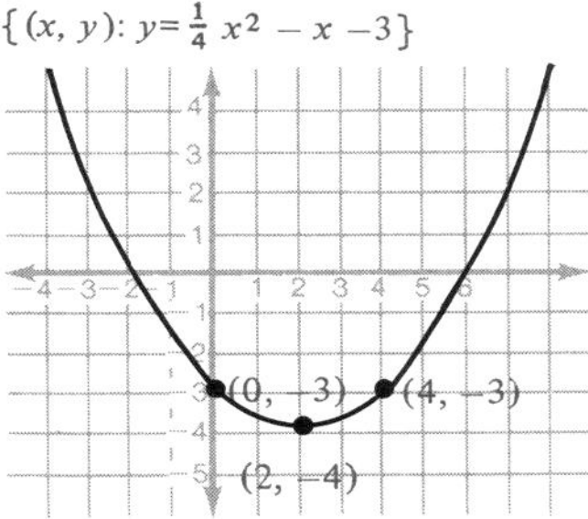

FIGURE 8

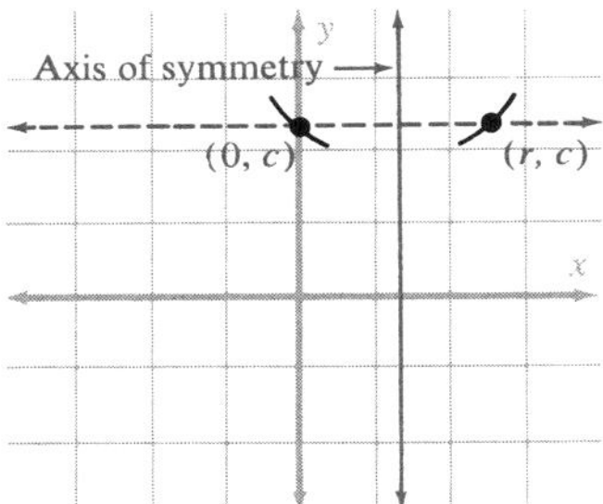

FIGURE 9

If $a < 0$, then the graph is "∩" shaped with the opening downward. If $a > 0$, then the graph is "∪" shaped with the opening upward.

$$\frac{1}{4}r^2 - r = 0$$

$$r(r - 4) = 0$$

$$r = 0 \quad \text{or} \quad r = 4$$

and the point (4, −3) is the point on the parabola corresponding to the point (0, −3). The axis of symmetry is the line for the equation $x = \frac{1}{2}(4)$. The vertex is the point $(2, s)$ that is on the parabola $y = \frac{1}{4}x^2 - x - 3$ [Figure 7]. Hence

$$s = \frac{1}{4}(2)^2 - (2) - 3$$

$$s = -4.$$

and the vertex of the parabola is (2, −4). With this information we can sketch the graph for the parabola [Figure 8].

In summary, to graph $y = ax^2 + bx + c$ where $b \neq 0$ find two points on the parabola with second coordinate c [Figure 9]. One of these is the point $(0, c)$ since $a \cdot (0)^2 + b \cdot 0 + c = c$. If the other point is (r, c), then

$$ar^2 + br + c = c$$

$$r(ar + b) = 0$$

$$r = 0 \quad \text{or} \quad r = -\frac{b}{a}.$$

Hence, the other point must be $\left(-\frac{b}{a}, c\right)$. From this information we find the axis of symmetry, the line for $x = -\frac{b}{2a}$, and the vertex of the parabola, $\left(-\frac{b}{2a}, \frac{-b^2 + 4ac}{4a}\right)$.

What is the significance of $a < 0$ versus $a > 0$ relative to the graphs for equations of the type $y = ax^2 + bx + c$?

In many cases the x-axis intersects the parabola in two points. Locating these points presents an alternative approach to sketching the parabola. For example, the solutions in R for the equation

$$\frac{1}{4}x^2 - x - 3 = 0$$

are the first coordinates for the two points on the x-axis. In solving this equation we find equivalent equations

$$x^2 - 4x - 12 = 0,$$

$$(x - 6)(x + 2) = 0,$$

and

$$x = 6 \quad \text{or} \quad x = -2.$$

Thus, (6, 0) and (-2, 0) are the two points on the parabola we seek. These two points can be used to locate the axis of symmetry, the line

$$x = \frac{1}{2}[(-2) + 6] = 2$$

which in turn, can be used to find the vertex, (2, -4).

Exercises

1. What type of geometric figure is the graph for each equation?

   a. $y = 2x^2$ b. $y = \frac{1}{2}x$ c. $y = 2x$

   d. $y = -2x$ e. $y = \frac{1}{2}x^2$ f. $y = -2x^2$

2. Sketch the graph for each part of Exercise 1.

3. Which parabolas have a vertex which is a "high" point? A "low" point?

   a. $y = -\frac{1}{3}x^2$ b. $3y - x^2 = 0$ c. $y = 3x^2$

   d. $y = 10x^2$ e. $y + 3x^2 = 0$ f. $10y = x^2$

4. Sketch the graph for each equation in Exercise 3.

5. In the equation $y = ax^2$, what is the significance of $a < 0$ versus $a > 0$ in sketching the graph of its solution set?

6. In the equations $y = ax^2$, $y = bx^2$ where $a \cdot b > 0$, what is the significance of $|a| > |b|$ in sketching the graph of each solution set?

7. Use the parabola for $y = 4x^2$ as a reference. Find the axis of symmetry and vertex of the parabola for each equation. Sketch the graph.

   a. $y = 4x^2 + 2$ b. $y = 4x^2 - 3$

   c. $y = 4x^2 + 5$ d. $y = 4x^2 - 7$

8. Replace $y = 4x^2$ by $y = -\frac{1}{3}x^2$ in the preceding instruction and use this reference.

   a. $y = -\frac{1}{3}x^2 + 1$ b. $y = -\frac{1}{3}x^2 - 1$

   c. $y = -\frac{1}{3}x^2 + 3$ d. $y = -\frac{1}{3}x^2 - 3$

9. Use one set of coordinate axes to find the axis of symmetry and vertex of the parabola for each equation. Sketch the graph.
 a. $y = 2x^2$ b. $y = 2x^2 + 1$ c. $y = 2x^2 + 2$
 d. $y = 2x^2 + 3$ e. $y = 2x^2 - 1$ f. $y = 2x^2 - 2$
 g. $y = 2x^2 - 3$ h. $y = 2x^2 + 5x - 3$

10. Find the points of the parabola for $y = -\frac{1}{4}x^2 + x + 1$ that are on the x-axis. Then find the vertical axis of symmetry, the vertex, and sketch the parabola.

11. Work Exercise 10 by finding the point of the parabola that is on the y-axis.

12. Show that the method in Exercise 10 does not apply to the equation $y = \frac{1}{4}x^2 - x + 3$.

Find the axis of symmetry and vertex of the parabola for each equation in Exercises 13–24. Sketch the graph.

13. $y = x^2 + x - 6$
14. $y = -x^2 + 7x - 10$
15. $y = 2x^2 - x - 6$
16. $y = -4x^2 - 8x + 5$
17. $y = \frac{1}{6}x^2 + \frac{1}{6}x - 1$
18. $y = -\frac{1}{2}x^2 - \frac{3}{2}x + 2$
19. $y = 4x^2 - 8x + 5$
20. $y = -4x^2 + 8x - 3$
21. $y = 3x^2 - 6x + 1$
22. $y = -3x^2 + 6x - 5$
23. $2y = x^2 + 4x + 6$
24. $2y + x^2 + 4x + 2 = 0$

25. For what real number r, is r^2 a minimum value? Is $-r^2$ a maximum value? Why?

26. For what real number r is $2(r - 4)^2$ a minimum value? Is $-2(r + 4)^2$ a maximum value? Is $2(r + 4)^2 + 5$ a minimum value?

27. Justify each step in the following derivation, where $r \in R$:

$$2r^2 - 4r - 1 = 2\left(r^2 - 2r - \frac{1}{2}\right) = 2\left[r^2 - 2r + (1 - 1) - \frac{1}{2}\right]$$

$$= 2\left[(r^2 - 2r + 1) + \left(-\frac{3}{2}\right)\right] = 2\left[(r - 1)^2 + \left(-\frac{3}{2}\right)\right]$$

$$= 2(r - 1)^2 - 3.$$

Use this information to find the axis of symmetry and vertex of the parabola for the equation $y = 2x^2 - 4x - 1$.

5-7 EQUATIONS OF CIRCLES AND ELLIPSES

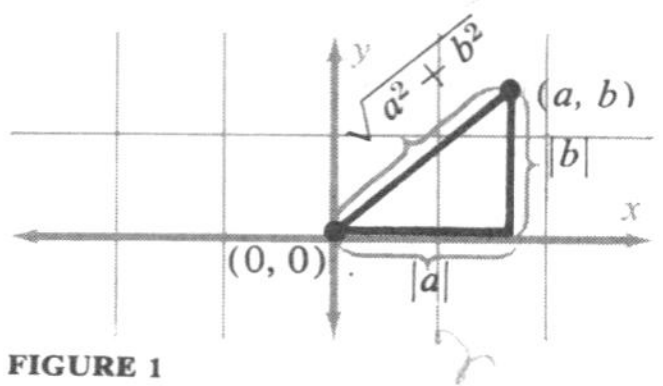

FIGURE 1

A **circle** is a set of points, each of which is the same distance from a given point called the **center** of the circle. Let us deduce the equation of a circle with its center at the origin.

From the Pythagorean theorem we know that the square of the distance from the origin to (a, b) in the coordinate plane is $a^2 + b^2$ [Figure 1]. Hence, the set of all points at a distance r from the origin corresponds to the solution set for

$$x^2 + y^2 = r^2.$$

Thus, the solution set for $x^2 + y^2 = r^2$ has a graph which is a circle with radius r and center $(0, 0)$ [Figure 2].

What is the solution set for $x^2 + y^2 = d$ if $d = 0$? If $d < 0$?

If $d = 0$, the solution set is $\{(0, 0)\}$;
• If $d < 0$, then for any real numbers r and s, $r^2 \geqslant 0$, $s^2 \geqslant 0$, and $r^2 + s^2 \geqslant 0$. Hence, no ordered pair of real numbers (r, s) is a member of the solution set and the solution set is $\emptyset$.

If (a, b) and (h, k) are two points [Figure 3], then the square of the distance between them is

$$(a - h)^2 + (b - k)^2.$$

The set of all points at a distance r from (h, k) is the solution set of

$$(x - h)^2 + (y - k)^2 = r^2.$$

That is, a circle with center at (h, k) and radius of length r is the graph for this equation.

$\{(x, y): x^2 + y^2 = c^2\}$

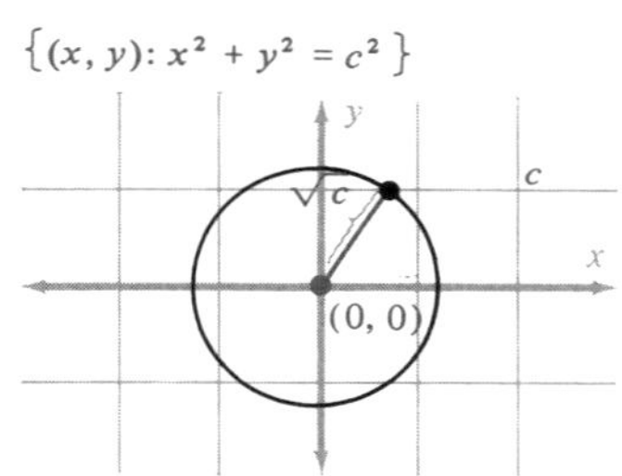

FIGURE 2

Consider the graph for the equation

$$x^2 + y^2 - 4x + 6y - 3 = 0.$$

Using two applications of completing the square, we find the equivalent sentences

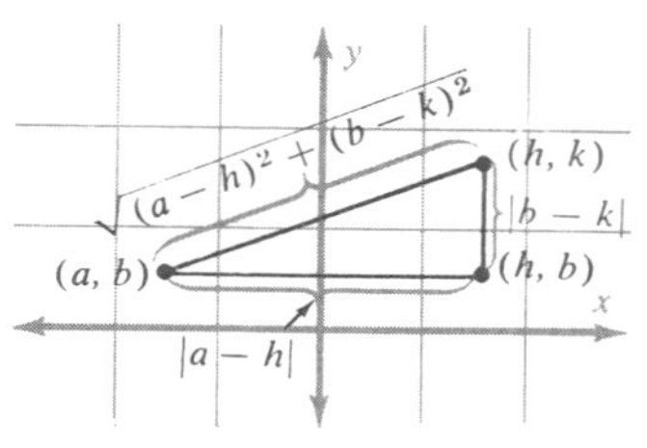

FIGURE 3

$$(x^2 - 4x + 4) + (y^2 + 6y + 9) = 3 + 4 + 9$$

and

$$(x - 2)^2 + [y - (-3)]^2 = 4^2.$$

The graph is a circle with radius 4 and center $(2, -3)$.

Completing the square in the equation

$$x^2 + y^2 + dx + ey + f = 0$$

we have

$$\left(x^2 + dx + \frac{d^2}{4}\right) + \left(y^2 + ey + \frac{e^2}{4}\right) = \frac{d^2}{4} + \frac{e^2}{4} - f$$

and

$$\left[x - \left(-\frac{d}{2}\right)\right]^2 + \left[y - \left(-\frac{e}{2}\right)\right]^2 = a$$

where a is the number $\frac{d^2}{4} + \frac{e^2}{4} - f$. If $a > 0$, the graph of the solution set is a circle with radius $\sqrt{a}$ and center $\left(-\frac{d}{2}, -\frac{e}{2}\right)$. What is the solution set if $a = 0$? If $a < 0$?

Another type of graph for quadratic equations in two variables is an **ellipse.** We will only consider ellipses that are symmetric with respect to the x-axis, the y-axis, and the origin. One way of characterizing an ellipse is as a set of points in a plane such that the sum of the distances between each point and two fixed points is constant. For example, if $(-3, 0)$ and $(3, 0)$ are the fixed points and 10 is the sum, then $(5, 0)$ and $(-5, 0)$ are on this ellipse [Figure 4] as are $(0, 4)$ and $(0, -4)$ [Figure 5]. This ellipse is sketched in Figure 6.

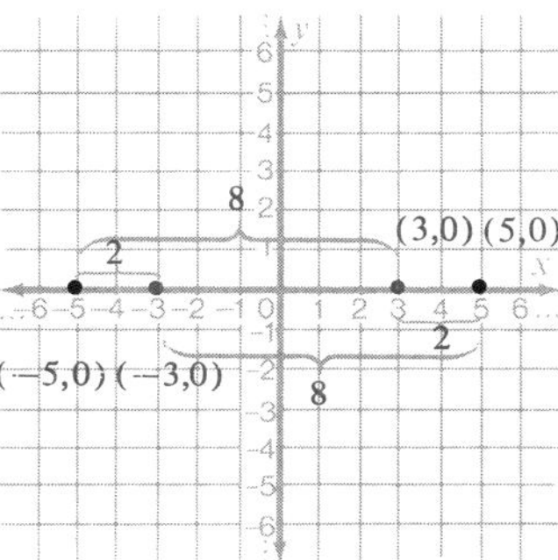

FIGURE 4

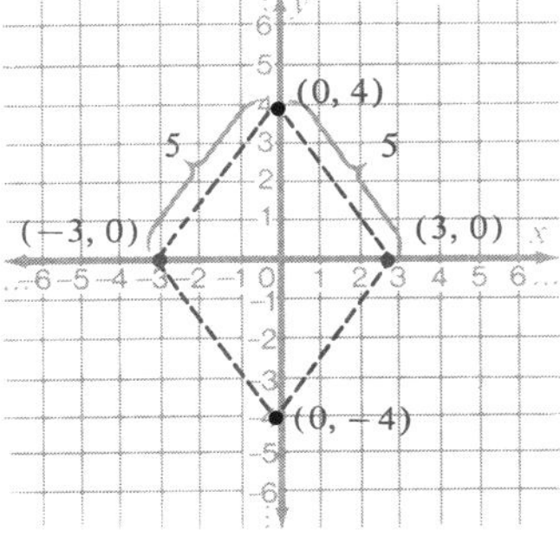

FIGURE 5

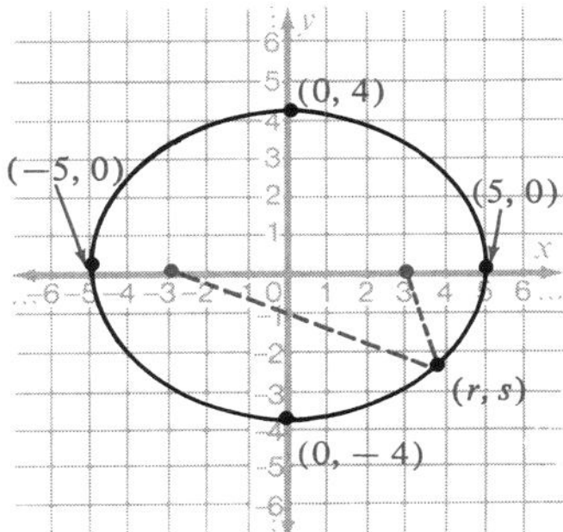

FIGURE 6

The points on this ellipse correspond to the members of the solution set for the equation

$$\sqrt{(x-3)^2+y^2}+\sqrt{[x-(-3)]^2+y^2}=10.$$

By subtracting and squaring we get an equivalent equation

$$(x-3)^2+y^2=100-20\sqrt{(x+3)^2+y^2}+(x+3)^2+y^2.$$

Collecting terms yields

$$5\sqrt{(x+3)^2+y^2}=25+3x.$$

Squaring and collecting terms gives

$$16x^2+25y^2-400=0.$$

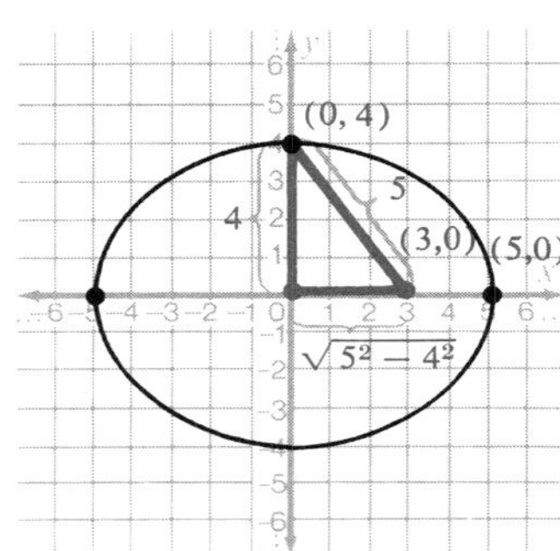

FIGURE 7

If we now divide by 400, if follows that

$$\frac{x^2}{25}+\frac{y^2}{16}=1$$

and finally

$$\frac{x^2}{5^2}+\frac{y^2}{4^2}=1.$$

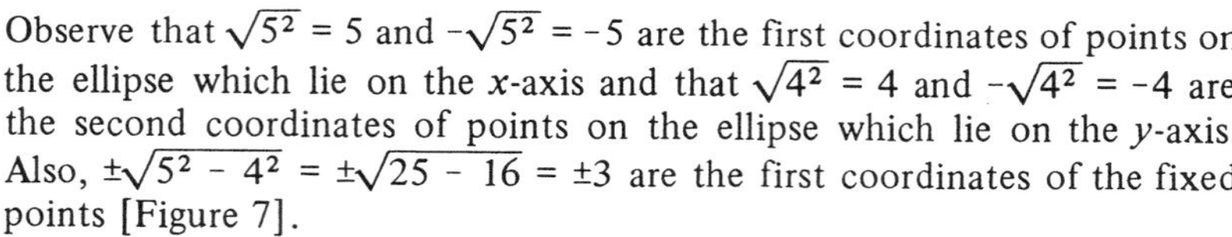

Observe that $\sqrt{5^2}=5$ and $-\sqrt{5^2}=-5$ are the first coordinates of points on the ellipse which lie on the x-axis and that $\sqrt{4^2}=4$ and $-\sqrt{4^2}=-4$ are the second coordinates of points on the ellipse which lie on the y-axis. Also, $\pm\sqrt{5^2-4^2}=\pm\sqrt{25-16}=\pm3$ are the first coordinates of the fixed points [Figure 7].

$\{(x, y): \frac{x^2}{a^2}+\frac{y^2}{b^2}=1\}$

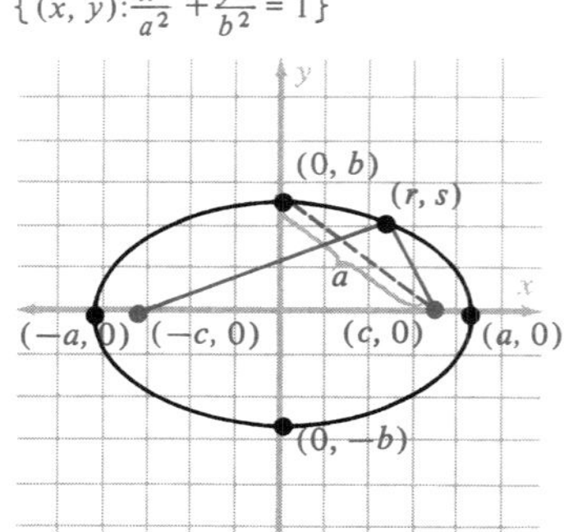

FIGURE 8

In general, if $a > b > 0$ then the graph for the quadratic equation

$$\frac{x^2}{a^2}+\frac{y^2}{b^2}=1$$

is an ellipse [Figure 8]. If $c^2=a^2-b^2$, the points $(c, 0)$ and $(-c, 0)$ are fixed points **(foci)**. The constant sum is $2a$. The segment with endpoints $(a, 0)$ and $(-a, 0)$ is called the **major axis**, and the segment with endpoints $(0, b)$ and $(0, -b)$ is called the **minor axis**.

Example Locate the foci, the endpoints of the major and minor axes, and the constant sum of the ellipse for

$$x^2+4y^2-100=0$$

and sketch the ellipse.

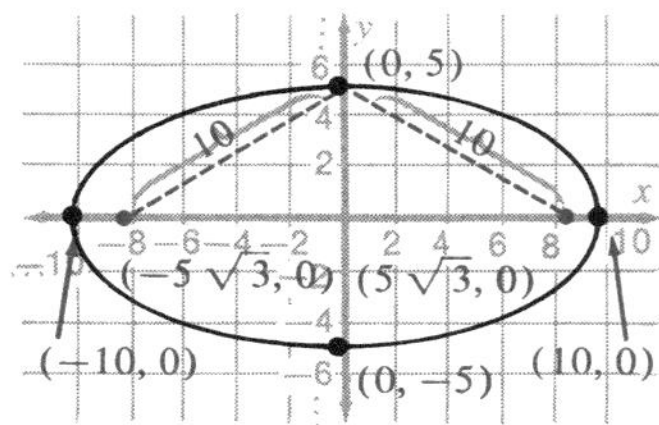

FIGURE 9

Solution: The equation $x^2 + 4y^2 - 100 = 0$ is equivalent to each of the following:

$$x^2 + 4y^2 = 100$$

$$\frac{x^2}{100} + \frac{y^2}{25} = 1.$$

$$\frac{x^2}{10^2} + \frac{y^2}{5^2} = 1.$$

Therefore, the points (−10, 0) and (10, 0) are endpoints of the major axis and the points (0, 5) and (0, −5) are endpoints of the minor axis. The points $(-\sqrt{100-25}, 0)$ and $(\sqrt{100-25}, 0)$ are the foci, and the constant sum is $2a$ where $a = 10$. The ellipse is sketched in Figure 9.

Standard form: $\frac{x^2}{(7)^2} + \frac{y^2}{(4)^2} = 1$

Endpoints of
major axis: (7, 0), (−7, 0)
minor axis: (0, 4), (0, −4)

Foci: $(\sqrt{33}, 0)$, $(-\sqrt{33}, 0)$

Constant sum: $2 \cdot 7 = 14$

Graph: Ellipse containing (7, 0), (−7, 0), (0, 4), and (0, −4)

Apply the directions for the preceding example to the ellipse for the equation $16x^2 + 49y^2 - 784 = 0$.

If the quadratic equation is equivalent to

$$\frac{x^2}{b^2} + \frac{y^2}{a^2} = 1$$

where $a > b > 0$ and $c^2 = a^2 - b^2$, then the points $(0, c)$ and $(0, -c)$ are the foci, the points $(0, a)$ and $(0, -a)$ are the endpoints of the major axis, and the points $(b, 0)$ and $(-b, 0)$ are the endpoints of the minor axis [Figure 10].

Exercises

The solution set for each equation in Exercises 1–16 is a circle, a singleton (set containing only one member), or the empty set. Determine which of the three types applies in each case. If the solution set is a circle, find the length of its radius and its center.

1. $x^2 + y^2 - 25 = 0$
2. $x^2 + y^2 - 10 = 0$
3. $x^2 + y^2 - 49 = 0$
4. $4x^2 + 4y^2 - 1 = 0$
5. $x^2 + y^2 + 4y + 5 = 0$
6. $x^2 + x + y^2 - 6 = 0$
7. $x^2 + y^2 - 10x + 24 = 0$
8. $x^2 + y^2 - 6x - 4y + 4 = 0$
9. $x^2 + y^2 - 10x + 6y + 30 = 0$
10. $x^2 + y^2 - 2x + 2y - 14 = 0$
11. $x^2 + y^2 + 14x - 10y + 74 = 0$
12. $x^2 + y^2 + 14x - 10y + 73 = 0$
13. $x^2 + y^2 + 14x - 10y + 75 = 0$
14. $x^2 + y^2 - 4x + 6y + 9 = 0$
15. $x^2 + y^2 - 4x + 6y + 13 = 0$
16. $2x^2 + 2y^2 - 2x - 2y - 7 = 0$

$\{(x, y): \frac{x^2}{b^2} + \frac{y^2}{a^2} = 1\}$

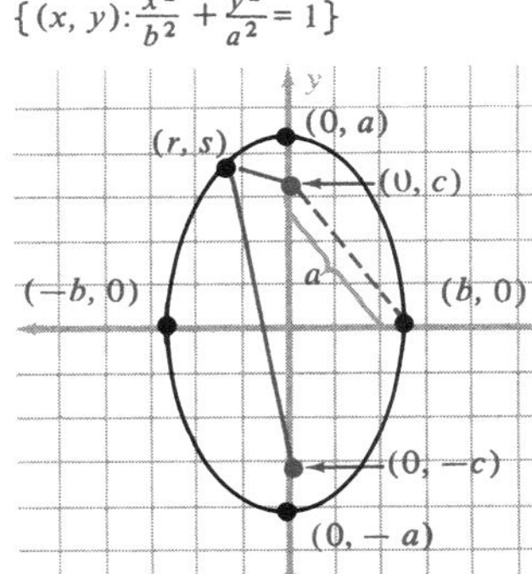

FIGURE 10

Sketch the graph for each sentence in Exercises 17-52.

17. $x^2 + y^2 - 4x + 10y + 13 = 0$
18. $4x^2 + 4y^2 - 4x + 12y - 15 = 0$
19. $2x^2 + 6x - y + 5 = 0$
20. $4x^2 + 4y^2 - 28y + 33 = 0$
21. $y^2 - 9 = 0$
22. $x^2 + y^2 - 4x + 10y + 29 = 0$
23. $3x^2 + 10x + y + 8 = 0$
24. $x^2 - 4 = 0$
25. $x^2 - y^2 = 0$
26. $x^2 + y^2 - 2x + 6y + 14 = 0$
27. $x^2 + y^2 - 16 \leqslant 0$
28. $x^2 + y^2 - 25 > 0$
29. $x^2 + y^2 - 2x + 6y + 6 < 0$
30. $x^2 + y^2 + 3y \geqslant 0$
31. $9x^2 + 25y^2 - 225 = 0$
32. $16x^2 + 9y^2 - 144 = 0$
33. $25x^2 + 9y^2 - 225 = 0$
34. $25x^2 + y^2 - 25 = 0$
35. $6x^2 + y^2 - 36 = 0$
36. $x^2 + 6y^2 - 36 = 0$
37. $100x^2 + y^2 - 100 = 0$
38. $x^2 + 100y^2 - 100 = 0$
39. $x^2 + 16y^2 - 16 = 0$
40. $16x^2 + y^2 - 16 = 0$
41. $16x^2 + 121y^2 - 1936 = 0$
42. $121x^2 + 16y^2 - 1936 = 0$
43. $100x^2 + 36y^2 - 225 = 0$
44. $36x^2 + 100y^2 - 225 = 0$
45. $4x^2 + 4y^2 - 9 = 0$
46. $9x^2 + 16y^2 = 0$
47. $9x^2 + 16y^2 - 144 = 0$
48. $9x^2 + 9y^2 - 25 = 0$
49. $6x^2 + y^2 - 36 < 0$
50. $6x^2 + y^2 - 36 \geqslant 0$
51. $x^2 + 16y^2 - 16 \geqslant 0$
52. $x^2 + 16y^2 - 16 < 0$

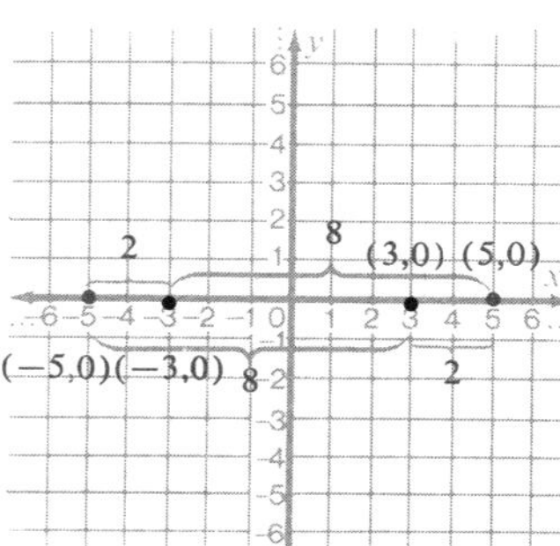

FIGURE 1

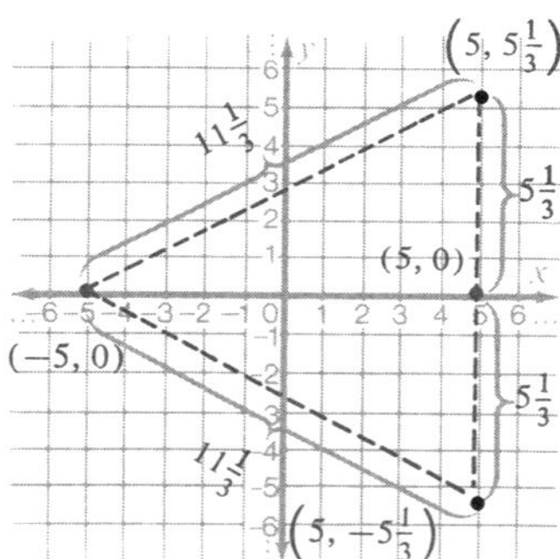

FIGURE 2

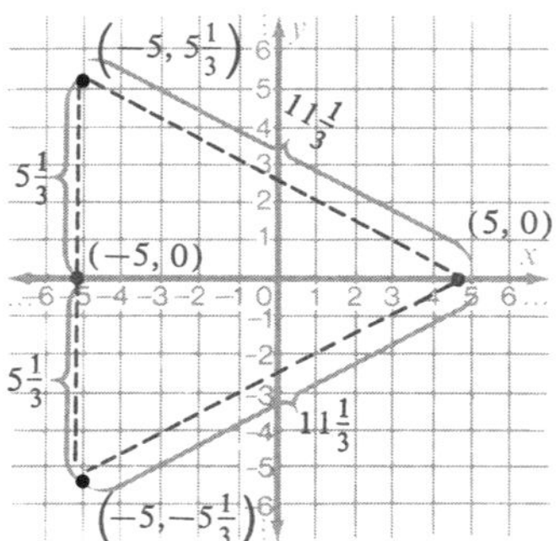

FIGURE 3

5-8 EQUATIONS OF HYPERBOLAS

Another geometric figure which is the graph for certain quadratic equations in two variables is the hyperbola. A **hyperbola** is the set of points in a plane such that the difference of the distance between each point and two fixed points is constant. As in the case for the ellipse, we will limit our treatment of hyperbolas to those that are symmetric with respect to both coordinate axes and the origin. To achieve this symmetry, the fixed points in the above definition are on one of the coordinate axes and are equidistant from the origin. For example, if the two fixed points are (5, 0) and (-5, 0) and the difference is 6, then (3, 0) and (-3, 0) are on the hyperbola [Figure 1] as are $\left(5, 5\frac{1}{3}\right)$ and $\left(5, -5\frac{1}{3}\right)$ [Figure 2] and the points $\left(-5, 5\frac{1}{3}\right)$ and $\left(-5, -5\frac{1}{3}\right)$ [Figure 3]. This hyperbola is sketched in Figure 4.

The points of this hyperbola correspond to members of the solution

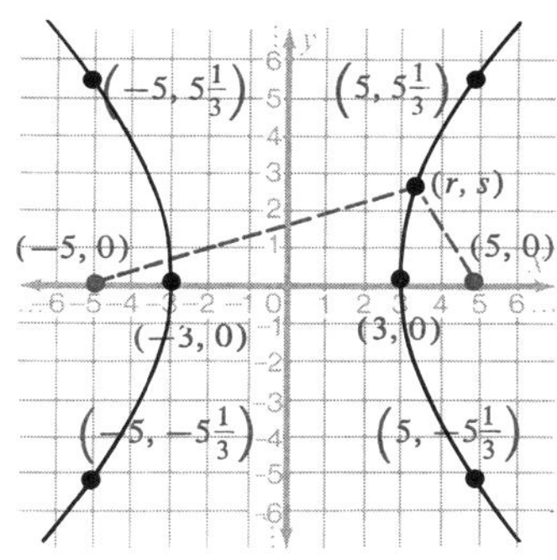

FIGURE 4

set for

$$\left|\sqrt{[x - (-5)]^2 + (y - 0)^2} - \sqrt{(x - 5)^2 + (y - 0)^2}\right| = 6.$$

This sentence is equivalent to

$$(1) \qquad \sqrt{(x + 5)^2 + y^2} - \sqrt{(x - 5)^2 + y^2} = 6$$

or

$$(2) \qquad \sqrt{(x + 5)^2 + y^2} - \sqrt{(x - 5)^2 + y^2} = -6.$$

Equation (1) is equivalent to

$$\sqrt{(x + 5)^2 + y^2} = 6 + \sqrt{(x - 5)^2 + y^2}.$$

Squaring both members, collecting terms, and dividing by 4 yields

$$5x - 9 = 3\sqrt{(x - 5)^2 + y^2}.$$

Again, squaring both members, we derive the following sequence of equivalent equations:

$$16x^2 - 9y^2 = 144$$

$$\frac{x^2}{9} - \frac{y^2}{16} = 1$$

$$\frac{x^2}{3^2} - \frac{y^2}{4^2} = 1.$$

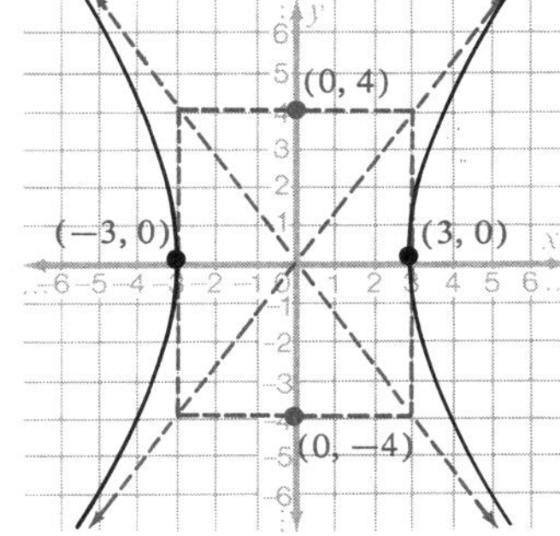

FIGURE 5

$\{(x, y): \frac{x^2}{a^2} - \frac{y^2}{b^2} = 1\}$

FIGURE 6

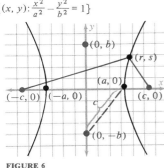

Equation (2) can be treated similarly. Observe that $\sqrt{3^2} = 3$ and $-\sqrt{3^2} = -3$ are the x-coordinates of points on the hyperbola which lie on the x-axis and that $\pm\sqrt{3^2 + 4^2} = \pm\sqrt{9 + 16} = \pm 5$ are the x-coordinates of the fixed points. Also notice in Figure 5 that $\sqrt{4^2} = 4$ and $-\sqrt{4^2} = -4$ are the y-coordinates of points on the y-axis that do not lie on the hyperbola. In general, if $c > a > 0$ and $c^2 - a^2 = b^2$, the graph of the quadratic equation $\frac{x^2}{a^2} - \frac{y^2}{b^2} = 1$ is a hyperbola with the points for $(c, 0)$ and $(-c, 0)$ as the fixed points and the constant difference of $2a$ [Figure 6]. The points $(a, 0)$ and $(-a, 0)$ are called **vertices** of this hyperbola and are the endpoints of the transverse axis. The points $(0, b)$ and $(0, -b)$ are the endpoints of the conjugate axis. The rectangle with vertices $(-a, b)$, (a, b), $(a, -b)$, and $(-a, -b)$ and with length $2a$ and width $2b$ [Figure 7] is an aid in sketching the hyperbola. The line determined by $(a, -b)$ and (a, b) and the line determined by $(a, -b)$ and $(-a, b)$ are called **asymptotes** for the hyperbola and have the equations

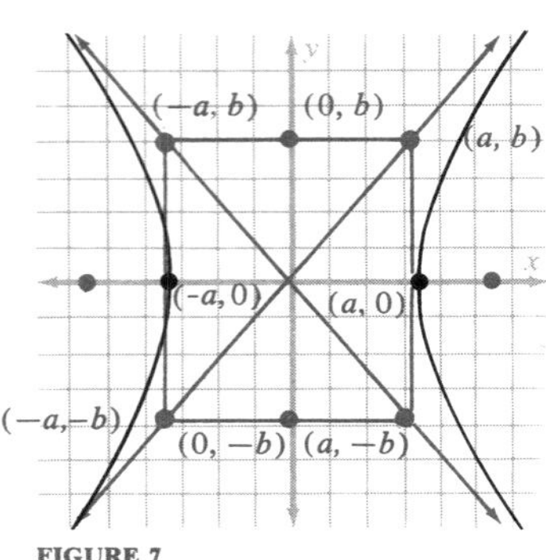

FIGURE 7

$$y = \frac{b}{a}x \quad \text{and} \quad y = -\frac{b}{a}x,$$

respectively. When $|x|$ is quite large, the points on the hyperbola are "close to" the corresponding points on these lines.

The hyperbola for the equation

$$\frac{y^2}{b^2} - \frac{x^2}{a^2} = 1$$

is called the **conjugate hyperbola** for the preceding hyperbola. This conjugate hyperbola has vertices at $(0, b)$ and $(0, -b)$ and fixed points at $(0, c)$ and $(0, -c)$ [Figure 8].

Example Locate the vertices, the points that determine the conjugate axis, and the fixed points of the hyperbola for

$$x^2 - 4y^2 + 100 = 0$$

and sketch this hyperbola.

Solution: The equation $x^2 - 4y^2 + 100 = 0$ is equivalent to each of the following:

$$x^2 - 4y^2 = -100$$

$$\frac{y^2}{25} - \frac{x^2}{100} = 1$$

$$\frac{y^2}{5^2} - \frac{x^2}{10^2} = 1.$$

Therefore, the vertices are $(0, 5)$ and $(0, -5)$ and the segment with points $(10, 0)$ and $(-10, 0)$ as endpoints is the conjugate axis. The fixed points are $(0, \sqrt{10^2 + 5^2})$ and $(0, -\sqrt{10^2 + 5^2})$, and the hyperbola is sketched in Figure 9.

Apply the directions in the preceding example to the hyperbola for the equation $x^2 - 4y^2 - 100 = 0$. How does this hyperbola relate to the hyperbola in Figure 9?

Standard form: $\frac{x^2}{(10)^2} - \frac{y^2}{(5)^2} = 1$

Vertices: $(10, 0)$, $(-10, 0)$

Endpoints of conjugate axis: $(0, 5)$, $(0, -5)$

Fixed points: $(5\sqrt{5}, 0)$, $(-5\sqrt{5}, 0)$

Graph: Hyperbola with same asymptotes as Figure 9 and with vertices at $(10, 0)$ and $(-10, 0)$

Relationship to Figure 9: Conjugate hyperbola

Exercises

Sketch the graph for each sentence in Exercises 1-22.

1. $9x^2 - 4y^2 - 36 = 0$
2. $4y^2 - 9x^2 - 36 = 0$
3. $9x^2 - y^2 + 9 = 0$
4. $y^2 - 9x^2 + 9 = 0$
5. $49x^2 - 4y^2 - 196 = 0$
6. $4x^2 - 49y^2 - 196 = 0$
7. $100x^2 - y^2 - 100 = 0$
8. $49x^2 - 4y^2 + 196 = 0$

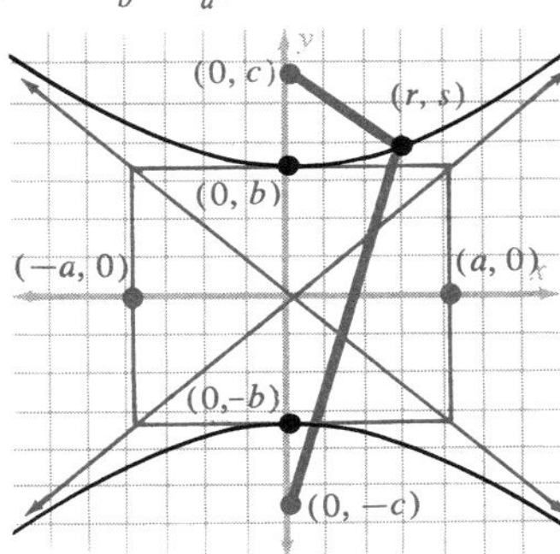

FIGURE 8

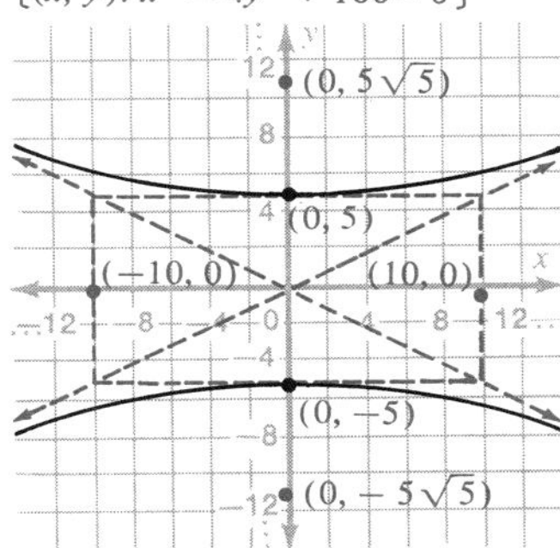

FIGURE 9

9. $4x^2 - 49y^2 + 196 = 0$
10. $x^2 - y^2 + 25 = 0$
11. $16x^2 - 25y^2 - 100 = 0$
12. $25y^2 - 16x^2 - 100 = 0$
13. $x^2 - 2y^2 - 20 = 0$
14. $y^2 - 2x^2 - 20 = 0$
15. $4x^2 - 5y^2 - 80 = 0$
16. $4y^2 - 5x^2 - 80 = 0$
17. $4x^2 + 5y^2 - 80 = 0$
18. $x^2 - y^2 - 25 = 0$
19. $4x^2 - y^2 - 16 \leqslant 0$
20. $4x^2 - y^2 - 16 > 0$
21. $4x^2 - y^2 + 16 < 0$
22. $x^2 - 64y^2 - 64 \geqslant 0$
23. Prove that

$$\frac{x^2}{a^2} - \frac{y^2}{b^2} = 1$$

is equivalent to the sentence

$$y = \frac{b}{a}\sqrt{x^2 - a^2} \quad \text{or} \quad y = -\frac{b}{a}\sqrt{x^2 - a^2}.$$

If $|x|$ is quite large, discuss the difference between

$$\frac{b}{a}x \quad \text{or} \quad -\frac{b}{a}x, \quad \text{and} \quad \frac{b}{a}\sqrt{x^2 - a^2} = \frac{b}{a}|x|\sqrt{1 - \frac{a^2}{x^2}}.$$

24. Prove that the hyperbola with fixed points at $(\sqrt{2}, \sqrt{2})$ and $(-\sqrt{2}, -\sqrt{2})$ and constant difference of $2\sqrt{2}$ is the graph of the solution set in $R \times R$ for the sentence $xy = 1$.
25. Derive the sentence for the hyperbola with fixed points $(-\sqrt{2}, \sqrt{2})$ and $(\sqrt{2}, -\sqrt{2})$ and constant difference of $2\sqrt{2}$.

Review Exercises

1. Solve the quadratic equation $2x^2 + 3x = 15 - 4x$ by the method of factoring. **5-1**
2. Solve $2x^2 + 5x - 3 = 0$ by the method of completing the square. **5-2**
3. Solve $8x^2 + 26x - 99 = 0$ by using the quadratic formula.
4. Find the members of the solution set in R and sketch the graphs of the solution set for each inequality. **5-3**
   a. $x^2 + 2x - 63 < 0$
   b. $x^2 - 2x - 63 \geqslant 0$
5. Use absolute value notation in writing a sentence equivalent to each sentence. **5-4**
   a. $-4 \leqslant x \leqslant 6$
   b. $x < -4$ or $x > 6$.

6. Find the solution set for each equation if R is the replacement set for the variables. 5-5
   a. $3u = \sqrt{u^2 + 32}$
   b. $\sqrt{2z - 1} = 1 + \sqrt{z - 1}$

Sketch the graph of the solution set for each sentence in Exercises 7-11.

7. $6y = x^2$ 5-6

8. $y = \frac{1}{6}x^2 - \frac{4}{3}x + 2$

9. $x^2 + y^2 - 6x + 8y = 0$ 5-7

10. $49x^2 + 9y^2 = 441$

11. $49x^2 - 9y^2 = 441$ 5-8

6. POLYNOMIAL AND RATIONAL FUNCTIONS

6-1 RELATIONS AND FUNCTIONS

When we make statements such as

Two is less than 8
Five is not equal to 8
Line L_1 is parallel to line L_2
John is enrolled in American history

we are working with a concept that involves a first object paired with a second object–that is, an ordered pair. To illustrate this notion, we begin with two sets: a set of "first" objects, $A = \{1, 2, 3\}$ and a set of second objects, $B = \{0, 2, 4, 6\}$. We define

$$A \times B = \{(a, b)\colon\ a \in A \text{ and } b \in B\}$$

as the cross product of A and B. In this example $A \times B$ contains twelve elements: (1, 0), (1, 2), (1, 4), (1, 6), (2, 0), (2, 2), (2, 4), (2, 6), (3, 0), (3, 2), (3, 4), and (3, 6). Which of these ordered pairs has a first number that is less than its second number?

This criterion yields a subset of $A \times B$, namely

$$n = \{(1, 2), (1, 4), (1, 6), (2, 4), (2, 6), (3, 4), (3, 6)\}.$$

Such a set is called a relation from A into B.

Definition If S and T are sets, then a **relation from S into T** is a subset of $S \times T$.

When denoting relations, it is customary to use lower case letters rather than capitals. We will follow this convention.

If the elements of the sets forming the cross product are real numbers, then we can represent the relation on a coordinate plane. This representation is called the **graph of the relation**.

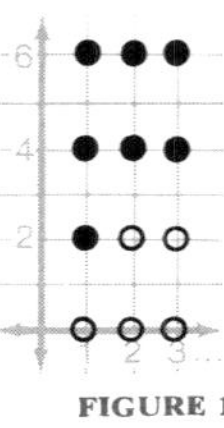
FIGURE 1

In Figure 1, the circles make up the graph of $A \times B$ and the filled-in circular regions make up the graph of n. The elements in a relation are ordered pairs. The set of first coordinates is called the **domain** of the relation. The set of second coordinates is called the **range** of the relation.

In the example $n \subset A \times B$, the domain of n is $\{1, 2, 3\}$ and the range of n is $\{2, 4, 6\}$.

If the elements in each ordered pair of a relation are interchanged, this process yields another set of ordered pairs.

Definition Let u be a relation from S into T. The relation

$$\{(t, s)\colon\ (s, t) \in u\}$$

from T into S is called the **inverse relation** of u and is denoted by u^{-1}.

The inverse relation of

$$n = \{(1, 2), (1, 4), (1, 6), (2, 4), (2, 6), (3, 4), (3, 6)\}$$

is

$$n^{-1} = \{(2, 1), (4, 1), (6, 1), (4, 2), (6, 2), (4, 3), (6, 3)\}.$$

The elements of n are identified by asking the question, "Is the first number less than the second number?" What question can be used to identify the elements in n^{-1}?

If $A = \{1, 2, 3\}$, $B = \{0, 2, 4, 6\}$, and $k = \{(1, 6), (2, 0), (3, 4)\}$ is a relation from A into B, sketch the graph of k. What is the domain of k? The range of k? What are the elements of the inverse relation k^{-1}? What are the domain and range of k^{-1}? Sketch the graph of k^{-1}.

k	$\{(1, 6), (2, 0), (3, 4)\}$
Domain	$\{1, 2, 3\} = A$
Range	$\{0, 4, 6\}$
k^{-1}	$\{(6, 1), (0, 2), (4, 3)\}$
Domain	$\{0, 4, 6\}$
Range	$\{1, 2, 3\}$

Graph of k

Graph of k^{-1}

A particular type of relation is called a function. To illustrate which relations are functions, we consider the relations

$$g = \{(a_1, b_1), (a_2, b_1), (a_3, b_2)\}$$

and

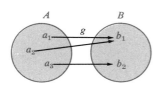

$$h = \{(a_1, b_1), (a_1, b_2), (a_2, b_2)\}$$

from $A = \{a_1, a_2, a_3\}$ into $B = \{b_1, b_2\}$ and their corresponding diagrams [Figures 2 and 3]. The important distinction between g and h is that each element in the domain of g has one and only one corresponding element in the range of g, whereas this statement does not hold for the relation h. The element a_1 of the domain of h has two corresponding elements in the range of h, namely b_1 and b_2.

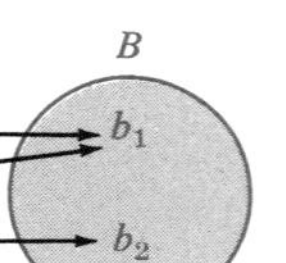
FIGURE 2

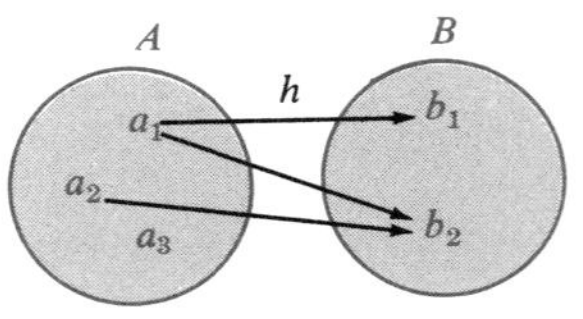
FIGURE 3

Definition A relation f from S into T is a **function** from S into T if and only if for each element s in S there is one and only one corresponding t in T (called the **functional value** of f at s) such that $(s, t) \in f$.

Examples The diagram in Figure 4 illustrates that the relation

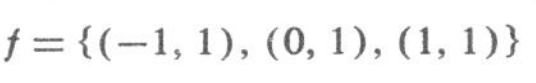

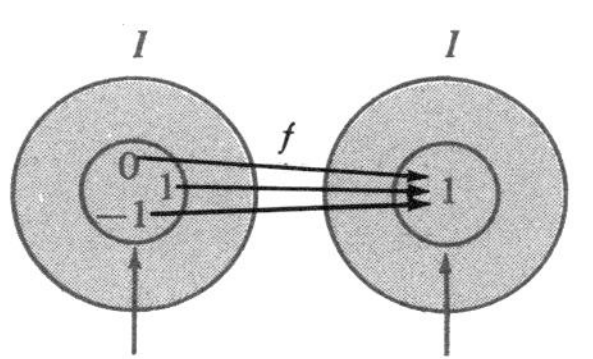

FIGURE 4

$$f = \{(-1, 1), (0, 1), (1, 1)\}$$

is a function. The relation

$$\{(x, y) \in I \times I: \ y \leqslant 2x\}$$

is not a function from I into I since at least one integer in its domain has more than one corresponding integer in its range [Figure 5].

Are the relations n, n^{-1}, k, k^{-1}, functions? Is f^{-1} in the above example, a function?

The relations k and k^{-1} are functions. The others are not since there exist at least two ordered pairs in each with identical first coordinates and distinct second coordinates.

$\{(x, y) \in I \times I: y \leq 2x\}$

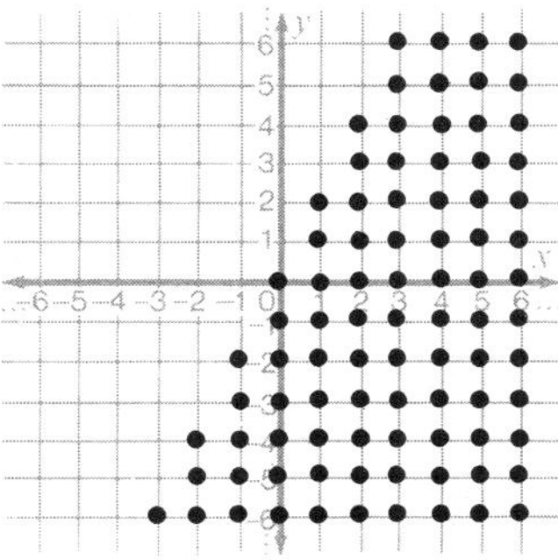

FIGURE 5

The fact that f is a function from S into T is sometimes denoted by f: $S \to T$. Since the relation $f = \{(x, y) \in I \times I: \ y = 2x + 1\}$ is a function from I into I, we write, "f: $I \to I$". Also, an alternative notation is often used to indicate that an ordered pair (s, t) belongs to a function f.

The numbers 1, 3, and 5 are the unique functional values of f at 0, 1, and 2, respectively. The alternative notation

$$f(0) = 1$$

$$f(1) = 3$$

$$f(2) = 5$$

displays this association. A symbol such as $f(0)$ is read "f of zero" or "the functional value of f at 0."

In general, if s is any element of the domain of a function f from S into T and t is the unique element in T such that $(s, t) \in f$, we write $t = f(s)$.

If $A = \{a_1, a_2, a_3\}$, $B = \{b_1, b_2\}$, $g = \{(a_1, b_1), (a_2, b_1), (a_3, b_2)\}$, and g: $A \to B$, what elements of B are $g(a_1)$, $g(a_2)$, and $g(a_3)$?

$g(a_1) = b_1, g(a_2) = b_1, g(a_3) = b_2$

The special relation

$$f_1 = \{(x, y): \ y = |x|\}$$

is called the **absolute value function**. The graph of f_1 [Figure 6] extends indefinitely to the right and left but contains no points below the x-axis. What is the domain of f_1? The range of f_1?

Another interesting function may be developed by first recalling that if r is a nonnegative real number, then $\sqrt{r}$ is the unique nonnegative real number such that $(\sqrt{r})^2 = r$. Thus, $\sqrt{25} = 5$ since $5^2 = 25$, $\sqrt{0} = 0$ since $0^2 = 0$, and $\sqrt{(-6)^2} = \sqrt{36} = 6$ since $6^2 = 36$.

A symbol such as $\sqrt{-4}$ does not denote a real number. With this agreement, we can define a function

$$f_2 = \{(x, y):\ y = \sqrt{x}\}$$

whose graph is Figure 7. It is important that the function f_2 be distinguished from the relation $f_3 = \{(x, y):\ y^2 = x\}$. Why is f_3 not a function?

	f_2	f_3
Graph	Figure 7	A parabola with vertex at the origin, with x-axis for axis of symmetry, and containing (4, 2) and (4, −2)
Domain	x: $x \geqslant 0$	$\{x:\ x \geqslant 0\}$
Range	x: $x \geqslant 0$	R

Sketch the graph of the relation f_3 and compare the graph with the graph for the function f_2. What is the domain of f_2? The domain of f_3? The range of f_2? The range of f_3?

Exercises

1. Given that $A = \{1, 2, 3, 4\}$, $B = \{3, 4, 5, 6\}$, and $c = \{(x, y) \in A \times B$: x and y are both even numbers$\}$:
   a. How many elements are in $A \times B$?
   b. List the elements of c.
   c. What is the domain of c?
   d. What is the range of c?
   e. Sketch the graph of c.
   f. List the elements of c^{-1}.
   g. What is the domain of c^{-1}?
   h. What is the range of c^{-1}?
2. What are the domain and range of each relation from R into R graphed in Figures 8–11?
3. Given that $C = \{2, 4, 6\}$, $D = \{3, 5, 7\}$, and

$$s_1 = \{(2, 3), (4, 7), (6, 5)\}$$
$$s_2 = \{(2, 5), (4, 5), (6, 5)\}$$
$$s_3 = \{(2, 5), (2, 3), (2, 7)\}$$
$$s_4 = \{(2, 7), (4, 7), (6, 3)\}$$
$$s_5 = \{(2, 3), (4, 7), (4, 5)\}$$
$$s_6 = \{(2, 3), (6, 5)\}$$

are relations from C into D:

$f_1 = \{(x, y):\ y = |x|\}$

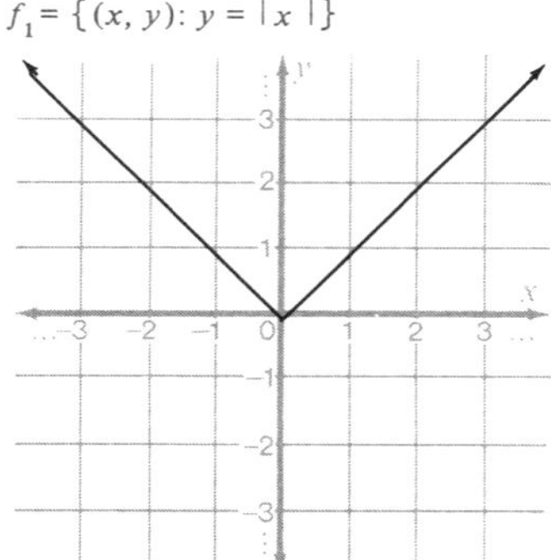

FIGURE 6

$f_2 = \{(x, y):\ y = \sqrt{x}\}$

FIGURE 7

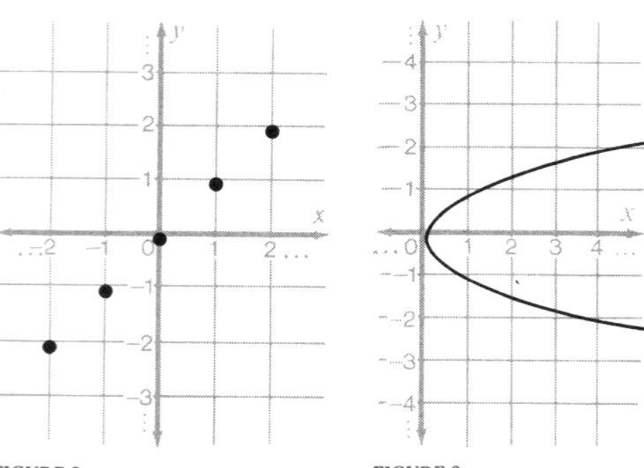

FIGURE 8 FIGURE 9

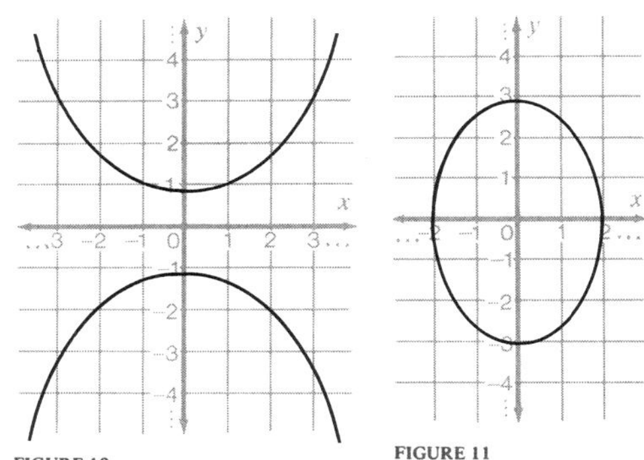

FIGURE 10

FIGURE 11

a. Give the domain for each relation.
b. Give the range for each relation.
c. Sketch the graph of each relation.
d. Sketch the graph of the inverse relation for each of these relations.
e. Which relations are functions?
f. Which inverse relations are functions?

For each relation in Exercises 4–15, give the domain and range, sketch the graph, find the inverse relation, and sketch the graph of the inverse relation.

4. $\{(x, y) \in I \times I\colon\ x \geqslant 0 \text{ and } y = x + 1\}$
5. $\{(x, y) \in I \times I\colon\ 0 \leqslant x \leqslant 4 \text{ and } -1 \leqslant y \leqslant 2\}$
6. $\{(x, y) \in I \times I\colon\ x \geqslant y$
7. $\{(x, y) \in I \times I\colon\ x^2 + y^2 = 4\}$
8. $\{(x, y)\colon\ x = 1\}$
9. $\{(x, y)\colon\ y = \pi\}$
10. $\{(x, y)\colon\ y = |x|\}$
11. $\{(x, y)\colon\ |y| = x\}$
12. $\{(x, y) \in I \times I\colon\ 0 \leqslant x \leqslant 4 \text{ or } -1 \leqslant y \leqslant 2\}$
13. $\{(x, y) \in I \times I\colon\ x^2 + y^2 = 25\}$
14. $\{(x, y)\colon\ -1 \leqslant x \leqslant 1, \text{ and } y > 1 \text{ or } y < -1\}$
15. $\{(x, y)\colon\ -1 \leqslant x \leqslant 1 \text{ and } y > 1, \text{ or } y < -1\}$

State the domain and range and sketch the graph of each function in Exercises 16–21.

16. $g_1 = \left\{(x, y)\colon\ y = \dfrac{|x|}{x}\right\}$
17. $g_2 = \{(x, y)\colon\ y = |x - 1|\}$
18. $g_3 = \{(x, y)\colon\ y = |x + 1|\}$

19. $g_4 = \{(x, y):\ y = \sqrt{x + 1}\}$
20. $g_5 = \{(x, y):\ y = \sqrt{x - 1}\}$
21. $g_6 = \{(x, y):\ y = x - [x]\}$
22. Refer to the functions in Exercises 16–19 and find each real number below

a. $g_1(1)$	**b.** $g_1(-1)$	**c.** $g_1(2)$
d. $g_2(1)$	**e.** $g_2(-1)$	**f.** $g_2(3)$
g. $g_3(-1)$	**h.** $g_3(1)$	**i.** $g_3(-3)$
j. $g_4(-1)$	**k.** $g_4(0)$	**l.** $g_3(3)$
m. $g_4(8)$	**n.** $g_4\left(\frac{1}{2}\right)$	**o.** $g_4\left(\frac{3}{4}\right)$

23. A relation f is called a **one-to-one correspondence** if and only if f is a function and the inverse relation f^{-1} is also a function. Which of the functions in Exercises 4–15 are one-to-one correspondences?
24. Give a geometric interpretation for the condition that a relation from R into R is a function.

6-2 LINEAR FUNCTIONS

$f = \{(x, y) : y = x^2 - 4x + 3)\}$

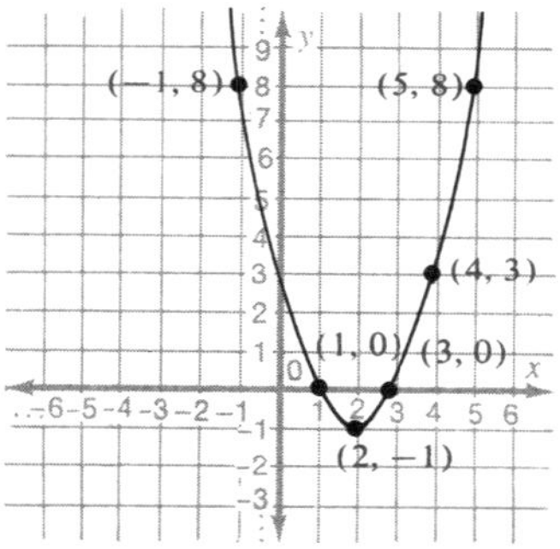

FIGURE 1

$\{(x, f(x)): f(x) = 3x - 5\}$

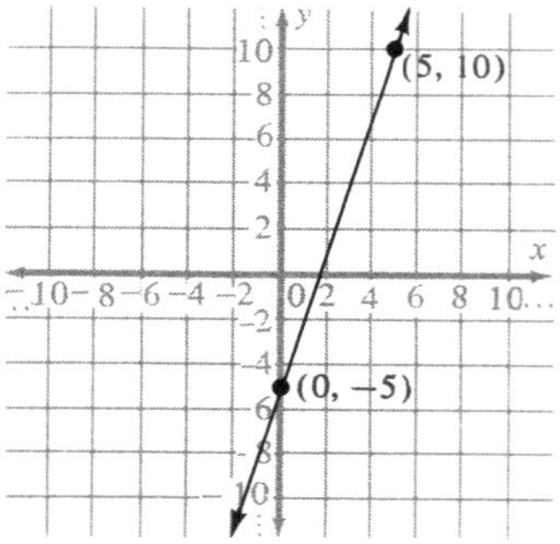

FIGURE 2

If we replace x by the number 2 in the polynomial $x^2 - 4x + 3$ we find a corresponding number

$$2 \rightarrow 2^2 - 4 \cdot 2 + 3 = -1$$

and an ordered pair $(2, -1)$. If x is replaced by 4 the corresponding number is 3 and the ordered pair is $(4, 3)$. Thus given a number we can find a second number by replacing x in the polynomial by the given number. This process yields the relation

$$f = \{(x, y):\ y = x^2 - 4x + 3\}.$$

Clearly, this relation is a function since a uniquely determines the number $(a \cdot a) - 4a + 3$. This relation is called the polynomial function from R into R for $x^2 - 4x + 3$.

The graph for this function is also the graph of the solution set for the equation

$$y = x^2 - 4x + 3.$$

We know that this graph is a parabola with a vertical axis of symmetry. By employing the methods developed in Chapter 5, Section 6, we are able to sketch the graph of f [Figure 1].

The graph of

$\{(x, y)\colon\ y = -x^2 - 2x + 15\}$

is a parabola with the vertex at (-1, 16), with vertical axis of symmetry (the line for $x = -1$), and containing (-5, 0) and (3, 0).

Sketch the graph of the function from R into R for the polynomial $-x^2 - 2x + 15$. What type of geometric figure is the graph?

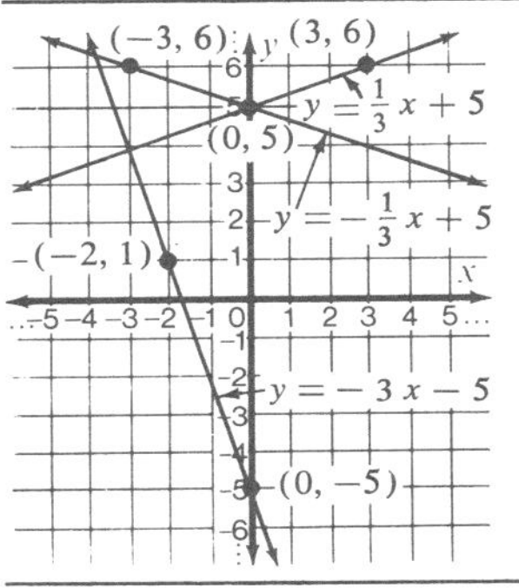

An important subset of the polynomial functions from R into R are called **linear functions.** Each linear function can be represented by an equation of the form $y = mx + b$. For example, the linear function $f = \{(x, y)\colon\ y = 3x - 5\}$ corresponds to the polynomial $3x + (-5)$. The graph of f is also the graph of the solution set for the linear equation $y = 3x - 5$. This graph is a line having slope 3 and y-intercept -5 [Figure 2]. Note that the equation is in the slope-y-intercept form.

Sketch the graph of the function from R into R for the polynomial $-3x - 5$; for $\frac{1}{3}x + 5$; for $-\frac{1}{3}x + 5$.

Exercises

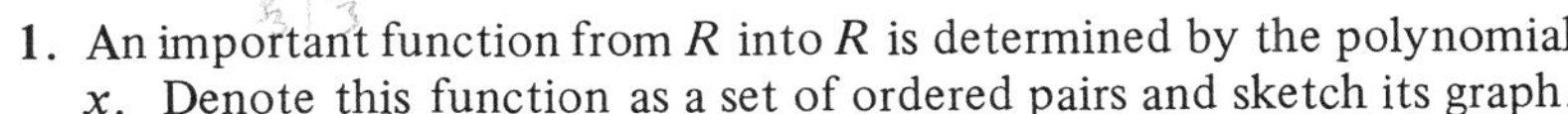

1. An important function from R into R is determined by the polynomial x. Denote this function as a set of ordered pairs and sketch its graph.
2. Sketch the graph of the function from R into R for each polynomial
   a. $2x$ b. $x^2 + 1$
   c. -2 d. x^{-2}
3. What set is the range of each function in Exercise 2?
4. Which of these polynomials have polynomial functions from R into R which are linear functions?
   a. x b. $-5x + 3$ c. x^2 d. 0 e. $x + x^2$ f. -4
5. Sketch the graph of the linear function from R into R determined by each polynomial. State the y-intercept and slope for the line.
   a. $2x + 5$ b. $-2x + 5$ c. $2x - 5$
   d. $-2x - 5$ e. $\frac{1}{2}x$ f. $\frac{1}{2}x + 4$
   g. $-\frac{1}{2}x$ h. $-\frac{1}{2}x - 4$ i. -4
   j. 0 k. 5 l. $x + 2$
6. Sketch the graph of the solution set for each equation if $R \times R$ is the replacement set for (x, y). Then find a polynomial such that the function from R into R for the polynomial has the same graph.
   a. $2x + 3y = 5$ b. $2x - 3y = 5$ c. $2x = 3y + 5 = 0$
   d. $2y = 5$ e. $3x - 2 = 0$ f. $10x + y = 5$
7. Find the collection of polynomials which have linear functions from R into R with graphs parallel to the graph of the linear function for $mx + b$.

8. Find the collection of polynomials which have linear functions from R into R with graphs containing the point for $(0, 0)$.
9. Perform Exercise 8 with (c, d) replacing $(0, 0)$.
10. Is the set of all lines in the plane equal to the set of graphs of solution sets for linear equations in two variables, x and y, with $R \times R$ as the replacement set for (x, y)? Explain.
11. Is the set of all lines in the plane equal to the set of graphs of linear functions from R into R? Explain.

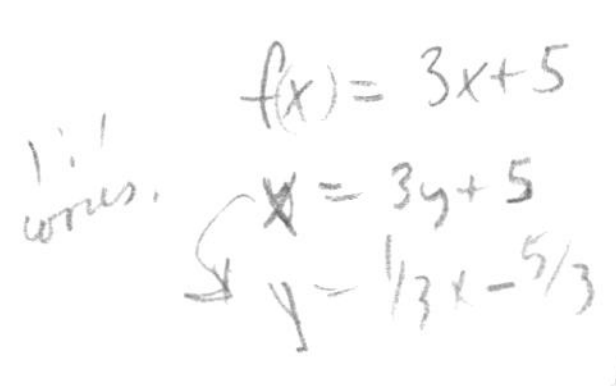

Let f be the function from R into R for each polynomial $mx + b$ in Exercises 12–14. Sketch the graph for each inverse relation f^{-1}.

12. $3x + 5$ 13. $-2x + 1$ 14. 3

15. Which inverse relations in Exercises 12–14 are functions?

6-3 QUADRATIC FUNCTIONS

Polynomial functions from R into R corresponding to polynomials of degree two are called **quadratic functions.** Each quadratic function can be represented by an expression of the form

$$\{(x, y): \ y = ax^2 + bx + c\}$$

where $ax^2 + bx + c$ is a polynomial of degree two. The function

$$f = \{(x, y): \ y = x^2\}$$

is associated with the polynomial x^2 and is the simplest quadratic function. The graph of this function is the graph of the solution set for the quadratic equation $y = x^2$ [Figure 1].

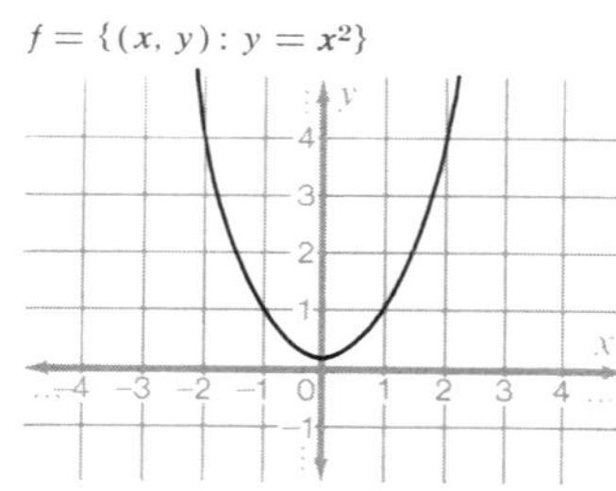

FIGURE 1

Sketch the graph of the quadratic function for the polynomial $-x^2$: $g = \{(x, y): \ y = -x^2\}$.

The graph of g is a parabola opening "downward" with the vertex at the origin, with the y-axis for the axis of symmetry, and containing $(-1, -1)$, $(1, -1)$, $(-2, -4)$, and $(2, -4)$.

In general, the graph of the quadratic function from R into R of the form $\{(x, y): \ y = ax^2 + bx + c\}$ is the parabola which is also the graph of the solution set for the equation

$$y = ax^2 + bx + c$$

provided that $a \neq 0$. Recall that the axis of symmetry for this parabola is a vertical line. If we can find two points on the parabola that are endpoints of a horizontal segment, then the midpoint is on the axis of symmetry.

For example

$$h = \{(x, y)\colon\ y = -x^2 + 2x - 5\}$$

is the quadratic function for the polynomial $-x^2 + 2x - 5$. Since

$$h(0) = -(0)^2 + 2 \cdot 0 - 5 = -5$$

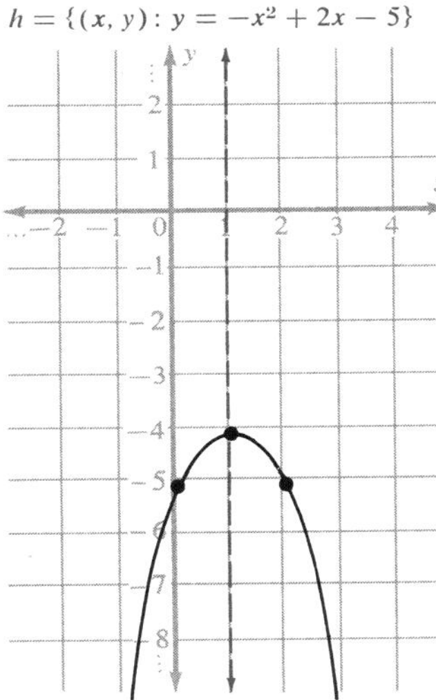

FIGURE 2

the point for $(0, -5)$ is on the parabola. If $(r, -5)$ corresponds to another point on this parabola then $h(r) = -5$; that is

$$h(r) = -r^2 + 2r - 5 = -5.$$

Therefore,

$$-r^2 + 2r = 0$$

and

$$r = 0 \quad \text{or} \quad r = 2.$$

Since $r \neq 0$, $r = 2$ and $h(2) = -5$. The horizontal segment has endpoints $(0, -5)$ and $(2, -5)$. The line for $x = 1$ is the vertical axis of symmetry; the vertex is the point for $(1, h(1)) = (1, -4)$; the graph of h is sketched in Figure 2.

Exercises

Sketch the graph of the quadratic function from R into R for each polynomial in Exercises 1–14. Find the vertex and axis of symmetry for each parabola.

1. $5x^2$
2. $-5x^2 + 1$
3. $-\frac{1}{5}x^2$
4. $\frac{1}{5}x^2 - 2$
5. $x^2 - 4x$
6. $-x^2 + 3x$
7. $2x^2 + 4$
8. $-2x^2 + 4$
9. $x^2 - 7x + 3$
10. $\frac{1}{3}x^2 + \frac{1}{2}x - 15$
11. $-2x^2 + 5x - 4$
12. $x^2 + 7x + 3$
13. $4x^2 + 8x - 5$
14. $-2x^2 + 6x - 3$

If f is the quadratic function from R into R for each polynomial in Exercises 15–18, find the elements r of the domain of the function such that $f(r) = 0$.

15. $12x^2 - x - 20$
16. $x^2 - 3x - 2$
17. $\pi x^2 - \frac{1}{\pi}x$
18. $-3x^2 + 7x - 3$

Find a polynomial such that the quadratic function for this polynomial contains each of the elements in Exercises 19 and 20.

19. $(0, 2), (1, -1), (4, 1)$ **20.** $(4, 4), (6, 1), (-1, 0)$

6-4 RATIONAL FUNCTIONS

For a polynomial, such as $x^2 - 4x + 3$, a function from R into R is defined by replacing x by real numbers r and by finding the ordered pairs $(r, r^2 - 4r + 3)$.

In a similar way, rational expressions give rise to **rational functions.** Consider the rational expression

$$\frac{1}{x - 4}.$$

If we replace x by the number 2, we derive a corresponding number

$$2 \longrightarrow \frac{1}{2 - 4} = -\frac{1}{2}.$$

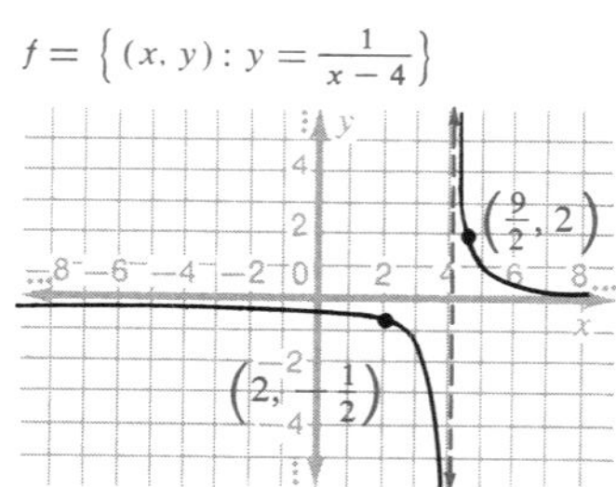

FIGURE 1

$g = \left\{ (x, y) : y = \frac{x^2 - x - 12}{x - 2} \right\}$

(0, 6)

(−3, 0)

(4, 0)

FIGURE 2

Similarly, if x is replaced by $\frac{9}{2}$, we derive

$$\frac{9}{2} \longrightarrow \frac{1}{\frac{9}{2} - 4} = 2.$$

This procedure gives us the two ordered pairs, $\left(2, -\frac{1}{2}\right)$ and $\left(\frac{9}{2}, 2\right)$. If a is any real number, with the exception of $a = 4$, then $\left(a, \frac{1}{a - 4}\right) \in R \times R$. The set of these ordered pairs defines a function which is the rational function corresponding to the rational expression $\frac{1}{x - 4}$. If we denote this function by f then

$$f = \left\{ (x, y): \ y = \frac{1}{x - 4} \right\}.$$

In sketching the graph for this function note, that if $|a|$ is quite large, then $\frac{1}{a - 4}$ is near zero $[f(404) = ?, f(-396) = ?]$.

If $|a - 4|$ is near zero (that is, a is near 4), then what is true of $f(a)$?

Find $f\left(\frac{401}{100}\right)$, $f\left(\frac{399}{100}\right)$, $f\left(\frac{4001}{1000}\right)$, and $f\left(\frac{3999}{1000}\right)$. Find $f(-2), f(-1), f(0), f(1), f(3), f(5)$, and $f(16)$; sketch the graph of f and compare your findings with Figure 1.

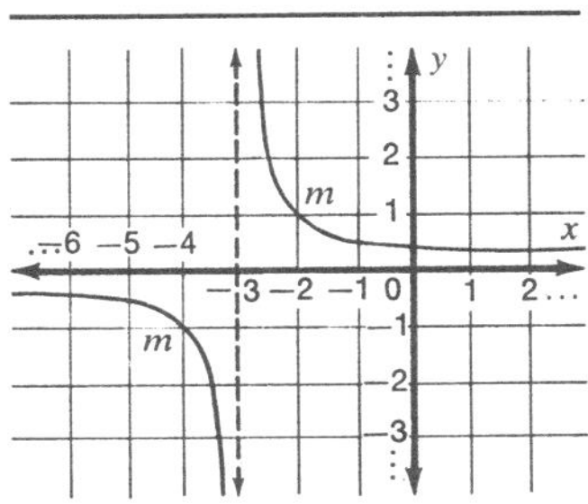

Sketch the graph for the function $m = \left\{(x, y)\colon\ y = \dfrac{1}{x+3}\right\}$.

Consider the function g for the rational expression

$$\frac{x^2 - x - 12}{x - 2} = \frac{(x+3)(x-4)}{x-2}.$$

For what two real numbers is it true that $g(a) = 0$? Notice that if $a = -3$ or if $a = 4$, $g(a) = 0$. Furthermore, if $a = 2$, $g(a)$ is undefined. As a approaches 2 but is less than 2, $g(a)$ becomes larger and larger. As a approaches 2 but is greater than 2, $g(a)$ becomes a smaller and smaller negative number. What is $g(a)$ if $|a|$ is quite large? As an aid in answering this question, consider

$h = \left\{(x, y)\colon y = \dfrac{x^2 + x - 6}{x + 3}\right\}$

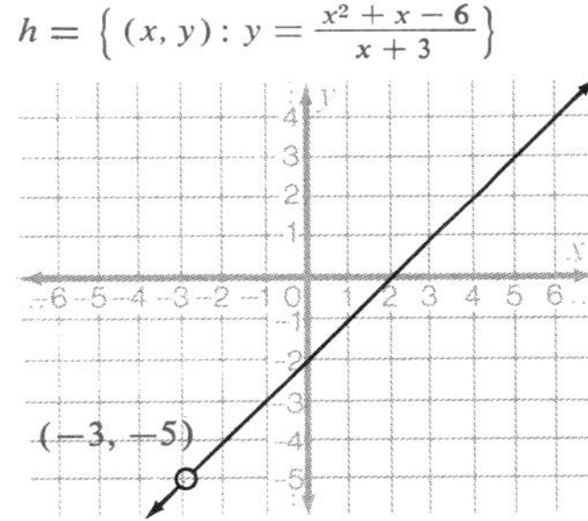

FIGURE 3

$$g(a) = \frac{a^2 - a - 12}{a - 2} = a + 1 + \frac{-10}{a - 2}.$$

We see that $g(a)$ is near $a + 1$ if $|a|$ is large. What is $g(1002)$? $g(-998)$? The graph of g is sketched in Figure 2.

An interesting type of function is illustrated by considering the rational function, call it h, for the rational expression

$$\frac{x^2 + x - 6}{x + 3} = \frac{(x+3)(x-2)}{(x+3)}.$$

If $a \neq -3$ then $h(a) = \dfrac{(a+3)(a-2)}{a+3} = a - 2$, and if $a = -3$ then no corresponding number $h(-3)$ is defined. Thus, the graph for h consists of points on a line for $y = x - 2$ with one point, $(-3, -5)$, removed [Figure 3].

Sketch the graph for the function

$$n = \left\{(x, y)\colon\ y = \frac{x^2 + 2x - 15}{x - 3}\right\}.$$

Let j be the function for the rational expression

$$\frac{x - 1}{x^2 - x - 6} = \frac{x - 1}{(x+2)(x-3)}.$$

The number $j(a)$ is defined for each real number a, except for which two real numbers? Discuss the nature of $j(a)$ for a near each of these two particular numbers. For large values of $|a|$ consider a second form of the equation

$$\frac{x-1}{x^2-x-6} = \frac{1-\dfrac{1}{x}}{x-1-\dfrac{6}{x}}$$

to see that the corresponding numbers $j(a)$ are near 0. The graph of j is shown as Figure 4.

$$j = \left\{ (x, y) : y = \frac{x-1}{x^2-x-6} \right\}$$

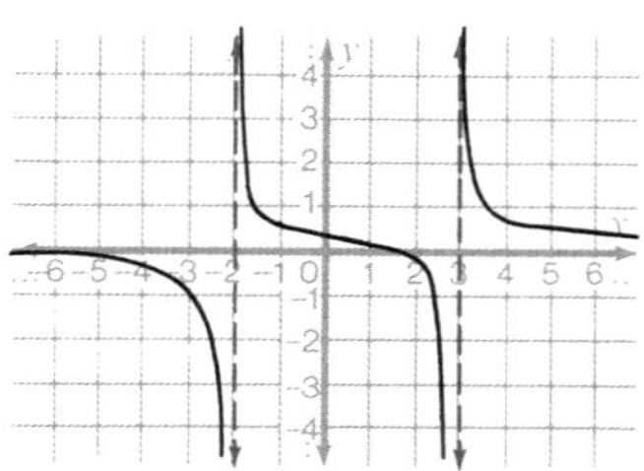

FIGURE 4

Let k be the function for the rational expression

$$\frac{3x^2+4x-4}{x^2+2x-15} = \frac{(3x-2)(x+2)}{(x+5)(x-3)}.$$

The number $k(a)$ is defined for each real number a, except for which two real numbers? Examine $k(a)$ for numbers a near each of these two numbers. For which numbers a is $k(a) = 0$? Note that for $|a| > 5$

$$k(a) = \frac{3a^2+4a-4}{a^2+2a-15} = \frac{3+\dfrac{4}{a}-\dfrac{4}{a^2}}{1+\dfrac{2}{a}-\dfrac{15}{a^2}}.$$

Use this to explain that $k(a)$ is near 3 when $|a|$ is quite large. Sketch the graph of k.

The function $k(a)$ is undefined for $a = -5$ and for $a = 3$. Approaching $a = -5$ from the left, $k(a)$ is positive and getting larger. Approaching $a = -5$ from the right, $k(a)$ is negative and getting smaller. Approaching $a = 3$ from the left, $k(a)$ is negative and getting smaller. Approaching $a = 3$ from the right, $k(a)$ is positive and getting larger. We see that $k(a) = 0$ for $a = -2$ or $a = \frac{2}{3}$. If $|a|$ is large,

$$k(a) = \frac{3+\dfrac{4}{a}-\dfrac{4}{a^2}}{1+\dfrac{2}{a}-\dfrac{15}{a^2}}$$

approaches $\frac{3}{1}$ because the terms involving a in both the numerator and the denominator approach zero.

$$k = \left\{ (x, y) : y = \frac{3x^2+4x-4}{x^2+2x-15} \right\}$$

Exercises

Discuss the function for each rational expression and sketch its graph in Exercises 1-20.

1. $\dfrac{1}{x+5}$

2. $\dfrac{x^2+2x-15}{x-3}$

3. $\dfrac{1}{2x-5}$

4. $\dfrac{x^2-2x-15}{x-1}$

5. $\dfrac{x-4}{x^2+5x-6}$

6. $\dfrac{2x^2-5x-12}{x-4}$

7. $\dfrac{-1}{x}$

8. $\dfrac{1}{x^2-3x-10}$

9. $\dfrac{x^2 - 5x - 24}{2x + 6}$

10. $\dfrac{1}{x^2}$

11. $\dfrac{4x^2 + 5x - 6}{x^2 - 2x - 24}$

12. $\dfrac{1}{x^2 - 2x + 1}$

13. $\dfrac{x^3 + 2x - 11x - 12}{x^2 + x - 12}$

14. $\dfrac{1}{x^2 + 4}$

15. $\dfrac{x^2 + 2x - 24}{x + 2}$

16. $\dfrac{x + 1}{x^2 + x - 20}$

17. $\dfrac{x^2 + x - 6}{x^2 + 2x - 48}$

18. $\dfrac{x^2 - 1}{x - 3}$

19. $\dfrac{x + 3}{x^2 + 3x - 4}$

20. $\dfrac{x^3 + x^2 + x + 1}{x^3 - 16x}$

21. Let f be the function for the rational expression

$$\frac{a_n x^n + a_{n-1} x^{n-1} + \cdots + a_1 x + a_0}{b_m x^m + b_{m-1} x^{m-1} + \cdots + b_1 x + b_0}$$

in $I(x)$. Discuss the nature of $f(a)$ for large values of $|a|$ if

a. $n < m$; b. $m < n$; c. $n = m$.

$f = \{(x, y) : y = 2x + 7\}$

FIGURE 1

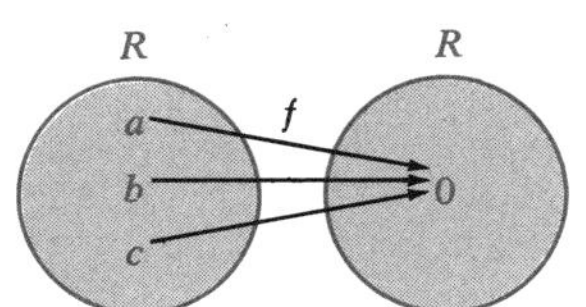

FIGURE 2

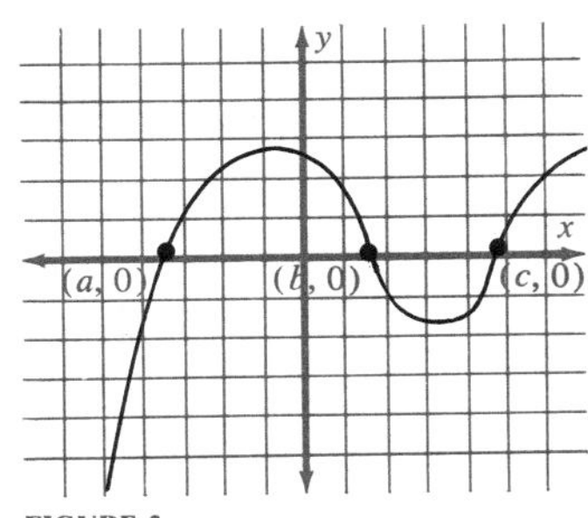

FIGURE 3

6-5 EQUATIONS

If we solve the equation $2x + 7 = 0$, we see that $-\frac{7}{2}$ is the one and only member of the solution set. Hence, $-\frac{7}{2}$ is the only real number mapped onto (matched with) zero by the function $\{(x, y): y = 2x + 7\}$ and $\left(-\frac{7}{2}, 0\right)$ is the point of the function which lies on the x-axis [Figure 1].

If f is any function from R into R, which elements of R, if any, are mapped onto the number 0 by the function f [Figure 2]? Or, if we consider the graph of f, what are the first numbers of ordered pairs for points of the graph on the x-axis [Figure 3]?

Definition If f is any function from R into R and $f(a) = 0$, then a is called a **zero of the function** f.

The function for the zero polynomial has an infinite number of zeros and the function for any polynomial of degree zero, e.g., 3, 100, $\frac{1}{2}$, has no zeros. The function for any polynomial of first degree has exactly one zero which can be found by solving an equation of the form $ax + b = 0$.

Find the zeros of the function from R into R for each polynomial: $4x + 5$, $-3x + 2$, $-\frac{1}{2}x + \frac{3}{4}$, $5x$, 0, and 6.

Real zeros: $-\frac{5}{4}, \frac{2}{3}, \frac{3}{2}, 0,$ $\{x\colon x \in R\}$, none, respectively.

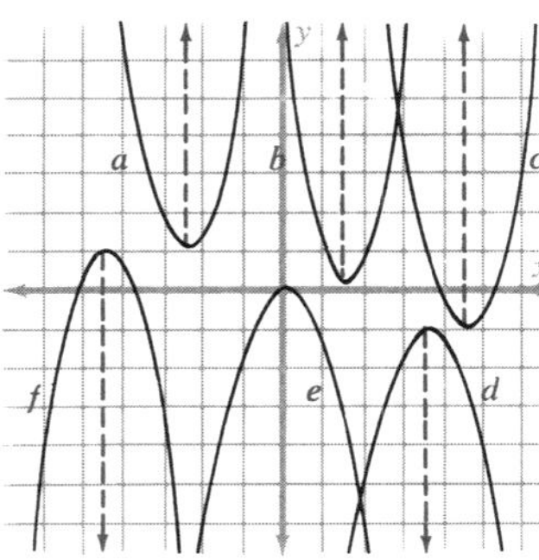

FIGURE 4

How many zeros may the function from R into R for a second degree polynomial have? The graph of any such function is a parabola with a vertical axis of symmetry. Figure 4 illustrates several such parabolas. How many points of each parabola in Figure 4 lie on the x-axis? How many zeros may a quadratic function from R into R have?

The set of zeros for the function $g = \{(x, y)\colon y = 2x^2 - x - 15\}$ is the solution set for the quadratic equation $2x^2 - x - 15 = 0$. Factoring, we have the equivalent equation

$$(2x + 5)(x - 3) = 0.$$

The set of zeros must be $\left\{-\frac{5}{2}, 3\right\}$, and $\left(-\frac{5}{2}, 0\right)$ and $(3, 0)$ are the two points on the graph of g that lie on the x-axis [Figure 5].

The zeros of the function from R into R for a second degree polynomial $ax^2 + bx + c$ are the members of the solution set for the quadratic equation

$$ax^2 + bx + c = 0.$$

Methods for solving equations of degree greater than two are developed in Chapter 8.

Equations whose members have the form of rational expressions are solved by finding equivalent equations that have obvious solution sets.

Example 1 Solve the following equation:

$$(1) \qquad \frac{3}{x - 3} + \frac{1}{x + 3} = \frac{17}{3x - 1}.$$

Solution: If x is replaced by $3, -3$, or $\frac{1}{3}$, then no true statement is found. If x is replaced by any other number, then (1) is true if and only if

$$(2) \qquad 3(x + 3)(3x - 1) + (x - 3)(3x - 1) = 17(x - 3)(x + 3)$$

is true with x replaced by the same number. Equation (2) was obtained by combining terms of the left member and then applying the definition of equality for rational expressions. Thus, (1) is equivalent to

$$(3) \qquad 3(3x^2 + 8x - 3) + (3x^2 - 10x + 3) = 17(x^2 - 9).$$

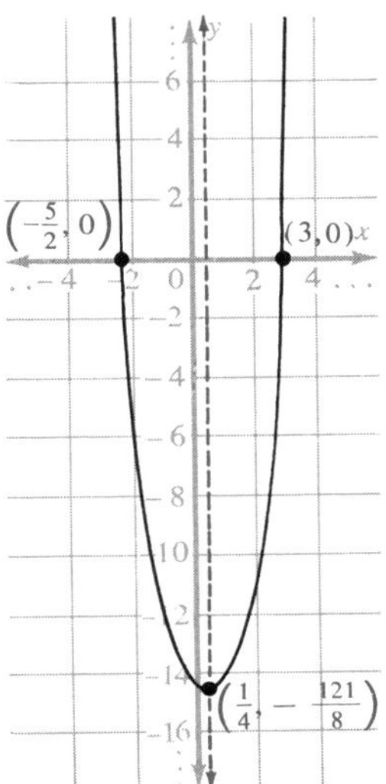

FIGURE 5

where

$$x \neq 3, \quad x \neq -3, \quad \text{and} \quad x \neq \frac{1}{3}$$

But (3) is equivalent to

$$5x^2 - 14x - 147 = 0$$

which is equivalent to

$$x = 7 \quad \text{or} \quad x = -\frac{21}{5}.$$

The solution set for (1) is $\left\{7, -\frac{21}{5}\right\}$.

Often a brief outline of such a discussion is presented and the details are understood. The members of Equation (1) are multiplied by the least common multiple of the denominators of the corresponding rational expressions. The resulting equation

$$5x^2 - 14x - 147 = 0$$

is solved. Any members of the solution set of this equation that do not satisfy the original sentence are cast aside. The remaining elements form the solution set of (1).

Example 2 Solve the following equation:

$$\text{(4)} \qquad \frac{7x - 23}{x - 4} = 6 + \frac{5}{x - 4}.$$

Solution: A replacement for x other than 4 satisfies the equation

$$\frac{(7x - 23)}{(x - 4)}(x - 4) = 6(x - 4) + \left(\frac{5}{x - 4}\right)(x - 4)$$

or

$$7x - 23 = 6(x - 4) + 5.$$

and conversely. Hence, (4) is equivalent to

$$7x - 23 = 6(x - 4) + 5$$

where

$$x \neq 4$$

which in turn, is equivalent to

$$x \neq 4 \quad \text{and} \quad x = 4.$$

The solution set is the empty set, ϕ.

Exercises

1. Find the set of zeros for the function from R into R corresponding to each polynomial.

   a. $2x - 3$ b. $\frac{1}{2}x + 5$ c. $-3x + 4$
   d. $-\frac{1}{3}x + 7$ e. $2x - 9$ f. $-7x + 3$
   g. $-3x + 13$ h. $5x - 3$ i. $-4x - 5$
   j. $\frac{1}{4}x - 3$ k. $-\frac{1}{2}x - 5$ l. $x^2 - x - 12$
   m. x^2 n. $x^2 - x - 20$ o. $x^2 + 4x - 21$
   p. $3x^2 + 7x - 6$ q. $3x^2 - 2x - 16$ r. $x^2 - 6x + 9$
   s. $5x^2 + 3x - 14$ t. $6x^2 - 5x - 6$ u. $6x^2 + 35x - 6$
   v. $x^2 + 5$ w. $\frac{1}{4}x^2 + 2x + 4$ x. $-x^2 + 4x - 5$

2. Sketch the graph of the polynomial function from R into R for each polynomial and indicate the points of the graph that lie on the x-axis.
   a. $x^2 + x + 4$ b. $x^2 - 6x + 9$ c. $-x^2 + 2x - 2$ d. $-x^2 - 4x - 4$

3. Find the set of zeros for the function corresponding to the first degree polynomial $mx + b$ $(m \neq 0)$.

4. If R is the replacement set for x, find the solution set for each equation.

   a. $\frac{x+3}{x+2} + \frac{x+7}{x+2} = 3$ b. $\frac{4x+3}{x+2} + \frac{4-x}{x+2} = 4$
   c. $\frac{4}{3x+1} + \frac{3}{x-1} = 1$ d. $\frac{5x+3}{x+2} = 3 + \frac{x-5}{x+2}$
   e. $\frac{1}{x} + x = \frac{5}{x}$ f. $\frac{4x-3}{x+1} = 5 - \frac{3}{x+1}$
   g. $\frac{2x}{x+2} + \frac{2}{x-4} = \frac{16-x}{x^2-2x-8}$ h. $3 + \frac{4}{2x+1} = \frac{4x+6}{2x+1}$
   i. $\frac{1}{x-1} + \frac{1}{x+3} = \frac{x-2}{x+3}$ j. $\frac{x}{x+1} = \frac{3}{x-1} - \frac{2x-12}{x^2-1}$
   k. $\frac{4x-1}{x+2} + \frac{3x-3}{x^2+3x+2} = \frac{3x}{x+1}$ l. $\frac{x}{x-3} + \frac{4}{x-2} = \frac{-2x}{x^2-5x+6}$
   m. $\frac{x}{3x+2} + \frac{x^2-2}{6x^2+x-2} = \frac{x}{1-2x}$ n. $\frac{x}{x-1} + \frac{x-1}{x-2} - \frac{8x+1}{3x+2} = 0$
   o. $\frac{x}{x-2} + \frac{1}{x+3} = \frac{3}{x-2} + \frac{5}{6-x-x^2}$
   p. $\frac{x}{x-1} + \frac{x-2}{x+2} = \frac{x}{x+2} + \frac{x^2+x+1}{x^2+x-2}$

5. Verify that the set of zeros for the function corresponding to a polynomial which has degree 0 is ϕ.

6. Verify that the set of zeros for the polynomial function corresponding to 0 is R.

7. The graph of a function from R into R for the second degree polynomial $ax^2 + bx + c$ is a parabola with a vertical axis of symmetry. What are the possibilities for the number of elements in the set of zeros for these quadratic functions? Consider the three cases:

$$b^2 - 4ac > 0, \quad b^2 - 4ac = 0, \quad \text{and} \quad b^2 - 4ac < 0.$$

Find the zeros of the function in each of these cases.

8. A water tank has two intake pipes. One pipe alone will fill the tank in 4 hours; the other alone will fill the tank in 6 hours. If both are opened, how long will it take to fill the tank?

9. The difference between two numbers is 14 and their quotient is $\frac{7}{9}$. Find the numbers.

10. The sum of the reciprocal of a number which is larger than 1, and twice the number is 5. Find the number.

11. The denominator of a fraction is 9 more than the numerator. If both the numerator and denominator are increased by 5 the resulting fraction is $\frac{1}{2}$. Find the original fraction.

12. A man and a boy, both working at a steady rate, can jointly complete a certain type of job in 40 minutes. If they worked alone on tasks of this type it would take the boy 60 minutes longer than the man to complete the work. How long does it take each one to complete the job when working alone?

13. If $\frac{1}{r} = \frac{1}{r_1} + \frac{1}{r_2} + \frac{1}{r_3}$, find r when
   a. $r_1 = 2, r_2 = 3, r_3 = 0.5$;
   b. $r_1 = 2.5, r_2 = 6.2, r_3 = 0.01$;
   c. $r_1 = 10, r_2 = 10, r_3 = 10$.

14. Let R be the replacement set for x and let a and b be real numbers. Prove that, for $x \neq a, b$:

$$\frac{1}{x - a} + \frac{1}{x - b} = 1$$

and

$$(x - b) + (x - a) = (x - a)(x - b)$$

are equivalent sentences.

Review Exercises

1. If $S = \{3, 4, 5\}$, $T = \{1, 2, 3\}$, and $w = \{(3, 1), (3, 2), (4, 2), (5, 1), (5, 2)\}$ what is the domain of the relation w from S into T? The range of w? What set is w^{-1}? 6-1

2. If $g = \{(x, y): y = |x - 2|\}$, what numbers are $g(0)$, $g(1)$, $g(-1)$, $g(2)$, and $g\left(-\frac{1}{2}\right)$? What is the domain of g? the range of g? Sketch the graph of g.

3. If f is the polynomial function from R into R for $4x^2 - 3x + 9$, what numbers are $f(0)$, $f(3)$, $f(-3)$, $f\left(\frac{1}{2}\right)$, and $f\left(-\frac{1}{2}\right)$? What is the domain of f?

4. Graph the function from R into R for each polynomial. 6-2

   a. $-4x + 3$ b. $-\frac{1}{4}x - 3$ c. $\frac{1}{4}x$ d. 4

5. Sketch the graph of the quadratic function from R into R for the polynomial $-4x^2 - 8x + 5$. 6-3

6. Sketch the graph of the rational function from R into R for each rational expression. 6-4

   a. $\frac{2}{x - 5}$ b. $\frac{2x^2 - x - 15}{x - 3}$ c. $\frac{x^2 + x - 6}{x - 1}$

7. Find the set of zeros of the function from R into R for each polynomial. 6-5

   a. $-4x + 7$ b. $x^2 + x - 2$ c. $x^2 + 4$

8. If R is the replacement set for x, find the solution set for the equation

$$\frac{2x - 5}{x + 3} = \frac{x}{x + 3}.$$

7. COMPLEX NUMBERS

7-1 SOLVING THE EQUATION $x^2 + 1 = 0$

Try to solve the following equation using real numbers as replacements for x:

$$x^2 + 1 = 0.$$

It cannot be done, because $r^2 \geqslant 0$ for any real number r, and therefore $r^2 + 1 > 0$. There is no real number whose square is a negative number. We would like to create a "number" system that contains solutions for $x^2 + 1 = 0$. We would also like to be able to add, subtract, multiply, and divide in this new number system, preserving what we do in the real number system.

Let us assume that we have a "number" i such that $i^2 + 1 = 0$, $i^2 = -1$, $i = \sqrt{-1}$, and $(x - i)(x + i) = x^2 + 1$. If we want the properties of multiplication to hold, we find that the following properties of i follow.

$$i^1 = i = \sqrt{-1}$$

$$i^2 = i \cdot i = -1$$

$$i^3 = i^2 \cdot i = (-1)i = -i$$

$$i^4 = i^2 \cdot i^2 = (-1)(-1) = 1$$

$$i^5 = i^4 \cdot i = 1 \cdot i = i$$

$$i^6 = i^4 \cdot i^2 = 1(-1) = -1$$

We would want i^n (where n is a positive integer) to be one of the elements $i, -1, -i$, or 1.

$i^7 = -i$; $i^8 = 1$; $i^9 = i$; $i^{10} = -1$; $i^{11} = -i$; $i^{12} = 1$

$i^{4m} = (i^4)^m = 1^m = 1$
$i^{4 \cdot q + r} = (i^4)^q \cdot i^r = 1^q \cdot i^r = i^r$, where $r \in \{0, 1, 2, 3\}$

Which elements of $\{1, -1, i, -i\}$ should i^n represent if $n = 7$? If $n = 8$? If $n = 9$? If $n = 10$? If $n = 11$? If $n = 12$? If n is a multiple of 4? If $n = 4q + r$ where q and r are integers and $0 \leqslant r < 4$?

Using the methods of Chapter 5, we see that the quadratic equation

$$x^2 - 4x + 13 = 0$$

has no real number solution. However, let us test the "number" $2 + 3i$ assuming, again, that we can apply the usual number properties and theorems. We have

$$(2 + 3i)^2 - 4(2 + 3i) + 13 = 2^2 + 2(2 \cdot 3i) + (3i)^2 - 8 - 12i + 13$$

$$= 4 + 12i + 9(-1) - 8 - 12i + 13 = 4 - 9 - 8 + 13 = 0.$$

Thus, the "number" $2 + 3i$ is a likely candidate for an element of the solution set for this quadratic equation.

$(2 - 3i)^2 - 4(2 - 3i) + 13$
$= 4 - 12i + 9i^2 - 8 + 12i + 13 = 0$

• The following equations are equivalent:

$$x^2 - 4x + 13 = 0$$
$$x^2 - 4x + 4 = -9$$
$$(x - 2)^2 = -9.$$

For each real number r,

$$(r - 2)^2 \geqslant 0;$$

hence, if R is the replacement set for x, the solution set of

$$x^2 - 4x + 13 = 0$$

is ϕ.

Show that the "number" $2 - 3i$ should also be in the solution set of the equation $x^2 - 4x + 13 = 0$. However, if R were the replacement set, then show that ϕ is the solution set of the quadratic equation.

An important benefit arises in having a number i such that $i^2 = -1$, and having the resulting properties for powers of i. If x in any polynomial with real coefficients is replaced by i, then we can derive a "number" of the type $a + bi$ where a and b are real numbers. For example, if we replace x in the polynomial

$$3x^4 - 2x^3 + 5x - 2$$

by the "number" i, then

$$3i^4 - 2i^3 + 5i - 2 = 3 \cdot 1 - 2(-i) + 5i - 2$$

$$= 3 + (-2)(-i) + 5i - 2$$

$$= 3 - 2 + 2i + 5i.$$

If a distributive property holds, then $2i + 5i = (2 + 5)i = 7i$ and

$$3i^4 - 2i^3 + 5i - 2 = (3 - 2) + (2 + 5)i = 1 + 7i.$$

In general, if p is a polynomial and if x is replaced by i, then we can derive a "number" of the type $a + bi$ where a and b are real numbers.

Exercises

1. Which of the "numbers" 1, -1, i, or $-i$ are the following powers of i?
   a. i^7 b. i^8 c. i^9 d. i^{10} e. i^{11}
   f. i^{12} g. i^{400} h. i^{411} i. i^{806} j. i^{1025}

Replace x by i in each of the following polynomials and derive a "number" of the type $a + bi$. (Either a or b, or both can be 0.)

2. $3x^5 - 7x^3 + 2x^2 + 5x + 2$
3. $x^6 + 5x^3 + 4x^2 - 3x + 7$
4. $x^5 - 2x^4 + x^3 - x^2 + 1$
5. $x^4 + 3x^3 + 3x^2 + 3x + 2$
6. $2x^6 - 5x^4 + 3x^2 - 7x + 3$
7. $3x^5 + 4x^4 - 7x^3 - 3x^2 + 5x - 8$
8. $3x^{49} - 2x^{20} + 1$
9. $5x^5 - 2x^4 + 7x^3 - 4x^2 + 3x - 1$
10. $17x^{47} - 3x^{21} + 3$
11. $2x^4 + x^3 + 2x^2 + x$

For each equation in Exercises 12–17 show that the indicated "numbers" are members of the solution set and find any real numbers that belong to the solution set.

12. $x^4 + x^3 - 11x^2 + x - 12 = 0;\ i,\ -i$
13. $x^2 - 6x + 25 = 0;\ 3 + 4i,\ 3 - 4i$
14. $x^2 - 4x + 5 = 0;\ 2 + i,\ 2 - i$
15. $x^5 - 3x^4 + 3x^3 - 9x^2 - 4x + 12 = 0;\ 2i,\ -2i$
16. $x^5 - 2x^4 + 6x^3 + 8x^2 - 40x = 0;\ 1 + 3i,\ 1 - 3i$
17. $x^4 - 2\sqrt{2}x^3 + 6\sqrt{2}x - 9 = 0;\ \sqrt{2} + i,\ \sqrt{2} - i$
18. Use the computational arithmetic for powers of i and for evaluating polynomials to suggest definitions of addition and multiplication for "numbers" $a + bi$ and $c + di$.

7-2 ADDITION

The speculative preceding section stemmed from desiring a nonempty solution set for the equation $x^2 + 1 = 0$. The discussion led to notation of the form $a + bi$ where a and b refer to real numbers and i, which denotes a "number" such that $i^2 = -1$.

Let us define a set of numbers that makes use of this notation.

Definition Let C be a set of numbers, called the set of **complex numbers**, where $C = \{a + bi\colon\ a \in R \text{ and } b \in R\}$.

The a is called the **real part** and the b is called the **imaginary part** of the complex number $a + bi$. Note that both the real and imaginary parts are real numbers.

If $a = 0$ then $a + bi$ is denoted by bi, and if $b = 0$ then we refer to $a + bi$ as a. If $a = 0$ and $b = 1$, the symbol i is used instead of $0 + 1i$. If $-b$ is a negative real number $a - bi$ is an alternative for $a + (-b)i$. For example, $2 - 3i = 2 + (-3)i$.

Denote each of the following complex numbers with a form $a + bi$, where a and b are real numbers: 3, $2i$, -4, $-i$, 0, $-2 - 3i$, and $2 - 3i$.

$3 = 3 + 0i$,
$2i = 0 + 2i$,
$-4 = (-4) + 0i$,
$-i = 0 + (-1)i$,
$0 = 0 + 0i$,
$-2 - 3i = (-2) + (-3)i$,
$2 - 3i = 2 + (-3)i$

Note that $a + bi$ and $c + di$ denote the same number in C if and only if $a = c$ and $b = d$ in the set of real numbers. Thus, we define equality of complex numbers as follows.

Definition In C, $a + bi = c + di$ if and only if $a = c$ and $b = d$ in R.

In defining a formal operation of addition for the set of complex numbers, we are consistent with the development presented in the first section of this chapter.

Definition For $a + bi$ and $c + di$ in C

$$(a + bi) + (c + di) = (a + c) + (b + d)i.$$

The sum of two complex numbers is also a complex number. For if $a + bi$ and $c + di$ are in C; a, b, c, d, $(a + c)$, and $(b + d)$ belong to the set of real numbers since addition on R is a closed operation. Thus,

$$(a + c) + (b + d)i$$

denotes a unique complex number.

Addition in the system of complex numbers satisfies properties that are analogous to the addition properties for the real number system. These properties follow from the definition of addition for complex numbers and properties of addition in the real number system.

Theorem 1 The operation of addition on C is associative. For $a + bi$, $c + di$, and $e + fi$ in C

$$[(a + bi) + (c + di)] + (e + fi) = (a + bi) + [(c + di) + (e + fi)].$$

Theorem 2 The operation of addition on C is commutative. For every $a + bi$ and $c + di$ in C,

$$(a + bi) + (c + di) = (c + di) + (a + bi).$$

Theorem 3 The number $0 + 0i = 0$ is the identity element for addition. For every $a + bi$ in C, $(a + bi) + 0 = a + bi$.

Theorem 4 Each number $a + bi$ in C has a corresponding additive inverse, $(-a) + (-b)i$. For each $a + bi$ in C,

$$(a + bi) + [(-a) + (-b)i] = 0.$$

We denote $(-a) + (-b)i$ as $-(a + bi)$.

Find the additive inverse of each of the following elements of C: $3 + 5i$, $3 - 5i$, $-3 + 5i$, $-3 - 5i$, $0 + 0i$, π, and $\sqrt{2}i$. Denote each of the following numbers in $a + bi$ form: $-(6i - 1)$, $-(2 - 9i)$, $-(-2i - 17)$, $-(1)$, and $-(-5i)$.

Element of C	*Additive inverse*
$3 + 5i$	$(-3) + (-5)i$
$3 - 5i$	$(-3) + 5i$
$-3 + 5i$	$3 + (-5)i$
$-3 - 5i$	$3 + 5i$
$0 + 0i$	$0 + 0i$
π	$(-\pi) + 0i$
$\sqrt{2}i$	$0 + (-\sqrt{2})i$

$-(6i - 1) = 1 + (-6)i$
$-(2 - 9)i = (-2) + 9i$
$-(-2i - 17) = 17 + 2i$
$-(1) = (-1) + 0i$
$-(-5i) = 0 + 5i$

Subtraction on C is defined as the inverse operation of addition.

Definition For $a + bi$ and $c + di$ in C, $(a + bi) - (c + di) = e + fi$ provided $e + fi$ is the unique element in C such that $(e + fi) + (c + di) = a + bi$.

For $(2 + 3i)$ and $(1 + 4i)$ in C, we find the difference, $(2 + 3i) - (1 + 4i)$ by subtracting real and imaginary parts:

$$(2 + 3i) - (1 + 4i) = (2 - 1) + (3 - 4)i = 1 - i$$

or by adding the additive inverse of $(1 + 4i)$ to $(2 + 3i)$:

$$(2 + 3i) + [(-1) + (-4)i] = 1 - i.$$

In each case the answer, $1 - i$, "checks" in the definition when added to $1 + 4i$. The difference of two complex numbers is found by either of these methods.

Theorem 5 For each $a + bi$ and $c + di$ in C

$$(a + bi) - (c + di) = (a - c) + (b - d)i = (a + bi) + [(-c) + (-d)i].$$

Since each element in C has an additive inverse, subtraction is a closed operation on C.

Exercises

State the real and imaginary parts for each of the following complex numbers:

1. $2 + 4i$
2. $-5 - 2i$
3. -5
4. $(6 + 2i) - (3 + 4i)$
5. $-3i$
6. $(2 + 2i) - (2 - 2i)$
7. $(2 + 3i) + [(5 + 2i) + (7 + i)]$
8. $[(2 + 3i) + (5 + 2i)] + (7 + i)$
9. $(4 + 3i) + (7 + 6i)$
10. $(7 + 6i) + (4 + 3i)$
11. $(4 + 3i) - (7 + 6i)$
12. $(7 + 6i) - (4 + 3i)$
13. $7 + 3 - 4$
14. $2i + 6i - 3i$

15. What is the additive inverse of each of these elements of C?
 a. $4 + 3i$ b. $2 - 3i$ c. $-5 + 2i$ d. -3 c. $-2i$
16. Show that for every $a, b \in R$, $a + bi = a + bi$.
17. Prove that in C, if $a + bi = c + di$ then $c + di = a + bi$.
18. Show that in C, if $a + bi = c + di$ and $c + di = e + fi$ then $a + bi = e + fi$.
19. Observe the three different uses of the '+' symbol in the example below:

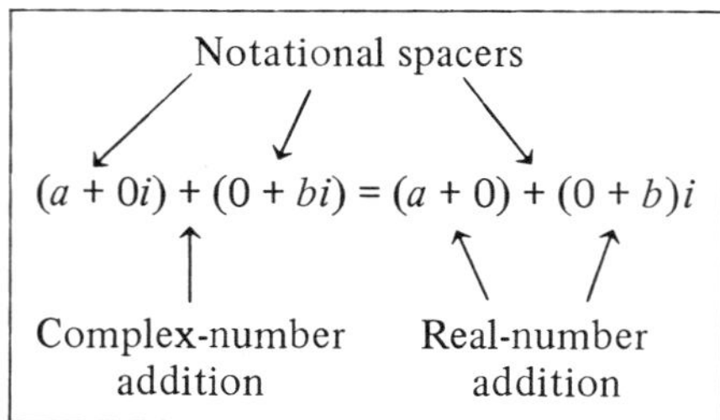

 Add the complex numbers a and bi, where $a \in R$ and $b \in R$, to convince yourself that the "+" symbol that separates addends is a symbol for complex-number addition as well as a spacing device.
20. Ordered pair notation (a, b) can be used for elements of C, instead of $a + bi$. Use ordered pairs to represent elements of C and define addition using this notation.

7-3 MULTIPLICATION

In Section 7-1 we proposed a few ideas about multiplication. For example, the product $i \cdot i$ should be $i^2 = -1$. If statements such as

$$i + i = 1i + 1i = (1 + 1)i = 2i$$

and

$$\begin{aligned}(2 + 3i)(4 + 5i) &= 2(4 + 5i) + 3i(4 + 5i)\\ &= 2(4) + 2(5i) + (3i)4 + (3i)(5i)\\ &= 2 \cdot 4 + (2 \cdot 5)i + (3 \cdot 4)i + (3 \cdot 5)i^2\\ &= 2 \cdot 4 + 3 \cdot 5(-1) + (2 \cdot 5)i + (3 \cdot 4)i\\ &= (8 - 15) + (10 + 12)i = -7 + 22i\end{aligned}$$

hold, we are assuming associativity and commutativity for multiplication and a distributive property for multiplication over addition.

In general, we want to be able to manipulate in the following way:

$$(a + bi)(c + di) = a(c + di) + bi(c + di)$$

$$= ac + adi + bci + bdi^2 = (ac - bd) + (ad + bc)i.$$

Thus, we use this end result as the definition of multiplication and in a sense, insure the presence of the desired properties.

Definition For $a + bi$ and $c + di$ in C

$$(a + bi)(c + di) = (ac - bd) + (ad + bc)i.$$

Using the definition of multiplication

$$(2 + 7i)(5 + 4i) = [(2 \cdot 5) - (7 \cdot 4)] + [(2 \cdot 4) + (7 \cdot 5)]i$$

$$= (-18) + 43i.$$

This agrees with the computation

$$(2 + 7i)(5 + 4i) = 2 \cdot 5 + [(7 \cdot 5) + (2 \cdot 4)]i + 28i^2$$

$$= (10 - 28) + 43i = (-18) + 43i.$$

Use the definition of multiplication in C to find each product: $(5 - 4i)(2 - 7i)$, $(0 + 1i)(0 + 1i)$, $(2 - 7i)^3$, $(0 + 0i)(0 + 1i)$, and $(4 + 3i)\left(\frac{4}{25} - \frac{4}{25}i\right)$.

$(5 - 4i)(2 - 7i) = -18 - 43i$
$(0 + 1i)(0 + 1i) = -1$
$(2 - 7i)^3 = (-45 - 28i)(2 - 7i)$
$= -286 + 259i$
$(0 + 0i)(0 + 1i) = 0$
$(4 + 3i)\left(\frac{4}{25} - \frac{3}{25}i\right) = 1$

The operation of multiplication in C has the following properties. Since real numbers are closed under multiplication, subtraction, and addition, $(ac - bd)$ and $(ad + bc)$ are unique real numbers. Thus,

$$(a + bi)(c + di) = (ac - bd) + (ad + bc)i$$

is a unique element of C. Multiplication in C is commutative. Also, multiplication in C is associative. As you would guess, the number $1 + 0i = 1$ is the identity element for multiplication. Every element of C has a multiplicative inverse. What is the multiplicative inverse of $3 = 3 + 0i$? Of $4i = 0 + 4i$? Of $3 + 4i$? It is relatively easy to see that

$$(3)\left(\frac{1}{3}\right) = 1 \quad \text{or} \quad (3 + 0i)\left(\frac{1}{3} + 0i\right) = 1 + 0i$$

and that

$$(4i)\left(-\frac{1}{4}i\right) = 1 \quad \text{or} \quad (0 + 4i)\left(0 - \frac{1}{4}i\right) = 1 + 0i.$$

However, the fact that $\left(\frac{3}{25} - \frac{4}{25}i\right)$ is the multiplicative inverse of $3 + 4i$ is not so easily seen. We will investigate the multiplicative inverse property further in the next section.

Exercises Perform the operations and state the real and imaginary parts for each resulting complex number in Exercises 1-16.

1. $(4 + 2i)(5 + 3i)$
2. $(3 - 2i)(3 + 2i)$
3. $(-2 - 3i)(7 + 2i)$
4. $(-4 + 3i)(4 + 3i)$
5. $(4 - 3i)(7 + 2i)$
6. $[(2 - 7i)(4 + 3i)](2 + 5i)$
7. $(7 - 3i)^2$
8. $(1 + 2i)\left(\frac{1}{5} - \frac{2}{5}i\right)$
9. $(1 + 3i)[(5 - 4i) + (1 + i)]$
10. $(8 - 3i)(8 + 3i)$
11. $(2 - 7i)[(4 + 3i)(2 + 5i)]$
12. $\left(\frac{1}{2} + \frac{2}{3}i\right)^2$
13. $(2 + 3i)^2$
14. $(1 - 2i)^4$
15. $(2 - 3i)\left(\frac{2}{13} + \frac{3}{13}i\right)$
16. $[(1 + 3i)(5 - 4i)] + [(1 + 3i)(1 + i)]$

Replace x by $2 + 3i$ in each of the polynomials in Exercises 17-20 and state the real and imaginary parts of the resulting complex number.

17. $x^2 - 4x + 13$
18. $2x^3 + 3x - 4$
19. $x^2 - x + 1$
20. $x^3 - 3x + 52$
21. Find the complex number which is the multiplicative inverse of each of the following:

    a. 3 b. $\frac{1}{5}$ c. $5i$ d. $\frac{1}{7}i$ e. $-4i$
22. State the distributive property for the system of complex numbers.

7-4 DIVISION

The operation of multiplication on the set of complex numbers has a corresponding inverse operation called division. If c_1 and c_2 are two complex numbers, what number is $c_1 \div c_2$? We know that this should be the unique complex number, say c_3, such that $c_3 \cdot c_2 = c_1$.

If $c_1 = (1 + 4i)$ and $c_2 = (2 + 3i)$, how do we find $c_1 \div c_2$? Division in the system of complex numbers is similar to division in the real number system, since every nonzero element has a multiplicative inverse. Furthermore, if c_2^{-1} is the multiplicative inverse of c_2, then

$$c_1 \div c_2 = c_1 \cdot c_2^{-1}.$$

Thus

$$(1 + 4i) \div (2 + 3i) = (1 + 4i)(2 + 3i)^{-1}.$$

Before multiplying, we will pursue the matter that each nonzero complex number has a multiplicative inverse. The number $2 + 3i$ is nonzero. What complex number $a + bi$ is its multiplicative inverse? What are the real numbers a and b such that $(2 + 3i)(a + bi) = 1$? We know that we want

$$(2 + 3i)(a + bi) = (2a - 3b) + (3a + 2b)i = 1 = 1 + 0i.$$

But this statement is true if and only if

$$2a - 3b = 1 \quad \text{and} \quad 3a + 2b = 0.$$

This statement holds if and only if

$$\frac{-3a}{2} = b \quad \text{and} \quad 2a - 3\left(\frac{-3a}{2}\right) = 1$$

or if and only if

$$a = \frac{2}{13} \quad \text{and} \quad b = \frac{-3}{13}.$$

It is easy to verify that

$$(2 + 3i)\left(\frac{2}{13} + \frac{-3}{13}i\right) = 1.$$

Therefore,

$$(1 + 4i) \div (2 + 3i) = (1 + 4i)(2 + 3i)^{-1}$$
$$= (1 + 4i)\left(\frac{2}{13} + \frac{-3}{13}i\right) = \frac{14}{13} + \frac{5}{13}i.$$

Find the multiplicative inverse of $6i$ and the quotient, $(4 + 3i) \div (6i)$; the multiplicative inverse of $4 - i$ and the quotient, $(3 + 5i) \div (4 - i)$.

$(6i)(a + bi) = 1$; hence

$$-6b + 6ai = 1 + 0i$$
$$-6b = 1 \quad \text{and} \quad 6a = 0$$
$$b = -\frac{1}{6} \quad \text{and} \quad a = 0$$

and $(6i)^{-1} = 0 + \left(-\frac{1}{6}i\right) = -\frac{1}{6}i$.

Thus,

$$(4 + 3i) \div 6i = (4 + 3i)\left(-\frac{1}{6}i\right)$$
$$= \frac{1}{2} + \left(-\frac{2}{3}\right)i.$$

• $(4 - i)(a + bi) = 1$; hence

$$(4a + b) + (4b - a)i = 1 + 0i$$
$$4a + b = 1 \quad \text{and} \quad 4b - a = 0$$
$$b = \frac{1}{17} \quad \text{and} \quad a = \frac{4}{17}$$

and $(4 - i)^{-1} = \frac{4}{17} + \frac{1}{17}i$. Thus

$$(3 + 5i) \div (4 - i)$$
$$= (3 + 5i)\left(\frac{4}{17} + \frac{1}{17}i\right) = \frac{7}{17} + \frac{23}{17}i.$$

Corresponding to the complex number $2 + 3i$ is the number $2 - 3i$ which is called the conjugate of $2 + 3i$. More generally, we have the following definition.

Definition The complex number $a - bi$ is the **conjugate** of the complex number $a + bi$.

We can use conjugates of complex numbers in computing multiplicative inverses. With the usual convention that $\frac{a + bi}{c + di}$ means $(a + bi) \div (c + di)$ we know that

$$\frac{1}{2 + 3i} = 1 \div (2 + 3i) = 1 \cdot (2 + 3i)^{-1} = (2 + 3i)^{-1}.$$

But what is the $a + bi$ form of $\frac{1}{2 + 3i}$? Using the corresponding properties for this notation, we have

$$\frac{1}{2 + 3i} = \frac{1}{2 + 3i} \cdot \frac{2 - 3i}{2 - 3i} = \frac{1(2 - 3i)}{(2 + 3i)(2 - 3i)}$$

$$= \frac{2 - 3i}{4 + 9} = \frac{2 - 3i}{13} = \frac{2}{13} + \frac{-3}{13}i.$$

We have the following general theorem.

Theorem 1 If $a + bi$ is a nonzero complex number, then

$$(a + bi)^{-1} = \frac{1}{a + bi} = \frac{1}{a + bi} \cdot \frac{a - bi}{a - bi} = \frac{a - bi}{a^2 + b^2}$$

$$= \frac{a}{a^2 + b^2} + \frac{-b}{a^2 + b^2}i.$$

We also make use of the conjugate in finding the real and imaginary part of the quotient of complex numbers. Another basic property that we use is expressed in the next theorem.

Theorem 2 The product of a complex number and its conjugate is real. (The imaginary part is 0.)

Note that the products of the denominators in the succeeding examples are applications of Theorem 2.

Example 1 Divide $2 + 5i$ by $5 + 3i$.

Solution:

$$\frac{2+5i}{5+3i}\cdot\frac{5-3i}{5-3i}=\frac{(10+15)+(-6+25)i}{25+9}=\frac{25+19i}{34}$$

$$=\frac{25}{34}+\frac{19}{34}i$$

Example 2 Express $\dfrac{7+2i}{3-4i}$ as a complex number in $a + bi$ form.

Solution:

$$\frac{7+2i}{3-4i}\cdot\frac{3+4i}{3+4i}=\frac{21+34i+8i^2}{9-16i^2}=\frac{13+34i}{25}=\frac{13}{25}+\frac{34}{25}i$$

Exercises

Find the conjugate and the multiplicative inverse of each complex number in Exercises 1-6.

1. $4 + 7i$ **2.** $-4 - 7i$ **3.** $-2 + 7i$ **4.** $3i$ **5.** $4 - 7i$ **6.** 5

State the real and imaginary parts of each complex number in Exercises 7-24.

7. $\dfrac{1}{6+3i}$ **8.** $\dfrac{7-4i}{4+3i}$ **9.** $\dfrac{4+2i}{-3i}$ **10.** $\dfrac{1}{4-3i}$

11. $\dfrac{-8-5i}{4-i}$ **12.** $\dfrac{2-5i}{3}$ **13.** $\dfrac{1}{2i}$ **14.** $\dfrac{1-i}{1+i}$

15. $\dfrac{1-3i}{2i}$ **16.** $\dfrac{1}{2}$ **17.** $\dfrac{3i}{1+2i}$ **18.** $\dfrac{3+4i}{1+\sqrt{2}i}$

(Assume $c, d, e, f \in R$, $c + di \neq 0$ and $e + fi \neq 0$.)

19. $\dfrac{(-3+4i)}{(7-2i)}\cdot\dfrac{(5-2i)}{(2+3i)}$ **20.** $\dfrac{2-i}{3+2i}+\dfrac{1+i}{4-i}$ **21.** $\dfrac{1}{-i}-\dfrac{2}{i}$

22. $\dfrac{c+di}{e+fi}$ **23.** $\dfrac{2}{1+3i}\cdot\dfrac{4i}{3-5i}$ **24.** $\dfrac{1}{c+di}+\dfrac{1}{e+fi}$

If z is a complex number, then $z = a + bi$ for some a and b in R. Denote the conjugate of z by $\bar{z} = \overline{a+bi}$; that is, $\bar{z} = a - bi$.

25. Show that $z \cdot \bar{z}$ is a real number.

26. Prove that $z + \bar{z}$ is a real number.

27. Show that $\bar{\bar{z}} = z$.

28. If $z = \bar{z}$, what is true of the real numbers a and b?

7-5 THE COMPLEX PLANE

The real number line is a geometric representation for the system of real numbers. Each real number is identified with a point on a line and each point on a line is identified with a real number.

What is the geometric representation for the complex number system? We know that each number $a + bi$ has a unique ordered pair of real numbers (a, b), where a is the real part and b is the imaginary part of the complex number. (See Exercise 20, p. 138.) We have already used the plane in representing the set $R \times R$.

If we use the natural correspondence between elements of $R \times R$ and complex numbers

$$(1, 3) \longleftrightarrow 1 + 3i \qquad (0, 2) \longleftrightarrow 0 + 2i = 2i$$

$$(3, 1) \longleftrightarrow 3 + 1i = 3 + i \qquad (-2, 0) \longleftrightarrow -2 + 0i = -2$$

$$(a, b) \longleftrightarrow a + bi$$

Complex Plane

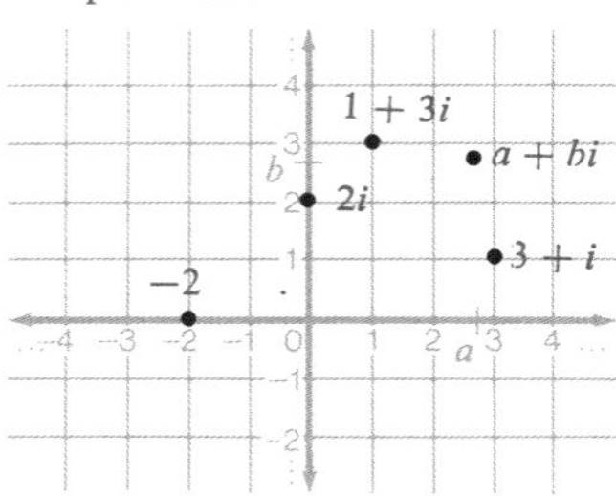

FIGURE 1

then we can identify the points in the plane with the corresponding complex numbers [Figure 1]. This one-to-one correspondence between complex numbers and points of a plane is referred to as the **complex number plane** or more briefly, the **complex plane.** The vertical axis is called the **imaginary axis** and the horizontal axis is referred to as the **real axis** for the complex plane.

How far is each point for the complex numbers $z_1 = 3 + i$, $z_2 = 1 + 3i$, $z_3 = 2i$, $z_4 = -2$, and $z = a + bi$ from the origin in the complex plane? The answers for z_3 and z_4 are both 2 [Figure 2]. The Pythagorean theorem is used to obtain the other answers [Figure 3 and Figure 4]. These distances are nonnegative real numbers and are referred to as the absolute values of the particular complex numbers.

Definition For the complex number $z = a + bi$, the real number $\sqrt{a^2 + b^2}$ is called the **absolute value** of z. The notation is $|z| = |a + bi| = \sqrt{a^2 + b^2}$.

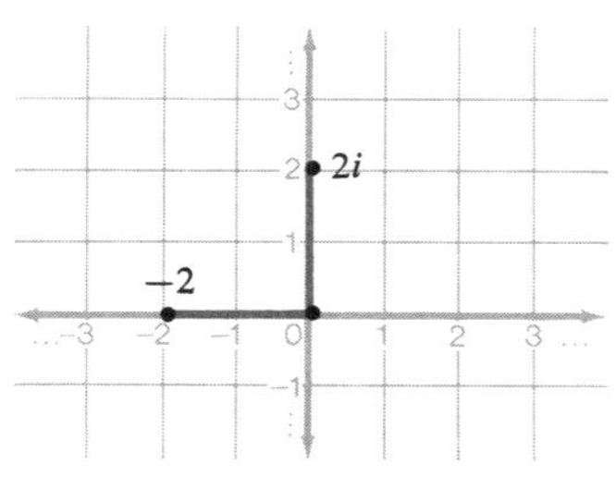

FIGURE 2

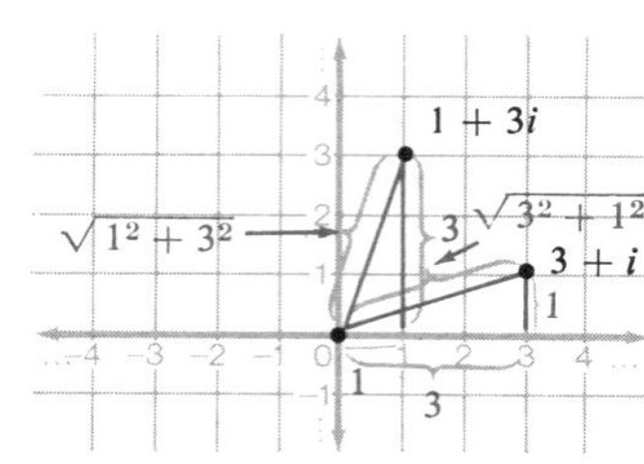

FIGURE 3

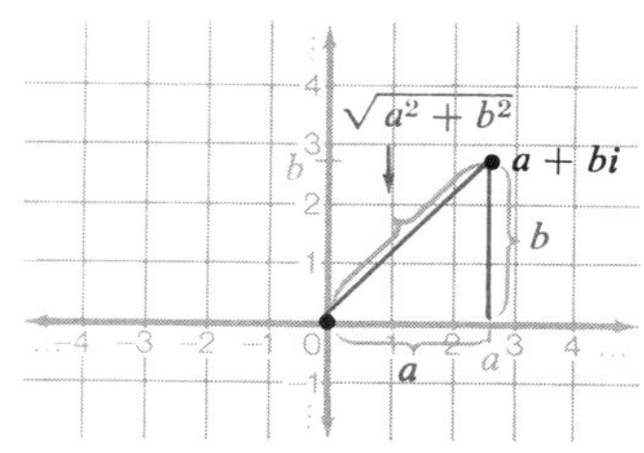

FIGURE 4

$5 = \sqrt{(4-0)^2 + (3-0)^2}$
$= \sqrt{(4-0)^2 + [(-3)+0]^2}$;

- $5 = |-4+3i| = |-4-3i|$;
- $5 = |5+0i| = |0+5i| = |-5+0i| = |0-5i| = |\sqrt{23}+\sqrt{2}i|$;
- If $r \in R$ and $r = 0$, only the complex number 0 has absolute value r. If $r > 0$, infinitely many complex numbers have absolute value r. (These complex numbers are represented by points on a circle with center at the origin and radius of length r.) If $r < 0$, no complex number has absolute value r.

Find the distance from $4 + 3i$ to the origin; from $4 - 3i$ to $0 + 0i$. Find the absolute value of $-4 + 3i$; of $-4 - 3i$. What nonnegative real number is $|5 + 0i|$? Is $|0 + 5i|$? Is $|-5 + 0i|$? Is $|0 - 5i|$? Is $|\sqrt{23} + \sqrt{2}i|$? How many complex numbers have the same absolute value?

The conjugate of $1 + 3i$ is $1 - 3i$. Note that the product

$$(1 + 3i)(1 - 3i) = 1^2 - 9i^2 = 1^2 + 3^2$$

is $(\sqrt{1^2 + 3^2})^2 = |1 + 3i|^2$. In general we have the following theorem.

Theorem 1 If $z = a + bi$ and $\bar{z} = a - bi$ is the conjugate of z, then $z \cdot \bar{z} = |z|^2$.

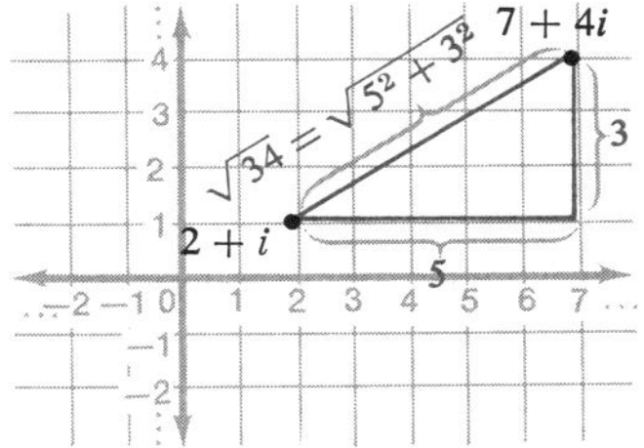

FIGURE 5

If we compare the absolute value concept for the real number system with the corresponding concept for complex numbers, both refer to a distance of a point from the origin. But, how can we describe the distance between two points where one of them is not necessarily the origin? On the real number line, for the real numbers r_1 and r_2, $|r_1 - r_2|$ served this purpose. If we consider the two points in the complex plane $2 + i$ and $7 + 4i$, then we can employ the Pythagorean theorem and find that the distance is $\sqrt{34}$ [Figure 5]. But, what number is $|(7 + 4i) - (2 + i)|$?

$$|(7 + 4i) - (2 + i)| = |5 + 3i| = \sqrt{5^2 + 3^2} = \sqrt{34}$$

In fact, if $z_1 = a + bi$ and $z_2 = c + di$ are two complex numbers, then $|z_1 - z_2|$ is the distance between the corresponding points in the complex plane.

Theorem 2 For z_1 and z_2 in C, $|z_1 - z_2| = |z_2 - z_1|$.

Exercises

Locate the point for each number in Exercises 1–4 in the complex plane.

1. $4 + 3i$, $3 + 4i$, $-5i$, 0, -3, $-4 - 3i$, $4 - 3i$
2. $3 + 5i$, $5 + 3i$, $-3 - 5i$, $-5 - 3i$
3. $3i$, $-4i$, -6, 4
4. Locate the conjugate of each number in Exercises 1-3.

Find the absolute value of each complex number in Exercises 5-8.

5. The numbers in Exercises 1–4.
6. $4 - i$, $1 + 3i$, $(4 - i)(1 + 3i)$, $\dfrac{1}{4 - i}$, $\dfrac{1}{1 + 3i}$, $\dfrac{1 + 3i}{4 - i}$
7. $-5 + 2i$, $2 - i$, $(-5 + 2i)(2 - i)$, $\dfrac{1}{-5 + 2i}$, $\dfrac{1}{2 - i}$, $\dfrac{-5 + 2i}{2 - i}$

8. $\frac{\sqrt{2}}{2} + \frac{\sqrt{2}}{2}i$, $2 + i$, $3 + 5i$, $(2 + i)(3 + 5i)$, $\frac{2 + i}{3 + 5i}$

Find the distance between each pair of points in the complex plane for Exercises 9–10.

9. $2 + 4i$, $4 + 7i$; $1 + 3i$, $-3 + 5i$; $-1 + 2i$, $7 - 5i$

10. i, -4; $1 + i$, $-1 - i$; $5 + 3i$, $5 - 3i$

11. Locate the points for each set of numbers in the complex plane.
    a. $6 + i$, $3 + 4i$, $(6 + i) + (3 + 4i)$, 0
    b. $4 + 7i$, $-1 - 5i$, $(4 + 7i) + (-1 - 5i)$, 0
    c. $a + bi$, $c + di$, $(a + bi) + (c + di)$, 0

12. a. In each of the parts of Exercise 11, the four points are the vertices of what type of geometric figure?
    b. Give a geometric interpretation for the addition operation of complex numbers.

13. a. Locate the point for each of the following numbers in the complex plane: $2 + 3i$, $4 - 2i$, $-3 - 4i$, $-2 + 5i$, $3i$, -4, $-5i$, 6, $a + bi$.
    b. Locate the conjugate of each of these numbers. Give a geometric description of the function $f\colon z \to \bar{z}$ for every $z \in C$.
    c. Locate the additive inverse of each of the above numbers. Give a geometric description for the function $f\colon z \to -z$ for every $z \in C$.

Sketch each set of points in Exercises 14–18 in the complex plane.

14. $\{z \in C\colon |z| = 1\}$

15. $\{z \in C\colon |z| < 3\}$

16. $\{z \in C\colon |z - (1 + i)| \leqslant 1\}$

17. $\{z \in C\colon |z + 3| > 2\}$
    (Hint: $z + 3 = z - (-3)$.)

18. $\{z \in C\colon |z - z_0| = r\}$ where $r \in R, r > 0$, and $z_0 \in C$.

19. Prove Theorem 1, page 145.

20. If z_1 and z_2 are complex numbers, give examples of each of the statements:
    a. $|z_1 \cdot z_2| = |z_1|\,|z_2|$ b. $\left|\frac{1}{z_1}\right| = \frac{1}{|z_1|}$, $(z_1 \neq 0)$ c. $|z_1 + z_2| \leqslant |z_1| + |z_2|$

Review Exercises

1. Replace x by i in the polynomial $2x^4 - 3x^3 - 5x^2 + x - 2$ and derive a "number" of the type $a + bi$ where $a, b \in R$. 7-1

2. State the real and imaginary parts for each complex number. 7-2
    a. $12 - 3i$ b. $(4 + 5i) + [8 + (-3)i]$
    c. $(7 + 3i) - (8 + 2i)$ d. $(4 + i) - (4 + i)$

3. State the real and imaginary parts for each complex number. 7-3
    a. $(3 + 2i)(5 + 7i)$ b. $(2 - 5i)^2$
    c. $(1 - 2i)\left(\frac{1}{5} + \frac{2}{5}i\right)$ d. $(-3 + 2i)^3$

4. Find the conjugate and multiplicative inverse of each complex number. 7-4
   a. $4i$ b. $3 - 7i$

5. State the real and imaginary parts of each complex number.
   a. $\dfrac{2 - 3i}{5 + 4i}$ b. $\dfrac{1 + 4i}{-3i}$

6. Locate the point corresponding to each number in the complex plane. 7-5
   a. $-7i$ b. $5 - 8i$ c. $-2 - 9i$

7. Find the absolute value of each complex number in Exercise 6.

8. What is the distance between $-1 + i$ and $5 - 7i$ in the complex plane?

8. HIGHER ORDER EQUATIONS

8-1 THE REMAINDER AND FACTOR THEOREMS

In this section we relate two notions which we have previously encountered: the function for a polynomial and the division algorithm for polynomials. Initially, we shall restrict our discussion to real numbers. If f is the function $x^3 - 19x + 30$, we write $f(x) = x^3 - 19x + 30$. Then $f(1) = 12$. On the other hand, if $x^3 - 19x + 30$ is divided by $x - 1$,

$$
\begin{array}{r|l}
 & x^2 + x - 18 \\
\hline
x - 1 & x^3 \qquad\quad - 19x + 30 \\
 & \underline{x^3 - x^2} \\
 & \quad x^2 - 19x \\
 & \quad \underline{x^2 - x} \\
 & \qquad - 18x + 30 \\
 & \qquad \underline{- 18x + 18} \\
 & \qquad\qquad\quad 12
\end{array}
$$

the remainder is 12.

Let us investigate the fact that $f(1)$ and the remainder upon dividing $x^3 - 19x + 30$ by $x - 1$ are the same number. We have observed that

$$x^3 - 19x + 30 = (x - 1)(x^2 + x - 18) + 12.$$

Replacing x by 1 on the right side of this equation gives

$$f(1) = (1 - 1)(1^2 + 1 - 18) + 12 = 0 + 12 = 12.$$

For a second example, we again use $x^3 - 19x + 30$ and its corresponding polynomial function.

$$f(-2) = (-2)^3 - 19(-2) + 30 = 60.$$

If $x^3 - 19x + 30$ is divided by $[x - (-2)]$ or $(x + 2)$, using synthetic division

we have

$$\begin{array}{r|rrrr} -2 & 1 & 0 & -19 & 30 \\ & & -2 & 4 & 30 \\ \hline & 1 & -2 & -15 & \boxed{60} \end{array}$$

The remainder is $f(-2) = 60$.

Use the factored form of $f(x) = x^3 - 19x + 30$ to verify that $f(-2) = 60$. If $g(x) = x^4 - 4$, find $g(3)$, $g(-3)$. Find the remainder on dividing $x^4 - 4$ by $x - 3$ and by $x + 3$.

$f(x) = [x - (-2)](x^2 - 2x - 15) + 60$

and

$$\begin{aligned} f(-2) &= [(-2) - (-2)] \\ &\quad \cdot [(-2)^2 - 2(-2) - 15] + 60 \\ &= 0[(-2)^2 - (-2) - 15] + 60 \\ &= 60 \end{aligned}$$

- $g(3) = (3)^4 - 4 = 77$
  $g(-3) = (-3)^4 - 4 = 77$

- $$\begin{array}{r|rrrrr} 3 & 1 & 0 & 0 & 0 & -4 \\ & & 3 & 9 & 27 & 81 \\ \hline & 1 & 3 & 9 & 27 & \boxed{77} \end{array}$$

$$\begin{array}{r|rrrrr} -3 & 1 & 0 & 0 & 0 & -4 \\ & & -3 & 9 & -27 & 81 \\ \hline & 1 & -3 & 9 & -27 & \boxed{77} \end{array}$$

The preceding examples illustrate a general result called the remainder theorem.

Theorem 1
(*Remainder theorem*)

If $f(x) = a_nx^n + a_{n-1}x^{n-1} + \cdots + a_1x + a_0$ then $f(a) = r$ where r is the remainder on dividing the polynomial by $x - a$.

Proof: By the division algorithm there is a unique pair of polynomials, a quotient q and a remainder r, such that

$$a_nx^n + a_{n-1}x^{n-1} + \cdots + a_1x + a_0 = q(x)(x - a) + r$$

and $\deg r < \deg (x - a)$ or $r = 0$. Since $\deg (x - a) = 1$, $\deg r = 0$ or $r = 0$. In either case, r is a real number. We now have

$$f(x) = q(x)(x - a) + r$$

and

$$f(a) = q(a)(a - a) + r = 0 + r = r$$

which is what we want to show.

If $f(x) = x^3 - 19x + 30$ then $f(3) = 0$, and by the remainder theorem when $x^3 - 19x + 30$ is divided by $x - 3$ the remainder is zero. Hence, $x - 3$ is a factor of $x^3 - 19x + 30$. Similarly, $f(2) = 0$ and $f(-5) = 0$; hence $(x - 2)$ and $[x - (-5)] = x + 5$ are also factors of $x^3 - 19x + 30$.

Theorem 2
(*Factor theorem*)

If $f(x) = a_nx^n + a_{n-1}x^{n-1} + \cdots + a_1x + a_0$ and $f(a) = 0$, then $(x - a)$ is a factor of the polynomial, and conversely.

Proof: There are unique polynomials, q and r, such that

$$a_nx^n + a_{n-1}x^{n-1} + \cdots + a_1x + a_0 = q(x - a) + r.$$

By the remainder theorem and the given condition, we know that

$$r = f(a) = 0.$$

Hence

$$a_nx^n + a_{n-1}x^{n-1} + \cdots + a_1x + a_0 = q(x - a)$$

and $(x - a)$ is a factor of the given polynomial.

Conversely, if $(x - a)$ is a factor of the polynomial then $r = 0$, but $r = f(a)$.

Use synthetic division to find $g(-2), g(-\sqrt{2}), g(-1), g(0), g(1), g(\sqrt{2})$, and $g(2)$ where $g(x) = x^4 - 3x^2 + 2$. Which elements of

$$\{-2, -\sqrt{2}, -1, 0, 1, \sqrt{2}, 2\}$$

are zeros of $g(x)$? Which elements of

$$\{(x + 2), (x + \sqrt{2}), (x + 1), x, (x - 1), (x - \sqrt{2}), (x - 2)\}$$

are factors of $x^4 - 3x^2 + 2$? Why?

$g(-2) = 6, g(-\sqrt{2}) = 0, g(-1) = 0$; $g(0) = 2$, $g(1) = 0$, $g(\sqrt{2}) = 0$, $g(2) = 6$;

- Zeros of $g(x)$: $-\sqrt{2}, -1, 1, \sqrt{2}$;
- Linear factors of $x^4 - 3x^2 + 2$: $(x + \sqrt{2})$, $(x + 1)$, $(x - 1)$, $(x - \sqrt{2})$;
- Apply the factor theorem.

To illustrate the converse consider the polynomial $x^4 - 7x^2 + 12$ which has $x - 2$ as a linear factor; that is,

$$x^4 - 7x^2 + 12 = (x - 2)(x^3 + 2x^2 - 3x - 6).$$

If $h(x) = x^4 - 7x^2 + 12$ then

$$h(2) = 0.$$

Exercises

1. If $f(x) = x^3 - x^2 + 2$ find:
   a. $f(1)$ b. $f(-2)$ c. $f(3)$ d. $f(-1)$
2. If $g(x) = x^4 + x^3 - 1$ find:
   a. $g(1)$ b. $g(-2)$ c. $g(3)$ d. $g(-1)$

Apply the division algorithm to find the remainder (by synthetic division) if the first polynomial in Exercises 3–10 is the divisor and the second is the dividend.

3. $x - 1$; $x^3 + 2x^2 - x + 1$
4. $x + 1$; $x^4 - x^2 + x - 3$
5. $x - 3$; $x^5 - x + 1$
6. $x + 2$; $x^5 + x^3 - 1$
7. $x - 1$; $x^3 + 4x - 3$
8. $x + 2$; $2x^3 - x^2 + 3x - 1$
9. $x - 2$; $x^4 + 3x^3 + x + 3$
10. $x + 2$; $x^4 - 3$
11. If the remainder theorem is applied to Exercise 1, what is the divisor in each case? Apply the division algorithm and compute the remainders using synthetic division.
12. Perform Exercise 11 with Exercise 1 replaced by Exercise 2.

For Exercises 13–15, let f be the polynomial function. Use the remainder theorem and synthetic division to compute the values. Check your results using a different procedure.

13. $x^3 - 7x + 6$; $f(1), f(2), f(3), f(-1), f(-2), f(-3)$

14. $x^3 - 3x^2 - x - 3$; $f(1), f(2), f(3), f(-1), f(-2), f(-3)$

15. $x^3 - \frac{3}{2}x^2 + \frac{3}{2}x - \frac{1}{2}$; $f(1), f\left(\frac{1}{2}\right), f\left(\frac{1}{4}\right), f(-1), f\left(-\frac{1}{2}\right), f\left(-\frac{1}{4}\right)$

16. Sketch the graph of the function $f(x) = 4x^4 + 4x^3 - 9x^2 - x + 2$ after finding the numbers $f\left(\frac{1}{2}\right)$, $f\left(\frac{3}{4}\right)$, $f(1)$, $f\left(\frac{3}{2}\right)$, $f(2)$, $f(3)$, $f\left(-\frac{1}{2}\right)$, $f(-1)$, $f\left(-\frac{3}{2}\right)$, $f(-2)$, and $f(-3)$ by synthetic division and the remainder theorem.

In Exercises 17 and 18

$$h(x) = x^3 - 6x^2 + 5x + 12 \quad \text{and} \quad k(x) = x^4 + 3x^3 - 11x^2 - 3x + 10.$$

17. Find each functional value.
a. $h(1)$ b. $h(-1)$ c. $h(3)$ d. $h(-5)$ e. $h(4)$ f. $h(2)$
g. $k(1)$ h. $k(-1)$ i. $k(3)$ j. $k(-5)$ k. $k(4)$ l. $k(2)$

18. Which of these polynomials are factors of $x^3 - 6x^2 + 5x + 12$? (Use the factor theorem and Exercise 17.)
a. $x - 1$ b. $x + 1$ c. $x - 3$ d. $x + 5$ e. $x - 2$ f. $x - 4$

19. Find which elements of the set $\{1, -1, 2, -2, 3, -3\}$ are zeros of the function for $x^4 + 2x^3 - 7x^2 - 8x + 12$. (Use synthetic division.)

20. Perform Exercise 19 with the set $\left\{1, -1, \frac{1}{3}, -\frac{1}{3}, \frac{1}{2}, -\frac{1}{2}, \frac{2}{3}, -\frac{2}{3}\right\}$ and the polynomial $6x^4 + x^3 - 8x^2 - x + 2$.

21. Let p and q be polynomials such that $p = q(x - a)$ where $a \in R$. Then prove that a zero of the function $q(x)$ is also a zero of the function $p(x)$.

22. In Exercise 21 let $p(x) = x^5 + x^4 - 7x^3 - 2x^2 + 5x + 6$ and $a = 2$. Find the polynomial q using synthetic division. Also show by synthetic division that $x + 3$ is a factor of the polynomial q.

23. Let p, q_1, q_2 be polynomials and a_1, a_2, r_1, r_2 belong to R such that $p = q_1(x - a_1) + r_1$ and $q_1 = q_2(x - a_2) + r_2$. Prove that $(x - a_2)$ is a factor of p if $(x - a_1)$ is a factor of p and $(x - a_2)$ is a factor of q_1.

8-2 COMPLEX NUMBERS AND THE QUADRATIC EQUATION

The remainder and factor theorems also hold when the complex numbers C are used to evaluate polynomials and define functions from C into

C. Since R is a subset of C, all polynomials with real number coefficients are also polynomials with complex number coefficients. If C is the replacement set, the solution set for the equation

$$a_n x^n + a_{n-1} x^{n-1} + \cdots + a_2 x^2 + a_1 x + a_0 = 0$$

is equal to the set of zeros of the function from C into C for the polynomial

$$a_n x^n + a_{n-1} x^{n-1} + \cdots + a_2 x^2 + a_1 x + a_0.$$

Consider the quadratic equation

$$x^2 + 1 = 0.$$

Let us compare the solution sets when R and C are used as the replacement sets. For example, if the replacement set is R, the solution set of the quadratic equation $x^2 + 1 = 0$ is the empty set. But, if the replacement set is C, the solution set of $x^2 + 1 = 0$ is $\{i, -i\}$.

Now consider the general quadratic equation

$$ax^2 + bx + c = 0$$

where a, b, and $c \in R$. If R is the replacement set for x, we know from the quadratic formula that the solution set has two, one, or no members. What is the nature of the solution set if C is the replacement set for x in the quadratic equation?

Let us examine the special case

$$x^2 + 9 = 0.$$

If C is the replacement set and we apply the quadratic formula to this equation ($a = 1, b = 0, c = 9$) then

$$x = \frac{\sqrt{-36}}{2} \quad \text{or} \quad x = -\frac{\sqrt{-36}}{2}.$$

In Section 7-1, we suggested letting $i = \sqrt{-1}$. We adopt that convention here. Thus:

$$\sqrt{-36} = \sqrt{36(-1)} = \sqrt{36}\sqrt{-1} = \sqrt{36}\,i = 6i.$$

Thus, the quadratic formula yields

$$x = \frac{\sqrt{-36}}{2} = 3i \quad \text{or} \quad x = -\frac{\sqrt{-36}}{2} = -3i$$

and possible solutions for $x^2 + 9 = 0$. If we test $3i$ in the equation we have

$$(3i)^2 + 9 = 3^2 i^2 + 9 = 9(-1) + 9 = 0.$$

Also

$$(-3i)^2 + 9 = (-3)^2 i^2 + 9 = 9(-1) + 9 = 0.$$

Hence, $3i$ and $-3i$ belong to the solution set.

Definition If a is a positive real number, then $\sqrt{-a}$ is the complex number $\sqrt{a}i = 0 + \sqrt{a}i$.

Recall that if a is a real number and $a \geqslant 0$ then $\sqrt{a}$ denotes the unique nonnegative real number b such that $b^2 = a$. Applying these definitions, we have

$$\sqrt{-49} = \sqrt{49}i = 7i, \quad \sqrt{-2} = \sqrt{2}i,$$

and

$$-\sqrt{-121} = -(\sqrt{121}i) = -(11i) = -11i.$$

Express each of the following complex numbers in $a + bi$ form: $\sqrt{144}$, $\sqrt{-144}$, $\sqrt{144} + \sqrt{-144}$, $-\sqrt{144}$, $-\sqrt{-144}$, $(\sqrt{144})(\sqrt{-144})$, $\frac{\sqrt{144}}{\sqrt{-144}}$, and $\frac{\sqrt{-144}}{\sqrt{144}}$.

$\sqrt{144} = 12 + 0i$,
$\sqrt{-144} = \sqrt{144}i = 0 + 12i$,
$\sqrt{144} + \sqrt{-144} = 12 + 12i$,
$-\sqrt{144} = (-12) + 0i$,
$-\sqrt{-144} = 0 + (-12i)$,
$(\sqrt{144})(\sqrt{-144}) = 0 + 144i$,
$\frac{\sqrt{144}}{\sqrt{-144}} = \frac{12}{12i} = 0 + (-1)i$,
$\frac{\sqrt{-144}}{\sqrt{144}} = \frac{12i}{12} = 0 + 1i$

With this new notation for complex numbers, the quadratic formula is

$$x = \frac{-b \pm \sqrt{b^2 - 4ac}}{2a},$$

where $a, b, c \in R$, $a \neq 0$, but C is the replacement set for x.

If we apply the quadratic formula to the equation $x^2 + x + 3 = 0$ the complex numbers $\frac{-1 + \sqrt{-11}}{2} = \frac{-1}{2} + \frac{\sqrt{11}}{2}i$ and $\frac{-1 - \sqrt{-11}}{2} = \frac{-1}{2} - \frac{\sqrt{11}}{2}i$ comprise the solution set. Note that one solution for the quadratic equation is the complex conjugate of the other. In fact, in any polynomial equation, if the coefficients are real numbers then if $a + bi$ is a complex number in the solution set, the conjugate $a - bi$ is also in the solution set.

Exercises

Find the real and imaginary parts for each complex number in Exercises 1–3.

1. $\sqrt{-64}$, $-\sqrt{-64}$, $\sqrt{64}$, $-\sqrt{64}$

2. $-\sqrt{\frac{4}{9}}$, $-\sqrt{-\frac{4}{9}}$, $\frac{6 + \sqrt{-144}}{8}$, $\frac{2 - \sqrt{-16}}{4}$, $\frac{10 - \sqrt{16}}{2}$

3. $\sqrt{-4} \cdot \sqrt{-9}$, $\sqrt{-4} \cdot \sqrt{9}$, $\sqrt{-25}(\sqrt{-16} + 1)$

4. Write a quadratic equation with each of the following solution sets.
   a. $\{3, -3\}$ b. $\{3i, -3i\}$ c. $\{3 + 3i, 3 - 3i\}$ d. $\{-3 + \sqrt{-3}, -3 - \sqrt{-3}\}$

5. Factor each polynomial using complex coefficients.
   a. $x^3 + 1$ b. $64x^3 - 27$ c. $x^4 + 4x + 3$

Find the solution set for each equation in Exercises 6-14 if R is the replacement set; if C is the replacement set.

6. $6x^2 - x - 12 = 0$
7. $4x^2 - 12x + 9 = 0$
8. $x^2 - 2x - 1 = 0$
9. $x^2 + 16 = 0$
10. $x^2 - 25 = 0$
11. $x^2 - 5 = 0$
12. $x^2 + 5 = 0$
13. $x^2 - 6x + 18 = 0$
14. $4x^2 - 4x + 21 = 0$

Factor each polynomial in Exercises 15-17 using integral coefficients; real coefficients; complex coefficients.

15. $6x^2 - x - 12$, $4x^2 - 12x + 9$, $x^2 - 2x - 1$
16. $x^2 + 16$, $x^2 - 25$, $x^2 - 5$
17. $x^2 + 5$, $x^2 - 6x + 18$, $4x^2 - 4x + 21$
18. If $2i$ is a given solution of $x^3 - \sqrt{2}x^2 + 4x - 4\sqrt{2} = 0$, factor $x^3 - \sqrt{2}x^2 + 4x - 4\sqrt{2}$ using complex coefficients.
19. If one solution of $2x^3 - 11x^2 + 20x + 13 = 0$ is $3 - 2i$, find the remaining solutions in C.
20. Show that if a complex number with nonzero imaginary part is a solution for the equation
$$ax^2 + bx + c = 0$$
where $a, b, c \in R$, then its conjugate is also a solution.
21. Prove that if $a + bi$ $(b \neq 0)$ is a solution of
$$a_n x^n + a_{n-1}x^{n-1} + \cdots + a_2 x^2 + a_1 x + a_0 = 0$$
where $a_n, a_{n-1}, \cdots, a_2, a_1, a_0 \in R$, then $a - bi$ is also a solution.
22. If a polynomial with real coefficients of degree 4 has an equation with $2 - i$ and $3i$ as solutions, write the equation.

8-3 RATIONAL AND COMPLEX ZEROS

In general, finding the zeros of a polynomial function is not an easy task. In this section we present a few basic results about polynomials which provide you some assistance. From the factor theorem the problem of finding zeros for a polynomial function is equivalent to finding linear factors of the polynomial. The important fundamental theorem of algebra follows.

Theorem 1 (*Fundamental theorem of algebra*) In the system of polynomials with complex number coefficients, every polynomial of degree $\geqslant 1$ has at least one linear factor.

Complex number coefficients for the linear factors are necessary as is shown by the polynomial

$$x^2 + 1 = (x - i)(x + i).$$

Real coefficients do not yield linear factors for $x^2 + 1$.

By repeated application of the fundamental theorem of algebra, a polynomial with complex coefficients of degree $n \geqslant 1$ has at least n linear factors. Also, the product of more than n linear polynomials has degree greater than n. Thus a polynomial with complex coefficients of degree $n \geqslant 1$ has exactly n linear factors. Since integers, rational numbers, and real numbers are imbedded in the system of complex numbers, the preceding statements are relevant in finding zeros for particular polynomial functions.

If we consider polynomials whose coefficients are integers, then information about rational zeros for the polynomial functions can be found. For example, $f(x) = 3x^3 - 2x^2 + 3x - 2$ has at most three zeros. (Why?) If a rational number $\frac{p}{q}$, where p and q are integers and have no common prime factor, is a zero of the function f, then

$$f\left(\frac{p}{q}\right) = 3\left(\frac{p}{q}\right)^3 - 2\left(\frac{p}{q}\right)^2 + 3\left(\frac{p}{q}\right) - 2 = 0.$$

This holds if and only if

$$3\left(\frac{p^3}{q^3}\right) - 2\left(\frac{p^2}{q^2}\right) + 3\left(\frac{p}{q}\right) - 2 = 0$$

and multiplying by q^3 yields

$$3p^3 - 2p^2q + 3q^2p - 2q^3 = 0.$$

Adding $2q^3$ to each member of the equation, and factoring p from the left side gives

$$p(3p^2 - 2pq + 3q^2) = 2 \cdot q^3.$$

Since p is a factor of the left member, p must be a factor of the right member, $2 \cdot q^3$. But by hypothesis p is not a factor of q^3 since p and q have no common prime factor. Hence, p is a factor of 2 and -2, the constant term of the polynomial. Hence, p must be a factor of the integer -2; that is, $p \in \{1, -1, 2, -2\}$.

Similarly, $3\left(\frac{p^3}{q^3}\right) - 2\left(\frac{p}{q}\right)^2 + 3\left(\frac{p}{q}\right) - 2 = 0$ if and only if:

$$3\left(\frac{p^3}{q^3}\right) - 2\left(\frac{p^2}{q^2}\right) + 3\left(\frac{p}{q}\right) - 2 = 0$$

$$3p^3 - 2p^2q + 3pq^2 - 2q^3 = 0$$

$$2p^2q - 3pq^2 + 2q^3 = 3p^3$$

and

$$q(2p^2 - 3pq + 2q^2) = 3p^3.$$

Since q is a factor of the left member, q must be a factor of the integer $3p^3$. We can conclude that q must be a factor of 3, the coefficient of the term with highest degree. Hence, $q \in \{1, -1, 3, -3\}$.

To summarize, if a rational number $\frac{p}{q}$ in simplest form is a zero of the polynomial function $f(x) = 3x^3 - 2x^2 + 3x - 2$, then p is a factor of -2, q is a factor of 3, and $\frac{p}{q}$ belongs to the set

$$\left\{\frac{1}{3}, \frac{2}{3}, 1, 2, -\frac{1}{3}, -\frac{2}{3}, -1, -2\right\}.$$

There may be no rational zero; however, if any rational zeros exist, they belong to this set. Trying $\frac{1}{3}$ and $\frac{2}{3}$ for this function

$$\begin{array}{c|cccc}
\frac{1}{3} & 3 & -2 & 3 & -2 \\
 & & 1 & -\frac{1}{3} & \frac{8}{9} \\
\hline
 & 3 & -1 & \frac{8}{3} & \boxed{-1\frac{1}{9}}
\end{array}
\qquad
\begin{array}{c|cccc}
\frac{2}{3} & 3 & -2 & 3 & -2 \\
 & & 2 & 0 & 2 \\
\hline
 & 3 & 0 & 3 & \boxed{0}
\end{array}$$

we find that $\frac{1}{3}$ is not a zero of f but that $\frac{2}{3}$ is a zero of this function. None of the other rational numbers are zeros of f since $f(x) = \left(x - \frac{2}{3}\right)(3x^2 + 3)$ and no rational numbers are zeros of $g(x) = 3(x^2 + 1)$.

Theorem 2 If $\frac{p}{q}$ is a rational number in simplest form and a zero of the function for a polynomial of degree n, $a_n x^n + a_{n-1}x^{n-1} + \cdots + a_1 + a_0$ having integral coefficients, then p must be a factor of a_0 and q must be a factor of a_n.

Use Theorem 2 to identify a set of rational numbers that contains all of the rational zeros of the function for $6x^3 + 7x^2 - 1$. Then find the rational zeros of the function.

Set of rational numbers containing the rational zeros:

$$\left\{-1, -\frac{1}{2}, -\frac{1}{3}, -\frac{1}{6}, \frac{1}{6}, \frac{1}{3}, \frac{1}{2}, 1\right\}$$

- Rational zeros: $-1, -\frac{1}{2}, \frac{1}{3}$.

According to Theorem 2, a rational zero of the function

$$f(x) = 6x^4 + 5x^3 + 4x^2 - 2x - 1$$

must be in the set

$$\left\{\frac{1}{6}, \frac{1}{3}, \frac{1}{2}, 1, -\frac{1}{6}, -\frac{1}{3}, -\frac{1}{2}, -1\right\}.$$

Using synthetic division, we have

$$\begin{array}{c|ccccc} \frac{1}{2} & 6 & 5 & 4 & -2 & -1 \\ & & 3 & 4 & 4 & 1 \\ \hline & 6 & 8 & 8 & 2 & \boxed{0} \end{array}$$

Hence, $\frac{1}{2}$ is a zero of f and $\left(x - \frac{1}{2}\right)$ is a factor of the polynomial:

$$6x^4 + 5x^3 + 4x^2 - 2x - 1 = \left(x - \frac{1}{2}\right)(6x^3 + 8x^2 + 8x + 2)$$

$$= \left(x - \frac{1}{2}\right)(2)(3x^3 + 4x^2 + 4x + 1).$$

Therefore any zero of f is a member of the solution set for

$$\left(x - \frac{1}{2}\right) = 0 \quad \text{or} \quad (3x^3 + 4x^2 + 4x + 1) = 0.$$

Each solution for these two equations will also be a zero of the original function $f(x)$. We know that the only zero of $x - \frac{1}{2}$ is $\frac{1}{2}$. What are the zeros of $g(x) = 3x^3 + 4x^2 + 4x - 1$? $\left(\text{In principle, } \frac{1}{2} \text{ could be a zero also, if } \left(x - \frac{1}{2}\right)^2 \text{ were a factor of the given polynomial.}\right)$

We now test the only candidates for rational zeros of $g(x)$, $\frac{1}{3}$ and $-\frac{1}{3}$.

$\frac{1}{3}$	3	4	4	1
		1	$\frac{5}{3}$	$\frac{17}{9}$
	3	5	$\frac{17}{3}$	$\frac{26}{9}$

$-\frac{1}{3}$	3	4	4	1
		-1	-1	-1
	3	3	3	0

Thus, $\frac{1}{3}$ is not a zero for $g(x)$ or $f(x)$, but $-\frac{1}{3}$ is a zero for both $g(x)$ and $f(x)$. Hence, if one zero of a function is found, we then try to factor the quotient polynomial which is of degree one less than the original polynomial and is simpler to handle. For, in general, if p and q are polynomials such that $p = q(x - a)$ where $a \in R$, then any zero of the function $q(x)$ is also a zero of the function $p(x)$.

Example 1 Find the rational zeros of $f(x) = x^5 - 11x^4 + 33x^3 + 11x^2 - 154x + 120$.

Solution: There are 16 positive factors of 120 and two factors of 1, the coefficient of x^5, hence there are 32 possible rational zeros to consider. Can you list them? We now show the synthetic-division computations that yield zeros for the function.

1	1	-11	33	11	-154	120
		1	-10	23	34	-120
5	1	-10	23	34	-120	0
		5	-25	-10	120	
-2	1	-5	-2	24	0	
		-2	14	-24		
	1	-7	12	0		

At this stage we know that

$$f(x) = (x - 1)(x - 5)(x + 2)(x^2 - 7x + 12)$$

and that

$$x^2 - 7x + 12 = (x - 4)(x - 3)$$

hence, 1, 5, -2, 3, and 4 are the zeros of the polynomial function.

Example 2 Factor $x^5 - 4x^4 - 8x^3 + 40x^2 - 17x - 60$ in the system of polynomials with complex coefficients.

Solution: Any rational solution of

$$x^5 - 4x^4 - 8x^3 + 40x^2 - 17x - 60 = 0$$

must belong to

$$\{\pm 1, \pm 2, \pm 3, \pm 4, \pm 5, \pm 10, \pm 12, \pm 15, \pm 20, \pm 30, \pm 60\}.$$

Using synthetic division we find that

$$\begin{array}{r|rrrrrr}
-1 & 1 & -4 & -8 & 40 & -17 & -60 \\
 & & -1 & 5 & 3 & -43 & 60 \\
\hline
4 & 1 & -5 & -3 & 43 & -60 & \boxed{0} \\
 & & 4 & -4 & -28 & 60 & \\
\hline
-3 & 1 & -1 & -7 & 15 & \boxed{0} & \\
 & & -3 & 12 & -15 & & \\
\hline
 & 1 & -4 & 5 & \boxed{0} & &
\end{array}$$

Hence,

$$x^5 - 4x^4 - 8x^3 + 40x^2 - 17x - 60$$
$$= (x + 1)(x - 4)(x + 3)(x^2 - 4x + 5).$$

The solution set for $x^2 - 4x + 5 = 0$ is $\left\{\dfrac{4 + \sqrt{-4}}{2}, \dfrac{4 - \sqrt{-4}}{2}\right\} = \{2 + i, 2 - i\}$; it follows that $x^2 - 4x + 5 = [x - (2 + i)]\,[x - (2 - i)]$ and the factorization we seek is

$$(x + 1)(x - 4)(x + 3)(x - 2 - i)(x - 2 + i).$$

Exercises

Use Theorem 2 of this section to list the set of possible rational zeros of the function for each polynomial in Exercises 1–4.

1. $2x^3 + 3x^2 + 3x + 1$

2. $3x^3 + 5x^2 + 4x - 2$

3. $4x^3 + 6x^2 - 10x + 3$

4. $x^5 + 5x^4 - 4x^3 - 11x^2 + 15x - 6$

5. Find one rational zero of each function in the preceding exercises.

Find the rational zeros of the function for each polynomial in Exercises 6–15.

6. $x^4 + x^3 - x^2 - 7x - 6$

7. $x^4 - 2x^3 - 4x^2 + 23x - 30$

8. $x^4 - 4x^3 - x^2 + 16x - 12$

9. $x^4 + 6x^3 - 9x^2 - 94x - 120$

10. $6x^4 - 11x^3 + 14x^2 + 7x - 6$

11. $15x^4 + 14x^3 + 8x^2 - 7x - 6$

12. $4x^5 - 8x^4 + 5x^3 - 5x^2 + x + 3$

13. $25x^4 + 30x^3 - 157x^2 - 42x + 72$

14. $6x^5 - 9x^4 + 14x^3 - 21x^2 + 4x - 6$

15. $40x^5 - 14x^4 - 121x^3 - 46x^2 + 39x + 18$

16. Prove that the function for $x^4 + 10x^2 + 9$ has no rational zeros.

17. Prove that any rational zero of the function for the polynomial $x^n + a_{n-1}x^{n-1} + \cdots + a_1x + a_0$, where the coefficients are integers, is an integer and is a factor of a_0.

In Exercises 18-23 factor each polynomial using complex coefficients.

18. $x^4 - 2x^3 - 13x^2 - 4x - 30$

19. $x^4 - 2x^3 - 2x^2 + 6x - 3$

20. $x^4 + x^2 - 20$

21. $x^4 + 9x^2 + 20$

22. $12x^4 + 8x^3 - 21x^2 - 5x + 6$

23. $x^5 - 5x^4 - 6x^3 + 53x^2 - 31x - 84$

24. Prove that the function for a polynomial of odd degree with real coefficients has at least one real zero.

25. Prove that every polynomial of degree $\geqslant 3$ with real coefficients can be expressed as a product of linear and quadratic polynomials with real coefficients.

8-4 IRRATIONAL ZEROS

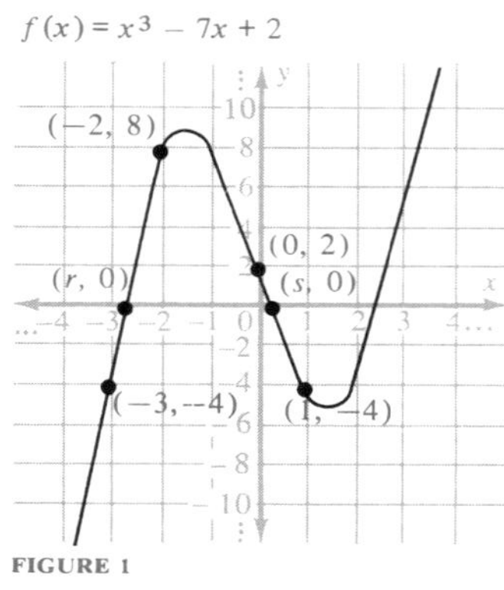

FIGURE 1

If we examine the polynomial function $f(x) = x^3 - 7x + 2$ for rational zeros, none of the factors of 2(1, 2, −1, −2) are zeros. Hence, no rational number is a zero for this function.

A standard method for approximating irrational zeros involves examining the graph of f. You might use synthetic division to calculate the coordinates of a few points on the graph of f. Then estimate a smooth curve between these points [Figure 1]. Note that the points (−3, −4) and (−2, 8) lie on the graph of f, that $f(-3) < 0$ and $f(-2) > 0$, and that the graph seems to cross the x-axis at the point $(r, 0)$ where $-3 < r < -2$. Therefore, $f(r) = 0$ and r is an irrational zero of f. Similarly, the smooth curve connecting the points (0, 2) and (1, −4) appear to intersect the x-axis at the point $(s, 0)$ where $0 < s < 1$. Therefore, another irrational zero seems to exist between 0 and 1.

The technique for locating zeros of polynomial functions by exploring the graph of the function is related to the following theorem.

Theorem If f is a polynomial function from R into R, $f(a) > 0$, and $f(b) < 0$, then there exists at least one real number c between a and b such that $f(c) = 0$.

Since $f(2) < 0$ and $f(3) > 0$, f has a zero between 2 and 3; ● This suggests that f has 3 *irrational* zeros (See text for other two.); ● Yes, $f(-4) = -34$ and $f(-2) = 8$; ● Yes, $f(-3) = -4$ and $f(-2.5) = 3.875$; ● No, $f(-2.5) > 0$ and $f(-2) = 8$.

Use the preceding theorem to locate any other zeros of the function $f(x) = x^3 - 7x + 2$ between consecutive integers. How many zeros does f seem to have? Are these real zeros rational or irrational? Does f have at least one zero between -4 and -2? Between -3 and -2.5? Between -2.5 and -2?

For the function $f(x) = x^3 - 7x + 2$ the number 0 is a very crude approximation for the irrational zero, s, between 0 and 1. One of the numbers 0.1, 0.2, 0.3, 0.4, 0.5, 0.6, 0.7, 0.8, 0.9, 1 might be a better approximation for s. Which one? If the graph of f between (0, 2) and (1, -4) were a line segment, then it would cross the x-axis at a point with first coordinate d between 0.3 and 0.4. From Figure 2, $\frac{d}{2} = \frac{1}{6}$, so $d = \frac{1}{3}$. What is $f(0.3)$?

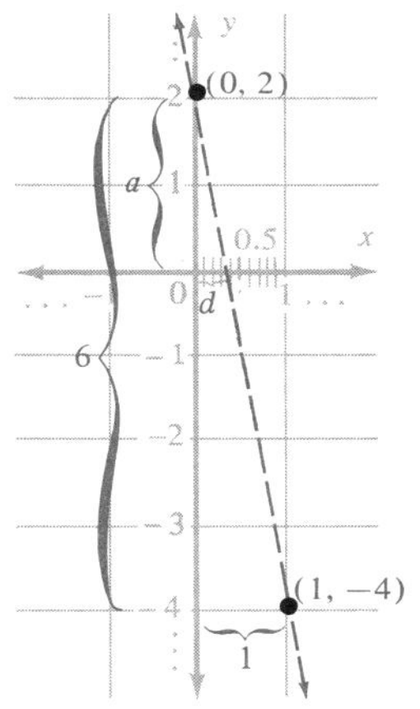

FIGURE 2

0.3	1	0	-7	2
		0.3	0.09	-2.073
	1	0.3	-6.91	-0.073 = $f(0.3)$

Since $f(0.3)$ is negative, the graph of f crosses at a point to the left of (0.3, 0). We examine $f(0.2)$:

0.2	1	0	-7	2
		0.2	0.04	-1.392
	1	0.2	-6.96	0.608 = $f(0.2)$

Since $f(0.2)$ is positive, the graph of f crosses at a point to the right of (0.2, 0). From this information, we judge that 0.3 is a good rational approximation of s to the nearest tenth. If $f(0.25)$ is a positive number, then we know that 0.3 is the best rational approximation for s to the nearest tenth. (Why?)

This process can be repeated until we reach the desired accuracy for approximating s. Figure 3 indicates the line segment connecting (0.2, 0.608) and (0.3, -0.073) that we use to approximate f. The details of approximating s to the nearest hundredth follow:

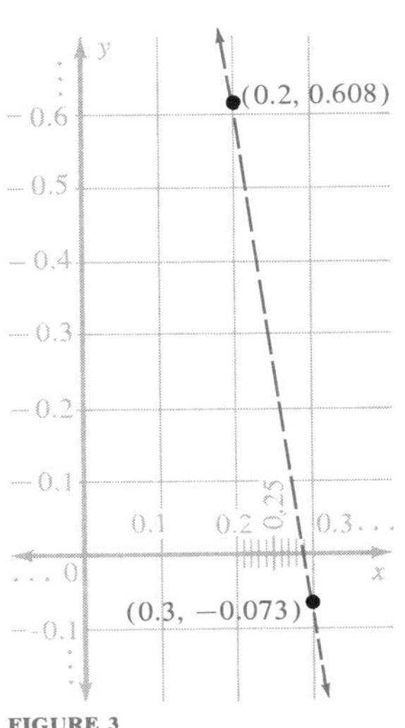

FIGURE 3

0.29	1	0	-7	2
		0.29	0.0841	-2.005301
	1	0.29	-6.9159	-0.005301

0.28	1	0	-7	2
		0.28	0.0784	-1.938048
	1	0.28	-6.9216	0.061952

Would you select 0.28 or 0.29 as the better approximation? If you have doubts about your choice, compute $f(0.285)$.

This procedure for finding a rational approximation of a real zero makes use of a line segment as an aid in guessing where the graph crosses the x-axis. As we continue the process, the line segment becomes a better approximation for the curve. In any event, we have at most nine points to consider at each stage of the process.

Exercises

Approximate, to the nearest hundredth, the indicated irrational zero of the function from R into R for each polynomial in Exercises 1–4.

1. $2x^3 - 5x + 1$, between 0 and 1
2. $x^3 - 5x^2 + 3$, between -1 and 0
3. $3x^4 - 3x - 1$, between 1 and 2
4. $x^3 - x + 2$, between -2 and -1
5. Approximate the other two irrational zeros for the example in this section $f(x) = x^3 - 7x + 12$ to the nearest tenth and to the nearest hundredth.

Find the zeros of the function for each polynomial in Exercises 6–7. (Approximate irrational zeros to the nearest hundredth.)

6. $x^3 - 6x + 1$
7. $x^3 - 3x^2 + 1$
8. Use the polynomial $x^2 - 2$ to define a function and approximate the irrational number $\sqrt{2}$ to the nearest thousandth using the method introduced in this section.
9. Follow the directions in Exercise 8 for the polynomial $x^2 - 3$ and the irrational number $\sqrt{3}$.

Review Exercises

1. If $h(x) = 2x^3 - 4x^2 + 3x - 5$ use the remainder theorem and synthetic division to find $h(1)$, $h(-1)$, $h(2)$, $h(-2)$, $h\left(\frac{3}{2}\right)$, $h\left(-\frac{3}{2}\right)$. 8-1
2. Which elements of $\{-12, -6, -4, -3, -2, -1, 1, 2, 3, 4, 6, 12\}$ are zeros of the function $f(x) = x^4 - 9x^2 - 4x + 12$?
3. Find the solution set if C is the replacement set. 8-2
   a. $x^2 - 4x + 13 = 0$
   b. $x^4 - 4x^2 - 5 = 0$
4. Find the rational zeros of the function 8-3

$$f(x) = 9x^4 - 12x^3 - 17x^2 + 8x + 4.$$

5. Find the solution set of the equation

$$x^4 - 2x^3 + x^2 - 8x - 12 = 0$$

   if the replacement set is R; if the replacement set is C.

6. Approximate, to the nearest hundredth, the irrational zero between 0 and 1 of the function 8-4

$$f(x) = 3x^3 + 2x^2 - 2x - 1$$

   from R into R.

9. EXPONENTIAL AND LOGARITHMIC FUNCTIONS

9-1 EXPONENTIAL FUNCTIONS

In this chapter we examine exponential functions and their inverses, the logarithmic functions. These classes of functions have several important and interesting applications.

$\{(x, y): y = b^x, x \text{ any rational number}\}$ where $b > 1$

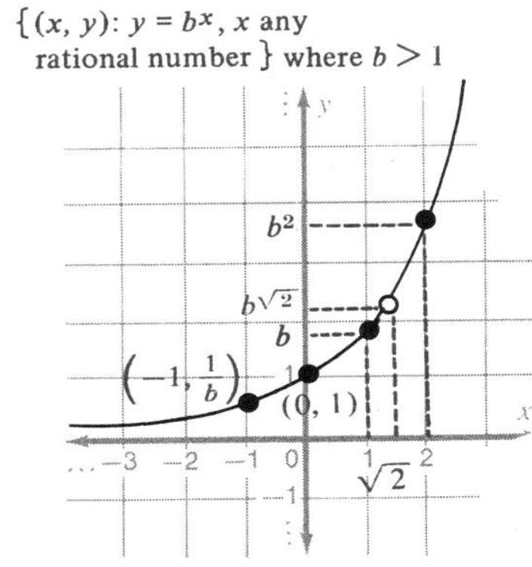

FIGURE 1

Let us first consider relations of the form $\{(x, y): y = b^x, x \in R\}$, where $b > 0$ and $b \neq 1$. If the real number exponent is a rational number m/n $(n > 0)$, then $b^x = b^{m/n} = (b^{1/n})^m$ where $b^{1/n}$ is the positive nth root of b. If the real number exponent is irrational a precise definition of b^x is beyond the scope of this book. However, the following geometric description should provide an intuitive feeling for the meaning of irrational exponents.

Our description begins with a discussion of the real number line. An infinite number of points on the real number line correspond to rational numbers and an infinite number of points correspond to irrational numbers. The points of each of these sets are distributed among the points of the other set in such a way that "arbitrarily close" to any point of either set we can find points that correspond to both rational and irrational numbers. Thus, if we think of removing the points corresponding to the irrational numbers, the remaining points which correspond to the rational numbers resemble a line with infinitely many "holes."

$\{(x, y): y = b^x, x \text{ any rational number}\}$ where $0 < b < 1$

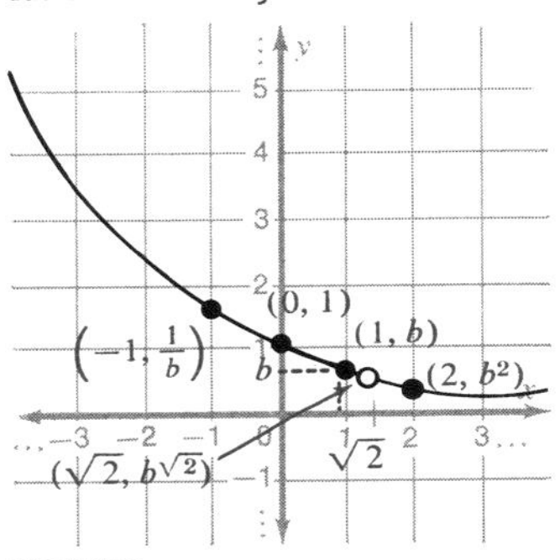

FIGURE 2

Consider the graphs of the relations $\{(x, y): y = b^x, x \text{ any rational number}\}$, where $b > 1$ [Figure 1] and where $0 < b < 1$ [Figure 2]. Since b^x is defined for each rational number x, the domain of each of these relations is the set of rational numbers. The graphs of the relations are sketched by plotting several points. Since the points on the x-axis with rational first coordinates are infinite in number and arbitrarily close together in any segment of the axis, the points of the graphs of the relations are infinite in number and arbitrarily close. Hence, the graphs of the relations cannot be pictured as anything but continuous curves, although in reality they are curves with holes above $(\sqrt{2}, 0)$ and above every other point on the x-axis with an irrational first coordinate.

We will define b^x, x irrational, to be the ordinate of the point that "fills" the "hole" above $(x, 0)$ in the graph of the relation $\{(x, y): y = b^x,$

x any rational number}, where $b > 0$ and $b \neq 1$. Thus we assume that b^x has meaning for any real number x if $b > 0$ and that b^x is a unique positive real number. We assume that the properties of rational exponents hold for irrational exponents as well.

From the assumption that b^x is a unique positive real number for each $b > 0$ and real number x, it follows that equations of the form $y = b^x$ define functions.

Definition Any function defined by $\{(x, y)\colon\ y = b^x\}$, where $b \neq 1$ is a positive real number, is called an **exponential function with base b.**

In the remainder of this book, the reader is expected to remember that the base of an exponential function is a positive real number not equal to 1.

Why is the function $\{(x, y)\colon\ y = b^x\}$, where $b = 1$, uninteresting? If $b < 0$, the relation $\{(x, y)\colon\ y = b^x\}$ is not an exponential function. Why is this case not included? What is the domain of an exponential function? The range?

It is a constant function; Because in real system even roots of negative numbers are not defined and irrational powers for negative numbers are not defined; Domain: R; Range: positive real numbers.

Figure 3 is the sketch of the exponential function $y = 10^x$. The ordered pairs for this function are of the form $(x, 10^x)$.

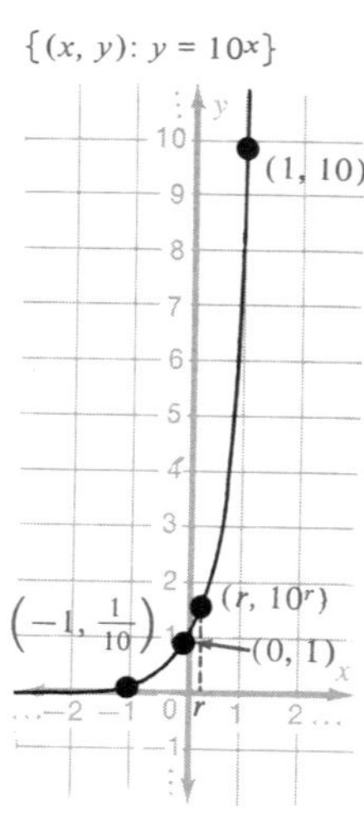

FIGURE 3

Exercises

1. Sketch the graph of the relation defined by each equation.
   a. $y = 10^x$ b. $y = 10^{-x}$ c. $y = \left(\frac{1}{10}\right)^{-x}$ d. $y = \left(\frac{1}{10}\right)^x$
2. Sketch the graphs of the relations a and b on the same set of axes. Do the same for relations c and d.
   $a = \{(x, y)\colon\ y = 3^x\}$ $c = \{(x, y)\colon\ y = 2^x\}$
   $b = \{(x, y)\colon\ y = 3^{-x}\}$ $d = \{(x, y)\colon\ y = 2^{-x}\}$
3. Write a statement that describes the relative positions of the graphs for the relations a and b in Exercise 2; for the relations c and d. (Hint: Consider the positions of the graphs relative to the y-axis.)
4. Follow the instructions in Exercise 2.
   $a = \{(x, y)\colon\ y = 4^x\}$ $c = \left\{(x, y)\colon\ y = \left(\frac{1}{4}\right)^x\right\}$
   $b = \{(x, y)\colon\ y = -4^x\}$ $d = \left\{(x, y)\colon\ y = -\left(\frac{1}{4}\right)^x\right\}$
5. Write a statement that describes the relative positions of the graphs for the relations a and b in Exercise 4; for the relations c and d. (Hint: Consider the positions of the graphs relative to the x-axis.)
6. Sketch the graph of each relation on the same set of axes.
   $a = \left\{(x, y)\colon\ y = \left(\frac{1}{3}\right)^x\right\}$ $b = \{(x, y)\colon\ y = 3^{-x}\}$
   What observation do you make about the graphs of these relations?

7. Sketch the graph of each relation on the same set of axes.

$$a = \{(x, y):\ y = 3^x\} \qquad b = \{(x, y):\ x = 3^y\}$$

Describe the relative positions of the graphs of these relations. (Hint: Sketch the graph of the line $y = x$.)

8. Plot as many points as are necessary to sketch each graph.

a. $y = 5^x$ b. $y = (0.1)^x$ c. $y = \left(\frac{3}{4}\right)^x$

d. $y = (\sqrt{2})^x$ e. $y = \pi^x$ f. $y = -2^x$

9-2 LOGARITHMIC FUNCTIONS

From the graphs of $\{(x, y):\ y = b^x\}$, where $b > 1$, [Figure 1, Section 9-1], where $0 < b < 1$, [Figure 2, Section 9-1] we see that each element in the range of an exponential function corresponds to only one element of the domain. Therefore, the inverse of an exponential function is also a function.

Definition The inverse of an exponential function $\{(x, y):\ y = b^x\}$ is called a **logarithmic function** and is denoted by

$$\{(y, x):\ x = \log_b y\}.$$

Again, $b > 0$ and $b \neq 1$. The expression $\log_b y$ is read "the logarithm of y to the base b."

Since the domain and range of exponential functions are, respectively, the range and domain of the logarithmic functions, the domain of a logarithmic function is the set of positive real numbers and the range is the set of all real numbers [Figure 1].

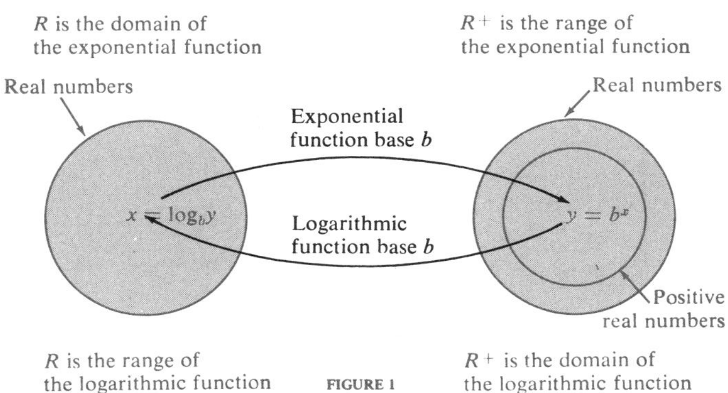
R is the domain of the exponential function
R+ is the range of the exponential function
Real numbers
Real numbers
Exponential function base b
x = log_b y
y = b^x
Logarithmic function base b
Positive real numbers
R is the range of the logarithmic function
R+ is the domain of the logarithmic function

FIGURE 1

By definition, a logarithmic function is the inverse of an exponential function

$$\{(x, y)\colon\ y = b^x\}.$$

Therefore, logarithmic functions are defined by:

$$\{(y, x)\colon\ y = b^x\}$$

using the definition of an inverse relation, and by:

$$\{(y, x)\colon\ x = \log_b y\}$$

using the notation introduced in the definition. Thus, equations

$$y = b^x \qquad \text{and} \qquad x = \log_b y$$

are equivalent. Thus, the logarithm of y to the base b, $\log_b y$, is the exponent of the power of b which is equal to y. In symbols, $y = b^{\log_b y}$.

Example Find n if $\log_3 27 = n$; if $\log_n 9 = -2$; if $\log_{81} n = \frac{1}{2}$.

Solution: If $\log_3 27 = n$, then $3^n = 27$; thus, $n = 3$.
If $\log_n 9 = -2$, then $n^{-2} = 9$; thus, $n = \frac{1}{3}$.
If $\log_{81} n = \frac{1}{2}$, then $81^{1/2} = n$; thus, $n = 9$.

Find m and n so that each of the following sentences will be true. If $25^{1/2} = 5$, then $\log_n 5 = \frac{1}{2}$; if $32^{1/5} = 2$, then $\log_{32} n = \frac{1}{5}$; if $5^2 = 25$, then $\log_5 25 = n$; if $9^{1/2} = 3$, then $\log_9 3 = n$. $\text{Log}_2\ 8 = n$ if and only if $m^3 = 8$; $\log_n 4 = 2$ if and only if $2^m = 4$; $\log_{10} n = 3$ if and only if $m^3 = 1000$. Since $a^1 = a$ and $a^0 = 1$, $\log_a a = n$ and $\log_a 1 = m$.

$n^{1/2} = 5$	$32^{1/5} = n$	$5^n = 25$	$9^n = 3$
$n = 25$	$n = 2$	$n = 2$	$n = \frac{1}{2}$

("Iff" is short for "if and only if.")
$2^n = 8$ ($n = 3$) iff $m^3 = 8$ ($m = 2$)
$n^2 = 4$ ($n = 2$) iff $2^m = 4$ ($m = 2$)
$10^3 = n$ ($n = 1000$) iff $m^3 = 1000$ ($m = 10$)
$a^n = a$ ($n = 1$) and $a^m = 1$ ($m = 0$)

Since logarithms are exponents, they have properties like those of exponents. In particular, the following property holds for logarithms.

Theorem If y_1 and y_2 are positive real numbers, then $y_1 = y_2$ if and only if

$$\log_b y_1 = \log_b y_2.$$

What does the graph of a logarithmic function look like? In general, the graph of a relation and the graph of its inverse relation are symmetric with respect to the line $y = x$. Two points, A and B, are said to be **symmetric with respect to a line** l if and only if l is the perpendicular bisector of the segment, $\overline{AB}$. In particular, it can be shown that the points (s, r) and (r, s) are symmetric with respect to the line $y = x$ [Exercise 9].

Since (s, r) is a member of the inverse of a relation if and only if (r, s) is a member of the relation, every point on the graph of the inverse relation is symmetric to some point on the graph of the relation with respect to the line $y = x$; that is, the graph of a relation and its inverse are symmetric with respect to the line $y = x$. Using this relationship, we can reflect the graph of the exponential function with base 10 [Figure 2] about the line $y = x$ to obtain the graph of the logarithmic function with base 10 [Figure 3].

Sketch the graph of $\{(x, y): y = 3^x\}$ and its inverse function $\{(y, x): y = 3^x\} = \{(y, x): x = \log_3 y\}$.

The graph $y = 3^x$ is a curve shaped like Figure 2 and containing the points $\left(-1, \frac{1}{3}\right)$, $(0, 1)$, and $(1, 3)$; The graph $x = \log_3 y$ is a curve shaped like Figure 3 and containing the points $\left(\frac{1}{3}, -1\right)$, $(1, 0)$, and $(3, 1)$.

In Figures 2 and 3 we adhere to the convention that the horizontal axis is the axis of the domain and the vertical axis is the axis of the range. Thus, in Figure 2 the horizontal axis is the x-axis and the vertical axis is the y-axis. However, x and y change roles for Figure 3; that is, the horizontal axis is the y-axis and the vertical axis is the x-axis. Therefore, this exponential function and its inverse cannot be pictured on the same coordinate system and at the same time maintain our convention. This dilemma can be resolved by interchanging the variables x and y:

$$\{(y, x): x = \log_{10} y\} = \{(x, y): y = \log_{10} x\}.$$

Now the graphs in Figures 2 and 3 can be pictured on the same coordinate system [Figure 4].

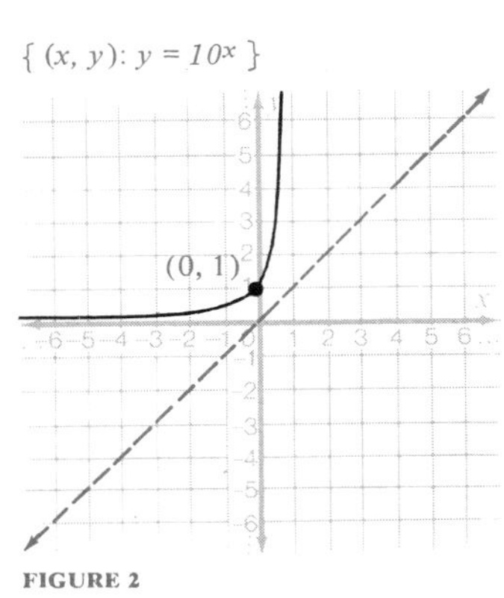

FIGURE 2

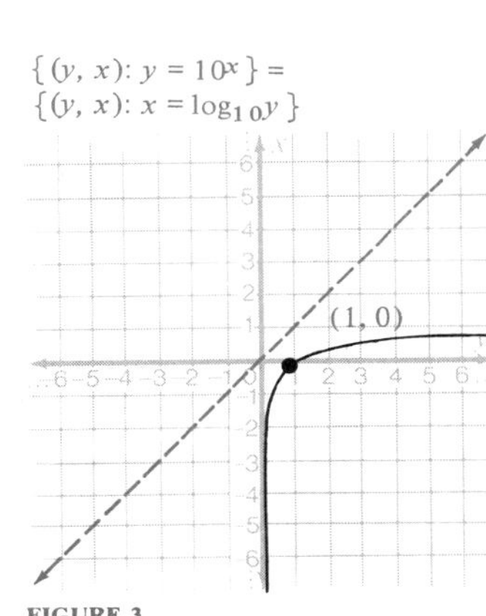

FIGURE 3

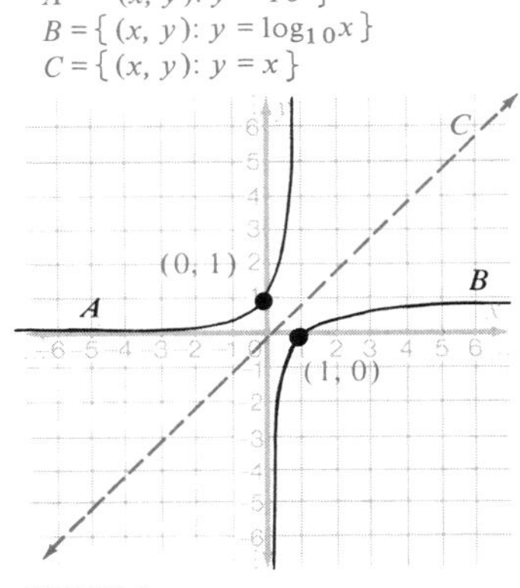

FIGURE 4

Exercises

1. Write an equivalent equation.

   a. $\log_2 8 = 3$ b. $4^3 = 64$ c. $25^{1/2} = 5$

   d. $\log_5 25 = 2$ e. $\log_4 \left(\frac{1}{16}\right) = -2$ f. $\log_{0.1} 100 = -2$

   g. $\log_2 2 = 1$ h. $(0.03)^2 = 0.0009$ i. $a^b = c$

   j. $\log_t m = z$ k. $(-2)^3 = -8$ l. $t^u = v$

2. Find b, x, or y.

   a. $\log_b 8 = 3$ b. $\log_8 2 = y$ c. $\log_{10} x = 2$

   d. $\log_b 16 = -4$ e. $\log_{27} x = -\frac{1}{3}$ f. $\log_\pi 1 = y$

   g. $\log_b 0.0049 = 2$ h. $\log_{0.5} x = 4$ i. $\log_b 0.25 = 2$

   j. $\log_b 10 = 0.1$ k. $\log_{1/3} x = -4$ l. $\log_{\sqrt{2}} \sqrt{2} = y$

3. Draw the graphs of

   $$f = \{(x, y): y = x^2, x \geqslant 0\}$$

   and

   $$g = \{(x, y): y^2 = x, y \geqslant 0\}$$

   on the same set of axes. Describe the relative positions of the graphs for f and g with respect to the line $y = x$. How are the functions f and g related?

4. Sketch the graphs of the functions defined by the equations $y = 6^x$ and $y = \log_6 x$ on the same set of axes.

5. Sketch the graph of the function defined by each equation.

   a. $y = \log_{10} x$ b. $y = \log_2 x$ c. $y = \log_5 x$

   d. $y = \log_3 x$ e. $y = \log_7 x$ f. $y = \log_4 x$

6. For each pair, determine whether the first number is less than, equal to, or greater than the second number.

   a. $\log_2 16$, $\log_3 9$ b. $\log_2 64$, $\log_2 4$ c. $\log_5 125$, $\log_4 65$

   d. $\log_6 36$, $\log_4 16$ e. $\log_7 1$, $\log_7 7$ f. $\log_2 17$, $\log_8 272$

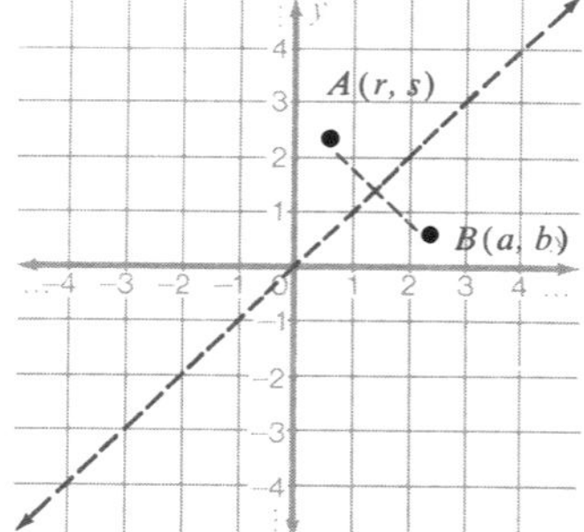

FIGURE 5

7. Which of the statements below are true statements?

   a. $\log_2 8 > \log_3 9$ b. $\log_3 27 < \log_4 256$

   c. $\log_5 25 > \log_7 49$ d. $\log_6 1 < \log_5 13$

   e. $\log_9 81 > \log_{13} 1$ f. $\log_7 7 < \log_8 2$

   g. $\log_{12} 10 > \log_9 1$ h. $\log_2 9 < \log_3 12$

   i. $\log_8 8 > \log_{11} 9$ j. $\log_4 12 < \log_2 7$

8. Prove that $\log_b 1 = 0$ and $\log_b b = 1$, for any base b such that $b > 0$ and $b \neq 1$.

9. Let $A(r, s)$ be an arbitrary point in the plane, and let $B(a, b)$ be the point in the plane symmetric to $A(r, s)$ with respect to the line $y = x$ [Figure 5]. Show that $a = s$ and $b = r$ by considering the equations for the slope of the line, $\overleftrightarrow{AB}$, and the midpoint of the segment, $\overline{AB}$.

9-3 PROPERTIES OF LOGARITHMS

Logarithms were invented to facilitate computation; however, since the advent of computers, the importance of logarithms as a computational aid has been greatly reduced. The theorems which follow provide a method for multiplying and dividing real numbers and for raising real numbers to a power or extracting roots of real numbers by using the logarithms of products, quotients, and powers. We have repeatedly said that logarithms are exponents; therefore, we use properties of exponents in proving these theorems.

Theorem 1 For any positive real numbers u, v, and b, $\log_b uv = \log_b u + \log_b v$.

Proof: If $z = \log_b uv$, $x = \log_b u$, and $y = \log_b v$, then $z = \log_b uv$ is equivalent to $b^z = uv$, $x = \log_b u$ is equivalent to $b^x = u$, and $y = \log_b v$ is equivalent to $b^y = v$ by the definition of logarithmic functions. Therefore

$$b^z = uv = b^x b^y = b^{x+y}$$

from which we see that $z = x + y$ (Why?) or

$$\log_b uv = \log_b u + \log_b v.$$

Example 1 If $\log_{10} 6 \approx 0.7782$, use Theorem 1 to approximate $\log_{10} 60$ and $\log_{10} 600$.

Solution:

$$\log_{10} 60 = \log_{10} (6 \cdot 10) = \log_{10} 6 + \log_{10} 10$$

$$\approx 0.7782 + 1 = 1.7782.$$

$$\log_{10} 600 = \log_{10} (6 \cdot 10 \cdot 10) = \log_{10} 6 + \log_{10} 10 + \log_{10} 10$$

$$\approx 0.7782 + 1 + 1 = 2.7782.$$

If $\log_{10} 7 \approx 0.8451$, approximate $\log_{10} 70$, $\log_{10} 7000$, and $\log_{10} 7{,}000{,}000$.

$\log_{10} 70 \approx 1.8451$
$\log_{10} 7000 \approx 3.8451$
$\log_{10} 7{,}000{,}000 \approx 6.8451$

The proof of a second property of logarithms uses the property $\dfrac{b^x}{b^y} = b^{x-y}$ of exponents.

Theorem 2 For any positive real numbers u, v, and b, $\log_b \left(\dfrac{u}{v}\right) = \log_b u - \log_b v$.

The proof of Theorem 2 is left as Exercise 7, page 172.

Example 2 If $\log_{10} 6 \approx 0.7782$, use Theorem 2 to approximate $\log_{10} 0.6$ and $\log_{10} 0.06$.

Solution:

$$\log_{10} 0.6 = \log_{10} \left(\frac{6}{10}\right) = \log_{10} 6 - \log_{10} 10$$

$$\approx 0.7782 - 1 = -0.2218.$$

$$\log_{10} 0.06 = \log_{10}\left(\frac{6}{100}\right) = \log_{10}\left(\frac{6}{10 \cdot 10}\right)$$

$$= \log_{10} 6 - \log_{10} 10 - \log_{10} 10$$

$$\approx 0.7782 - 1 - 1 = -1.2218.$$

If $\log_{10} 7 \approx 0.8451$, approximate $\log_{10} 0.7$, $\log_{10} 0.007$, and $\log_{10} 0.000007$.

$\log_{10} 0.7 \approx -0.1549$
$\log_{10} 0.007 \approx -2.1549$
$\log_{10} 0.000007 \approx -5.1549$

The next property is useful in evaluating the logarithm of powers and roots of a number.

Theorem 3 For any positive real numbers u and b and for any real number v, $\log_b u^v = v \log_b u$.

Proof: If $z = \log_b u^v$ and $w = \log_b u$, then

$$z = \log_b u^v \quad \text{is equivalent to} \quad b^z = u^v$$

$$w = \log_b u \quad \text{is equivalent to} \quad b^w = u.$$

Substituting b^w for u in $b^z = u^v$, we have

$$b^z = (b^w)^v = b^{vw} \quad \text{(Why?)}$$

and $\quad z = vw \quad$ (Why?)

or $\quad \log_b u^v = v \log_b u.$

Example 3 If $\log_{10} 6 \approx 0.7782$, use Theorem 3 to approximate $\log_{10} 1296$.

Solution:

$$\log_{10} 1296 = \log_{10} 6^4 = 4 \log_{10} 6$$

$$\approx (4)(0.7782) = 3.1128.$$

If $\log_{10} 7 \approx 0.8451$, approximate $\log_{10} 49$, $\log_{10} 2401$, and $\log_{10} \sqrt{7}$.

$\log_{10} 49 \approx 1.6902$
$\log_{10} 2401 \approx 3.3804$
$\log_{10} \sqrt{7} \approx 0.4226$

Example 4 Find $\log_3 \dfrac{(27)(81^{1/2})}{243}$.

Solution:

$$\log_3 \frac{(27)(81^{1/2})}{243} = \log_3 [(27)(81^{1/2})] - \log_3 243 \quad \text{(Theorem 2)}$$

$$= \log_3 27 + \log_3 81^{1/2} - \log_3 243 \quad \text{(Theorem 1)}$$

$$= \log_3 27 + \frac{1}{2}\log_3 81 - \log_3 243 \quad \text{(Theorem 3)}$$

$$= \log_3 3^3 + \frac{1}{2}\log_3 3^4 - \log_3 3^5$$

$$= 3 + \frac{1}{2}(4) - 5 = 0$$

Since $\log_3 \dfrac{(27)(81^{1/2})}{243} = 0$ in Example 4, what real number is the quotient $\dfrac{(27)(81^{1/2})}{243}$?

If $\log_3 x = 0$, then $3^0 = x$ and $1 = x$.

Exercises

1. What are the domain and range of a logarithmic function?
2. What numbers can be the base of a logarithmic function?
3. Use Theorems 1, 2, and 3 to rewrite each expression as a sum.
   a. $\log_b \dfrac{x^2}{y}$ b. $\log_b 7x$ c. $\log_b \dfrac{x^2y}{4z^{1/2}}$
   d. $\log_b x^2y$ e. $\log_b \dfrac{x^2y}{z^{1/2}}$ f. $\log_b \left(\dfrac{x}{y}\right)^{1/2}$
   g. $\log_b \dfrac{2}{xyz}$ h. $\log_b \dfrac{xy}{8z^2}$ i. $\log_b \dfrac{z^{1/2}x}{2}$
   j. $\log_b \dfrac{xy}{y^2z}$ k. $\log_b \dfrac{\sqrt{xy^3}}{x^2z^{3/2}}$ l. $\log_b \dfrac{(x^2y)^{1/3}}{z^{1/2}}$
4. Rewrite each of the following as the logarithm of a single term.
   a. $\log_2 n + \log_2 m - \log_2 7$
   b. $2 \log_4 x + 3 \log_4 y - \dfrac{1}{2}(\log_4 9 + \log_4 z)$
   c. $2 \log_a x - \log_a y$
   d. $\log_3 2 - \log_3 x - \log_3 y - \log_3 z$
   e. $\dfrac{1}{2}\log_6 a - \dfrac{1}{2}\log_6 b$
   f. $\dfrac{1}{2}\log_b z + \log_b x - \log_b 2$
   g. $\log_7 x + \log_7 y - 2 \log_7 y - \log_7 z$
   h. $\dfrac{1}{2}\log_3 x + 2 \log_3 y - \dfrac{1}{2}\log_3 z$
   i. $\dfrac{1}{2}(\log_5 x - \log_5 y) - 2 \log_5 4z$
   j. $-\log_2 x + 2 \log_2 5$

k. $\frac{1}{2}\log_b x + 2\log_b y - \log_b z$

l. $\frac{1}{2}\log_b x + \frac{1}{2}\log_b y$

m. $-\left(\frac{1}{2}\log_b x - \log_b z\right) - \log_b y$

5. Approximate n if $\log_{10} 2 \approx 0.3010$, $\log_{10} 4 \approx 0.6021$, $\log_{10} 5 \approx 0.6990$, and $\log_{10} 9 \approx 0.9542$.
   a. $\log_{10} 8 = n$ b. $\log_{10} 45 = n$ c. $\log_{10} 40 = n$
   d. $\log_{10} 25 = n$ e. $\log_{10} 81 = n$ f. $\log_{10} 3 = n$
   g. $\log_{10} 0.4 = n$ h. $\log_{10} 4.5 = n$ i. $\log_{10} 6 = n$
   j. $\log_{10} 400 = n$

6. Find the solution set of each equation.
   a. $\log_4 x - \log_4 3 = 4$ b. $\frac{1}{2}\log_3 x = 1$
   c. $\log_3 x + \log_3 6 = 3$ d. $\frac{1}{2}\log_8 x + \frac{1}{2}\log_8 x^3 = 2$
   e. $\frac{1}{2}\log_3 (x - 2) - \log_3 4 = 2$ f. $2\log_5 x - (\log_5 x + \log_5 4) = 3$
   g. $2\log_6 x - \frac{1}{2}\log_6 9 + 5 = 7$ h. $4\log_2 x \div \log_2 16 = 4$
   i. $2\log_6 x - \frac{1}{3}\log_6 27 = 2$ j. $\frac{3}{2}\log_7 x = 3$
   k. $3\log_4 x - \frac{1}{2}\log_4 36 + \frac{1}{3}\log_4 8 = 3$
   l. $\frac{1}{2}\log_2 (x^2 + x) - \frac{1}{2}\log_2 x = 2$

7. Prove Theorem 2, page 169.

8. Express each real number as the sum of a negative integer and a positive number between zero and one.
   a. -5.6789 b. -2.7753 c. -8.7772 d. -4.3215

9-4 COMMON LOGARITHMS

The logarithm of a number to the base 10 is called the **common logarithm** of the number. Henceforth, the common logarithm of a number y, $\log_{10} y$, will be denoted by $\log y$.

Table I (pages 372–373) is used to find the common logarithm of a number. The decimal points were omitted when Table I was printed, each numeral in the body of the table names a number between 0 and 1, and each numeral in the N-column (leftmost column) names a number between 1 and 10. Moreover, the numerals in the body of the table name the common logarithms of the numbers named by the numerals in the N-column. From this discussion it would seem that Table I can only be used to find the common logarithms of numbers between 1 and 10. To overcome the apparent limitations of Table I we introduce scientific notation.

Definition A positive real number is said to be expressed in **scientific notation** if and only if it is expressed as the product of a real number r such that $1 \leqslant r < 10$ and an integral power of 10.

Some examples of scientific notation are:

$$0.0376 = (3.76) \cdot 10^{-2}$$

$$0.376 = (3.76) \cdot 10^{-1}$$

$$3.76 = (3.76) \cdot 10^{0}$$

$$37.6 = (3.76) \cdot 10^{1}$$

$$376 = (3.76) \cdot 10^{2}.$$

Express each of the following in scientific notation: 9738, 0.0053, 92, 0.0000357.

$9738 = (9.738) \cdot 10^3$
$0.0053 = (5.3) \cdot 10^{-3}$
$92 = (9.2) \cdot 10$
$0.0000357 = (3.57) \cdot 10^{-5}$

Suppose $y = r \cdot 10^n$ when expressed in scientific notation. Then

$$\begin{aligned}\log y &= \log (r \cdot 10^n) && \text{(Why?)}\\ &= \log r + \log 10^n && \text{(Why?)}\\ &= \log r + n \log 10 && \text{(Why?)}\\ &= \log r + n. && \text{(Why?)}\end{aligned}$$

Since $1 \leqslant r < 10$, $\log r$ can be found from Table I. Thus, $\log y$ is the sum of a nonnegative number less than 1 and an integer. A logarithm is said to be in **standard form** when it is expressed in this form. The nonnegative number less than 1 is called the **mantissa** of the logarithm and the integer is called its **characteristic.**

Example 1 Express log 256 in standard form. What are the mantissa and characteristic of log 256?

Solution: The number is first expressed in scientific notation:

$$256 = (2.56) \cdot 10^2.$$

Therefore

$$\log 256 = \log (2.56) \cdot 10^2$$
$$= \log 2.56 + \log 10^2$$
$$= \log 2.56 + 2 \log 10$$
$$= \log 2.56 + 2.$$

Using Table I, we first find 25 in the column headed N. We then move to the right in the row containing 25 to the column headed 6. The table entry in the 25-row and 6-column is 4082. All the numerals in the body of Table I name numbers between 0 and 1; hence, $\log 2.56 \approx 0.4082$. Thus

$$\log 256 \approx 0.4082 + 2.$$

The mantissa and characteristic of log 256 are 0.4082 and 2. The symbol "$\approx$" is used to remind us that table values are usually approximations.

Example 2 Express log 0.000627 in standard form. What are the mantissa and characteristic of log 0.000627?

Solution: Expressed in scientific notation

$$0.000627 = (6.27) \cdot 10^{-4}.$$

Thus

$$\log 0.000627 = \log [(6.27) \cdot 10^{-4}] = \log 6.27 + \log 10^{-4}$$
$$\approx 0.7973 + (-4).$$

The mantissa and characteristic of log 0.000627 are 0.7973 and -4 respectively.

In Example 2 the characteristic is a negative integer. In fact, the characteristic of the logarithm of a positive number less than one is always negative. Since $0.7973 + (-4) = -3.2027$, we might write $\log 0.000627 \approx -3.2027$ also; however, neither the characteristic nor the mantissa are displayed and a transformation to the standard form is necessary to retrieve them. In a practical situation it is a good idea to keep the standard form until it is clear that the sum of the characteristic and mantissa is desired.

Example 3 Find the standard form of the logarithm, -2.8777.

Solution:

$$-2.8777 = 3 + (-2.8777) + (-3)$$
$$= 0.1223 + (-3)$$

The characteristic and mantissa of -2.8777 are -3 and 0.1223, respectively.

$-3.1500 = 0.8500 + (-4)$
$-5.4989 = 0.5011 + (-6)$
$-4.7352 = 0.2648 + (-5)$

The numbers -3.1500, -5.4989, and -4.7352 are common logarithms of three positive real numbers. Find the standard form of each logarithm.

The mantissa of a logarithm of a number depends only on the digits in the numeral for the number. It is independent of the position of the decimal point as indicated in the table below. The characteristic of the logarithm of a number depends on the position of the decimal point. With a little practice you should be able to determine the characteristic of the common logarithm of a number without pencil and paper calculations. The characteristic of the common logarithm of a number is the exponent of the power of ten when the number is expressed in scientific notation.

y	*Characteristic of* $\log y$	*Mantissa of* $\log y$
2.47	0	0.3927
24.7	1	0.3927
247	2	0.3927
2470	3	0.3927
0.247	−1	0.3927
0.0247	−2	0.3927
0.00247	−3	0.3927

Table I also may be used to find a number y given its common logarithm, x. If $x = \log y$, the number y is called the **antilogarithm** of x and is denoted by $y = \text{antilog}_{10}\, x$ or $y = \text{antilog}\, x$. Thus, $y = \text{antilog}\, x$ is equivalent to $x = \log y$.

Example 4 Find antilog 0.4713.

Solution: Since $y = \text{antilog}\ 0.4713$ is equivalent to $\log y = 0.4713$, $\log y$ has characteristic zero and mantissa 0.4713. Find 4713 in the body of Table I. The corresponding numeral in the N-column is 29; the numeral heading the column containing 4713 is 6. Thus the numeral naming antilog 0.4713 has the digits 2, 9, and 6, in that order. Since $\log y$ has characteristic zero, antilog $0.4713 \approx 2.96$.

Example 5 Evaluate antilog 3.6010.

Solution: Since $y = \text{antilog}\ 3.6010$ is equivalent to $\log y = 3.6010$, $\log y$ has characteristic 3 and mantissa 0.6010. Find 6010 in the body of Table I. The corresponding numeral in the N-column is 39 and the numeral heading the column containing 6010 is 9. Since $\log y$ has characteristic 3, antilog $3.6010 \approx 3990$.

Example 6 Evaluate antilog (−2.4789).

Solution: The equation $y = \text{antilog}\ (-2.4789)$ is equivalent to $\log y = -2.4789$. In standard form, $\log y = 0.5211 + (-3)$. Thus, $\log y$ has characteristic −3 and mantissa 0.5211. Find 5211 in the body of Table I, and locate the numerals 3, 3, and 2 naming antilog (−2.4789). Then, since the characteristic of log y is −3, antilog (−2.4789) ≈ 0.00332.

What is the antilogarithm of 0.4409? Of 0.4713? Find antilog 3.5717, antilog 5.7604, antilog (−1.1599), and antilog (−0.8097).

x	antilog x
0.4409	2.76
0.4713	2.96
3.5717	3730
5.7604	576,000
−1.1599	0.0692
−0.8097	0.155

Exercises

1. Express each number in scientific notation.
   a. 34.75 b. 8.953 c. 0.00321
   d. 21 e. 0.0000576 f. 93,000,000
   g. 186,000 h. 0.00000673 i. 1968
   j. 725.623 k. 32.789 l. 0.05992
   m. 397,000 n. 0.0009422 o. 7946.8

2. What is the characteristic of the common logarithm of each number?
   a. 3.75 b. 0.000247 c. 235
   d. 62.1 e. 8 f. 0.062
   g. 98.2 h. 1.35 i. 0.609
   j. 0.0000025 k. $(9.67) \cdot 10^8$ l. 0.00000327
   m. 2000 n. 0.0973 o. $(3.79) \cdot 10^{-3}$
   p. 96,200,000

3. What is the mantissa of the common logarithm of each number in Exercise 2?

4. Find the common logarithm of each number. Leave your answer in standard form.

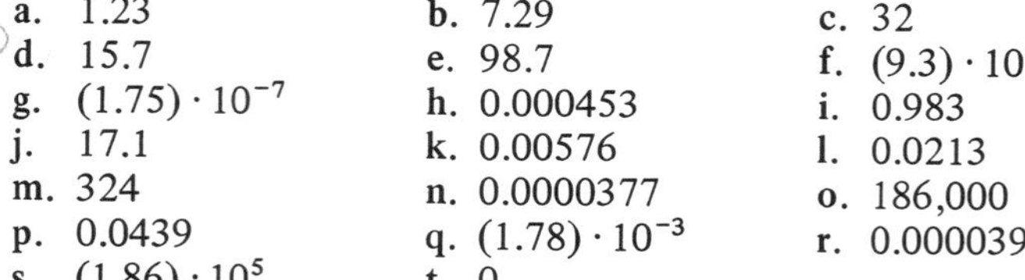

   a. 1.23 b. 7.29 c. 32
   d. 15.7 e. 98.7 f. $(9.3) \cdot 10^6$
   g. $(1.75) \cdot 10^{-7}$ h. 0.000453 i. 0.983
   j. 17.1 k. 0.00576 l. 0.0213
   m. 324 n. 0.0000377 o. 186,000
   p. 0.0439 q. $(1.78) \cdot 10^{-3}$ r. 0.000039
   s. $(1.86) \cdot 10^5$ t. 0

5. Interpret each number as the common logarithm of some number. Write the standard form of each logarithm.
   a. 2.7536 b. 5.8998 c. 3.5877
   d. −2.6271 e. −1.2175 f. −3.5331
   g. −6.8962 h. −2.2899 i. −5.7471
   j. −4.2007 k. −10.8633 l. −7.1918

6. Find the antilogarithm of each number.
   a. 0.8609 b. 0.5378 c. 1.1004
   d. 2.9420 e. 1.7332 f. 3.7634
   g. −1.0443 h. −2.4881 i. −3.6778
   j. −2.1226 k. 2.3979 l. 1.6848
   m. −2.2233 n. −5.1524 o. −1.4123

9-5 LINEAR INTERPOLATION

Table I also can be used to approximate the mantissas of the common logarithms of real numbers which require more than three digits to name the factor r $(1 \leqslant r < 10)$ in their scientific notation representations. The procedure for doing this is justified geometrically by the following observation.

A segment joining two points on the graph of a function may be thought to approximate the part of the curve that is between these points. For "nice" curves, the closer the points are together, the better the segment joining them approximates the part of the curve between them [Figure 1]. The graphs of the logarithmic functions are "nice" in this sense. Moreover, the entries in Table I name points on the graph of $\{(x, y)\colon\ y = \log x\}$ sufficiently close together to determine segments which closely approximate the curve between them. Approximating a small part of a logarithmic curve by a segment provides the rationale for using Table I to approximate the mantissa of the common logarithm of a number named by more than three digits. This procedure is called **linear interpolation.**

$\{(x, y)\colon y = \log_b x\}$

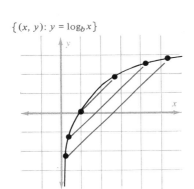

FIGURE 1

We illustrate the process of linear interpolation by finding log 897.6. The characteristic of log 897.6 is 2 and linear interpolation is used to approximate the mantissa.

We first observe that 897.6 is between Table I entries 897 and 898. We assume log 897.6 is between log 897 and log 898. Part of the graph of $\{(x, y)\colon\ y = \log x\}$ is pictured in Figure 2 and the points on the graph are of the form $(x, \log x)$. Thus, $P\,(897, 2.9528)$ and $Q\,(898, 2.9533)$ are on the graph, since from Table I $\log 897 \approx 2.9528$ and $\log 898 \approx 2.9533$. Point T $(897.6, \log 897.6)$ is also on the graph although a value for the second coordinate of T cannot be read directly from Table I. The best we can do is to use the second coordinate of $R\,(897.6, y)$ as an approximation for log 897.6 where R is the intersection of segment $\overline{PQ}$ and the line $x = 897.6$. The approximation error is the length of segment $\overline{TR}$.

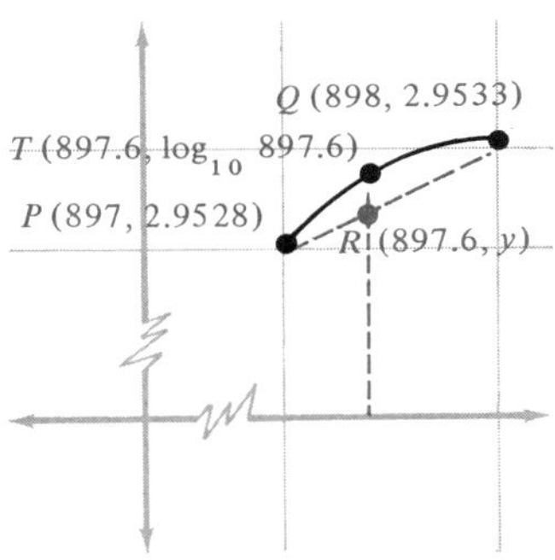

FIGURE 2

The slope of line $\overleftrightarrow{PQ}$ can be calculated using points P and R, points R and Q, or points P and Q. Using points P and R and points P and Q:

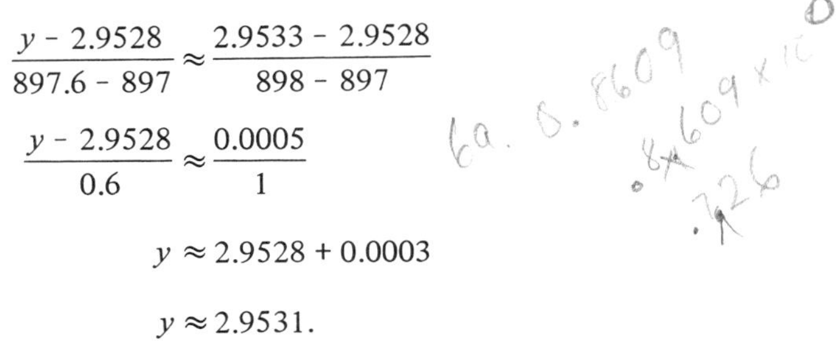

$$\frac{y - 2.9528}{897.6 - 897} \approx \frac{2.9533 - 2.9528}{898 - 897}$$

$$\frac{y - 2.9528}{0.6} \approx \frac{0.0005}{1}$$

$$y \approx 2.9528 + 0.0003$$

$$y \approx 2.9531.$$

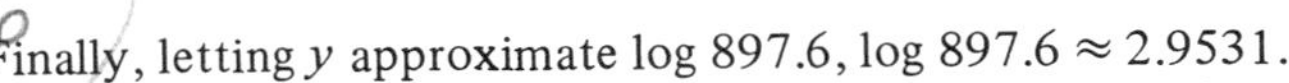

Finally, letting y approximate log 897.6, $\log 897.6 \approx 2.9531$.

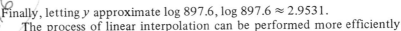

The process of linear interpolation can be performed more efficiently using the format below:

		x	$\log x$		
1	0.6	897	2.9528	d	0.0005
		897.6	y		
		898	2.9533		

We see that the appropriate slopes are

$$\frac{d}{0.6} \approx \frac{0.0005}{1}$$

where $d = y - 2.9528$. Thus $y \approx 2.9528 + d \approx 2.9528 + 0.0003 = 2.9531$ and $\log 897.6 \approx 2.9531$.

Example 1 Approximate log 82.15.

Solution: The nearest numbers whose logarithms can be found using Table I and between which is 82.15 are 82.1 and 82.2. Also, $\log 82.1 \approx 1.9143$ and $\log 82.2 \approx 1.9149$:

		x	$\log x$		
0.1	0.05	82.1	1.9143	d	0.0006
		82.15	y		
		82.2	1.9149		

Thus

$$\frac{d}{0.05} \approx \frac{0.0006}{0.1}$$

and $d \approx \dfrac{(0.05)(0.0006)}{0.1} = 0.0003$ where

$$d \approx y - 1.9143.$$

Solving for y, we have

$$y \approx 1.9143 + d \approx 1.9143 + 0.0003 = 1.9146$$

and

$$\log 82.15 \approx 1.9146.$$

$\log 321.2 \approx 2.5068$
$\log 9.414 \approx 0.9738$
$\log 0.05737 \approx 0.7587 + (-2)$

Use linear interpolation to approximate the common logarithm of each of the following: 321.2, 9.414, and 0.05737.

Example 2 Approximate a if $\log a \approx 2.6505$. This is equivalent to finding the antilogarithm of 2.6505.

Solution: The digits naming a are determined by 0.6505, the mantissa of $\log a$. In the body of Table I we see that 0.6505 is between 0.6503 ($\log 4.47$) and 0.6513 ($\log 4.48$):

	x	$\log x$	
0.01 { b [	4.47	0.6503] 0.0002	} 0.0010
	x	0.6505	
	4.48	0.6513	

Using slopes as in the earlier examples, we have

$$\frac{0.0002}{b} \approx \frac{0.0010}{0.01}$$

$$\frac{0.0002}{b} \approx 0.1$$

$$b \approx \frac{0.0002}{0.1} = 0.002$$

where $b \approx x - 4.47$. Thus, $x \approx 4.47 + b \approx 4.47 + 0.002 = 4.472$. Since the characteristic of $\log a$ is 2, it follows that $a \approx (x)(10^2) \approx 447.2$.

$\log x$	x
0.6620 − 3	0.004592
2.4619	289.7
0.1124 + 1	12.95
−1.7269 or 0.2731 − 2	0.01875

Approximate x if $\log x \approx 0.6620 - 3$; if $\log x \approx 2.4619$; if $\log x \approx 0.1124 + 1$; if $\log x \approx -1.7269$.

Exercises

1. Use linear interpolation to approximate the common logarithm of each number.

   a. 19.67 b. 0.01932 c. 1.798
   d. 197.2 e. 0.003257 f. 5763
   g. 7369 h. 98.83 i. 1,398,000
   j. 698.1 k. 0.001030 l. 9.007
   m. 27.34 n. 178.3 o. 0.0002006
   p. 32.71 q. 197.8 r. 0.05709
   s. 0.001403 t. 83265

2. Use linear interpolation to approximate the antilogarithm of each number.

   a. 0.5979 b. 1.1109 c. 2.6887

d. -1.3274 e. -2.5432 f. 3.2974
g. 4.8430 h. -3.7392 i. -4.4578
j. -2.3928 k. 1.3024 l. -3.1342
m. 2.0172 n. -5.5732 o. -10.6359
p. 0.2823 q. 2.6753 r. -1.5783
s. -2.3642 t. -8.8531

9-6 COMPUTATION USING LOGARITHMS

Computing with logarithms is less laborious than finding complicated products, quotients, powers and roots by arithmetic algorithms because the properties of logarithms enable us to perform addition instead of multiplication (Theorem 1, Section 9-3), subtraction instead of division (Theorem 2, Section 9-3), and multiplication instead of raising to a power (Theorem 3, Section 9-3).

Example 1 Compute (3.45)(0.000893) using common logarithms.

Solution: If $x = (3.45)(0.000893)$, then

$$\log x = \log (3.45)(0.000893) = \log 3.45 + \log 0.000893.$$

$$\begin{aligned} \log 3.45 &\approx 0.5378 \\ + \log 0.000893 &\approx 0.9509 - 4 \\ \hline \log x &\approx 1.4887 - 4 \\ \log x &\approx 0.4887 - 3 \end{aligned}$$

Thus far, we have found the logarithm of the desired product. From the definition of antilogarithm, antilog $(\log x) = x$; hence

$$x \approx \text{antilog}\,(0.4887 - 3) \approx 0.003081.$$

Use logarithms to find the product of 23.5 and 1750.

$$\begin{aligned} \log 23.5 &\approx 1.3711 \\ \log 1750 &\approx 3.2430 \\ \hline \log (23.5)(1750) &\approx 4.6141 \\ (23.5)(1750) &\approx 41{,}300 \end{aligned}$$

Example 2 Use logarithms to compute $3.45 \div 0.00976$.

Solution: If $x = 3.45 \div 0.00976$, then

$$\log x = \log \left[\frac{3.45}{0.00976}\right] = \log 3.45 - \log 0.00976$$

$$\log 3.45 \approx 0.5378$$
$$(-)\ \log 0.00976 \approx 0.9894 - 3$$
$$\log x \approx -0.4516 + 3$$
$$\log x \approx 2.5484$$
$$x \approx \text{antilog } 2.5484$$
$$x \approx 353.5$$

Use logarithms to find the quotient of 8.80 and 247.8.

$x = (8.80) \div (247.8)$
$\log 8.80 \approx 0.9445$
$(-)\ \log 247.8 \approx 2.3941$
$\log x \approx 0.5504 - 2$
$x \approx 0.03551$

Example 3 Use logarithms to compute $(3750)^{2/5}$.

Solution: If $x = (3750)^{2/5}$, then

$$\log x = \log (3750)^{2/5} = \frac{2}{5} \log 3750$$
$$\approx \left(\frac{2}{5}\right)(3.5740)$$
$$\log x \approx 1.4296$$
$$x \approx 26.89.$$

Evaluate $57^{1/2}$, $89^{1/3}$, and $121^{1/2}$ using logarithms. Use a table of square roots and cube roots to check your answers.

$\frac{1}{2} \log 57 \approx 0.8779$
$57^{1/2} \approx 7.55$
$89^{1/5} \approx 4.465$, $121^{1/2} \approx 11.0$

When calculating using logarithms, fewer errors are committed if the steps of the processes are planned and performed systematically. The formats illustrated in Examples 1–3 can be extended to more complicated examples.

Example 4 Compute $\left[\dfrac{(0.057)(98.75)}{0.395}\right]^{1/3}$.

Solution: If x is equal to the desired power, then

$$\log x = \log \left[\frac{(0.057)(98.75)}{0.395}\right]^{1/3} = \frac{1}{3} \log \left[\frac{(0.057)(98.75)}{0.395}\right]$$
$$= \frac{1}{3} \left[(\log 0.057 + \log 98.75) - \log 0.395\right]$$
$$\log 0.057 \approx 0.7559 - 2$$
$$(+)\ \log 98.75 \approx 0.9945 + 1$$
$$\log 0.057 + \log 98.75 \approx 1.7504 - 1 = 0.7504$$

$$\begin{array}{rl}
\log 0.057 \ + \ \log 98.75 & \approx 0.7504 \\
(-)\log 0.395 & \approx 0.5966 - 1 \\
\hline
(\log 0.057 \ + \ \log 98.75) - \ \log 0.395 & \approx 0.1538 + 1 \\
\log x & \approx \left(\frac{1}{3}\right)(1.1538) \\
\log x & \approx 0.3846 \\
x & \approx 2.424.
\end{array}$$

Exercises

1. Use logarithms to find the following. Interpolate when necessary.
   a. $(989)(14.68)$
   b. $(0.00679)(3.78)$
   c. $(32.7)(0.062)^2$
   d. $(9385)^3(1.73)^4$
   e. $(16.89)^{1/4}(0.2689)^2$
   f. $(0.006834)^{0.4}$
   g. $(75{,}250)^{2/3}$
   h. $(64.75)^{1/2}(27.65)^{1/3}$
   i. $0.629 \div 13.75$
   j. $(0.575)^{0.5} \div (0.4732)^3$
   k. $\dfrac{(0.257)(98.7)}{36.72}$
   l. $(18.7)(4.789)^2$
   m. $\dfrac{(0.00347)(67{,}320)^3}{13{,}250}$
   n. $\dfrac{(13.75)(624)}{(27.5)(7.32)}$
   o. $\dfrac{(13.7)(14.6)(0.00653)}{(18.7)(321)^{-2}}$
   p. $\left[\dfrac{(149)(1.73)^2(1890)}{(1.414)(21.7)^{0.3}}\right]^{1/5}$

If P dollars are invested at an annual interest rate r and interest is compounded k times a year, the amount A that the investment is worth at the end of n years is given by the formula

$$A = P\left(1 + \frac{r}{k}\right)^{kn}.$$

If the interest is compounded continuously at an annual rate r for n years, the amount is

$$A = Pe^{nr}$$

where e is an irrational number. Use 2.718 as an approximation for e and the formulas above when working Exercises 2, 3, and 4.

2. If \$10,000 is invested at 8% interest and interest is compounded annually, find the value of the investment at the end of 7 years.
3. On his tenth birthday, Johnny deposited \$10 in a bank at 4% interest compounded quarterly. Today is Johnny's seventeenth birthday. Assuming that he made no further deposits or withdrawals, how much is in his account today?
4. Two banks, A and B, pay per annum interest rates of 5% and $8\frac{1}{8}$%, respectively. Bank A compounds interest continuously whereas bank B compounds interest quarterly. Which bank pays more interest at the end of a year?
5. An object dropped in a vacuum free falls s feet in t seconds according

to the relationship $s = \frac{1}{2}gt^2$, where $g = 32$ ft/sec^2. If a man skydives form an airplane and opens his parachute 15.673 seconds later, how far does he free fall if air resistance is neglected?

6. Assume that the maximum speed at which an automobile may round an unbanked curve is

$$v = (\mu gr)^{1/2}$$

where μ is the coefficient of friction, $g = 32$ ft/sec^2, and r is the radius of curvature. If the coefficient of friction, μ, is 0.525 and the radius of curvature is 252.6 feet, what is the maximum speed in miles per hour at which the automobile may round the curve?

7. Ignoring air resistance, an object thrown downward with a velocity (v_0) falls s feet in t seconds according to the formula

$$s = v_0 t + \frac{1}{2}gt^2$$

where $g = 32$ ft/sec^2. How far does an object thrown downward with a velocity of 88 ft/sec fall in 6.3257 seconds?

8. Newton's law of universal gravitation indicates that every particle of mass attracts every other particle of mass and that the force with which they attract each other is directly proportional to the product of their masses and inversely proportional to the square of the distance between them; that is

$$F = \frac{m_1 m_2}{r^2} G$$

where F denotes the force, m_1 and m_2 denote masses, r denotes distance, and G is the constant of proportionality. If the mass of the earth is $(4.1) \cdot 10^{23}$ slugs, the mass of the moon is approximately 0.0123 of the mass of the earth, and the constant $G = (3.44) \cdot 10^{-8}$ (lb/ft^2)/slug^2, what is the force of attraction between the earth and the moon? Use 250,000 miles as an approximation for r. [Note: 1 slug = 1 lb/(ft/sec^2).]

9. The velocity a satellite requires to hold it in an orbit with radius d (when measured from the center of the earth) is

$$v = \sqrt{\frac{Gm_e}{d}}$$

where m_e is the mass of the earth and $G = (3.44) \cdot 10^{-8}$ (lb ft^2)/slug^2 is a constant. (Note that the velocity required to hold a satellite in a prescribed orbit does not depend on the mass of the satellite.) If the mass of the earth is $(4.1) \cdot 10^{23}$ slugs and a space capsule is to be put into orbit 200 miles above the earth, what velocity (in mph) must it attain to stay in orbit? Assume that the length of the radius of the earth is 4000 miles.

10. According to the theory of relativity, the mass of a body depends on the speed at which it is moving. The relationship between the mass of the body at rest, m_0, and its mass m at speed v is

$$m = \frac{m_0}{\left(1 - \frac{v^2}{c^2}\right)^{1/2}}$$

where c is the speed of light, approximately $(3) \cdot 10^8$ meters/second. If the rest mass of a body is 0.00076 grams, what is its mass when its speed is $(2.345) \cdot 10^7$ meters/second?

9-7 LOGARITHMS TO BASES OTHER THAN 10

The common logarithm of a number can be approximated using Table I. In fact, Table I can be used to find the logarithm of a number to any desired base. For example, to find $\log_3 73.9$, we recall that $\log_3 73.9 = x$ if and only if $3^x = 73.9$. Therefore

$$\log 3^x = \log 73.9$$

$$x \log 3 = \log 73.9$$

$$x = \frac{\log 73.9}{\log 3}$$

and

$$\log_3 73.9 = \frac{\log 73.9}{\log 3} \approx \frac{1.8686}{0.4771}$$

where approximations for log 73.9 and log 3 are found in Table I.

Using the above example as a model, find the quotient that approximates $\log_4 739$.

$$x = \frac{\log 739}{\log 4} \approx \frac{2.8686}{0.6021}$$

The following theorem provides a formula for using Table I to compute the logarithm of a number to any base.

Theorem For any positive real numbers x and b, $\log_b x = \dfrac{\log x}{\log b}$.

Example What real number is $\log_5 6.91$?

Solution: Using the above theorem and Table I,

$$\log_5 6.91 = \frac{\log 6.91}{\log 5} \approx \frac{0.8395}{0.6990}.$$

At this point, we can calculate the desired quotient by the division algorithm or with logarithms. If

$$y = \frac{0.8395}{0.6990}, \quad \text{then}$$

$$\log y = \log\left(\frac{0.8395}{0.6990}\right)$$

$$\log y = \log 0.8395 - \log 0.6990$$

$$\log 0.8395 \approx 0.9240 - 1$$

$$(-)\ \log 0.6990 \approx 0.8445 - 1$$

$$\log y \approx 0.0795$$

$$y \approx \text{antilog}\,(0.0795)$$

$$y \approx 1.201.$$

Therefore, $\frac{0.8395}{0.6990} \approx 1.201$ and $\log_5 6.91 \approx 1.201$.

Exercises Use Table I to approximate each real number in Exercises 1–8.

1. $\log_3 4.57$
2. $\log_6 7.5$
3. $\log_4 21.3$
4. $\log_{1.4} 215$
5. $\log_4 1814$
6. $\log_{3/2} 33.76$
7. $\log_2 1.356$
8. $\log_3 359.2$

Use Table I to approximate each real number in Exercises 9–11.

9. $\log_\pi \sqrt{2}$
10. $\log_{2.4} (417)(0.229)$
11. $\log_3 (729.8)^3$

9-8 EXPONENTIAL EQUATIONS

Equations in which a variable appears in an exponent are called **exponential equations.** The equations

(1) $4^{3x+1} = 128$

(2) $0.5^x = 19$

are examples of exponential equations. In Equation (1), 4 and 128 are both powers of 2. Hence, (1) is equivalent to

$$(2^2)^{3x+1} = 2^7$$

$$2^{6x+2} = 2^7$$

$$6x + 2 = 7.$$

The solution set is $\left\{\frac{5}{6}\right\}$.

Solve the following exponential equations: $27^{x-1} = 81$; $-8^x = 16$; $25^{3x+2} = 625$.

$$27^{x-1} = 81 \qquad 25^{3x+2} = 625$$

$$3^{3(x-1)} = 3^4 \qquad 5^{2(3x+2)} = 5^4$$

$$3(x-1) = 4 \qquad 2(3x+2) = 4$$

$$x = \frac{7}{3} \qquad x = 0$$

Since -8^x is negative for all real numbers x, there is no real number for which $-8^x = 16$; that is, the solution set is ϕ.

In Equation (2) we cannot easily express both sides of the equation in the same base. Logarithms are useful for working with such exponential equations as is demonstrated below:

$$0.5^x = 19$$

$$\log 0.5^x = \log 19$$

$$x \log 0.5 = \log 19$$

$$x = \frac{\log 19}{\log 0.5} \approx \frac{1.2788}{0.6990 - 1}$$

$$x \approx \frac{1.2785}{-0.3010}.$$

Involved divisions, such as the one above, can be performed efficiently using logarithms. Because the domain of the logarithmic functions is the set of positive real numbers and x is negative, we use logarithms to find $-x$, a positive number. Then the quotient x is the negative of $-x$.

$$-x \approx \frac{1.2785}{0.3010}$$

$$\log(-x) \approx \log\left(\frac{1.2785}{0.3010}\right)$$

$$= \log 1.2785 - \log 0.3010$$

$$\log 1.2785 \approx 0.1067$$

$$\underline{(-) \log 0.3010 \approx 0.4786 - 1}$$

$$\log(-x) \approx 0.6281$$

$$-x \approx 4.247$$

$$x \approx -4.247.$$

The solution of $0.5^x = 19$ is approximately -4.247.

Equation	Approximate solution
$3^x = 5$	1.465
$4^{2x} = 12$	0.896
$0.6^x = 2$	-1.357
$0.3^{2x} = 7$	-0.808

Use logarithms to approximate the real solutions of each of the following exponential equations: $3^x = 5; 4^{2x} = 12; 0.6^x = 2; 0.3^{2x} = 7$.

Exercises

1. Find the solution set of each exponential equation.
   a. $5^{3x+2} = 125$ b. $36^{x+2} = 216$ c. $7^{-x} = \frac{1}{49}$
   d. $\left(\frac{2}{3}\right)^x = \frac{81}{16}$ e. $3^x - 2 = 6$ f. $5^{2x+2} - 9 = 45$
   g. $6^x = 25$ h. $13^x = 9.3$ i. $7^x = 7.5$
   j. $(0.4)^{2x} = 17$ k. $(0.7)^{x-1} = 0.6$ l. $3^x = 2^{x+1}$
   m. $4^{2x+1} = 7^x$ n. $(0.3)^x = 4^{2x}$ o. $64^{(1/2)x-2} = 16$
2. Find the solution set of each logarithmic equation by solving the equivalent exponential equation.
   a. $x = \log_2 3$ b. $y = \log_5 7$ c. $z = \log_3 8$
   d. $r = \log_7 13$ e. $t = \log_4 9$ f. $u = \log_6 2$
   g. $w = \log_9 13$ h. $x = \log_7 5$ i. $z = \log_2 100$
3. Find the solution set of each exponential equation.
   a. $5^{2x}(3^{x+1}) = 16$ b. $5^{\log x} = 7$ c. $5^{2^{x+1}} = 2$ d. $3^{4^{x+1}} = 11$
4. Find the solution set of each inequality.
   a. $2^x > \frac{7}{3}$ b. $5^x < \frac{5}{3}$ c. $(0.04)^{2x} > \left(\frac{1}{3}\right)^3$ d. $6^{x+1} < (0.5)^3$
5. Justify your method for solving the inequalities in Exercise 4.

9-9 APPLICATIONS

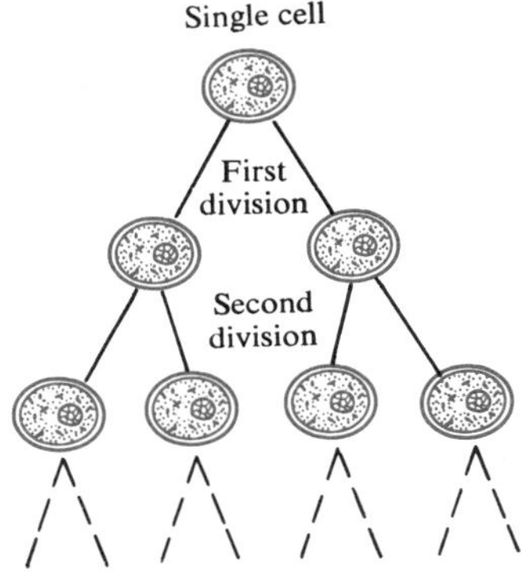

FIGURE 1

Growth and Decay: The growth in a living organism due to cell divisions, the growth of a bacteria culture, population increase, and the decomposition of radioactive materials are examples of processes during which the amount of change that takes place at a given time is proportional to the amount (number of cells, bacteria, people, or radioactive atoms) present at the time. In this section, we consider two of these problems, human growth and radioactive decay. Each of these problems involves exponential equations.

An adult human body is the result of successive divisions of cells. The process begins with a single cell. Two new cells are formed by the division of a single cell. Then each of the two new cells divides into two cells, and so on. Biologists call this method of cell division **mitosis**. The first two stages of cell division in the growth process are illustrated in Figure 1.

The body of a mature human contains approximately 10^{14} cells. Considering the large number of cells in the human body and the method of forming them from a single cell by successive divisions, a natural conclusion is that a large number of cell divisions are necessary to produce the mature body. How many cell divisions do you think are necessary? If we let n be the number of cell divisions necessary to grow from a single cell into an adult human, n is the solution of the equation

$$2^n = 10^{14}.$$

Find the number of cell divisions necessary to grow from a single cell into an adult human by solving this exponential equation.

Over a period of time, radioactive elements change into substances with different chemical and physical properties. Physicists assume that this phenomenon occurs because radioactive atoms are unstable and break. As time passes, the amount of radioactive substance present decreases according to the formula

$$(1) \qquad A = A_0 a^{ct}$$

where A is the amount of radioactive substance present at time t, A_0 is the amount of radioactive substance present when $t = 0$, a is any convenient base, and c is a constant associated with the substance being considered. Notice that A_0, a and c are constants; hence, the equation is an exponential equation.

Example The **half-life** of a radioactive substance is the length of time in which half a given amount will be gone. The radioactive material radium F has a half-life of 140 days. Starting with 20 milligrams, how much radium F is left after t days? After 280 days? In how many days will there be 4 milligrams left?

Solution: Since the half-life of radium F is 140 days, there will be only 10 of the 20 milligrams left after 140 days. That is, if $A_0 = 20$ milligrams, then $A = \frac{A_0}{2} = \frac{20}{2} = 10$ when $t = 140$. Substituting these values into equation (1), we obtain

$$10 = 20a^{140c}$$

$$\frac{1}{2} = a^{140c}.$$

Since a is any base, it seems convenient to let a be $\frac{1}{2}$. In fact, whenever the half-life of a substance is known, it is convenient to let a be $\frac{1}{2}$ because $a^{ct} = \frac{A}{A_0} = \frac{1}{2}$ when t equals the half-life of the substance. If $a = \frac{1}{2}$, we obtain the equation $\frac{1}{2} = \left(\frac{1}{2}\right)^{140c}$. Thus, $1 = 140c$ and $c = \frac{1}{140}$. After finding c, we can conclude that the amount of radium F left (out of 20 milligrams) at any time t (in days) is given by the formula

$$A = 20\left(\frac{1}{2}\right)^{t/140}.$$

To find out how much radium F is left after 280 days, we let t be 280. Then

$$A = 20\left(\frac{1}{2}\right)^{280/140} = 20\left(\frac{1}{2}\right)^{2} = 5.$$

The number of days until there are 4 milligrams of radium F left is found by solving the equation

$$4 = 20\left(\frac{1}{2}\right)^{t/140}$$

$$\frac{4}{20} = \left(\frac{1}{2}\right)^{t/140}$$

Using logarithms, we have

$$\log 0.2 = \log (0.5)^{t/140}$$

$$\log 0.2 = \left(\frac{t}{140}\right) \log 0.5$$

$$t = \frac{140 (\log 0.2)}{\log 0.5} \approx 325.07 \text{ (days)}.$$

The Slide Rule: Logarithms were invented by John Napier in 1614. Less than a decade later, logarithmic scales were constructed on lines and were used to facilitate computations. Over the centuries, the idea of using logarithmic scales lead to the modern slide rule.

Before considering a slide rule's design and operation, we diagram and briefly describe the parts of a slide rule—the stator, the slide, the indicator or cursor, and the hairline [Figure 2].

The **stator** consists of two bars rigidly held together by plates. A third bar, called the **slide**, fits tongue-in-groove between the two bars of the stator but fits loosely enough to move back and forth. There are scales on the slide and on each bar of the stator. The **indicator** (or **cursor**) has a **hairline** which is used to read the scales or to position the slide.

In order to understand the design of several scales on the slide rule, it is necessary to consider the properties of logarithms which provide the

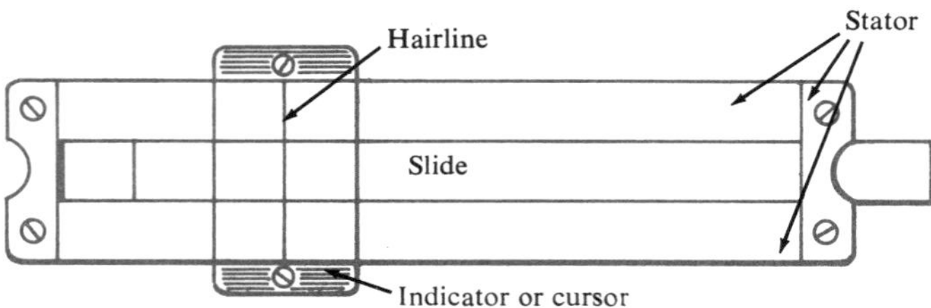

FIGURE 2

rationale for the design and operation of the slide rule. Recall that $\log_b uv = \log_b u + \log_b v$ for any positive real numbers u, v, and b.

This property suggests the design for the *C*- and *D*-scales of a slide rule. The *C*- and *D*-scales are identical, adjacent scales with the *C*-scale on the slide and *D*-scale on the stator as pictured. The points labeled 1 on the ends of each scale are called the **left index** and **right index**. Figure 3 shows

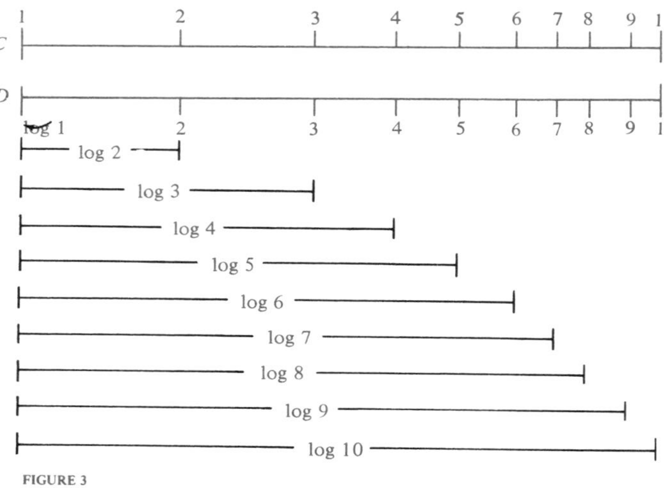

FIGURE 3

that the distance from the left index to the points labeled 1, 2, 3, and so on are log 1, log 2, log 3, · · · , log 10. Hence, the numeral 2 on each scale names the antilogarithm of the distance between the left index and the point 2 or antilog (log 2).

This design is ingenious as we can see by considering the problem:

$$\log(2 \cdot 3) = \log 2 + \log 3.$$

If we position the *C*-scale so that the left index is above the point 2 on the *D*-scale, we observe that the number of the point on the *D*-scale corresponding to 3 on the *C*-scale is the product $2 \cdot 3 = 6$ [Figure 4].

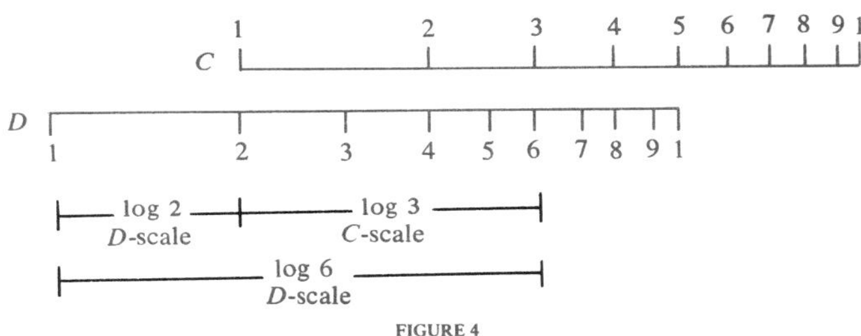

FIGURE 4

In the procedure described, we added the distance, log 2 (*D*-scale), to the distance, log 3 (*C*-scale), and located the point which is log 6 units from the left index of the *D*-scale. Labeling the points with the antilogarithms of the lengths of the segments enables us to read the product directly from the slide rule.

Exercises

1. If the process of cell division illustrated in Figure 1 is continued, how many cells are there after the fifth division? The seventh division? The nth division, where n is a positive integer?

2. Starting with a single cell, after how many cell divisions will there be at least 1,000,000 cells?

3. Suppose that each cell divides into five cells at each stage in the growth of an organism. How many cell divisions of this kind would be necessary to grow from a single cell into a mature human?

4. The half-life or uranium X is 24 days. Starting with 100 milligrams of uranium X, how much is left in t days? In 96 days? In how many days will there be 10 milligrams left?

5. Discuss why we can talk about the amount of radioactive substance remaining after a period of time which is more than twice the half-life of the substance.

6. The growth of a bacteria culture as time passes is also described by Equation (1). If we start with 100 bacteria and find that there are 1000 bacteria three hours later, what is the number of bacteria after t hours? After 6 hours? In how many hours will there be 10^6 bacteria?

7. Cut two strips of stiff paper one-inch wide and ten-inches long. Look up log 1, log 2, log 3, . . . , log 9, and log 10 in Table I. If the length of each strip corresponds to one unit, then each inch is one tenth of a unit. Locate the points on the strips corresponding to log 1 units from the left end of the strip, log 2 units from the left end of the strip, and so on. Put the scale on the bottom edge of one strip (*C*-scale) and on the top edge of the other (*D*-scale).

8. Find the procedure for using the *C*- and *D*-scales to divide on a slide rule. Does your procedure work for $9 \div 3$, $8 \div 2$, $5 \div 3$?

9. The *C*- and *D*-scales of a slide rule are similar to Table I since only the logarithms of numbers between one and ten are on the scales. Thus,

we must keep track of the characteristics of the logarithms of the numbers involved in a problem just as we do when using Table I. How would you use your C- and D-scales to multiply 200 by 300?

10. Design a scale (A-scale) such that the square root of a number on the D-scale is the corresponding number on the A-scale when the left indices of the A- and D-scales are aligned. How can the A- and D-scales be used to square a number?

Review Exercises

1. Sketch the graph of the relation $\left\{(x, y): y = \left(\frac{1}{2}\right)^x\right\}$. What integers (not necessarily consecutive) does $\left(\frac{1}{2}\right)^{-\pi}$ appear to be between? 9-1
2. Sketch the graph of the function $\{(x, y): y = (0.2)^x\}$.
3. Find the values of m and n for which $\log_n 16 = 4$ if and only if $2^m = 16$. 9-2
4. Find a real number x such that $\log_{0.5} x = 4$.
5. Write an equation equivalent to $x^y = z$ where x, y, and z are real numbers, $x > 0$, and $x \neq 1$.
6. Sketch the graph of the function $r = \{(x, y): y = \log_3 x\}$ and its inverse. What set of ordered pairs is r^{-1}?
7. Write $\frac{1}{2} \log_4 a + 2 \log_4 b - \frac{2}{3} \log_4 c$ as the logarithm of a single term. 9-3
8. Find the solution set of $\frac{1}{2} \log z - 2 = 0$.
9. Express 37,654.692 in scientific notation.
10. Find the antilogarithm of 3.9222. 9-4
11. Find log 0.000174.
12. Use linear interpolation to approximate log 0.4021.
13. Use linear interpolation to approximate antilog (−1.9503). 9-5
14. Use logarithms to find $\dfrac{8^{1/2}(18.7)}{32.45}$. 9-6
15. An object thrown downward with a velocity v_0 falls s feet in t seconds according to the formula $s = v_0 t + \frac{1}{2}gt^2$, if $g = 32$ ft/sec^2. Ignoring air resistance, how far does an object thrown downward with a velocity of 100 ft/sec fall in 3.457 seconds?
16. Use Table I to approximate $\log_3 0.00438$. 9-7

17. Find the solution set of the equation $(0.25)^{x+2} = 64$ without using logarithms. 9-8

18. Find the solution set of each equation.

   a. $4^x = 5^{x+1}$

   b. $\left(\frac{1}{9}\right)^{2x-1} = 27$

19. Find the solution set of the equation $v = \log_7 9$ by solving the equivalent exponential equation.

20. The half-life of a certain radioactive substance is 100 days. Starting with 10 milligrams, how much of the substance is left after 125 days? In how many days will there be 3 milligrams left? 9-9

10. MATRICES, DETERMINANTS, AND LINEAR PROGRAMMING

10-1 MATRICES

To solve a linear equation such as

$$5x = 4$$

where the set of real numbers R is the replacement set, we recall that the multiplicative inverse of 5 is $\frac{1}{5}$. The following equations are equivalent:

$$(5^{-1})(5x) = (5^{-1})(4)$$

$$(5^{-1} \cdot 5)x = (5^{-1})(4)$$

$$1 \cdot x = \frac{4}{5}$$

$$x = \frac{4}{5}.$$

In this section mathematical objects called matrices are introduced which allow systems of linear equations to be treated as one linear matrix equation. To change the system of linear equations in Section 3-5 to matrix form,

$$(1) \qquad \begin{cases} 2x + 3y = 11 \\ x - 2y = 5 \end{cases}$$

can be written as

$$(2) \qquad \begin{cases} 2x_1 + 3x_2 = 11 \\ x_1 - 2x_2 = 5 \end{cases}$$

where x and y are replaced by x_1 and x_2, respectively. Using capital letters to denote matrices, we will be able to express (2) as

$$(3) \qquad AX = B.$$

Then, if a "matrix" multiplicative inverse A^{-1} exists for A, this system can be solved by

$$\begin{aligned} A^{-1} \cdot (AX) &= A^{-1} \cdot B \\ (A^{-1} \cdot A)X &= A^{-1} \cdot B \\ I \cdot X &= A^{-1} \cdot B \\ X &= A^{-1} \cdot B \end{aligned}$$

Of course, we will need to understand matrix operations and have an identity matrix for multiplication. This technique can be applied to systems of three linear equations in three variables, such as

$$\begin{cases} x_1 - 3x_2 - 2x_3 = 9 \\ 3x_1 + 2x_2 + 6x_3 = 20 \\ 4x_1 - x_2 + 3x_3 = 25 \end{cases}$$

and, in general, to systems of n linear equations in n variables.

In order to represent systems of equations in matrix form we need to identify these objects and discuss the arithmetic of their operations of multiplication and addition.

Definition A **matrix** is a rectangular array of elements from a number system.

Some examples of matrices are

$$C = \begin{bmatrix} 5 & 2 \\ 6 & -1 \end{bmatrix}, \quad D = \begin{bmatrix} 2 & 4 & 3 \\ -1 & 0 & 8 \end{bmatrix}, \quad E = \begin{bmatrix} 1 \\ 4 \\ 3 \end{bmatrix},$$

$$F = \begin{bmatrix} 4 & -1 & -5 \\ 2 & 3 & 4 \end{bmatrix}.$$

The rows of a matrix are "numbered" from top to bottom, the columns from left to right. Matrix C has 2 rows and 2 columns; matrix E has 3 rows and 1 column.

Definition A matrix with only one row is called a **row vector**. A matrix with only one column is called a **column vector**.

Definition The **dimension** of a matrix with m rows and n columns is $m \times n$, read "m by n." A matrix of dimension $n \times n$ is said to have **order** n.

2×2, 2×3, 3×1, 2×3

What is the dimension of matrix C? D? E? F?

We can let a capital letter represent a matrix, or at times we will consider the entries in a matrix. An "address" or double subscript notation is used to identify an element. The first subscript gives the row position; the second, the column address. For example, if

$$A = \begin{bmatrix} a_{11} & a_{12} & a_{13} \\ a_{21} & a_{22} & a_{23} \\ a_{31} & a_{32} & a_{33} \\ a_{41} & a_{42} & a_{43} \end{bmatrix}$$

then a_{42} is the element of matrix A appearing in the 4th row and 2nd column.

Definition Two matrices A and B are **equal** ($A = B$) if and only if they have the same dimension and the elements in all corresponding positions are equal.

In the real number system, any two numbers have a sum. In a system of matrices, only matrices of the same dimension can be added. Matrix addition is a simple operation. Elements in the same relative position are added.

Example

$$\begin{bmatrix} 2 & 0 & 5 \\ -1 & 2 & 7 \end{bmatrix} + \begin{bmatrix} 4 & 3 & -2 \\ 7 & 6 & 2 \end{bmatrix} = \begin{bmatrix} 2+4 & 0+3 & 5+(-2) \\ (-1)+7 & 2+6 & 7+2 \end{bmatrix}$$

$$= \begin{bmatrix} 6 & 3 & 1 \\ 6 & 8 & 9 \end{bmatrix}$$

$$\begin{bmatrix} 6 & 3 & -2 \\ 1 & 3 & 12 \end{bmatrix}$$

Find the sum $D + F$.

Any matrix in which all of the elements are 0 is called a **zero matrix** and denoted by 0.

Exercises

The following matrices are referred to in the Exercises.

$$G = \begin{bmatrix} 1 & 5 & -2 & 3 \\ 4 & -1 & 0 & 6 \\ 2 & 5 & -3 & 8 \end{bmatrix} \quad H = \begin{bmatrix} 6 & -3 & 4 & -5 \\ 2 & 9 & -8 & 3 \\ 1 & 4 & -7 & -9 \end{bmatrix} \quad I = \begin{bmatrix} 1 & 0 & 0 \\ 0 & 1 & 0 \\ 0 & 0 & 1 \end{bmatrix}$$

$$J = \begin{bmatrix} 4 & -9 & 3 & 2 \\ 7 & 5 & -3 & 8 \\ 0 & 9 & 6 & 3 \end{bmatrix} \quad K = \begin{bmatrix} 4 & -1 & 2 \\ 5 & -3 & 8 \\ 7 & 6 & 3 \end{bmatrix} \quad L = \begin{bmatrix} 1 & 1 & 2 \\ 0 & 4 & 0 \\ 3 & 9 & -2 \end{bmatrix}$$

1. State the dimension of each matrix.
2. Find the sums of all pairs where addition is possible.
3. Find 3 nonzero row vectors whose sum is [3 4 2].
4. Verify that $G + H = H + G$ and $K + L = L + K$. What property for matrix addition does this suggest?
5. Compare $(G + H) + J$ and $G + (H + J)$. What property does this result suggest?
6. Find $G + 0$ and $0 + K$.
7. Find a matrix M such that $L + M = 0$. M is called the **additive inverse** of L and can be denoted by $(-L)$. Find the additive inverse for G, J, and K.

10-2 MATRIX MULTIPLICATION

Checking a solution of a linear equation provides us with a key to understanding matrix multiplication. If (2, −3, 4) is a proposed solution for the equation

$$3x_1 + 7x_2 + 5x_3 = 5$$

we check the sum of the products $3 \cdot 2$, $7 \cdot (-3)$, and $5 \cdot 4$; which is the number 5. Let the coefficients of the left member form a row vector and the numbers of the proposed solution form a column vector. Then the "product" of these two vectors (in the stated order) is:

$$\begin{bmatrix} 3 & 7 & 5 \end{bmatrix} \begin{bmatrix} 2 \\ -3 \\ 4 \end{bmatrix} = 3 \cdot 2 + 7 \cdot (-3) + 5 \cdot 4 = 5.$$

Find the "products."

−64, 34, −11

$$\begin{bmatrix} 4 & -3 & 7 \end{bmatrix} \begin{bmatrix} 1 \\ 4 \\ -8 \end{bmatrix}, \quad \begin{bmatrix} 2 & -5 \end{bmatrix} \begin{bmatrix} 7 \\ -4 \end{bmatrix}, \quad \begin{bmatrix} 1 & -3 & 2 & 6 \end{bmatrix} \begin{bmatrix} 4 \\ 3 \\ -9 \\ 2 \end{bmatrix}$$

The preceding technique of finding sums of products of the elements is the basis for determining the elements of the product of two matrices.

Each of the numbers in the product matrix is determined by sums of products using a row vector in the first matrix with a column vector in the second matrix. The number p_{23} (in row 2, column 3) of the following example is the "product" of the row 2 vector of M and the column 3 vector of N.

$$\underset{M}{\begin{bmatrix} 3 & 6 & 4 \\ -2 & 8 & 7 \end{bmatrix}} \underset{N}{\begin{bmatrix} -1 & 4 & 9 \\ 2 & 6 & 7 \\ 4 & 8 & 3 \end{bmatrix}} = \underset{P}{\begin{bmatrix} p_{11} & p_{12} & p_{13} \\ p_{21} & p_{22} & p_{23} \end{bmatrix}}$$

Hence,

$$p_{23} = (-2) \cdot 9 + 8 \cdot 7 + 7 \cdot 3 = 59.$$

$$p_{12} = [3 \quad 6 \quad 4] \begin{bmatrix} 4 \\ 6 \\ 8 \end{bmatrix} = 3 \cdot 4 + 6 \cdot 6 + 4 \cdot 8 = 80.$$

(1st row vector of M) (2nd column vector of N)

Show that $p_{11} = 25, p_{13} = 81, p_{21} = 46$, and $p_{22} = 96$.

In our example M is a 2×3 matrix; N is a 3×3 matrix. From the method of multiplication the product MN must have the same number of rows as the first matrix, M, and the same number of columns as the second, N. Thus P must be a 2×3 matrix. Also the method cannot be applied unless the number of columns in the first matrix is equal to the number of rows in the second factor (M has 3 columns and N has 3 rows).

Definition (*Matrix multiplication*) The product of a $k \times m$ matrix A and a $m \times n$ matrix B is the $k \times n$ matrix C where the elements c_{ij} of C are

$$c_{ij} = a_{i1}b_{1j} + a_{i2}b_{2j} + \cdots + a_{im}b_{mj}$$

for integers $1 \leqslant i \leqslant k$ and $1 \leqslant j \leqslant n$.

In the definition, c_{ij} is the "product" of the ith row vector of A and the jth column vector of B.

Example

$$\begin{bmatrix} 4 & -1 & 5 \\ 6 & 2 & -3 \end{bmatrix} \begin{bmatrix} 3 & -2 \\ 1 & 7 \\ -4 & 0 \end{bmatrix} = \begin{bmatrix} 4 \cdot 3 + (-1)1 + 5 \cdot 4 & 4(-2) + (-1)7 + 5 \cdot 0 \\ 6 \cdot 3 + 2 \cdot 1 + (-3)4 & 6(-2) + 2 \cdot 7 + (-3)0 \end{bmatrix}$$

$$= \begin{bmatrix} 31 & -15 \\ 8 & 2 \end{bmatrix}$$

Special square matrices possess the matrix multiplication identity property. (A **square matrix** is one with the same number of rows and columns.) If the previous matrix M is multiplied by the 2×2 matrix with 1's on the main diagonal and 0's elsewhere then the product is M, itself:

$$\begin{bmatrix} 1 & 0 \\ 0 & 1 \end{bmatrix} \begin{bmatrix} 3 & 6 & 4 \\ -2 & 8 & 7 \end{bmatrix} = \begin{bmatrix} 1 \cdot 3 + 0(-2) & 1 \cdot 6 + 0 \cdot 8 & 1 \cdot 4 + 0 \cdot 7 \\ 0 \cdot 3 + 1(-2) & 0 \cdot 6 + 1 \cdot 8 & 0 \cdot 4 + 1 \cdot 7 \end{bmatrix}$$

$$= \begin{bmatrix} 3 & 6 & 4 \\ -2 & 8 & 7 \end{bmatrix}$$

Thus, the matrix $\begin{bmatrix} 1 & 0 \\ 0 & 1 \end{bmatrix}$ displays the identity property, and is denoted by I_2.

Definition The $n \times n$ **identity matrix** is the matrix with 1's in the main diagonal positions (positions with equal row and column addresses) and 0's at each of the other positions; and is denoted by I_n.

Show that $MI_3 = M$.

Exercises

1. Find the following matrix products:

   a. $[-2 \quad 5 \quad 3] \begin{bmatrix} 4 \\ -2 \\ -7 \end{bmatrix}$

   b. $[4 \quad -2 \quad 3] \begin{bmatrix} 0 \\ 1 \\ 0 \end{bmatrix}$

   c. $[2 \quad -1 \quad 5 \quad 7] \begin{bmatrix} 6 \\ 3 \\ -5 \\ 8 \end{bmatrix}$

   d. $[a_1 \quad a_2 \quad a_2] \begin{bmatrix} b_1 \\ b_2 \\ b_3 \end{bmatrix}$

2. The director of an intramural athletic program purchased 3 basketballs, 4 volleyballs, 24 softballs, and 8 softball bats. Each basketball cost \$16.00, each volleyball \$9.50, each softball \$2.40, and each bat \$2.10. Represent the number of each item purchased by a row vector and represent the cost per type of item by a column vector. Find the total cost of this athletic equipment as the product of the row and column vectors.

Use the matrices below in Exercises 3–6.

$$A = \begin{bmatrix} a_{11} & a_{12} & a_{13} \\ a_{21} & a_{22} & a_{23} \\ a_{31} & a_{32} & a_{33} \\ a_{41} & a_{42} & a_{43} \end{bmatrix}, \quad B = \begin{bmatrix} b_{11} & b_{12} & b_{13} & b_{14} \\ b_{21} & b_{22} & b_{23} & b_{24} \\ b_{31} & b_{32} & b_{33} & b_{34} \end{bmatrix}, \quad C = [4 \quad -3 \quad 2],$$

$$D = \begin{bmatrix} -5 \\ 0 \\ 4 \end{bmatrix}, \quad E = \begin{bmatrix} 4 & 5 & 1 \\ 2 & -10 & 3 \\ -1 & 0 & 7 \end{bmatrix}, \quad F = \begin{bmatrix} 2 & -4 \\ 4 & 3 \\ 1 & -1 \end{bmatrix}, \quad K = \begin{bmatrix} 5 & -3 & -2 \\ 2 & 1 & 4 \\ 1 & -1 & 0 \end{bmatrix}.$$

3. Find the products:

   a. CD b. EF c. EK d. KE e. I_3E f. FI_2 g. AB h. BA

4. Find examples in Exercise 3 which show that matrix multiplication is *not* commutative.

5. Illustrate the distributive property for matrix multiplication over matrix addition by showing that $(E+K)F$ is the same matrix as $EF+KF$.

6. Find the products:

   a. $\begin{bmatrix} 1 \\ 3 \end{bmatrix} [2 \quad 5]$ b. DC c. $\begin{bmatrix} a_1 \\ a_2 \\ a_3 \\ a_4 \end{bmatrix} [b_1 \quad b_2 \quad b_3]$

10-3 GAUSSIAN ELIMINATION

The methods for solving systems of linear equations which were discussed in Chapter 3 can be expressed with matrix notation. A special sequencing of steps from these methods is called **Gaussian elimination.** Recall that if the original system is

$$\begin{cases} a_{11}x_1 + a_{12}x_2 = c_1 \\ a_{21}x_1 + a_{22}x_2 = c_2 \end{cases}$$

the "addition-subtraction" method is based on the principle that the equation system

$$\begin{cases} a_{11}x_1 + a_{12}x_2 = c_1 \\ a_{21}x_1 + a_{22}x_2 + k(a_{11}x_1 + a_{12}x_2) = c_2 + kc_1 \end{cases}$$

has the same solution set as the original system. The second equation was replaced. We make use of three principles in solving equations by the addition-subtraction method:

1. An equation may be replaced by the sum of its members with a multiple of the members of another equation;
2. An equation may be replaced by an equation whose members are a nonzero multiple of the given members; that is, multiplying an equation by a nonzero number does not change the solution set;
3. The order of two equations in a system can be interchanged with no effect on the solution set.

These three principles are called **elementary transformations.**

In order to apply these principles using matrix notation, the system of equations is represented by the **augmented matrix** which is the coefficient matrix augmented on the right with the constant column vector. For the system

$$\begin{cases} 2x_1 - 3x_2 = 14 \\ 3x_1 + 5x_2 = 8 \end{cases}$$

the augmented matrix is

$$\left[\begin{array}{cc|c} 2 & -3 & 14 \\ 3 & 5 & 8 \end{array}\right].$$

In solving this equation the 1st row vector is multiplied by $\frac{1}{2}$ (a type 2 elementary transformation) in order that the coefficient of x_1 be 1 in the resulting equation. Thus we have the matrix

$$\left[\begin{array}{cc|c} \frac{1}{2}\cdot 2 & \frac{1}{2}\cdot(-3) & \frac{1}{2}\cdot 14 \\ 3 & 5 & 8 \end{array}\right] = \left[\begin{array}{cc|c} 1 & -\frac{3}{2} & 7 \\ 3 & 5 & 8 \end{array}\right].$$

Now the x_1 variable is "eliminated" in the other equation by adding -3 times the 1st row vector to the second row vector (a type 1 elementary transformation). Thus we have

$$\left[\begin{array}{cc|c} 1 & -\frac{3}{2} & 7 \\ 3+(-3)\cdot 1 & 5+(-3)\cdot\left(-\frac{3}{2}\right) & 8+(-3)\cdot 7 \end{array}\right] = \left[\begin{array}{cc|c} 1 & -\frac{3}{2} & 7 \\ 0 & \frac{19}{2} & -13 \end{array}\right]$$

The second row vector in this matrix is multiplied by $\frac{2}{19}$ (a type 2 transformation) to yield a coefficient of 1 for x_2.

$$\left[\begin{array}{cc|c} 1 & -\frac{3}{2} & 7 \\ 0 & \frac{19}{2}\cdot\frac{2}{19} & (-13)\frac{2}{19} \end{array}\right] = \left[\begin{array}{cc|c} 1 & -\frac{3}{2} & 7 \\ 0 & 1 & -\frac{26}{19} \end{array}\right]$$

The solution set for the given system is readily obtained from this matrix. The second row vector corresponds to the equation $x_2 = -\frac{26}{19}$. The first row vector represents $x_1 - \frac{3}{2}x_2 = 7$. Since $-\frac{26}{19}$ must replace x_2 in a solution for the system we substitute back to the preceding row:

$$x_1 - \frac{3}{2}\left(-\frac{26}{19}\right) = 7.$$

In solving for x_1 we find that $x_1 = \frac{94}{19}$ and the ordered pair $\left(\frac{94}{19}, -\frac{26}{19}\right)$ is the only solution for the system. Note that this last step, in effect, is another type 1 elementary transformation—we multiply the second row vector by $\frac{3}{2}$ and add it to the first row vector.

Apply similar steps using an augmented matrix to solve the system:

$$\begin{cases} 3x_1 + 5x_2 = 8 \\ 2x_1 - 3x_2 = 14 \end{cases} \qquad \left[\begin{array}{cc|c} 3 & 5 & 8 \\ 2 & -3 & 14 \end{array}\right]$$

$$\left[\begin{array}{cc|c} 1 & \frac{5}{3} & \frac{8}{3} \\ 0 & 1 & -\frac{26}{19} \end{array}\right]$$

$$\left(\frac{94}{19}, -\frac{26}{19}\right)$$

This method is readily applied to a system of 3 equations in 3 variables, or larger systems. The number of calculations is increased, but the basic principles are the same. Applying the method to the system

$$\begin{cases} 6x_1 - 8x_2 + 3x_3 = 6 \\ 3x_1 + 2x_2 - 4x_3 = 4 \\ 5x_1 - 6x_2 + 9x_3 = -2 \end{cases}$$

we have the sequence of matrices:

$$\left[\begin{array}{ccc|c} 6 & -8 & 3 & 6 \\ 3 & 2 & -4 & 4 \\ 5 & -6 & 9 & -2 \end{array}\right]$$

$$\left[\begin{array}{ccc|c} 1 & -\frac{4}{3} & \frac{1}{2} & 1 \\ 3 & 2 & 4 & 4 \\ 5 & -6 & 9 & -2 \end{array}\right]$$

(1st row vector times $\frac{1}{6}$, in order that the first element of the row be 1)

$$\left[\begin{array}{ccc|c} 1 & -\frac{4}{3} & \frac{1}{2} & 1 \\ 0 & 6 & -\frac{11}{2} & 1 \\ 0 & \frac{2}{3} & \frac{13}{2} & -7 \end{array}\right]$$

(−3 times 1st row vector added to 2nd row vector; −5 times 1st row vector added to 3rd row vector)

$$\left[\begin{array}{ccc|c} 1 & -\frac{4}{3} & \frac{1}{2} & 1 \\ 0 & 1 & -\frac{11}{12} & \frac{1}{6} \\ 0 & \frac{2}{3} & \frac{13}{2} & -7 \end{array}\right]$$ $\left(\frac{1}{6}\right.$ times 2nd row vector, in order that the 2nd element–first nonzero element–in that row be $\left.1\right)$

$$\left[\begin{array}{ccc|c} 1 & -\frac{4}{3} & \frac{1}{2} & 1 \\ 0 & 1 & -\frac{11}{12} & \frac{1}{6} \\ 0 & 0 & \frac{64}{9} & -\frac{64}{9} \end{array}\right]$$ $\left(-\frac{2}{3}\right.$ times the 2nd row vector added to the 3rd row vector$\left.\right)$

$$\left[\begin{array}{ccc|c} 1 & -\frac{4}{3} & \frac{1}{2} & 1 \\ 0 & 1 & -\frac{11}{12} & \frac{1}{6} \\ 0 & 0 & 1 & -1 \end{array}\right]$$ $\left(\frac{9}{64}\right.$ times the 3rd row vector, in order that the 3rd element–first nonzero element–be $\left.1\right)$

The solution is recovered from this last matrix. From the third row vector we have

$$x_3 = -1.$$

Using this information along with the 2nd row vector yields

$$x_2 - \frac{11}{12}(-1) = \frac{1}{6}$$

$$x_2 = \frac{1}{6} + \frac{11}{12}(-1)$$

$$x_2 = -\frac{3}{4}.$$

Finally, using these numbers in the first row vector, called **back substituting**, we have

$$x_1 - \frac{4}{3}\left(-\frac{3}{4}\right) + \frac{1}{2}(-1) = 1$$

$$x_1 = 1 + \frac{4}{3}\left(-\frac{3}{4}\right) - \frac{1}{2}(-1)$$

$$x_1 = \frac{1}{2}$$

and the solution set is $\left\{\left(\frac{1}{2}, -\frac{3}{4}, -1\right)\right\}$

Exercises Write the augmented matrix for each system and use the Gaussian elimination method to solve the system.

1. $\begin{cases} x_1 - 3x_2 - 2x_3 = 9 \\ 3x_1 + 2x_2 + 6x_3 = 20 \\ 4x_1 - x_2 + 3x_3 = 25 \end{cases}$

2. $\begin{cases} x_1 + x_2 + x_3 = -2 \\ x_1 - x_2 + x_3 = 12 \\ x_1 + 2x_2 + 3x_3 = -3 \end{cases}$

3. $\begin{cases} 2x_1 + 3x_2 + x_3 = 1 \\ x_1 - 2x_2 - x_3 = -10 \\ -x_1 + 4x_2 + 2x_3 = 15 \end{cases}$

4. $\begin{cases} x_1 + x_2 + x_3 = -1 \\ -x_1 + x_2 + x_3 = 7 \\ 4x_1 - x_2 + 3x_3 = 0 \end{cases}$

5. $\begin{cases} x_1 + x_2 - x_3 = -3 \\ 5x_1 - x_2 - x_3 = 1 \\ 2x_1 - 6x_2 + 3x_3 = 1 \end{cases}$

6. $\begin{cases} x_1 + 4x_2 + x_3 = -5 \\ -5x_1 + x_2 + x_3 = 2 \\ 11x_1 - 2x_2 - x_3 = -1 \end{cases}$

7. $\begin{cases} 5x_1 + x_2 + x_3 = 4 \\ x_1 + 3x_2 + x_3 = 3 \\ x_1 + x_2 - x_3 = -1 \end{cases}$

8. $\begin{cases} 4x_1 + 4x_2 + 4x_3 = 7 \\ 8x_1 - 4x_2 + 2x_3 = -7 \\ x_1 + x_3 = 3 \end{cases}$

9. $\begin{cases} x_1 - x_2 + x_3 - x_4 = 0 \\ 3x_1 + 3x_2 + 3x_3 + 3x_4 = -1 \\ -2x_1 + 2x_2 + x_3 + 2x_4 = -1 \\ -x_1 + x_2 - 4x_3 + 4x_4 = -1 \end{cases}$

10. $\begin{cases} x_1 + x_2 + 5x_3 + x_4 = 0 \\ 2x_1 + 7x_2 - 3x_3 + 2x_4 = -13 \\ x_1 + 2x_2 + x_3 - x_4 = 0 \\ 3x_1 - x_2 + 4x_3 - 3x_4 = 1 \end{cases}$

10-4 SYSTEMS IN MATRIX NOTATION

The system of linear equations,

$$\begin{cases} 2x_1 + 3x_2 = 11 \\ x_1 - 2x_2 = 5 \end{cases}$$

can now be expressed in matrix style. Let

$$A = \begin{bmatrix} 2 & 3 \\ 1 & -2 \end{bmatrix}, \quad X = \begin{bmatrix} x_1 \\ x_2 \end{bmatrix}, \quad \text{and} \quad B = \begin{bmatrix} 11 \\ 5 \end{bmatrix}.$$

Then

$$\begin{bmatrix} 2 & 3 \\ 1 & -2 \end{bmatrix} \begin{bmatrix} x_1 \\ x_2 \end{bmatrix} = \begin{bmatrix} 11 \\ 5 \end{bmatrix} \quad \text{or} \quad AX = B$$

denotes this system. Since the matrix product

$$\begin{bmatrix} \frac{2}{7} & \frac{3}{7} \\ \frac{1}{7} & -\frac{2}{7} \end{bmatrix} \begin{bmatrix} 2 & 3 \\ 1 & -2 \end{bmatrix} = \begin{bmatrix} 1 & 0 \\ 0 & 1 \end{bmatrix},$$

the matrix $\begin{bmatrix} \frac{2}{7} & \frac{3}{7} \\ \frac{1}{7} & -\frac{2}{7} \end{bmatrix}$ is called the (multiplicative) **inverse** of A, and is denoted by A^{-1}. We have $A^{-1}A = A(A^{-1}) = I_2$. Thus

$$A^{-1}AX = A^{-1}B$$

$$I_2X = A^{-1}B$$

$$X = A^{-1}B$$

$$\begin{bmatrix} 1 & 0 \\ 0 & 1 \end{bmatrix} \begin{bmatrix} x_1 \\ x_2 \end{bmatrix} = \begin{bmatrix} \frac{2}{7} & \frac{3}{7} \\ \frac{1}{7} & -\frac{2}{7} \end{bmatrix} \begin{bmatrix} 11 \\ 5 \end{bmatrix} = \begin{bmatrix} \frac{37}{7} \\ \frac{1}{7} \end{bmatrix}$$

$$\begin{bmatrix} x_1 \\ x_2 \end{bmatrix} = \begin{bmatrix} \frac{37}{7} \\ \frac{1}{7} \end{bmatrix}$$

or

$$\begin{cases} x_1 = \frac{37}{7} \\ x_2 = \frac{1}{7} \end{cases}$$

The inverse for matrix A was actually found by using the following information. If matrix $A = \begin{bmatrix} a_{11} & a_{12} \\ a_{21} & a_{22} \end{bmatrix}$ has an inverse $A^{-1} = \begin{bmatrix} b_{11} & b_{12} \\ b_{21} & b_{22} \end{bmatrix}$ then

$$A(A^{-1}) = \begin{bmatrix} 1 & 0 \\ 0 & 1 \end{bmatrix}$$

and

$$\begin{cases} b_{11}a_{11} + b_{12}a_{21} = 1 & b_{21}a_{11} + b_{22}a_{21} = 0 \\ b_{11}a_{12} + b_{12}a_{22} = 0 & b_{21}a_{12} + b_{22}a_{22} = 1 \end{cases}$$

must hold. From this it follows that

$$b_{11} = \frac{a_{22}}{a_{11}a_{22} - a_{12}a_{21}} \qquad b_{21} = \frac{-a_{21}}{a_{11}a_{22} - a_{12}a_{21}}$$

$$b_{12} = \frac{-a_{12}}{a_{11}a_{22} - a_{12}a_{21}} \qquad b_{22} = \frac{a_{11}}{a_{11}a_{22} - a_{12}a_{21}}.$$

The denominator for each fraction, $a_{11}a_{22} - a_{12}a_{21}$, is called the **determinant** of the 2×2 matrix A, and is denoted by $|A|$. Note that the determinant is a number. Using this information in the preceding example, we find that

$$A = \begin{bmatrix} 2 & 3 \\ 1 & -2 \end{bmatrix} \text{ and } |A| = 2(-2) - 1 \cdot 3 = -7.$$

The diagonal elements in A are interchanged, additive inverses of the other elements are found, and then all of these elements are divided by $|A| = -7$ to yield

$$A^{-1} = \begin{bmatrix} \frac{-2}{-7} & \frac{-3}{-7} \\ \frac{-1}{-7} & \frac{2}{-7} \end{bmatrix} = \begin{bmatrix} \frac{2}{7} & \frac{3}{7} \\ \frac{1}{7} & -\frac{2}{7} \end{bmatrix}.$$

If $|A| = 0$ for a 2×2 matrix then this procedure fails, but in this case there is no inverse for A and the system does not have a unique solution. The system is either inconsistent or dependent.

The example and exercises in this section involve systems of two linear equations in two variables. The general case of a system of n linear equations in n variables can also be examined in matrix form

$$AX = B$$

where A is an $n \times n$ matrix, $X = \begin{bmatrix} x_1 \\ x_2 \\ \cdot \\ \cdot \\ \cdot \\ x_n \end{bmatrix}$, and B is a column vector with n

elements. If $n \geqslant 3$ and matrix A has an inverse, its calculation is beyond the development presented in this text.

Exercises Write each system of linear equations in matrix form $AX = B$, compute A^{-1}, and solve the system by finding the product $A^{-1}B$.

1. $\begin{cases} 5x_1 - 3x_2 = 21 \\ 2x_1 + 7x_2 = -8 \end{cases}$

2. $\begin{cases} 18x_1 - 30x_2 = -11 \\ -4x_1 + 3x_2 = 0 \end{cases}$

3. $\begin{cases} 0.4x_1 + 0.1x_2 = -0.1 \\ 0.3x_1 + 0.7x_2 = 0.3 \end{cases}$

4. $\begin{cases} 9x_1 + 15x_2 = -5 \\ -9x_1 - 5x_2 = 5 \end{cases}$

5. $\begin{cases} 4x_1 - 3x_2 + 43 = 0 \\ 2x_1 - 7x_2 + 71 = 0 \end{cases}$

6. $\begin{cases} 7x_1 - 9x_2 = 13 \\ 4x_1 - 3x_2 = 9 \end{cases}$

7. $\begin{cases} 12x_1 + 16x_2 + 7 = 0 \\ 13 + 12x_1 + 24x_2 = 0 \end{cases}$

8. $\begin{cases} 6x_1 + 2x_2 - 17 = 0 \\ 3x_2 + 4 - 5x_1 = 0 \end{cases}$

Find the solution set for each of the following systems. (The method of this section fails. Why?)

9. $\begin{cases} 4x_1 + 5x_2 + 2 = 0 \\ 10x_2 + 4 = -8x_1 \end{cases}$

10. $\begin{cases} 3x_1 + 5x_2 = 7 \\ 6x_1 + 9 = -10x_2 \end{cases}$

10-5 DETERMINANTS

The determinant for a 2×2 matrix

$$A = \begin{bmatrix} a_{11} & a_{12} \\ a_{21} & a_{22} \end{bmatrix}$$

was introduced in Section 10-4. The determinant of matrix A, denoted by

$$|A| = \begin{vmatrix} a_{11} & a_{12} \\ a_{21} & a_{22} \end{vmatrix},$$

is the number $a_{11}a_{22} - a_{12}a_{21}$.

Examples If

$$B = \begin{bmatrix} 3 & 7 \\ -2 & 5 \end{bmatrix} \quad \text{and} \quad C = \begin{bmatrix} -1 & 0 \\ 0 & 4 \end{bmatrix}$$

then

$$|B| = \begin{vmatrix} 3 & 7 \\ -2 & 5 \end{vmatrix} = 3 \cdot 5 - 7(-2) = 29,$$

and

$$|C| = \begin{vmatrix} -1 & 0 \\ 0 & 4 \end{vmatrix} = (-1)4 - 0 \cdot 0 = -4.$$

A determinant is associated with every square matrix, and this determinant is quite useful in many situations. In this section we will find out how to calculate the determinant of a square matrix, and also develop a few properties which facilitate computation.

Consider the matrix

$$\begin{bmatrix} 4 & -3 & 9 \\ 8 & 7 & -4 \\ 1 & 5 & 3 \end{bmatrix}$$

and the element 5 in the third row and second column. Now delete the third row and the second column to form the 2 X 2 matrix

$$\begin{bmatrix} 4 & -3 & 9 \\ 8 & 7 & -4 \\ 1 & (5) & 3 \end{bmatrix} = \begin{bmatrix} 4 & 9 \\ 8 & -4 \end{bmatrix}.$$

Now multiply the determinant of this matrix by $(-1)^{3+2}$

$$(-1)^{3+2} \begin{vmatrix} 4 & 9 \\ 8 & -4 \end{vmatrix} = (-1)\,[(-16) - 72] = 88.$$

The exponent of -1 is the sum of the row and column numbers. This 2 X 2 matrix is called the **minor**, and the product is called the **cofactor**, of the element in the third row and second column.

In general, the cofactor of an element a_{ij} in the ith row and jth column of an $n \times n$ matrix A is the product of $(-1)^{i+j}$ and the determinant of the $(n-1) \times (n-1)$ matrix formed by deleting the ith row and jth column of A.

Find the cofactors of the elements 1 and 3 in the preceding 3 X 3 matrix.

$(-1)^{3+1}(12 - 63) = -51$;
$(-1)^{3+3}(28 - (-24)) = 52$

We now define determinants in terms of cofactors.

Definition The **determinant** of a square matrix is the sum of the products of elements and corresponding cofactors where the elements form one row or one column of the matrix.

Example Using the third row in the preceding matrix we have

$$\begin{vmatrix} 4 & -3 & 9 \\ 8 & 7 & -4 \\ 1 & 5 & 3 \end{vmatrix} = 1(-1)^{3+1}\begin{vmatrix} -3 & 9 \\ 7 & -4 \end{vmatrix} + 5(1)^{3+2}\begin{vmatrix} 4 & 9 \\ 8 & -4 \end{vmatrix}$$

$$+ 3(-1)^{3+3}\begin{vmatrix} 4 & -3 \\ 8 & 7 \end{vmatrix}$$

$$= 1(-51) + 5 \cdot 88 + 3 \cdot 52 = 545.$$

Find the determinant of this same matrix using the 1st row; 1st column; 2nd row; 3rd column.

$4 \cdot 41 + (-3)(-28) + 9 \cdot 33 = 545;$
$4 \cdot 41 + 8 \cdot 54 + 1(-51) = 545;$
$8 \cdot 54 + 7 \cdot 3 + (-4)(-23) = 545;$
$9 \cdot 33 + (-4)(-23) + 3 \cdot 52 = 545.$

The following theorems aid in the computation of determinants. The basis for some of these results appear in a few of the exercises. The matrices A and B are square matrices of the same dimension.

Theorem 1 If all of the elements in one row or one column of A are zeros then $|A| = 0$.

Theorem 2 If matrix B is formed from A by adding a multiple on one row vector of A to another row vector of A (a multiple of one column vector to another column vector) then $|B| = |A|$.

To show that Theorem 2 holds for 2 X 2 matrices consider the matrices

$$A = \begin{bmatrix} a_{11} & a_{12} \\ a_{21} & a_{22} \end{bmatrix} \quad \text{and} \quad B = \begin{bmatrix} a_{11} & a_{12} \\ a_{21} + ka_{11} & a_{22} + ka_{12} \end{bmatrix}$$

where k is a real number. Then

$$|A| = \begin{vmatrix} a_{11} & a_{12} \\ a_{21} & a_{22} \end{vmatrix} = a_{11}a_{22} - a_{12}a_{21}$$

and

$$|B| = a_{11}(a_{22} + ka_{12}) - a_{12}(a_{21} + ka_{11}) = a_{11}a_{22} - a_{12}a_{21}.$$

Theorem 3 If matrix B is formed by multiplying each element of one row vector (or one column vector) of A by the nonzero number c then $|B| = c\,|A|$.

To illustrate this theorem for 2 X 2 matrices, let

$$A = \begin{bmatrix} a_{11} & a_{12} \\ a_{21} & a_{22} \end{bmatrix} \quad \text{and} \quad B = \begin{bmatrix} a_{11} & 3 \cdot a_{12} \\ a_{21} & 3 \cdot a_{22} \end{bmatrix}.$$

We see that $|B| = a_{11} \cdot 3a_{22} - 3a_{12}a_{21} = 3(a_{11}a_{22} - a_{12}a_{21}) = 3\,|A|$.

A strategy used in computing a determinant involves deriving a matrix with an element 1 in some location. Then a matrix with 0's in the other row (or column) positions is derived. Finally, the determinant is computed using this row (or column). The following example illustrates this strategy.

$$\begin{vmatrix} 3 & 4 & -2 \\ 5 & 12 & -5 \\ 1 & 6 & 4 \end{vmatrix} = 2 \cdot \begin{vmatrix} 3 & 2 & -2 \\ 5 & 6 & -5 \\ 1 & 3 & 4 \end{vmatrix}$$ (Theorem 3; factor 2 from the elements of the 2nd column vector)

$$= 2 \cdot \begin{vmatrix} 0 & -7 & -14 \\ 5 & 6 & -5 \\ 1 & 3 & 4 \end{vmatrix}$$ (Theorem 2; add −3 times the third row vector to the 1st row vector)

$$= 2 \cdot \begin{vmatrix} 0 & -7 & -14 \\ 0 & -9 & -25 \\ 1 & 3 & 4 \end{vmatrix}$$ (Theorem 2; add −5 times the 3rd row vector to the second row vector)

$$= 2 \cdot 1 \cdot (-1)^{3+1} \begin{vmatrix} -7 & -14 \\ -9 & -25 \end{vmatrix} = 2(175 - 126) = 98.$$

In special cases determinants can be used to solve a system of n linear equations in n variables. The result is known as Cramer's rule.

Theorem (*Cramer's rule*) In a system of n linear equations in $x_1, x_2, \cdots, x_n$ if C is the coefficient matrix, A_i is the matrix formed by replacing column i of C by the last column of the augmented matrix for $1 \leqslant i \leqslant n$, and if $|C| \neq 0$ then $\left\{\left(\frac{|A_1|}{|C|}, \frac{|A_2|}{|C|}, \cdots, \frac{|A_n|}{|C|}\right)\right\}$ is the solution of the system.

Example In the system from Section 10-3

$$\begin{cases} 2x_1 - 3x_2 = 14 \\ 3x_1 + 5x_2 = 8 \end{cases}$$

$$C = \begin{bmatrix} 2 & -3 \\ 3 & 5 \end{bmatrix}, \quad A_1 = \begin{bmatrix} 14 & -3 \\ 8 & 5 \end{bmatrix}, \quad \text{and} \quad A_2 = \begin{bmatrix} 2 & 14 \\ 3 & 8 \end{bmatrix}.$$

The solution is $\left\{\left(\frac{|A_1|}{|C|}, \frac{|A_2|}{|C|}\right)\right\}$.

Compute $|C|$, $|A_1|$, and $|A_2|$, and solve the preceding system.

19, 94, −26, $\left(\frac{94}{19}, \frac{-26}{19}\right)$

Exercises

1. Compute the determinant for each of the following matrices:

a. $\begin{bmatrix} 7 & 11 \\ 9 & 13 \end{bmatrix}$ b. $\begin{bmatrix} 12 & -17 \\ 13 & 8 \end{bmatrix}$ c. $\begin{bmatrix} 23 & -11 \\ 15 & -28 \end{bmatrix}$

d. $\begin{bmatrix} 4 & 0 & -7 \\ 9 & 1 & 11 \\ -6 & 0 & 3 \end{bmatrix}$ e. $\begin{bmatrix} 12 & -7 & 19 \\ 0 & 0 & 5 \\ -9 & 3 & 43 \end{bmatrix}$ f. $\begin{bmatrix} 2 & 5 & 1 \\ -3 & 7 & -8 \\ 11 & -4 & 15 \end{bmatrix}$

g. $\begin{bmatrix} 1 & -35 & 63 \\ 3 & 95 & 6 \\ 1 & -35 & 63 \end{bmatrix}$ h. $\begin{bmatrix} 4 & 0 & 3 & 7 \\ 5 & 2 & 4 & 9 \\ -6 & -1 & 5 & 8 \\ 3 & 0 & 8 & 2 \end{bmatrix}$ i. $\begin{bmatrix} 7 & 0 & 0 & 0 \\ 4 & -3 & 0 & 0 \\ 8 & 2 & 2 & 0 \\ -9 & 14 & 3 & 5 \end{bmatrix}$

2. Employ Property 3 in computing the determinant of each of the matrices:

a. $\begin{bmatrix} 1 & 3 & 4 \\ 5 & 15 & 10 \\ -1 & 5 & 2 \end{bmatrix}$ b. $\begin{bmatrix} 9 & 12 & -6 & 3 \\ 1 & -7 & 8 & 0 \\ -1 & 3 & -4 & 5 \\ 2 & 4 & -8 & 6 \end{bmatrix}$

3. Apply Cramer's rule to solve each of the systems:

a. $\begin{cases} 5x_1 - 3x_2 = 21 \\ 2x_1 + 7x_2 = -8 \end{cases}$ b. $\begin{cases} 2x_1 - 3x_2 + x_3 = 4 \\ x_1 + 2x_2 - 5x_3 = -2 \\ 3x_1 - x_2 - x_3 = 3 \end{cases}$

4. Show that the determinant of a square matrix with two rows (or two columns) alike is zero.

5. If the determinant of a 1×1 matrix $[a]$ is defined to be a then use this information and the definition of the determinant of a square matrix given in this section to derive the number

$$\begin{vmatrix} a_{11} & a_{12} \\ a_{21} & a_{22} \end{vmatrix}.$$

10-6 LINEAR PROGRAMMING

The topic of linear programming in mathematics is related to linear systems. Each of the sentences in the following system corresponds to a half-plane together with the boundary line.

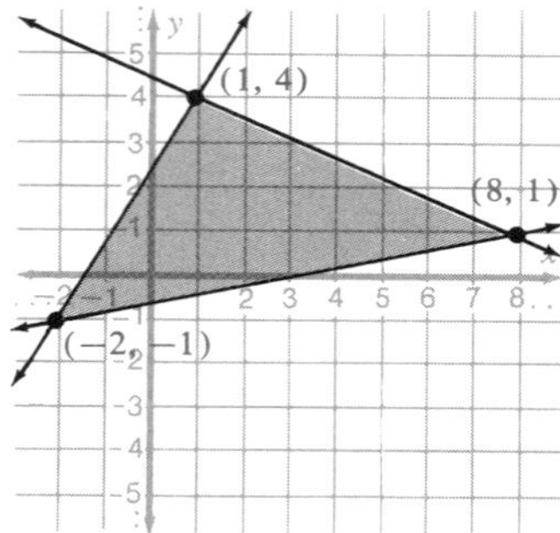

FIGURE 1

$$\begin{cases} 3x + 7y \leqslant 31 \\ x - 5y \leqslant 3 \\ 5x - 3y \geqslant -7 \end{cases}$$

The points common to the solution set for each sentence comprise the triangular region sketched in Figure 1, assuming that $R \times R$ is the replacement set for (x, y). Verify that the region represents the graph of the solution set for the system.

Graph the solution set for the system of inequalities

$$\begin{cases} 2x - y \leqslant 4 \\ 2x + y \geqslant -4 \\ y \leqslant 4 \end{cases}$$

where $R \times R$ is the replacement set for (x, y).

The graph of the solution set is a triangular region with vertices (−4, 4), (4, 4), and (0, −4).

A polygonal region (the set of points on or in the interior of a polygon) which is the graph of the solution set of a system of linear sentences plays an important role in linear programming.

Example A small manufacturing concern manufactures two types of products. Three machines A, B, and C are used in making each product. Machines A and B are available for 8 hours each day and machine C is available for 10 hours. Each product of the first type requires 1 hour, $\frac{1}{2}$ hour, and 1 hour on machines A, B, C, respectively. Each product of the second type requires $\frac{1}{3}$ hour, 1 hour, and 1 hour on machines A, B, and C. If the profit of the first product is \$20 each while the profit on the second product is \$15 each, how many articles of each type should be manufactured each day to realize a maximum profit?

Let x and y be variables such that replacements for x and y correspond to possible numbers of first and second type products, respectively. Clearly we are seeking numbers such that

$$x \geqslant 0$$
$$y \geqslant 0.$$

Machines A and B are available for at most 8 hours; hence, the number of articles of each type must be in the solution sets of

$$x + \frac{1}{3}y \leqslant 8 \qquad (3x + y \leqslant 24).$$

$$\frac{1}{2}x + \ y \leqslant 8 \qquad (\ x + 2y \leqslant 16).$$

Since machine C is available for 10 hours and each article of either type requires 1 hour, we have

$$x + y \leqslant 10.$$

The number of articles of each type must yield true statements for each of the five sentences. We are interested in the solution set for the system:

$$\begin{cases} x \geqslant 0 \\ y \geqslant 0 \\ 3x + \ y \geqslant 24 \\ x + 2y \geqslant 16 \\ x + \ y \geqslant 10 \end{cases}$$

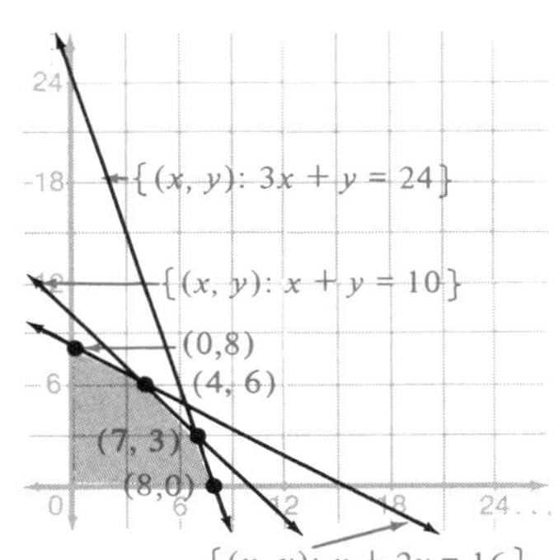

FIGURE 2

The polygonal region sketched in Figure 2 is the graph of the solution set for this system.

From the original information given, the profit in dollars from the articles manufactured during one day is given by the expression

$$20x + 15y$$

where the polygonal region is the replacement set for (x, y). The next theorem, a fundamental principle of linear programming, is used to find the maximum profit.

Theorem If P is a polygonal region, a and b are real numbers, and $f(x, y) = ax + by$ is a function from P into R, then the function f assumes a maximum (and minimum) value at a vertex of P.

Table 1

x	y	$20x + 15y$
0	0	0
0	8	120
4	6	170
7	3	185
8	0	160

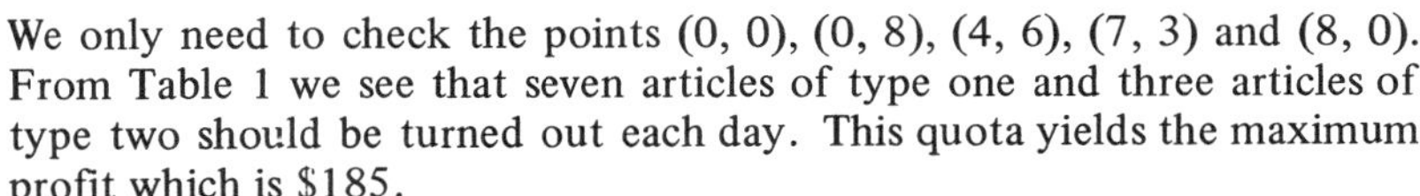

We only need to check the points (0, 0), (0, 8), (4, 6), (7, 3) and (8, 0). From Table 1 we see that seven articles of type one and three articles of type two should be turned out each day. This quota yields the maximum profit which is \$185.

This problem in linear programming involved maximizing profits subject to certain conditions called constraints. In our example the system of inequalities describes the constraints.

Exercises

Graph the solution set for each system in Exercises 1–6.

1. $\begin{cases} 11x - 4y \leqslant 27 \\ 3x - 8y \geqslant -41 \\ 2x + \ y \geqslant -2 \end{cases}$

2. $\begin{cases} x \geqslant 0 \\ y \geqslant 0 \\ x + y \leqslant 10 \\ 3x + y \leqslant 15 \end{cases}$

3. $\begin{cases} x \geqslant 0 \\ y \geqslant 0 \\ 3x + 2y \leqslant 12 \end{cases}$

4. $\begin{cases} x \geqslant 0 \\ y \geqslant 0 \\ 4x + y \leqslant 24 \\ x + 3y \leqslant 15 \end{cases}$

5. $\begin{cases} 10x + 4y \leqslant 32 \\ 10x - 4y \geqslant -32 \\ y \geqslant -2 \end{cases}$

6. $\begin{cases} x + y \geqslant 0 \\ x + 2y \leqslant 12 \\ x - y \leqslant 8 \end{cases}$

In Exercises 7–10 find the maximum and minimum value for each expression in each polygonal region described in Exercises 1–6.

7. $2x + y$ 8. $5x + 7y$ 9. $x - 2y$ 10. $3y - 2x$

11. Use the example in the text, but assume a profit of \$10 for each article of the first type and a profit of \$15 for each article of the second type. How many articles of each type should be turned out each day to realize a maximum profit?

12. Bill and Jim manufacture two different types of toys, T_1 and T_2. Each type requires one hour of work by Bill. Jim needs to work 2 hours on each T_1 toy and 1 hour on each type T_2 toy. Bill is able to work 8 hours a day, while Jim works 10 hours. If the profit is \$4 for each T_1 toy and \$3 for each T_2 toy, determine the most profitable number of T_1 and T_2 toys to manufacture each day.

13. Work Exercise 12 if the profit is \$2 for each T_1 toy and \$3 for each T_2 toy.

14. A manufacturing company makes two types of items, I_1 and I_2. Three machines, M_1, M_2, and M_3, are used in the process. Item I_1 requires 3 hours work on M_1, 1.5 hours on M_2, and 2.4 hours on M_3. Item I_2 requires 2 hours on M_1, 4 hours on M_2, and 3.2 hours on M_3. If the profit is \$50 for each item I_1 and each item I_2, how many articles of each type should be manufactured each day to realize a maximum profit? Each machine can be operated for 24 hours a day.

15. Work Exercise 14 if the profit is \$40 for each I_1 item and \$80 for each I_2 item.

10-7 LINEAR INEQUALITIES AND ABSOLUTE VALUE

Consider the sentence

$$|x| - x \leqslant 0$$

where R is the replacement set for x. From the meaning of absolute value if $x \geqslant 0$, then $|x| = x$; if $x < 0$, then $|x| = -x$. Thus, the original sentence is equivalent to

(1) $(x \geqslant 0 \text{ and } x - x \leqslant 0) \text{ or } (x < 0 \text{ and } -x - x \leqslant 0)$

The sentence $|x| - x \leqslant 0$ is equivalent to $|x| \leqslant x$ whose solution set in $R \times R$ is $\{(x, y): x \geqslant 0\}$:

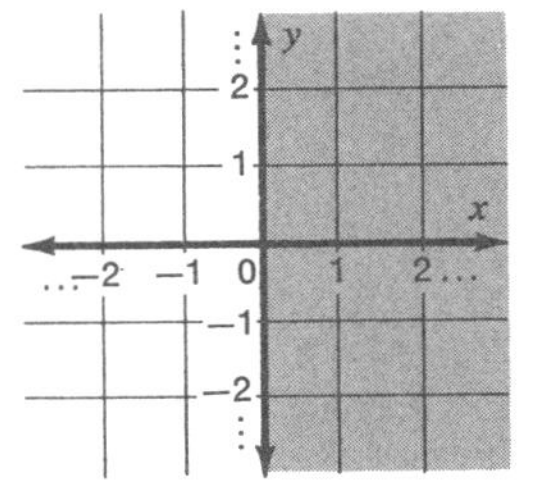

$$(2) \quad \begin{cases} x \geqslant 0 \\ x - x \leqslant 0 \end{cases} \text{ or } \begin{cases} x < 0 \\ x \geqslant 0 \end{cases}$$

The first component is true for all nonnegative real numbers; no real number belongs to the solution set of the second component. Combining these results, we conclude that the solution set of the original sentence is the set of nonnegative real numbers.

If $R \times R$ is the replacement set for (x, y), what is the solution set for the sentence $|x| - x \leqslant 0$? Sketch its graph.

If $R \times R$ is the replacement set for (x, y), what is the solution set for the sentence

$$|x| - |y| > 0?$$

By considering each of the cases

$$x \geqslant 0 \quad \text{and} \quad y \geqslant 0$$

or

$$x < 0 \quad \text{and} \quad y \geqslant 0$$

or

$$x \geqslant 0 \quad \text{and} \quad y < 0$$

or

$$x < 0 \quad \text{and} \quad y < 0$$

we can write the following compound sentences which are equivalent to the original sentence:

$$(3) \quad \begin{cases} x \geqslant 0 \\ y \geqslant 0 \\ x - y > 0 \end{cases} \text{ or } \begin{cases} x < 0 \\ y \geqslant 0 \\ -x - y > 0 \end{cases} \text{ or } \begin{cases} x \geqslant 0 \\ y < 0 \\ x + y > 0 \end{cases} \text{ or } \begin{cases} x < 0 \\ y < 0 \\ -x + y > 0 \end{cases}$$

$$(4) \quad \begin{cases} x \geqslant 0 \\ y \geqslant 0 \\ y < x \end{cases} \text{ or } \begin{cases} x < 0 \\ y \geqslant 0 \\ y < -x \end{cases} \text{ or } \begin{cases} x \geqslant 0 \\ y < 0 \\ y > -x \end{cases} \text{ or } \begin{cases} x < 0 \\ y < 0 \\ y > x \end{cases}$$

The graph of the solution set for each component is indicated in Figures 1, 2, 3, and 4, respectively.

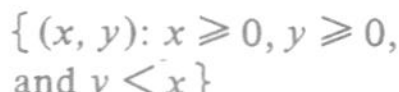

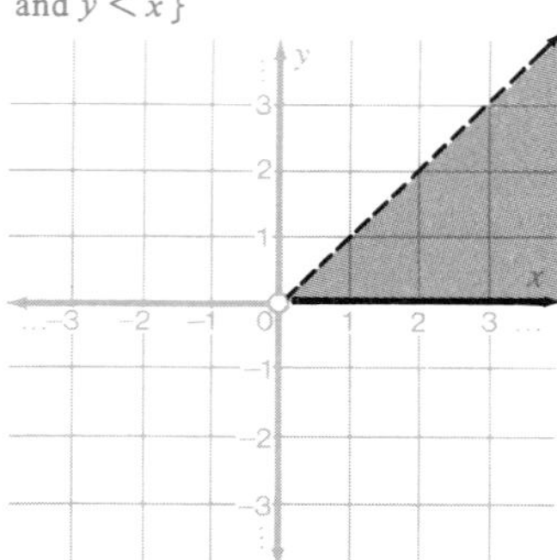

FIGURE 1

$\{(x, y): x < 0, y < 0, \text{ and } y > x\}$

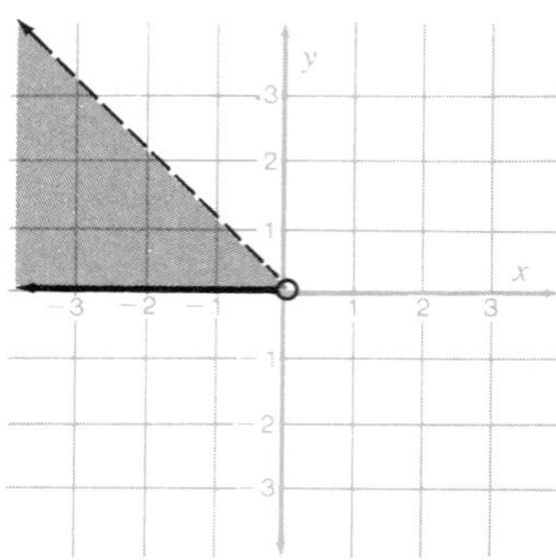

FIGURE 2

$\{(x, y): x < 0, y \geqslant 0, \text{ and } y < -x\}$

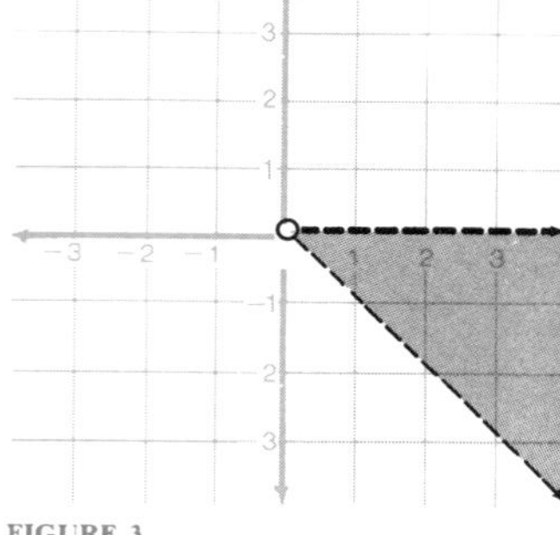

FIGURE 3

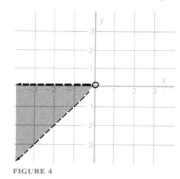

$\{(x, y): x \geqslant 0, y < 0, \text{ and } y > -x\}$

FIGURE 4

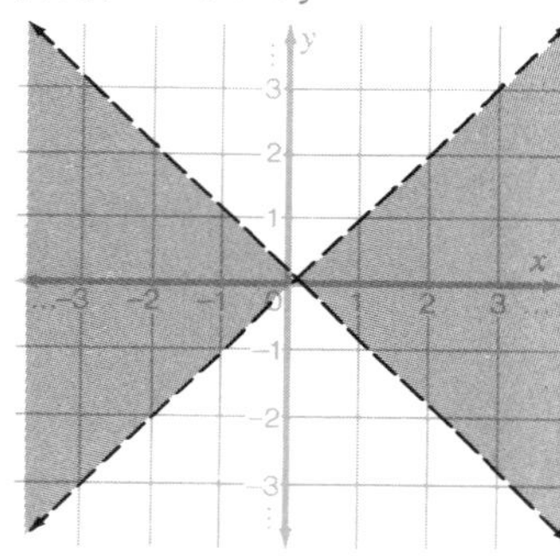

FIGURE 5

The solution set for the original sentence is the union of the solution sets for each component, and its graph is Figure 5.

Exercises

1. If R is the replacement set for x, write an equivalent compound sentence whose components are linear sentences for each inequality.

   a. $|2x| - x < 4$ b. $|x| + x > 0$ c. $|x| - \frac{1}{2}x > 2$

   d. $|x| < 5$ e. $|x - 2| < 4$

2. Graph the solution set for each inequality in Exercise 1.
3. Graph the solution set for each inequality in Exercise 1 if $R \times R$ is the replacement set for (x, y).

Sketch the graph of the solution set for each sentence, in Exercises 4–9, if $R \times R$ is the replacement set for (x, y):

4. $|x| + y = 1$ 5. $|x - 2| < 1$ 6. $|x| + |y| \leqslant 1$

7. $x - |y| \leqslant 0$ 8. $|y + 3| > 2$ 9. $|x - y| > 1$

In Exercises 10–12, assume that a is a real number, $a \geqslant 0$, and the replacement set for x is R.

10. Prove that $|x| = a$ is equivalent to $x = a$ or $x = -a$.

11. Prove that $|x| < a$ is equivalent to $-a < x < a$.

12. Prove that $|x| > a$ is equivalent to $x < -a$ or $x > a$.

Review Exercises

1. If 10-1

$$A = \begin{bmatrix} 1 & 4 & 3 \\ -2 & 1 & 7 \\ 5 & 3 & 2 \end{bmatrix}, \quad B = \begin{bmatrix} -5 & 3 & 1 \\ 2 & -7 & 3 \\ 4 & 8 & -1 \end{bmatrix}, \quad \text{and} \quad I = \begin{bmatrix} 1 & 0 & 0 \\ 0 & 1 & 0 \\ 0 & 0 & 1 \end{bmatrix}$$

find the matrices $A + B$, and $(A + B) + I$.

2. Using the matrices in Exercise 1, find AB, BA, and AI. 10-2

3. Write the augmented matrix for the system 10-3

$$\begin{cases} x_1 + 5x_2 + 7x_3 = -6 \\ 3x_1 - 4x_2 - 3x_3 = 15 \\ -5x_1 + 2x_2 - 4x_3 = -20 \end{cases}$$

and solve by using the Gaussian elimination method.

4. Write the system 10-4

$$\begin{cases} 14x_1 + 6x_2 = -5 \\ 12x_1 + 3x_2 = -5 \end{cases}$$

in matrix form $AX = B$, compute A^{-1}, and solve the system by finding $A^{-1}B$.

5. Find $|A|$, $|B|$, and $|I|$ for the matrices in Exercise 1. 10-5

6. Graph the solution set of the system below if the replacement set for (x, y) is $R \times R$. 10-6

$$\begin{cases} x + 2y \leqslant 8 \\ x - 2y \leqslant 8 \\ x + 1 \geqslant 0 \end{cases}$$

7. Graph the solution set of the inequality $|x| - 2x > 2$ if the replacement set for x is R; if the replacement set for (x, y) is $R \times R$. 10-7

11. TRIGONOMETRIC FUNCTIONS

11-1 ANGLES AND DEGREE MEASURE

The word *trigonometry* is a combination of two Greek words meaning "triangle measurement." Isolated instances of the use of trigonometric ideas have been traced back as far as the pyramids of Egypt. However, the early systemic study of trigonometry is attributed to the ancient Greek mathematicians who were interested in trigonometry as a tool for astronomy and surveying. The study of trigonometric relations as functions originated in the seventeenth century. This development became possible only after the invention of adequate algebraic symbolism. In this chapter we present the characteristics of the trigonometric functions. Trigonometry as triangle measurement will be considered in Chapter 14.

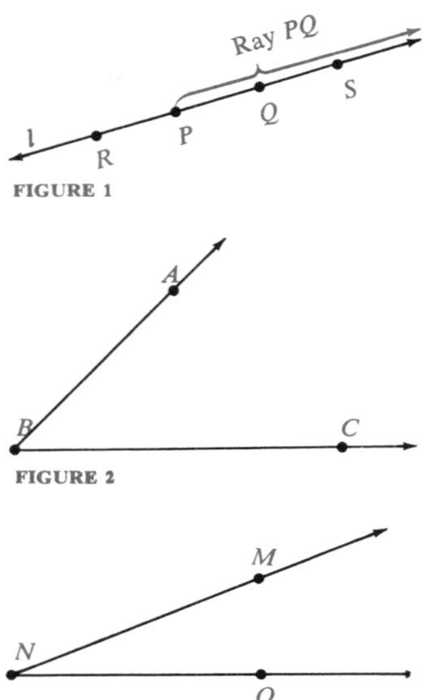

FIGURE 1

FIGURE 2

FIGURE 3

Before introducing the trigonometric relations it is helpful to recall several geometric concepts. Consider a line l and a point P on line l [Figure 1]. The union of P and the set of all points on one "side" of P is a **ray** and P is the **endpoint** of the ray. A ray is named by first indicating its endpoint and then any other point on the ray. In Figure 1, we have indicated ray PQ, which in symbols may be written $\overrightarrow{PQ}$.

In geometry, an **angle** is defined to be the union of two rays with a common endpoint called the **vertex**. Each of the rays is called a **side** of the angle. The angle pictured in Figure 2 is the union of rays $\overrightarrow{BA}$ and $\overrightarrow{BC}$. Point B is the vertex of the angle; the rays, $\overrightarrow{BA}$ and $\overrightarrow{BC}$, are the sides of the angle. The angle is named angle ABC, angle CBA, or since there is only one angle with point B as vertex, angle B (in symbols $\angle ABC$, $\angle CBA$, or $\angle B$).

In trigonometry it is helpful to think of an angle as a figure obtained by **rotating** a ray about its endpoint. The original ray is called the **initial side** of the angle, and the ray into which the initial side is rotated is called the **terminal side** of the angle. Thus, in trigonometry, we think of an angle as the union of two rays with a common endpoint (a set of points) and a rotation. For example, in geometry $\angle MNO = \overrightarrow{NM} \cup \overrightarrow{NO}$ [Figure 3]; however, in trigonometry $\angle MNO$ is the set of points $\overrightarrow{NM} \cup \overrightarrow{NO}$ and a rotation

$\overrightarrow{PQ}$, $\overrightarrow{UV}$; $\overrightarrow{PR}$, $\overrightarrow{UT}$; Counterclockwise, clockwise; less than one revolution, more than one revolution; $\angle QPR$, $\angle VUT$

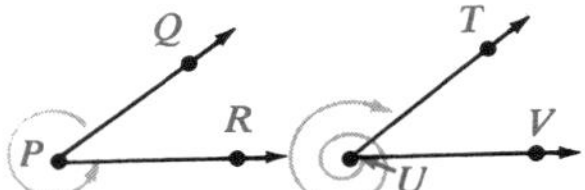

which takes $\overrightarrow{NM}$ into $\overrightarrow{NO}$. Thus, the same set of points pictures a different angle for each rotation that generates $\angle MNO$ [Figure 4].

In Figure 5, $\overrightarrow{EF}$ is the initial side, $\overrightarrow{ED}$ is the terminal side, and the arrow-tipped arc indicates that $\angle FED$ was obtained by rotating $\overrightarrow{EF}$ counterclockwise about its endpoint. To indicate an angle in trigonometry name a point on the initial ray, the vertex, and a point on the terminal ray, in that order.

In the figure what is the initial side in each case? The terminal side? Is the rotation clockwise or counterclockwise? More than one revolution or less than one revolution? Name the angle in each case.

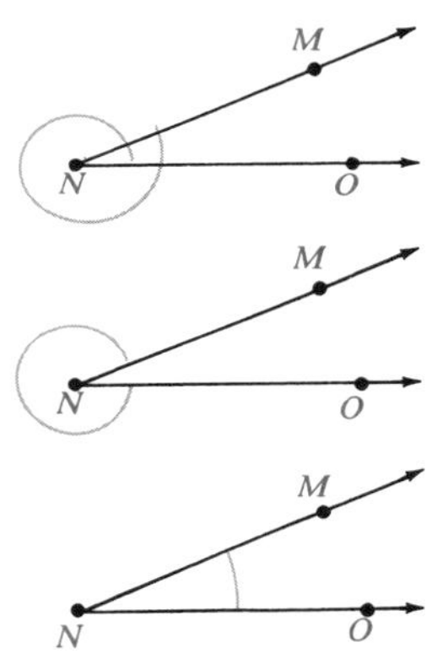

FIGURE 4

To measure $\angle ABC$, place its vertex at the center of a circle. Let the intersection of the initial side of the angle and the circle correspond to zero and divide the circle into 360 congruent arcs called **degrees** [Figure 6]. Then the number of degrees through which the initial side of an angle rotates is the degree measure of the angle, denoted $m_d \angle ABC$. More precise angle measurement may be obtained by dividing each degree into 60 congruent arcs. Each of these arcs is called a **minute**. The division of a minute into 60 congruent arcs gives a subunit of a degree called a **second**. Degrees are symbolized by °, minutes by ′, and seconds by ″. Thus, 43°17′45″ means 43 degrees, 17 minutes, and 45 seconds.

By convention, if an angle is the result of rotating the initial side in a counterclockwise direction, it is assigned a positive measure; if the rotation is clockwise, the angle is assigned a negative measure [Figure 7]. This concept of an angle is sometimes called a **directed angle**. An angle is said to be in **standard position** if its vertex is at the origin and if its initial side is on the positive-horizontal axis of a rectangular coordinate system [Figure 8].

The degree measure of an angle is a real number. To see this suppose an angle has measure 3°7′5″. The measure of this angle in degree units only is $3 + 7\left(\frac{1}{60}\right) + 5\left(\frac{1}{60}\right)^2 = \frac{11225}{3600}$. Degree measure can be interpreted as a

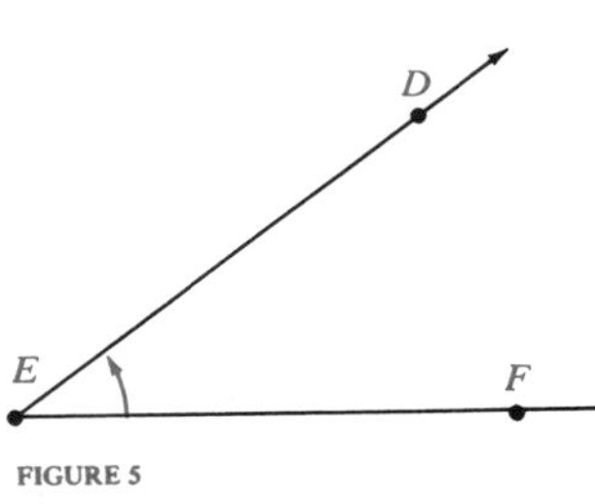

FIGURE 5

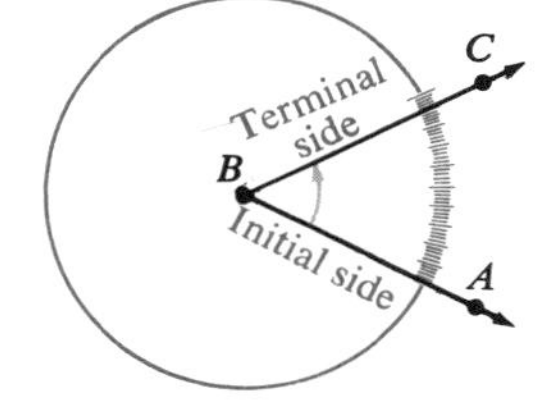

FIGURE 6

Angle with positive measure

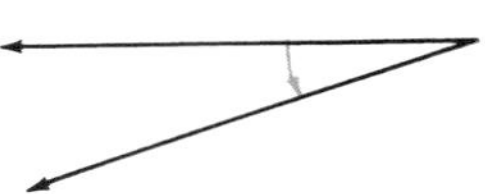

Angle with negative measure

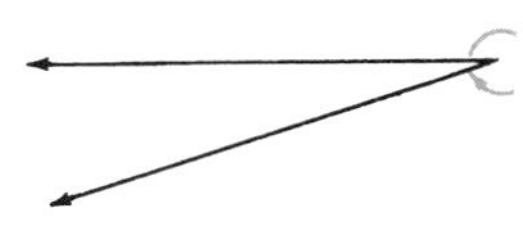

FIGURE 7

Angle in standard position

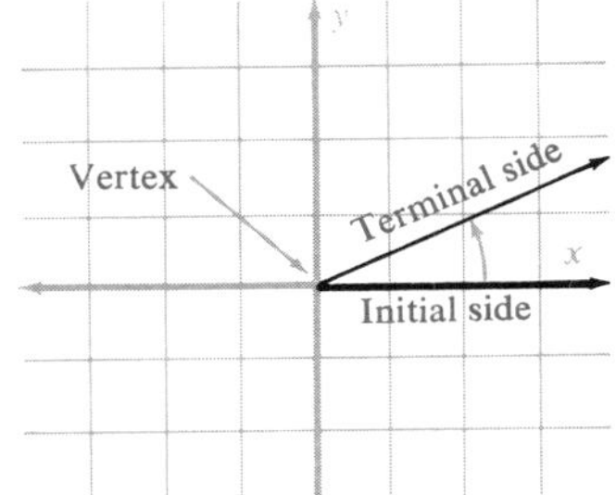

FIGURE 8

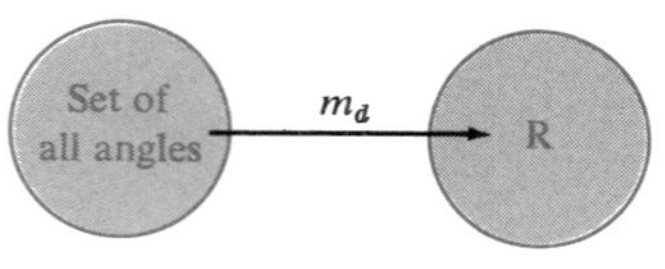

FIGURE 9

function m_d from the set of all angles into the set of real numbers [Figure 9]. Negative real numbers correspond to degree measures of angles that were formed by rotating the initial side in a clockwise direction. The notation $\angle A$ can be used to represent a trigonometric angle, with rotation being understood from the sign of its measure.

$m_d \angle B = \frac{121}{4}$, $m_d \angle C = 5.3433$, $m_d \angle D = -\frac{2101}{600}$

If $\angle B$ has measure $30°15'$, find $m_d \angle B$. If $\angle C$ has measure $5°20'36''$, find a decimal approximation for $m_d \angle C$ to the nearest ten thousandth. If $m_d \angle D = -m_d \angle E$ and $m \angle E = 3°30'6''$, find $m_d \angle D$.

Exercises

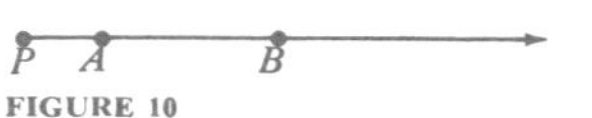

FIGURE 10

1. Let points P, A, and B be as shown in Figure 10. Consider the rays, $\overrightarrow{PA}$ and $\overrightarrow{PB}$. Are they the same set of points?
2. State whether the angles pictured have positive or negative measure.

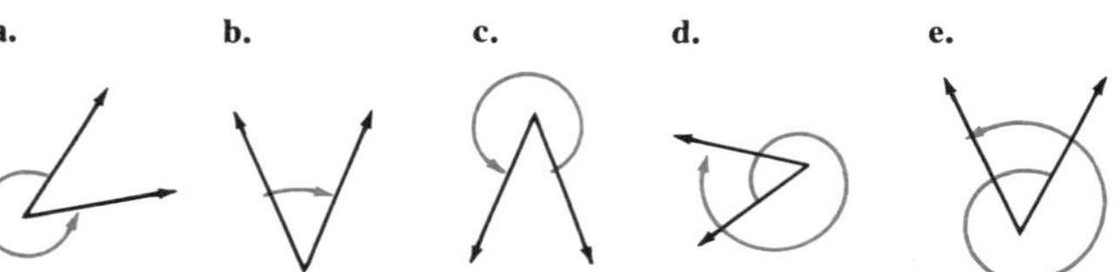

3. The rotation generating the angles pictured is indicated by an arrow-tipped arc. Copy each drawing and label its initial and terminal sides. Indicate whether the measure of the angle is positive or negative.

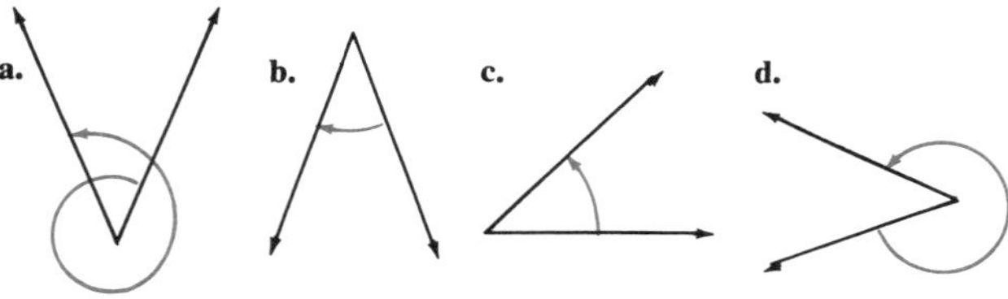

4. a. Name the angles shown in Figure 11 which are in standard position.
   b. Name the initial and terminal sides of each angle named in part a.
5. Draw a picture of an angle in standard position with positive measure whose initial side has rotated through the negative x-axis.
6. Three distinct points of a line are labeled X, Y, and Z. How many different rays can be named using the names of these points?
7. Let A, B, and C be three noncollinear points. How many rays containing two of the points (one as the endpoint) are determined by the three points?
8. a. What is the intersection of the rays in Exercise 7?
   b. What is the union of the rays in Exercise 7?

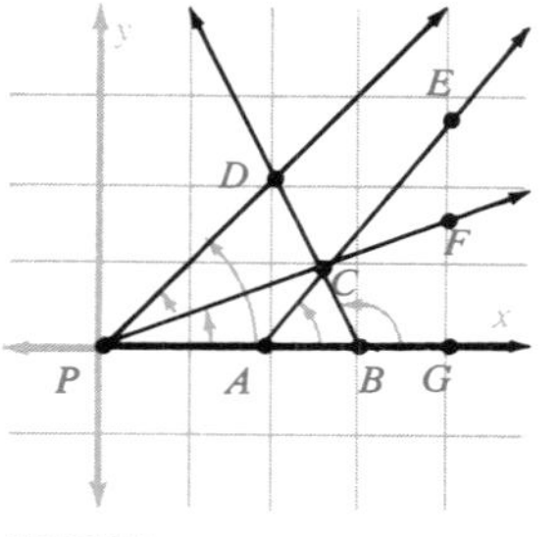

FIGURE 11

9. If five points of a line are labeled, how many different rays can be named using the names of these points?

11-2 RADIAN MEASURE

Mathematicians often use a different unit to measure angles, called radian measure, to simplify many formulas found in mathematics and physics.

Definition Let $\angle A$ be an angle with vertex at the center of a circle of radius t. Let the "length" of the arc that the angle subtends be s [Figure 1]. The **radian measure** of $\angle A$ is $\frac{s}{t}$. In symbols: $m_r \angle A = \frac{s}{t}$.

We will frequently use the Greek letter θ (theta) to represent the radian measure of an angle; thus $\theta = \frac{s}{t}$.

FIGURE 1

Three assumptions underlie the definition of the radian measure of an angle:

1. Both the length of the radius and the "length" of the arc subtended by the angle are measured in the same linear units.
2. The "length" is positive or negative depending on whether the rotation is counterclockwise or clockwise.
3. The radius t is always positive.

From the first assumption, $\frac{s}{t}$, the radian measure of the angle, is a real number independent of the unit of the linear measure for s and t. The last two assumptions justify assigning a positive number to an angle formed by a counterclockwise rotation and a negative number to an angle formed by a clockwise rotation.

$m_r \angle E = \frac{2}{3}$.

If the vertex of $\angle E$ is at the center of a circle of radius 1 foot and $\angle E$ subtends an arc of 8 inches, find $m_r \angle E$. Sketch two pairs of angles, $\angle F$ and $\angle G$, such that $m_r \angle F = -m_r \angle G$.

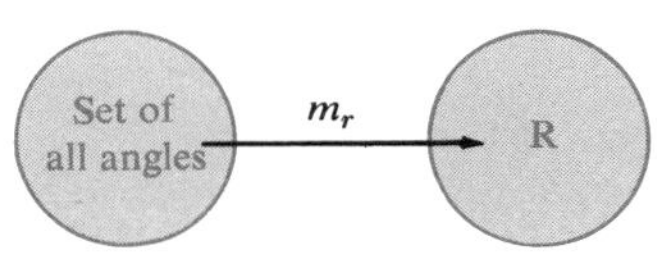

FIGURE 2

As is true of degree measure for angles, radian measure for angles is a function from the set of all angles into the set of real numbers [Figure 2] and is independent of the length of the radius of the circle. If we know any two of the three numbers—the radian measure of an angle, the "length" s of the subtended arc, and the length t of the radius of the circle—we can compute the other.

Example 1 Find the length t of the radius of a circle if a central angle of the circle with measure 2 radians subtends an arc of 10 inches.

Solution: Since $\theta = \frac{s}{t}$, we have

$$t = \frac{s}{\theta}$$

$$t = \frac{10\text{ (inches)}}{2} = 5\text{ (inches)}$$

Example 2 If the length of the radius of a circle is 8 feet, find the radian measure θ of a central angle which subtends an arc of 2 feet.

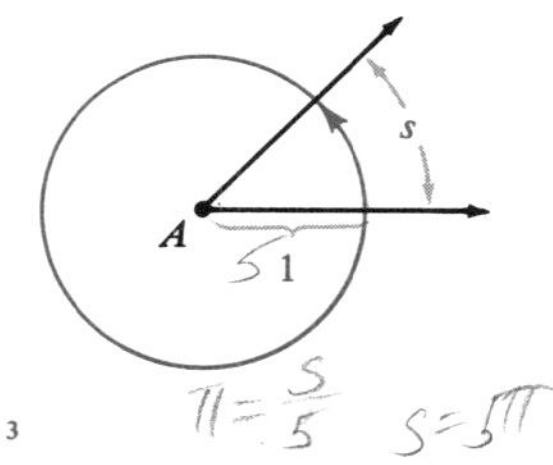

FIGURE 3

Solution: Since $\theta = \frac{s}{t}$, we have

$$\theta = \frac{2\text{ (feet)}}{8\text{ (feet)}} = \frac{1}{4}$$

$s = 5\pi$ (meters)

If the length of the radius of a circle is 5 meters, find the length of the arc subtended by a central angle with measure π radians.

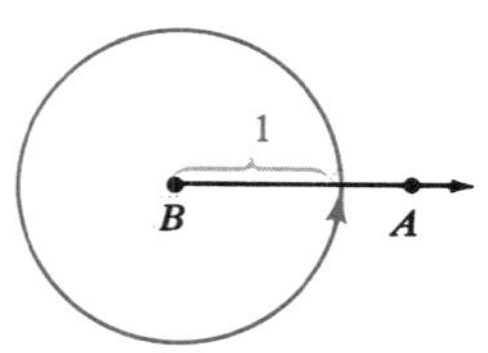

FIGURE 4

Notice in Figure 3 that if $t = 1$ then $m_r \angle A = \frac{s}{1} = s$. That is, the radian measure of a central angle of a circle with unit radius is equal to the "length" of the arc subtended. Moreover, the length of the arc subtended by $\angle ABA$ [Figure 4] which is obtained by a counterclockwise rotation of $\overrightarrow{BA}$ around to its original position is equal to 2π, the circumference of the circle. Therefore,

$$m_r \angle ABA = \frac{s}{t} = \frac{2\pi(1)}{1} = 2\pi.$$

$m_r \angle ABD = \pi$, $m_r \angle ABC = \frac{\pi}{2}$, $m_r \angle ABE = \frac{3\pi}{2}$

If each angle in the figure is in standard position, what is $m_r \angle ABD$? $m_r \angle ABC$? $m_r \angle ABE$?

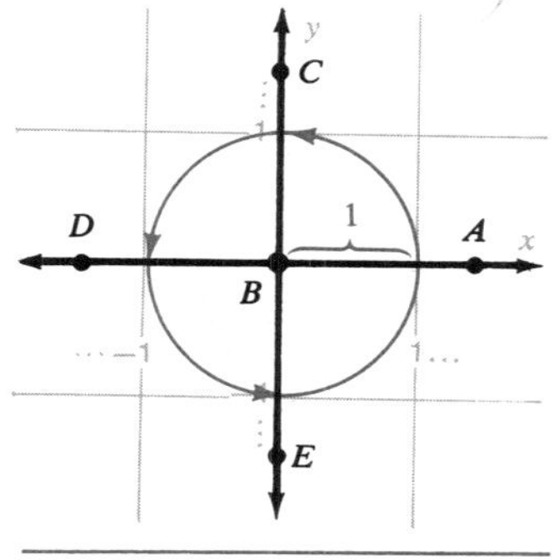

As we have seen, m_d and m_r are both functions from the set of all angles into the set of real numbers. However, each function assigns a different number to the same angle with one exception: If $m_r \angle A = 0$, then $m_d \angle A = 0$. For example, if $m_r \angle B = -\frac{3\pi}{2}$, then $m_d \angle B = -270$; and if $m_r \angle C = \pi$, then $m_d \angle C = 180$.

$m_r \angle B = -\frac{\pi}{2}$ and $m_d \angle B = -90$; $\frac{-\frac{\pi}{2}}{2\pi} = -\frac{1}{4}$ and $-\frac{90}{360} = -\frac{1}{4}$; $\frac{-\frac{\pi}{2}}{-90} = \frac{\pi}{180}$; $\frac{-\frac{3\pi}{2}}{-270} = \frac{\pi}{180}$. The ratio of the radian measure to the degree measure of each of the angles is $\frac{\pi}{180}$.

Find the radian measure and the degree measure of an angle, $\angle B$, in standard position formed by a clockwise rotation with terminal side the negative y-axis; the ratio of the radian measure to 2π and the ratio of the degree measure to 360; the ratio of the radian measure to the degree measure; the ratio of the radian measure of $\angle B$ and $\angle C$ mentioned above to the degree measure of the same angle. What do you observe about these ratios?

Thus, if the radius of a circle is t, the length of each degree is $\frac{1}{360}$ of the circumference of the circle or $\frac{2\pi t}{360}$. The degree measure of an angle is the number of degree units in the arc on the circle subtended by the angle with vertex at the center. If the subtended arc has "length" s, then the degree measure of the angle is

$$\frac{s}{\frac{2\pi t}{360}}.$$

On the other hand, the radian measure of the angle that subtends an arc of a circle of "length" s is $\frac{s}{t}$. We can now derive the following theorem.

Theorem 1 The ratio of the nonzero radian measure of any angle, $\angle A$, to the degree measure of the same angle is the constant $\frac{180}{\pi}$.

Proof: Since $m_r \angle A \neq 0$, $m_r \angle A \neq m_d \angle A$. Using the observations in the paragraph preceding the statement of the theorem, we see that

$$m_d \angle A = \frac{s}{\frac{2\pi t}{360}} = \frac{s}{t} \cdot \frac{360}{2\pi} = m_r \angle A \cdot \frac{180}{\pi}.$$

Hence, $\frac{m_d \angle A}{m_r \angle A} = \frac{180}{\pi}$.

Theorem 2 For any angle, $\angle A$,

$$m_r \angle A = \frac{\pi}{180} m_d \angle A \quad \text{and} \quad m_d \angle A = \frac{180}{\pi} m_r \angle A.$$

Thus, if $m_d \angle A = 1$, then $m_r \angle A = \frac{\pi}{180}$ (approximately 0.017453 radians); if $m_r \angle A = 1$, then $m_d \angle A = \frac{180}{\pi}$ (approximately $57°17'45''$).

Example 1 Find the radian measure of $\angle A$ if $m_d \angle A = 120$.

Solution: $m_r \angle A = \dfrac{\pi}{180} m_d \angle A = \dfrac{\pi}{180}(120) = \dfrac{2\pi}{3}$.

Example 2 Find the degree measure of $\angle B$ if $m_r \angle B = 3\pi$.

Solution: $m_d \angle B = \dfrac{180}{\pi} m_r \angle B = \dfrac{180}{\pi}(3\pi) = 540$.

Exercises

1. Let t be the length of the radius of a circle, θ the radian measure of a central angle of the circle, and s the "length" of the arc subtended by the central angle. Find t, s, or θ for each of the following sets of data.

a. $t = 0.75$, $\theta = \dfrac{3\pi}{4}$ b. $t = 10$, $\theta = -\dfrac{\pi}{10}$

c. $\theta = \dfrac{3}{2}$, $s = 5\pi$ d. $\theta = -3.5$, $s = -7$

e. $s = -\dfrac{3\pi}{5}$, $t = \pi$ f. $s = \pi$, $t = 1$

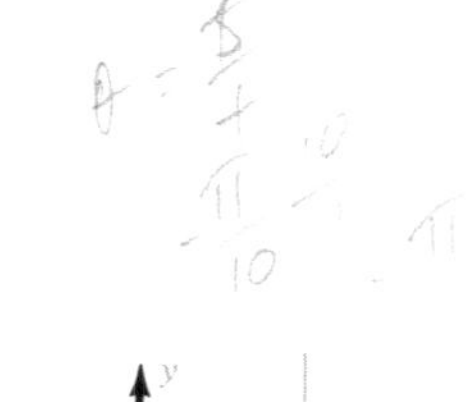
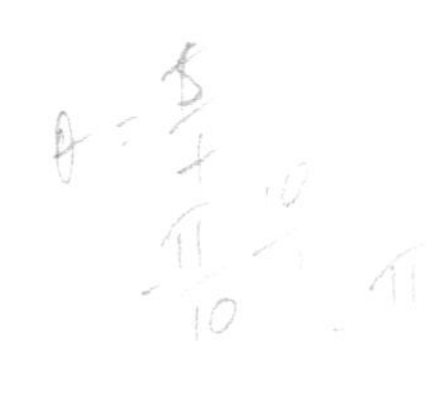

FIGURE 5

2. Find the radian measure of each of the following angles in standard position [Figure 5].

a. $m_r \angle ABA$ b. $m_r \angle ABD$

c. $m_r \angle ABE$ d. $m_r \angle ABC$

3. Find the radian measure corresponding to each angle measure.

a. $30°$ b. $90°$ c. $(-15)°$

d. $270°$ e. $300°$ f. $(-135)°$

g. $450°$ h. $6°30'$ i. $31°30'$

4. Find the degree measure corresponding to each radian measure.

a. 3π b. $-\dfrac{\pi}{3}$ c. -1.876

d. $-\dfrac{\pi}{4}$ e. $\dfrac{\pi}{12}$ f. $\dfrac{2\pi}{3}$

g. 2 h. $\dfrac{7\pi}{4}$ i. $-\dfrac{5\pi}{4}$

5. In what quadrant is the terminal side of the central angles in standard position with the following measures?

a. $135°$ b. $-\dfrac{15\pi}{4}$ radians c. $525°$

d. $\dfrac{5\pi}{4}$ radians e. $300°$ f. $-\dfrac{4\pi}{3}$ radians

g. $(-30)°$ h. $\dfrac{\pi}{12}$ radians i. $-\dfrac{3\pi}{4}$ radians

6. If the radius of a wagon wheel is 2 feet find the distance the wagon moves if the wheel travels through an angle with measure 7 radians.

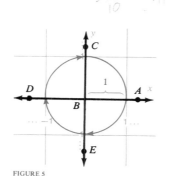

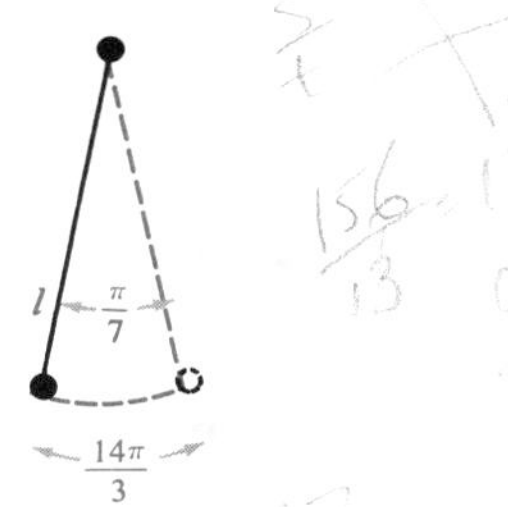

FIGURE 6

7. A bicycle with 26 inch wheels travels 13 feet. What is the radian measure of the angle through which a wheel of the bicycle has turned?

8. Find the length l of a pendulum if the tip of the pendulum traces an arc of length $\frac{14\pi}{3}$ and the measure of the angle this arc subtends is $\frac{\pi}{7}$ radians [Figure 6].

9. Friction gears use friction to transmit their motion from one to another. If a friction gear with radius 6 inches that is in contact with a friction gear with radius 4 inches moves through an angle of measure 5 radians, through how many radians does the friction gear with radius 4 inches move?

10. If the hour hand of a clock points to 6 and you move the hand back (counterclockwise) to 2, what is the radian measure of the angle through which you turn it? If you move the hour hand ahead (clockwise) from 6 to 2, what is the radian measure of the angle through which you turn it? Can you check your work by considering your answers to the previous two questions? Is the check foolproof or could you have made a mistake in answering the first two questions?

11. If the second hand on a wristwatch is $\frac{1}{2}$ inch long, how far does the tip of the hand travel in 20 seconds? In 75 seconds?

12. The moon subtends an angle of approximately 32′ at the center of the earth. If the approximate distance from the center of the earth to the moon is 240,000 miles, approximate the diameter of the moon.

13. New Orleans and Port Arthur, Canada, have approximately the same longitude. If the latitude of New Orleans is about 30°N and of Port Arthur is about 48°N, approximate the distance between these cities. Use 4000 miles for the length of the radius of the earth.

11-3 THE WRAPPING FUNCTION

Since the radian measure of an angle is independent of the length of the radius of the circle chosen, we can simplify calculations by choosing a circle whose radius has length 1 (called the **unit circle**). For when a unit circle is centered at the vertex of an angle

$$\theta = \frac{s}{t} = \frac{s}{1} = s,$$

that is, the "length" of the arc subtended by an angle in the unit circle and the radian measure of the angle are equal.

To see that every real number is the radian measure of some angle, let θ be a real number. Start at point $(1, 0)$ and move along the unit circle $|\theta|$

TABLE 1

θ	$P(\theta) = P(\theta + 2\pi n)$
0	(1, 0)
$\frac{\pi}{2}$	(0, 1)
π	(−1, 0)
$\frac{3\pi}{2}$	(0, −1)

units in a clockwise direction if θ is negative and in a counterclockwise direction if θ is positive [Figure 1]. In this way a unique point (x, y) on the unit circle is located. The points (x, y) and $(0, 0)$ determine the terminal side of an angle in standard position with radian measure θ. A function whose domain is the set of real numbers and whose range is the set of all ordered pairs for points on the unit circle is defined by this procedure. We call this function the **wrapping function *P*.** Note that

$$P(\theta) = (x, y)$$

where (x, y) are the coordinates of the point on the unit circle associated with the real number θ. The real number θ can be interpreted as the radian measure of an angle in standard position with the point $P(\theta)$ on its terminal side. Given a real number θ the coordinates $P(\theta)$ are, in general, difficult to find. However, there are many real numbers θ for which the coordinates $P(\theta)$ can be found easily. Consider Figure 2. If $\theta = 0$ or $\theta = 2\pi$ (the circumference of the unit circle), then $P(\theta) = (1, 0)$. P associates $(0, 1)$ with $\theta = \frac{\pi}{2}$. (Why?) If $\theta = -\frac{\pi}{2}$, then $P(\theta) = (0, -1)$.

θ	π	$-\pi$	$\frac{3\pi}{2}$	$-\frac{3\pi}{2}$
$P(\theta)$	(−1, 0)	(−1, 0)	(0, −1)	(0, 1)

$(1, 0) = P(4\pi) = P(6\pi) = P(8\pi)$
$= P(-2\pi) = P(-4\pi) = P(-8\pi)$

What is $P(\theta)$ if θ is equal to $\pi, -\pi, \frac{3\pi}{2}$, or $-\frac{3\pi}{2}$? What is $P(\theta)$ if θ is equal to $4\pi, 6\pi, 8\pi, -2\pi, -4\pi$, or -8π?

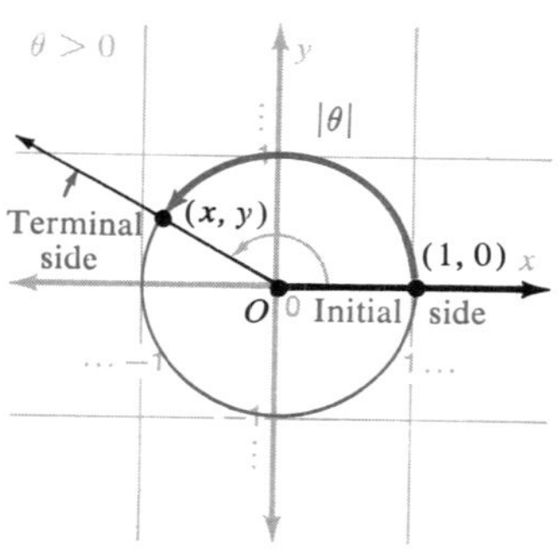

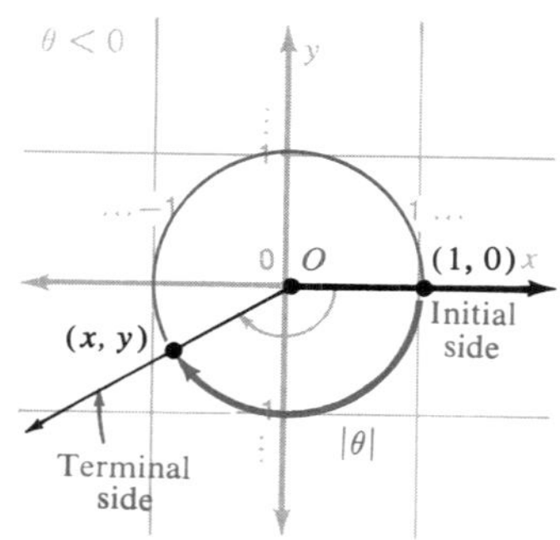

FIGURE 1

Notice that if θ is equal to $\frac{5\pi}{2}$ $\left(\text{or } \frac{\pi}{2} + 2\pi\right)$, $-\frac{3\pi}{2}$ $\left(\text{or } \frac{\pi}{2} - 2\pi\right)$, $\frac{9\pi}{2}$ $\left(\text{or } \frac{\pi}{2} + 4\pi\right)$, or $-\frac{7\pi}{2}$ $\left(\text{or } \frac{\pi}{2} - 4\pi\right)$, then $P(\theta) = (0, 1)$ in each case. Since the circumference of the unit circle is 2π, if we move 2π units along the circle from the point $P(\theta)$ in either a counterclockwise or clockwise direction, we return to the point $P(\theta)$. Therefore, *the coordinates of the point that P associates with any real number θ are also the coordinates of the point associated with any one of the infinite set of real numbers $\{\theta + 2\pi n$: n is an integer$\}$*. Interpreting a real number as the radian measure of an angle, the previous statement implies that there are infinitely many angles in standard position with radian measures that differ by an integral multiple of 2π that have the same terminal side. Hereafter, we will denote $\{\theta + 2\pi n$: n is an integer$\}$ by $\{\theta + 2\pi n\}$. The coordinates of the points on the unit circle associated with the real numbers considered thus far are summarized in Table 1.

In the following examples we show how the coordinates $P(\theta)$ for several special real numbers θ can be found. The point $P(\theta) = (x, y)$ is on the unit circle [Figure 3]; hence, by the Pythagorean theorem its coordinates satisfy the equation

$$x^2 + y^2 = 1.$$

For special θ, we can find a second relationship between the coordinates $P(\theta)$ and solve for x and y.

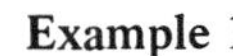

Example 1 Find the coordinates $P(\theta)$ when $\theta = \dfrac{\pi}{4}$.

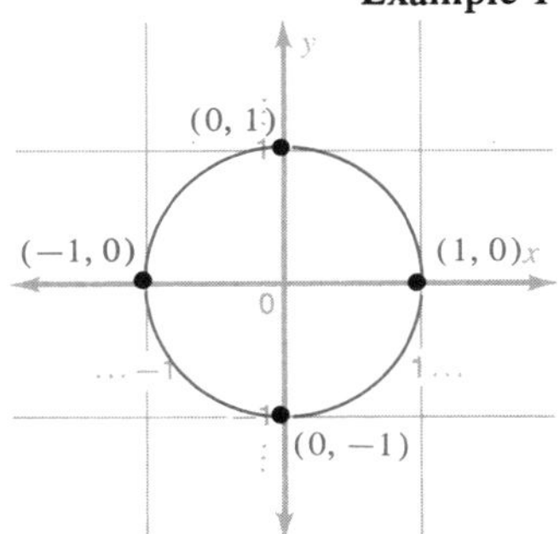

FIGURE 2

Solution: Since $\dfrac{\pi}{4}$ units is one-eighth of the circumference of the circle, the point $P\left(\dfrac{\pi}{4}\right)$ is the midpoint of the arc of the circle joining the points (1, 0) and (0, 1) [Figure 4]. Thus, the point $P\left(\dfrac{\pi}{4}\right)$ is equidistant from the x- and y-axes and so $x = y$. Solving the system

$$\begin{cases} x^2 + y^2 = 1 \\ x = y \end{cases}$$

we have: $$x = \frac{\sqrt{2}}{2} \quad \text{or} \quad x = -\frac{\sqrt{2}}{2}$$

Since x and y are both positive, $P\left(\dfrac{\pi}{4}\right) = \left(\dfrac{\sqrt{2}}{2}, \dfrac{\sqrt{2}}{2}\right)$.

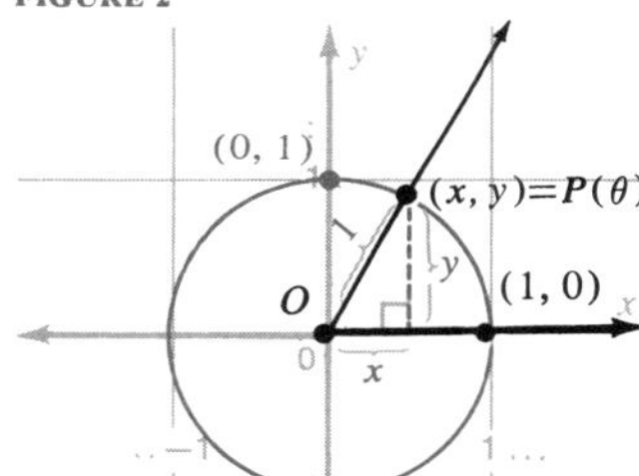

FIGURE 3

What are the coordinates $P(\theta)$ when θ is equal to $\dfrac{9\pi}{4}$ $\left(\text{or } \dfrac{\pi}{4} + 2\pi\right)$, $\dfrac{17\pi}{4}$ $\left(\text{or } \dfrac{\pi}{4} + 4\pi\right)$, $-\dfrac{7\pi}{4}$ $\left(\text{or } \dfrac{\pi}{4} - 2\pi\right)$, and $-\dfrac{15\pi}{4}$ $\left(\text{or } \dfrac{\pi}{4} - 4\pi\right)$? Was it necessary to do a great deal of computation to answer the previous question?

$P\left(\dfrac{9\pi}{4}\right) = P\left(\dfrac{17\pi}{4}\right) = P\left(-\dfrac{7\pi}{4}\right) = P\left(-\dfrac{15\pi}{4}\right) = \left(\dfrac{\sqrt{2}}{2}, \dfrac{\sqrt{2}}{2}\right)$; No

---

Example 2 What are the coordinates $P(\theta)$ when $\theta = \dfrac{3\pi}{4}$?

Solution: In Figure 5 $\angle QOR \cong \angle TOS$. Thus, $\triangle QOR$ and $\triangle TOS$ are congruent right triangles. (Why?) Then $\overline{OR} \cong \overline{OS}$ and $\overline{QR} \cong \overline{TS}$, which implies that the pairs of segments have equal measure. But $P\left(\dfrac{3\pi}{4}\right)$ corresponds to Q, a point in the second quadrant, and $P\left(\dfrac{\pi}{4}\right)$ corresponds to T, a point in the first quadrant. Hence, the x-coordinate of $P\left(\dfrac{3\pi}{4}\right)$ is the negative of the x-coordinate of $P\left(\dfrac{\pi}{4}\right)$; the y-coordinates of $P\left(\dfrac{3\pi}{4}\right)$ and $P\left(\dfrac{\pi}{4}\right)$ are equal. Since $P\left(\dfrac{\pi}{4}\right) = \left(\dfrac{\sqrt{2}}{2}, \dfrac{\sqrt{2}}{2}\right)$, the coordinates $P\left(\dfrac{3\pi}{4}\right)$ are

$$x = -\frac{\sqrt{2}}{2} \quad \text{and} \quad y = \frac{\sqrt{2}}{2}.$$

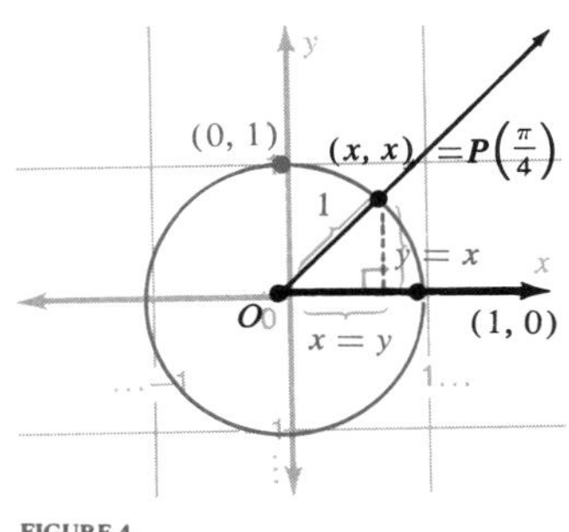

FIGURE 4

$$P(\theta) = P\left(\frac{3\pi}{4} + 2\pi n\right) = \left(-\frac{\sqrt{2}}{2}, \frac{\sqrt{2}}{2}\right)$$

What are the coordinates $P(\theta)$ when $\theta \in \left\{\frac{3\pi}{4} + 2\pi n\right\}$?

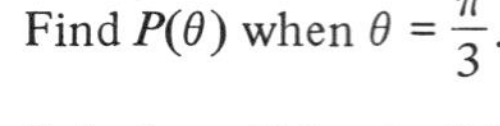

Example 3 Find $P(\theta)$ when $\theta = \frac{\pi}{3}$.

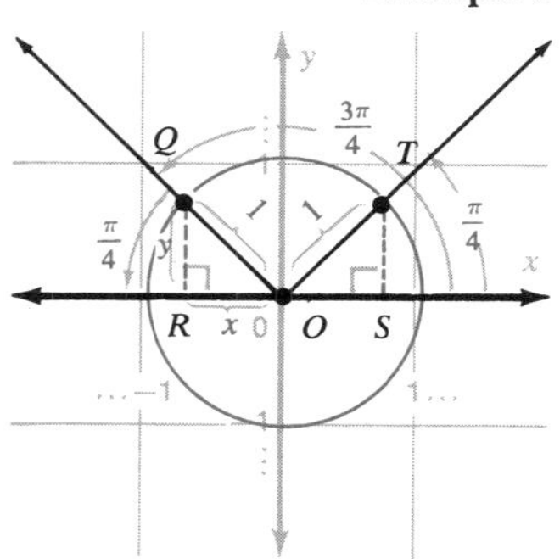

FIGURE 5

Solution: Triangle RQO is a 30-60-right triangle with $m_d\ \angle ORQ = 30$ and $m_d\ \angle ROQ = 60$ [Figure 6]. (Why?) Since the length of the side opposite the 30° angle in a 30-60-right triangle is equal to one-half the length of the hypotenuse, the measure of $\overline{OQ} = x = \frac{1}{2}$. Substituting $\frac{1}{2}$ for x in the equation $x^2 + y^2 = 1$ we find

$$y = \frac{\sqrt{3}}{2} \quad \text{or} \quad y = -\frac{\sqrt{3}}{2}.$$

From Figure 6 it is apparent that y is positive; hence, $P\left(\frac{\pi}{3}\right) = \left(\frac{1}{2}, \frac{\sqrt{3}}{2}\right)$.

The coordinates $P(\theta)$ associated with $\theta = \frac{\pi}{6}$ can be determined by similar means. Verify that $P\left(\frac{\pi}{6}\right) = \left(\frac{\sqrt{3}}{2}, \frac{1}{2}\right)$. Coordinates of points on the unit circle corresponding to the values of θ which we have considered are listed in Table 2. The coordinates of the points that P associates with other real numbers θ which have been computed by more complicated means are available in tables.

$$P\left(\frac{\pi}{3} + 2\pi n\right) = \left(\frac{1}{2}, \frac{\sqrt{3}}{2}\right)$$
$$P\left(\frac{\pi}{6} + 2\pi n\right) = \left(\frac{\sqrt{3}}{2}, \frac{1}{2}\right)$$

What are the coordinates of the point(s) associated with the numbers $\frac{\pi}{3} + 2\pi n$? With the numbers $\frac{\pi}{6} + 2\pi n$?

Exercises

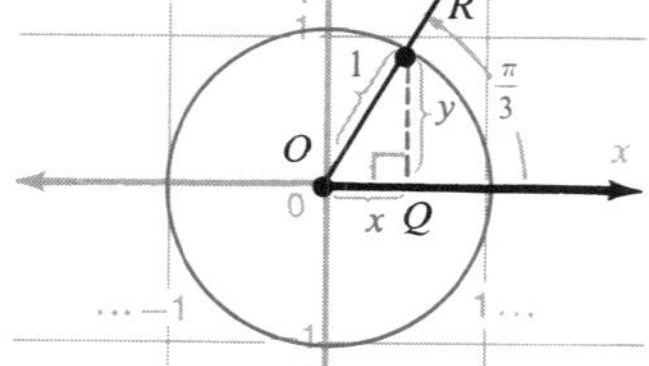

FIGURE 6

1. Find the coordinates $P(\theta)$ for any number θ other than 0, $\frac{\pi}{2}$, or $\frac{3\pi}{2}$ in each of the sets.

   a. $\{0 + 2\pi n\}$ b. $\left\{\frac{\pi}{2} + 2\pi n\right\}$ c. $\left\{\frac{3\pi}{2} + 2\pi n\right\}$

2. Find the approximate location of the point $P(\theta)$ on a graph of the unit circle which is associated with each value of θ.

   a. $\frac{5\pi}{3}$ b. $m_d(\theta) = 30$ c. $\frac{2\pi}{3}$

   d. $-\frac{7\pi}{3}$ e. $m_d(\theta) = 135$ f. $\frac{\pi}{2}$

   g. $m_d(\theta) = 720$ h. $m_d(\theta) = 90$ i. $\frac{3\pi}{2}$

TABLE 2

θ	$P(\theta) = P(\theta + 2\pi n)$
0	(1, 0)
$\frac{\pi}{6}$	$\left(\frac{\sqrt{3}}{2}, \frac{1}{2}\right)$
$\frac{\pi}{4}$	$\left(\frac{\sqrt{2}}{2}, \frac{\sqrt{2}}{2}\right)$
$\frac{\pi}{3}$	$\left(\frac{1}{2}, \frac{\sqrt{3}}{2}\right)$
$\frac{\pi}{2}$	(0, 1)
$\frac{3\pi}{4}$	$\left(-\frac{\sqrt{2}}{2}, \frac{\sqrt{2}}{2}\right)$
π	(-1, 0)
$\frac{3\pi}{2}$	(0, -1)

j. 4π k. $-\frac{17\pi}{3}$ l. -7

m. $m_d(\theta) = -135$ n. $-\frac{\pi}{4}$ o. $m_d(\theta) = 300$

3. List the five smallest positive numbers and the five largest negative numbers in each set (n is an integer).

a. $\{\pi + 2\pi n\}$ b. $\left\{-\frac{\pi}{3} + n\pi\right\}$ c. $\left\{\frac{\pi}{2} + n\frac{\pi}{2}\right\}$

d. $\{7 + 2n\}$ e. $\left\{-\frac{\pi}{2} + 2\pi n\right\}$ f. $\left\{-3\pi + \left(\frac{5\pi}{2}\right)n\right\}$

g. $\{3\pi + 2\pi n\}$

4. List all multiples of $\frac{\pi}{2}$ that satisfy the following inequalities.

a. $|\theta| < \frac{\pi}{2}$ b. $|\theta| \leq \frac{\pi}{2}$ c. $-\frac{3\pi}{2} < \theta \leq \frac{5\pi}{2}$ d. $-\frac{\pi}{3} \leq \theta < \frac{\pi}{6}$

5. Determine which of the following correspond to points on the unit circle.

a. $\left(\frac{1}{2}, \frac{\sqrt{3}}{2}\right)$ b. $\left(\frac{\sqrt{3}}{2}, \frac{2}{2}\right)$ c. $(-1, 0)$

d. $\left(\frac{\sqrt{2}}{2}, \frac{\sqrt{2}}{2}\right)$ e. $\left(-\frac{1}{2}, \frac{1}{2}\right)$ f. $\left(\frac{1}{2}, \frac{1}{2}\right)$

g. $(0, 1)$ h. $(1, 1)$ i. $\left(-\frac{1}{2}, \frac{5}{8}\right)$

j. $(0, -1)$ k. $(-1, -1)$ l. $\left(\frac{\sqrt{3}}{2}, \frac{1}{2}\right)$

6. Find the set of real numbers that the wrapping function P associates with each ordered pair of Exercise 5 which corresponds to a point on the unit circle.

7. Find $P(\theta)$ for the following values of θ.

a. -8π b. $-\frac{15\pi}{4}$ c. $\frac{13\pi}{3}$ d. $\frac{19\pi}{6}$

e. $\frac{25\pi}{6}$ f. $-\frac{11\pi}{3}$ g. $\frac{11\pi}{4}$ h. $-\frac{23\pi}{6}$

i. $\frac{7\pi}{2}$ j. $-\frac{4\pi}{3}$ k. $\frac{7\pi}{4}$ l. $\frac{17\pi}{3}$

8. Find $P(\theta)$ for each value of θ.

a. $\pi + \frac{\pi}{2}$ b. $\frac{\pi}{3} + \frac{\pi}{6}$ c. $\frac{\pi}{2} + \frac{\pi}{4}$

d. $\frac{2\pi}{3} + \frac{3\pi}{2}$ e. $\frac{\pi}{4} - 32\pi$ f. $\pi + \frac{3\pi}{2}$

g. $\frac{5\pi}{2} + \frac{\pi}{4}$ h. $\frac{7\pi}{4} - 3\pi$ i. $16\pi - \frac{3\pi}{2}$

11-4 TRIGONOMETRIC FUNCTIONS

From the wrapping function we define six functions called the **trigonometric functions**: the sine, cosine, tangent, cotangent, secant, and cosecant functions. The abbreviations **sin, cos, tan, cot, sec,** and **csc** are used for these words. The six trigonometric functions are determined by the equations below.

Definition If $P(\theta) = (x, y)$ are the coordinates of the point on the unit circle associated with the real number θ by the wrapping function P, then

$$\sin\theta = y \qquad \csc\theta = \frac{1}{y} \quad (\text{if } y \neq 0)$$

$$\cos\theta = x \qquad \sec\theta = \frac{1}{x} \quad (\text{if } x \neq 0)$$

$$\tan\theta = \frac{y}{x} \quad (\text{if } x \neq 0) \qquad \cot\theta = \frac{x}{y} \quad (\text{if } y \neq 0).$$

Thus, the domain of the sine function is R and its range is the set of all second coordinates of points on the unit circle associated with these real numbers by the wrapping function P. Hence, if $\theta \in R$,

$$\text{sine } \theta = \sin\theta = y$$

where y is the second coordinate of $P(\theta)$. In a similar way,

$$\text{cosine } \theta = \cos\theta = x$$

where x is the first coordinate of $P(\theta)$.

Verify that the point for $\left(\frac{1}{4}, \frac{\sqrt{15}}{4}\right)$ is on the unit circle. If $\theta \in R$ such that $P(\theta) = \left(\frac{1}{4}, \frac{\sqrt{15}}{4}\right)$, find $\sin\theta$, $\cos\theta$, $\tan\theta$, $\frac{\sin\theta}{\cos\theta}$, $\cot\theta$, $\frac{\cos\theta}{\sin\theta}$, $\sec\theta$, $\csc\theta$, $(\sin\theta)(\csc\theta)$, $(\cos\theta)(\sec\theta)$, and $(\tan\theta)(\cot\theta)$.

$\left(\frac{1}{4}\right)^2 + \left(\frac{\sqrt{15}}{4}\right)^2 = 1$; ●

$\sin\theta$	$\cos\theta$	$\tan\theta$	$\frac{\sin\theta}{\cos\theta}$
$\frac{\sqrt{15}}{4}$	$\frac{1}{4}$	$\sqrt{15}$	$\sqrt{15}$

$\cot\theta$	$\frac{\cos\theta}{\sin\theta}$	$\sec\theta$	$\csc\theta$
$\frac{\sqrt{15}}{15}$	$\frac{\sqrt{15}}{15}$	4	$\frac{4\sqrt{15}}{15}$

$(\sin\theta)(\csc\theta) = (\cos\theta)(\sec\theta) = (\tan\theta)(\cot\theta) = 1$

Several relationships among the trigonometric functions are apparent from their definitions. For example, for those values of θ for which a denominator of zero is not obtained

$$\csc\theta = \frac{1}{y} = \frac{1}{\sin\theta} \quad \text{or} \quad \csc\theta \cdot \sin\theta = 1$$

$$\sec\theta = \frac{1}{x} = \frac{1}{\cos\theta} \quad \text{or} \quad \sec\theta \cdot \cos\theta = 1$$

$$\cot\theta = \frac{x}{y} = \frac{1}{\tan\theta} \quad \text{or} \quad \cot\theta \cdot \tan\theta = 1.$$

Two relations, f and g, whose domains are the same (except at values of x where $f(x) = 0$ or $g(x) = 0$) are **reciprocals** of each other if and only if $f(x) \cdot g(x) = 1$. Thus, the sine and cosecant functions are reciprocal functions; the cosine and secant functions are reciprocal functions; the tangent and cotangent functions are reciprocal functions.

Another useful relationship is that

$$\tan\theta = \frac{y}{x} = \frac{\sin\theta}{\cos\theta}$$

if θ is a number such that $\cos\theta \neq 0$. A similar relationship exists among the cotangent, sine, and cosine functions:

$$\cot\theta = \frac{x}{y} = \frac{\cos\theta}{\sin\theta}$$

if θ is a number such that $\sin\theta \neq 0$.

Since the values of the trigonometric functions for θ are defined in terms of the coordinates of a point on the unit circle, the values of the trigonometric functions for a real number θ are related to the quadrant in which the point $P(\theta)$ lies [Figure 1]. The coordinates of a point in the first quadrant are both positive. Hence, if the point corresponding to $P(\theta)$ is in the first quadrant, $\sin\theta$, $\cos\theta$, $\tan\theta$, $\cot\theta$, $\sec\theta$, and $\csc\theta$ are all positive. The first coordinate of a point in the second quadrant is negative; its second coordinate is positive. If θ is a real number such that $P(\theta)$ corresponds to a point in the second quadrant, $\sin\theta$ is a positive real number and $\cos\theta$ is negative. It is now possible to determine if $\tan\theta$, $\cot\theta$, $\sec\theta$, and $\csc\theta$ are positive or negative when the point $P(\theta)$ is in the second quadrant. The relationship between the values of the trigonometric functions for a real number θ and the quadrant of the point $P(\theta)$ is summarized in Table 1.

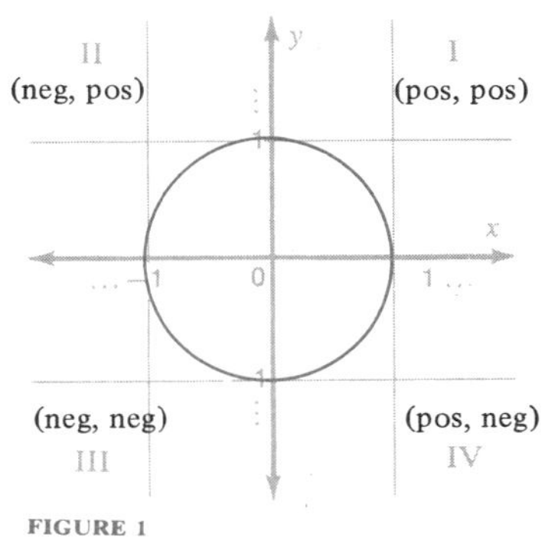

FIGURE 1

TABLE 1

Quadrant of $P(\theta)$	$\sin\theta$	$\cos\theta$	$\tan\theta$	$\cot\theta$	$\sec\theta$	$\csc\theta$
I	pos	pos	pos	pos	pos	pos
II	pos	neg	neg	neg	neg	pos
III	neg	neg	pos	pos	neg	neg
IV	neg	pos	neg	neg	pos	neg

In Section 11-3 the coordinates $P(\theta)$ for certain real numbers θ were determined. Since the values of the trigonometric functions for a real number θ are defined in terms of the coordinates of $P(\theta)$, the values of the trigonometric functions for these special real numbers can be obtained from Table 2 in Section 11-3. A summary of the values of the trigonometric functions for these special real numbers may be obtained by completing Table 2. Notice that since the function P assigns the same ordered pair to the real number θ and any number of the form $(\theta + 2\pi n)$, the values of the trigonometric functions for all these numbers are the same.

TABLE 2

θ	$\sin(\theta + 2\pi n)$	$\cos(\theta + 2\pi n)$	$\tan(\theta + 2\pi n)$	$\cot(\theta + 2\pi n)$	$\sec(\theta + 2\pi n)$	$\csc(\theta + 2\pi n)$
0	0	1	0	undefined	1	undefined
$\frac{\pi}{6}$	$\frac{1}{2}$	$\frac{\sqrt{3}}{2}$				
$\frac{\pi}{4}$	$\frac{\sqrt{2}}{2}$		1			$\sqrt{2}$
$\frac{\pi}{3}$		$\frac{1}{2}$		$\frac{\sqrt{3}}{3}$		$\frac{2\sqrt{3}}{3}$
$\frac{\pi}{2}$			undefined		undefined	
$\frac{3\pi}{4}$	$\frac{\sqrt{2}}{2}$	$-\frac{\sqrt{2}}{2}$	-1			
π		-1				undefined
$\frac{3\pi}{2}$				0		

Exercises

1. Complete Table 2.
2. List the trigonometric functions, if any, that are undefined for each number.

   a. $-\pi$ b. $-\frac{\pi}{4}$ c. $-\frac{3\pi}{2}$ d. π

   e. 0 f. $\frac{\pi}{2}$ g. $\frac{3\pi}{4}$ h. -2π

3. Simplify each expression.

   a. $\sin\frac{\pi}{3} \cdot \cot\frac{\pi}{2}$ b. $\cot\frac{\pi}{4} + \sec\frac{\pi}{6} - \cos\frac{\pi}{3}$

   c. $\dfrac{\tan\frac{\pi}{6} \cdot \cot\frac{3\pi}{2}}{\csc\frac{\pi}{3}}$

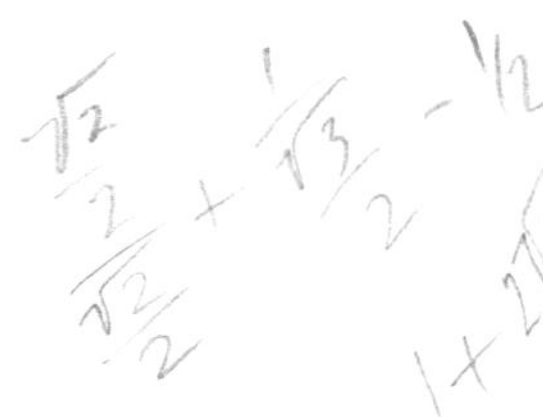

4. If $P(\theta) = \left(\frac{\sqrt{3}}{2}, \frac{1}{2}\right)$ find the values of the trigonometric functions for θ. Find a value of θ for which $P(\theta) = \left(\frac{\sqrt{3}}{2}, \frac{1}{2}\right)$.

5. Given $\sin\theta = \frac{\sqrt{3}}{2}$ and $\cos\theta = \frac{1}{2}$, find $\tan\theta$, $\cot\theta$, $\sec\theta$, and $\csc\theta$. Find a value for θ for which the values of the trigonometric functions are those you have just computed.

6. Given $\tan\theta = \sqrt{3}$ and $\sec\theta = 2$, find $\sin\theta$, $\cos\theta$, $\cot\theta$, and $\csc\theta$. Find a value for θ for which the values of the trigonometric functions are those you have just computed.

7. Prove that the point $\left(\frac{\sqrt{2}}{2}, \frac{\sqrt{2}}{2}\right)$ is on the unit circle. What is θ if $P(\theta) = \left(\frac{\sqrt{2}}{2}, \frac{\sqrt{2}}{2}\right)$ and $0 \leqslant \theta \leqslant 2\pi$? Find the values of the six trigonometric functions for θ.

8. Use the table in Exercise 1 to find a value of θ that makes each sentence true.

   a. $\sin\frac{\pi}{3} = \cos\theta$ b. $\cos\frac{\pi}{4} = \sin\theta$ c. $\tan\frac{\pi}{4} = \cot\theta$

   d. $\sec\frac{\pi}{6} = \csc\theta$ e. $\csc\frac{\pi}{4} = \sec\theta$ f. $\cot\frac{\pi}{6} = \tan\theta$

9. Show that each sentence is a true statement.

   a. $\tan\frac{\pi}{3} \neq \tan\frac{\pi}{6} + \tan\frac{\pi}{6}$ b. $\cos\frac{\pi}{2} \neq \cos\frac{\pi}{4} + \cos\frac{\pi}{4}$

10. Show that each sentence is a true statement.

   a. $\dfrac{\sin\frac{\pi}{3}\cdot\sec\frac{\pi}{3}}{\cos\frac{\pi}{4}} = \sec\frac{\pi}{4}\cdot\tan\frac{\pi}{3}$

   b. $\sin\frac{\pi}{3} = \tan\frac{\pi}{4}\cdot\cot\frac{\pi}{6}\cdot\sin\frac{\pi}{6}$

11. In the designated quadrant, which of the corresponding statements are true, false, or may be either depending on θ?

   a. Quadrant I: $\sin\theta > \tan\theta$ b. Quadrant II: $\cot\theta < \sin\theta$
   c. Quadrant III: $\tan\theta < \cos\theta$ d. Quadrant IV: $\cos\theta > \sin\theta$
   e. Quadrant I: $\cot\theta > \csc\theta$ f. Quadrant II: $\sec\theta < \csc\theta$
   g. Quadrant III: $\cot\theta > \sec\theta$ h. Quadrant IV: $\cot\theta < \sec\theta$

12. Determine the quadrant(s) of the point $P(\theta)$ in each condition.

   a. $\tan\theta > 0$ and $\cot\theta > 0$ b. $\sec\theta < 0$ and $\sin\theta > 0$
   c. $\cos\theta > 0$ and $\tan\theta > 0$ d. $\sin\theta > 0$ and $\cos\theta > 0$
   e. $\sec\theta > 0$ and $\sin\theta < 0$ f. $\cos\theta < 0$ and $\tan\theta > 0$
   g. $\csc\theta < 0$ and $\cos\theta < 0$ h. $\sin\theta < 0$ and $\csc\theta < 0$

13. Find the number θ in the interval from 0 to 2π $(0 \leqslant \theta < 2\pi)$ that has the same trigonometric functional values as each of the following:

a. $\frac{19\pi}{3}$ b. $\frac{7\pi}{5}$ c. $-\frac{\pi}{2}$ d. -3

e. $\frac{15\pi}{2}$ f. $\frac{11\pi}{2}$ g. $-\frac{23\pi}{6}$ h. $\frac{9\pi}{4}$

i. 6 j. $-\frac{17\pi}{3}$ k. 7π l. $\frac{37\pi}{6}$

14. Simplify each expression.

a. $\sin \frac{\pi}{3} \cdot \cos \frac{11\pi}{2} - \tan \frac{\pi}{4}$

b. $\frac{1}{\cos \pi} - \sec 5\pi + \cot (-\pi)$

c. $\left[\tan (-3\pi) + \csc \frac{37\pi}{6}\right]\left(\sec \frac{9\pi}{4}\right)$

d. $\left(\cot \frac{25\pi}{6}\right)^{-1}\left(\tan \frac{13\pi}{6}\right)$

e. $\dfrac{\sin \frac{7\pi}{3}}{\cos\left(-\frac{11\pi}{3}\right)}$

f. $\sin 7\pi + \cos (-7\pi) - \cot (-6\pi) - \sec 6\pi$

11-5 REDUCTION FORMULAS

Any real number θ such that $0 \leqslant \theta \leqslant \frac{\pi}{2}$ is called a **reference number.** The wrapping function P associates each reference number with the coordinates of a point on the unit circle in the first quadrant, (0, 1) or (1, 0). This section discusses a method of expressing the values of the trigonometric functions for any real number in terms of the trigonometric functional values of a reference number. The formulas obtained are called **reduction formulas.**

For every real number ψ, $P(\psi) = P(\psi + 2\pi n)$ for any integer n. Thus there is a smallest nonnegative number φ such that $0 \leqslant \varphi < 2\pi$ and $P(\psi) = P(\varphi)$. For example, if $\psi = \frac{11\pi}{2}$, there is a real number $\varphi, 0 \leqslant \varphi < 2\pi$, such that $P(\psi) = P\left(\frac{11\pi}{2}\right) = P(\varphi)$. To find φ, refer to the set of real numbers $\left\{\frac{11\pi}{2} + 2\pi n\right\}$. Since $\frac{11\pi}{2}$ is greater than 2π we add a negative multiple of

2π to get the desired number:

$$\frac{11\pi}{2} - 2(2\pi) = \frac{11\pi}{2} - \frac{8\pi}{2} = \frac{3\pi}{2}$$

and $0 \leqslant \frac{3\pi}{2} < 2\pi$. If the given number is negative, we add a positive multiple of 2π to find the desired number.

What is the number σ such that $0 \leqslant \sigma < 2\pi$ and $P(\sigma) = P\left(-\frac{17\pi}{4}\right)$?

$\sigma = -\frac{17\pi}{4} + 3(2\pi) = \frac{7\pi}{4}$

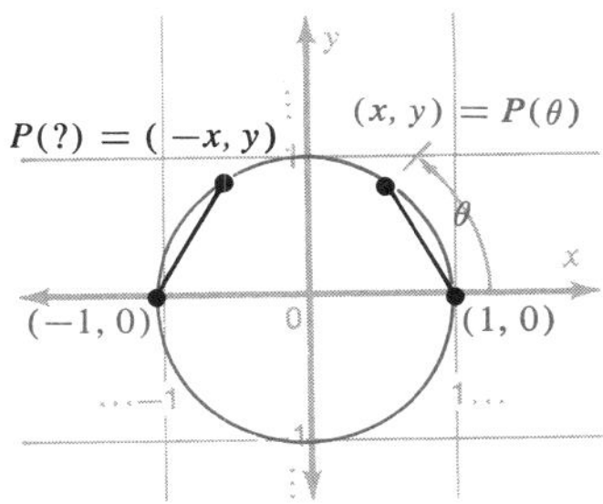

FIGURE 1

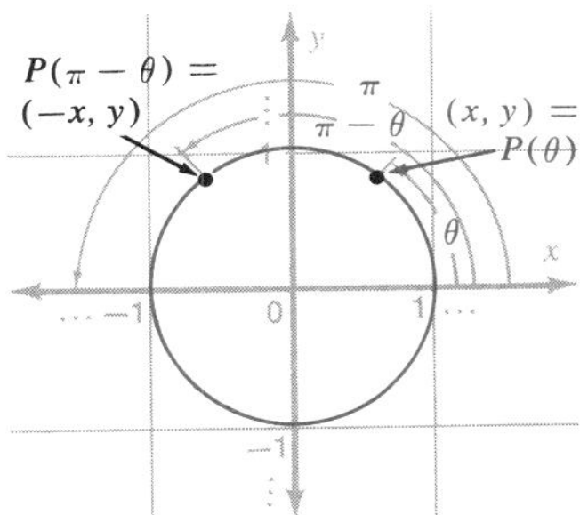

FIGURE 2

The next step is to express the values of the trigonometric functions for the numbers between 0 and 2π in terms of the trigonometric functional values of a corresponding reference number.

Let θ be a reference number $\left(\text{a number such that } 0 \leqslant \theta \leqslant \frac{\pi}{2}\right)$ so that the point $P(\theta) = (x, y)$ is in the first quadrant. The point $(-x, y)$ is also on the unit circle. Thus, the wrapping function P assigns some nonnegative number less than 2π to the point $(-x, y)$. What is this number? Since the point (x, y) is in the first quadrant, the point $(-x, y)$ is in the second quadrant [Figure 1]. By the Pythagorean theorem the distance (length of the chord) between the points $(-x, y)$ and $(-1, 0)$ is equal to the distance (length of the chord) between the points (x, y) and $(1, 0)$. Thus, the arcs of the unit circle between these pairs of points are congruent. Since the length of the arc between the points (x, y) and $(1, 0)$ is θ, the length of the arc of the unit circle between the points $(-x, y)$ and $(-1, 0)$ is also θ [Figure 2]. Therefore, the length of the arc of the unit circle between the points $(1, 0)$ and $(-x, y)$ is $\pi - \theta$. Thus, $P(\pi - \theta) = (-x, y)$.

From the definitions of the trigonometric functions, we know that

$$\cot(\pi - \theta) = -\frac{x}{y} \quad \text{and} \quad \cot\theta = \frac{x}{y}.$$

Hence,

$$\cot(\pi - \theta) = -\frac{x}{y} = -\left(\frac{x}{y}\right) = -\cot\theta.$$

Similarly,

$$\cos(\pi - \theta) = -x = -\cos\theta.$$

We can use the definition of Section 11-4 to find the remaining relationships between the trigonometric functional values of $\pi - \theta$ and the values of the same trigonometric function at θ. Thus, we obtain the following set of equations (called **reduction formulas**) which make up Theorem 1.

Theorem 1 If $0 \leqslant \theta \leqslant \frac{\pi}{2}$, then θ is the reference number associated with $(\pi - \theta)$ and

$$\sin(\pi - \theta) = \sin\theta \qquad \csc(\pi - \theta) = \csc\theta$$

$$\cos(\pi - \theta) = -\cos\theta \qquad \sec(\pi - \theta) = -\sec\theta$$

$$\tan(\pi - \theta) = -\tan\theta \qquad \cot(\pi - \theta) = -\cot\theta$$

θ	$\frac{4\pi}{6}$	$\frac{5\pi}{6}$	$\frac{6\pi}{6}$	$\frac{3\pi}{4}$	$\frac{4\pi}{4}$	$\frac{2\pi}{3}$	$\frac{3\pi}{3}$
$\pi - \theta$	$\frac{\pi}{3}$	$\frac{\pi}{6}$	0	$\frac{\pi}{4}$	0	$\frac{\pi}{3}$	0

$\frac{\pi}{2} \leqslant \pi - \theta \leqslant \pi,\ \pi \leqslant \pi + \theta \leqslant \frac{3\pi}{2},$
$-\frac{\pi}{2} \leqslant -\theta \leqslant 0$

Find the reference number associated with each of the following: $\frac{4\pi}{6}, \frac{5\pi}{6}, \frac{6\pi}{6}, \frac{3\pi}{4}, \frac{4\pi}{4}, \frac{2\pi}{3}, \frac{3\pi}{3}$. If $0 \leqslant \theta \leqslant \frac{\pi}{2}$, then _______ $\leqslant \pi - \theta \leqslant$ _______; _______ $\leqslant \pi + \theta \leqslant$ _______; and _______ $\leqslant -\theta \leqslant$ _______.

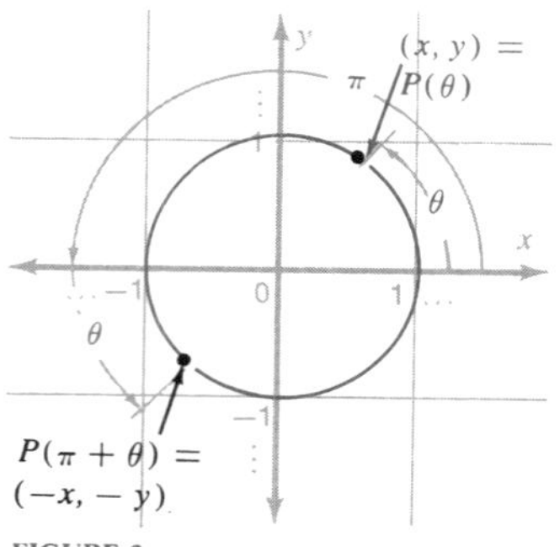

FIGURE 3

Similar reduction formulas can be obtained for any nonnegative number less than 2π that the wrapping function associates with points on the unit circle in quadrants III and IV. If $P(\theta) = (x, y)$, note that

$$P(\pi + \theta) = (-x, -y)$$

[Figure 3]. Then using the definitions of the values of the trigonometric functions as was done above we obtain Theorem 2.

Theorem 2 If $0 \leqslant \theta \leqslant \frac{\pi}{2}$, then θ is the reference number associated with $(\pi + \theta)$ and

$$\sin(\pi + \theta) = -\sin\theta \qquad \csc(\pi + \theta) = -\csc\theta$$

$$\cos(\pi + \theta) = -\cos\theta \qquad \sec(\pi + \theta) = -\sec\theta$$

$$\tan(\pi + \theta) = \tan\theta \qquad \cot(\pi + \theta) = \cot\theta$$

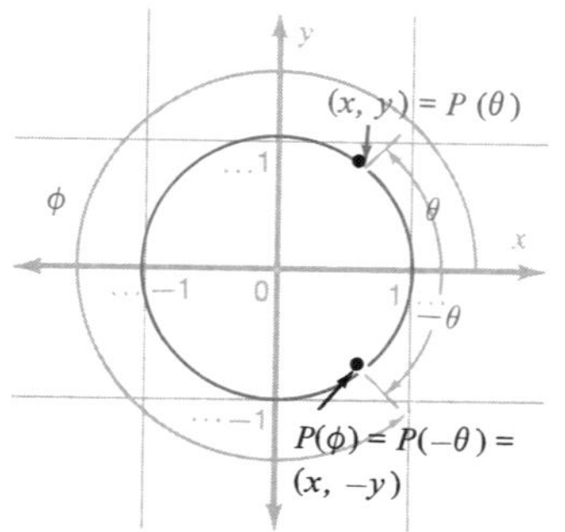

FIGURE 4

Before stating the theorem which gives the next set of reduction formulas, we observe that if φ is a nonnegative real number less than 2π and the point $P(\varphi)$ is in quadrant IV, then $P(\varphi) = P(-\theta)$ where θ is a reference number [Figure 4]. This observation justifies the form of the reduction formulas which constitute Theorem 3.

Theorem 3 If $0 \leqslant \theta \leqslant \frac{\pi}{2}$, then θ is the reference number associated with $(-\theta)$ and

$$\sin(-\theta) = -\sin\theta \qquad \csc(-\theta) = -\csc\theta$$

$$\cos(-\theta) = \cos\theta \qquad \sec(-\theta) = \sec\theta$$

$$\tan(-\theta) = -\tan\theta \qquad \cot(-\theta) = -\cot\theta$$

This theorem can be proved by observing that if $P(\theta) = (x, y)$, then $P(-\theta) = (x, -y)$ [Figure 4] and proceeding as in Theorem 1.

Example 1 Find the values of the trigonometric functions for $\varphi = \frac{5\pi}{6}$.

Solution: Since the point $P(\varphi)$ is in the second quadrant, we use Theorem 1. If $\pi - \theta = \frac{5\pi}{6}$, then θ, the reference number, is $\frac{\pi}{6}$ and

$$\sin\frac{5\pi}{6} = \sin\frac{\pi}{6} = \frac{1}{2} \qquad \csc\frac{5\pi}{6} = \csc\frac{\pi}{6} = 2$$

$$\cos\frac{5\pi}{6} = -\cos\frac{\pi}{6} = -\frac{\sqrt{3}}{2} \qquad \sec\frac{5\pi}{6} = -\sec\frac{\pi}{6} = -\frac{2\sqrt{3}}{3}$$

$$\tan\frac{5\pi}{6} = -\tan\frac{\pi}{6} = -\frac{\sqrt{3}}{3} \qquad \cot\frac{5\pi}{6} = -\cot\frac{\pi}{6} = -\sqrt{3}$$

Example 2 Find the trigonometric functional values of $\varphi = -\frac{13\pi}{3}$

Solution: The function P associates each number in the set $\left\{-\frac{13\pi}{3} + 2\pi n\right\}$ with the same point on the unit circle. One of these numbers is between 0 and 2π, namely

$$-\frac{13\pi}{3} + (2\pi)3 = -\frac{13\pi}{3} + \frac{18\pi}{3} = \frac{5\pi}{3}.$$

The number $\frac{5\pi}{3}$ corresponds to a point in the fourth quadrant and hence to the point associated with the negative of some reference number, in this case $-\frac{\pi}{3}$. Using Theorem 3, we have the following six statements.

$$\sin\left(-\frac{13\pi}{3}\right) = \sin\frac{5\pi}{3} = \sin\left(-\frac{\pi}{3}\right) = -\sin\frac{\pi}{3} = -\frac{\sqrt{3}}{2}$$

$$\cos\left(-\frac{13\pi}{3}\right) = \cos\frac{5\pi}{3} = \cos\left(-\frac{\pi}{3}\right) = \cos\frac{\pi}{3} = \frac{1}{2}$$

$$\tan\left(-\frac{13\pi}{3}\right) = \tan\frac{5\pi}{3} = \tan\left(-\frac{\pi}{3}\right) = \tan\frac{\pi}{3} = \sqrt{3}$$

$$\csc\left(-\frac{13\pi}{3}\right) = \csc\frac{5\pi}{3} = \csc\left(-\frac{\pi}{3}\right) = -\csc\frac{\pi}{3} = -\frac{2\sqrt{3}}{3}$$

$$\sec\left(-\frac{13\pi}{3}\right) = \sec\frac{5\pi}{3} = \sec\left(-\frac{\pi}{3}\right) = \sec\frac{\pi}{3} = 2$$

$$\cot\left(-\frac{13\pi}{3}\right) = \cot\frac{5\pi}{3} = \cot\left(-\frac{\pi}{3}\right) = \cot\frac{\pi}{3} = \frac{\sqrt{3}}{2}$$

Find the reference number for $\frac{3\pi}{4}$ and use Theorem 1 to find $\sin\frac{3\pi}{4}$, $\cos\frac{3\pi}{4}$, and $\tan\frac{3\pi}{4}$. Do your results agree with Table 2, Section 11-4?

Reference number $= \pi - \frac{3\pi}{4} = \frac{\pi}{4}$ and by Theorem 1

$$\sin\frac{3\pi}{4} = \sin\frac{\pi}{4} = \frac{\sqrt{2}}{2}$$

$$\cos\frac{3\pi}{4} = -\cos\frac{\pi}{4} = -\frac{\sqrt{2}}{2}$$

$$\tan\frac{3\pi}{4} = -\tan\frac{\pi}{4} = -1.$$

Find the reference number for $\frac{7\pi}{6}$; for $\frac{11\pi}{6}$; for $-\frac{\pi}{4}$. Evaluate $\cot\frac{7\pi}{6}$, $\csc\frac{11\pi}{6}$, and $\sec\left(-\frac{\pi}{4}\right)$.

θ	*Reference number*	*Value*
$\frac{7\pi}{6}$	$\frac{\pi}{6}$	$\cot\frac{7\pi}{6} = \sqrt{3}$
$\frac{11\pi}{6}$	$\frac{\pi}{6}$	$\csc\frac{11\pi}{6} = -2$
$-\frac{\pi}{4}$	$\frac{\pi}{4}$	$\sec\left(-\frac{\pi}{4}\right) = \sqrt{2}$

Because of their importance in triangle applications, we will derive another set of reduction formulas. Let θ be a reference number with $P(\theta) = (x, y)$. Then the number $\frac{\pi}{2} - \theta$ is a reference number and the coordinates $P\left(\frac{\pi}{2} - \theta\right)$ are related to the coordinates $P(\theta)$ [Figure 5] as follows. If $P(\theta) = (x, y)$, then $P\left(\frac{\pi}{2} - \theta\right) = (y, x)$. Assuming that $P\left(\frac{\pi}{2} - \theta\right) = (y, x)$, the definitions of the trigonometric functions justify the following set of reduction formulas.

Theorem 4 If $0 \leqslant \theta \leqslant \frac{\pi}{2}$, then

$$\sin\left(\frac{\pi}{2} - \theta\right) = \cos\theta \qquad \csc\left(\frac{\pi}{2} - \theta\right) = \sec\theta$$

$$\cos\left(\frac{\pi}{2} - \theta\right) = \sin\theta \qquad \sec\left(\frac{\pi}{2} - \theta\right) = \csc\theta$$

$$\tan\left(\frac{\pi}{2} - \theta\right) = \cot\theta \qquad \cot\left(\frac{\pi}{2} - \theta\right) = \tan\theta$$

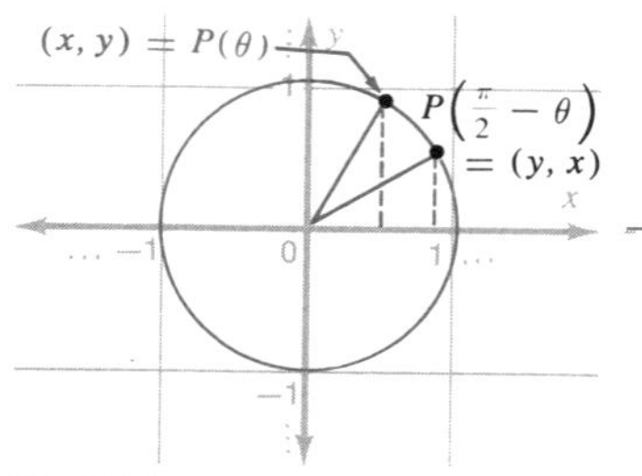

FIGURE 5

Similar sets of reduction formulas may be derived for such numbers as $\frac{\pi}{2} + \theta$, $\frac{3\pi}{2} + \theta$, and $\frac{3\pi}{2} - \theta$.

How are the coordinates $P\left(\frac{\pi}{2} + \theta\right)$, $P\left(\frac{3\pi}{2} + \theta\right)$, and $P\left(\frac{3\pi}{2} - \theta\right)$ related to $P(\theta) = (x, y)$, where θ is a reference number?

$P\left(\frac{\pi}{2} + \theta\right) = (-y, x)$,

$P\left(\frac{3\pi}{2} + \theta\right) = (y, -x)$,

$P\left(\frac{3\pi}{2} - \theta\right) = (-y, -x)$

Theorems 1–4 state that the reduction formulas hold when θ is a reference number. Actually, this restriction is unnecessary because the reduction formulas are valid for any real number for which the trigonometric functions involved are defined. This can be verified quite simply by using the results in Section 13-3 (see problem 11, Section 13-3).

Exercises

1. Find the number between 0 and 2π for which the six trigonometric functions have the same values as each number below.

   a. $\frac{17\pi}{3}$ b. $-\frac{22\pi}{6}$ c. $\frac{13\pi}{4}$
   d. $\frac{11\pi}{3}$ e. $-\frac{35\pi}{6}$ f. -2
   g. $-\frac{3\pi}{4}$ h. $\frac{18\pi}{5}$ i. $-\frac{21\pi}{9}$

2. Find the number of the form θ, $\pi - \theta$, $\pi + \theta$, or $-\theta$ (where θ is a reference number) that has the same trigonometric functional values as each of the following.

   a. $\frac{5\pi}{3}$ b. $\frac{\pi}{3}$ c. $\frac{\pi}{4}$ d. $\frac{7\pi}{6}$
   e. $\frac{\pi}{6}$ f. 3 g. $\frac{23\pi}{24}$ h. $\frac{24\pi}{14}$
   i. 5.5 j. $-\frac{7\pi}{3}$ k. 5 l. $-\frac{56\pi}{17}$

3. Use the results obtained in Exercises 1–2 (a–e) and the reduction formulas to evaluate each of the following.

   a. $\tan\frac{17\pi}{3}$ b. $\cos\frac{11\pi}{3}$ c. $\sin\left(-\frac{22\pi}{6}\right)$
   d. $\cot\frac{13\pi}{4}$ e. $\sec\left(-\frac{35\pi}{6}\right)$ f. $\cot\left(-\frac{35\pi}{6}\right)$
   g. $\cot\left(-\frac{22\pi}{6}\right)$ h. $\sec\frac{13\pi}{4}$ i. $\sin\left(-\frac{35\pi}{6}\right)$
   j. $\cos\frac{17\pi}{3}$ k. $\csc\left(-\frac{35\pi}{6}\right)$ l. $\tan\frac{13\pi}{4}$
   m. $\cot\frac{11\pi}{3}$ n. $\sec\left(-\frac{22\pi}{6}\right)$ o. $\sin\frac{11\pi}{3}$
   p. $\tan\left(-\frac{22\pi}{6}\right)$ q. $\cot\frac{17\pi}{3}$ r. $\cos\left(-\frac{22\pi}{6}\right)$
   s. $\tan\left(-\frac{35\pi}{6}\right)$ t. $\csc\frac{17\pi}{3}$ u. $\cos\frac{13\pi}{4}$

4. Express each of the following as $\sec\theta$, $\csc\theta$, or $\cot\theta$ where θ is the reference number.

   a. $\cot\frac{17\pi}{3}$ b. $\csc\left(-\frac{3\pi}{2}\right)$ c. $\csc\left(-\frac{7\pi}{2}\right)$

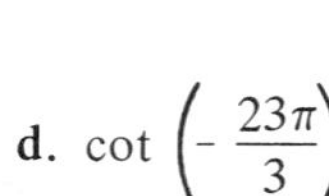

d. $\cot\left(-\frac{23\pi}{3}\right)$ e. $\sec\frac{11\pi}{2}$ f. $\sec\frac{19\pi}{3}$

g. $\cot\frac{13\pi}{8}$ h. $\sec\left(-\frac{11\pi}{6}\right)$ i. $\cot\frac{23\pi}{3}$

5. Express each of the following as $\sin\theta$, $\cos\theta$, or $\tan\theta$ where θ is the reference number.

a. $\cos\frac{17\pi}{3}$ b. $\sin\frac{18\pi}{2}$ c. $\tan\left(-\frac{5\pi}{2}\right)$

d. $\sin\left(-\frac{13\pi}{4}\right)$ e. $\tan\frac{13\pi}{11}$ f. $\cos\frac{7\pi}{12}$

g. $\tan\left(-\frac{\pi}{6}\right)$ h. $\sin 18$ i. $\cos\frac{85\pi}{7}$

6. Evaluate each expression.

a. $\cot\frac{13\pi}{6}\sin\left(-\frac{13\pi}{4}\right) - \sec\frac{11\pi}{2}$ b. $\dfrac{\sec\frac{14\pi}{3}}{\cot\frac{23\pi}{3} - 1}$

c. $\cos\left(-\frac{3\pi}{4}\right)\left(\sin\frac{9\pi}{4}\right) - 1$ d. $1 + \tan^2\left(-\frac{\pi}{6}\right)$

e. $2\sin\frac{17\pi}{3}\cos\frac{17\pi}{3}$ f. $\tan\frac{17\pi}{6} + \sec\left(-\frac{14\pi}{6}\right)$

Review Exercises

1. Discuss the ways that mathematicians use the word *angle*. **11-1**

2. a. Name the angles in Figure 1 that are in standard position.
   b. What are the initial and terminal sides of each angle named in part a?

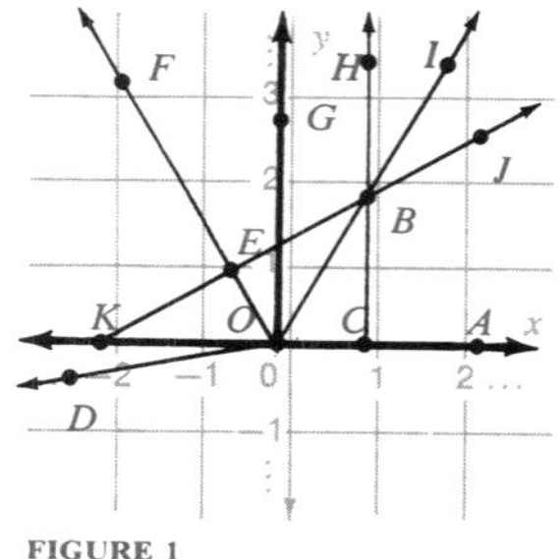

FIGURE 1

3. Determine if the measure of each angle is positive or negative.

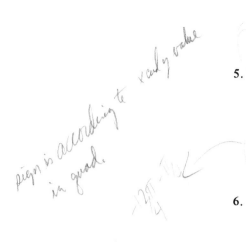

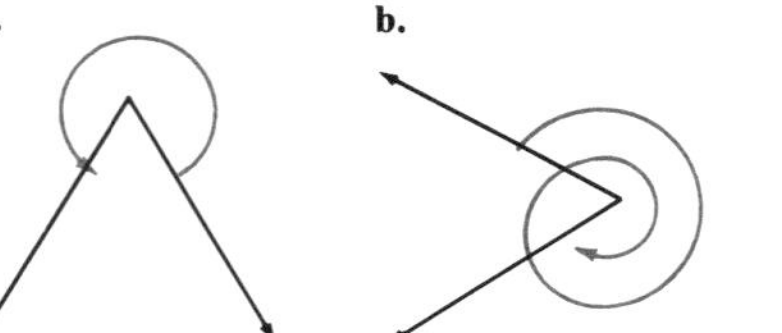

4. The length of each chain supporting the seat of a child's swing is 6 feet. When the swing is at its highest points forward and backward, the radian measure of the angle determined by these chains is $\frac{3\pi}{4}$. How far does the child travel in one trip between these high points? **11-2**

5. Find the radian measure of an angle if its degree measure is $\frac{90}{\pi}$.

6. Find the degree measure of an angle if its radian measure is $\frac{\pi}{5}$.

7. Find the smallest positive number and the largest negative number in the set, $\left\{-\frac{3\pi}{2} + 2\pi n\right\}$. 11-3

8. Find the coordinates $P(\theta)$ associated with each of the following values of θ.

a. -11π b. $\frac{13\pi}{6}$ c. $\frac{18\pi}{4}$ d. $-\frac{47\pi}{6}$

9. What relationship must exist between x and y in order for the point (x, y) to be on the unit circle with center at the origin?

10. If $P(\theta) = \left(\frac{1}{5}, \frac{2\sqrt{6}}{5}\right)$, find each trigonometric functional value.

a. $\sin\theta$
b. $\sin\theta \cdot \cos\theta$
c. $\cos\theta$
d. $\cos\theta \cdot \sec\theta$
e. $\tan\theta$
f. $\sin^2\theta + \cos^2\theta$

11. Given $\sin\psi = \frac{2}{3}$ and $\tan\psi = \frac{2\sqrt{3}}{9}$, find $\sec\psi$. 11-4

12. Find φ if $\tan\varphi = \sqrt{3}$ and $\sec\varphi = 2$.

13. In which quadrant(s) are the following statements always true? Always false? Either true or false depending on θ?
a. $\sin\theta > \cos\theta$ b. $\tan\theta > \cot\theta$ c. $\sec\theta > \csc\theta$

14. Determine the quadrant of $P(\theta)$ in each condition.
a. $\csc\theta > 0$ and $\cos\theta < 0$
b. $\sec\theta > 0$ and $\tan\theta < 0$
c. $\sin\theta < 0$ and $\cot\theta > 0$

15. Simplify $4\left(\cos\frac{13\pi}{6}\right)^2 + \sin\frac{\pi}{6} - \tan\pi$.

16. Find the number between 0 and 2π for which the six trigonometric functions have the same values as each number below.

a. $-\frac{7\pi}{4}$ b. $\frac{25\pi}{6}$ c. 3

17. Find the reference number associated with each real number. 11-5

a. $-\frac{7\pi}{4}$ b. $\frac{25\pi}{6}$ c. 5π

18. Evaluate each expression.

a. $\cos\left(-\frac{7\pi}{4}\right)$ b. $\tan\frac{25\pi}{6}$ c. $\sec 5\pi$

12. GRAPHS OF THE TRIGONOMETRIC FUNCTIONS

12-1 DOMAIN, RANGE, AND PERIODICITY

What are the domain and range of the sine function? Recall that the domain of the wrapping function P is the set of real numbers R and that its range is the set of ordered pairs for points on the unit circle. Since for any real number θ, $\sin \theta = y$ for $P(\theta) = (x, y)$, the coordinates of a point on the unit circle, the domain of the sine function is the set of real numbers. Moreover, the y-coordinates of the points on the unit circle are between -1 and 1 inclusive and so the range of the sine function is $\{-1 \leqslant y \leqslant 1\}$.

What are the domain and range of the cosine function?

Domain: set of real numbers; Range: set of real numbers between and including -1 and 1.

The tangent function

$$\tan \theta = \frac{y}{x}, \quad \text{for} \quad P(\theta) = (x, y)$$

is defined for all y and all nonzero x, so the domain of the tangent function is the set of all real numbers θ such that the first coordinate of $P(\theta)$ is nonzero. The only points on the unit circle with first coordinates equal to zero are the points $(0, 1)$ and $(0, -1)$. These points correspond to the set

$$\left\{\frac{\pi}{2} + 2\pi n\right\} \cup \left\{\frac{3\pi}{2} + 2\pi n\right\} = \left\{\frac{\pi}{2} + \pi n\right\}.$$

The domain of the tangent function is $\left\{\theta\colon\ \theta \neq \frac{\pi}{2} + \pi n\right\}$.

What is the domain of the secant function?

Domain: $\{\theta\colon\ \theta \neq \frac{\pi}{2} + n\pi\}$

The domain and range of the tangent function can also be determined by considering the unit circle and a line l tangent to the circle at $(1, 0)$

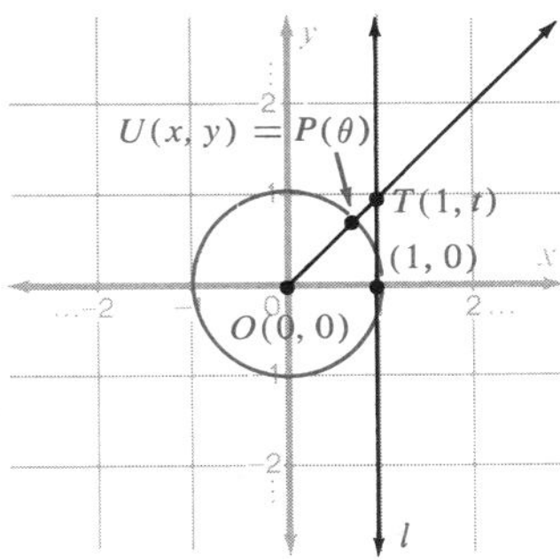

FIGURE 1

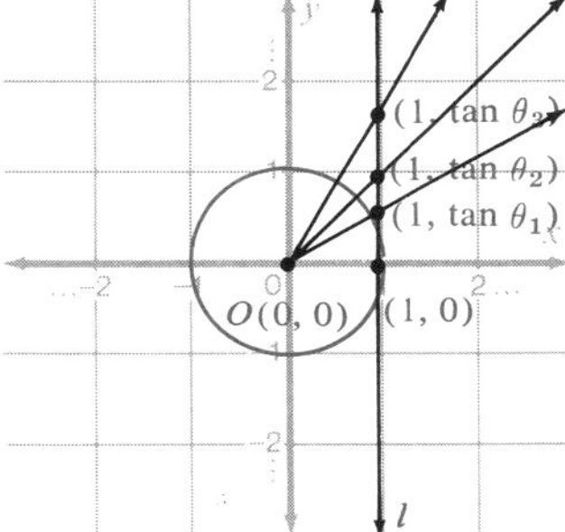

FIGURE 2

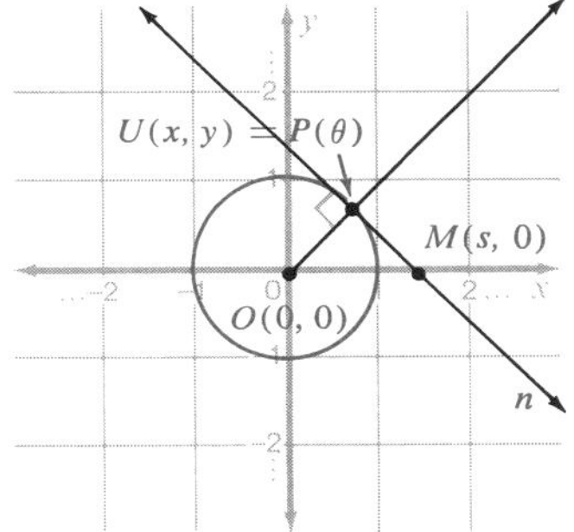

FIGURE 3

[Figure 1]. Let U be the point $P(\theta)$ on the unit circle, T be the intersection of $\overleftrightarrow{OU}$ with line l, and t denote the second coordinate of T. Using points T and O and points U and O to determine the slope of line OU, we have

$$\frac{t}{1} = \frac{y}{x}.$$

Since $\frac{y}{x} = \tan\theta$, $t = \tan\theta$ and $T(1, t)$ is the point $T(1, \tan\theta)$. If $\theta = 0$, then $P(\theta) = (1, 0)$, the point $P(\theta)$ is identical to T, and $\tan\theta = 0$. As θ increases from 0 to $\frac{\pi}{2}$, T moves up line l; that is, $\tan\theta$ assumes all nonnegative real values [Figure 2]. When $\theta = \frac{\pi}{2}$, $\overleftrightarrow{OU}$ is parallel to line l; hence, $\overleftrightarrow{OU}$ does not intersect line l and $\tan\frac{\pi}{2}$ does not exist, a result obtained earlier. If θ decreases from 0 to $-\frac{\pi}{2}$, the point T moves down line l; that is, $\tan\theta$ assumes all negative real values. $\left(\text{What is } \tan\left(-\frac{\pi}{2}\right)\text{? Why?}\right)$ Hence, the range of the tangent function is the set of real numbers.

The range of the secant function can be determined using Figure 3 where n is the line tangent to the unit circle at $U(x, y) = P(\theta)$ and where $M(s, 0)$ is the point where n intersects the x-axis. Line n is perpendicular to line OU. Hence, their slopes, $\frac{y}{x - s}$ and $\frac{y}{x}$, are negative reciprocals of each other; that is

$$\frac{y}{x - s} = -\frac{x}{y}, \quad xs = x^2 + y^2, \quad \text{and} \quad s = \frac{x^2 + y^2}{x}.$$

Since $U(x, y)$ is on the unit circle, $x^2 + y^2 = 1$,

$$s = \frac{1}{x} = \sec\theta,$$

and $M(s, 0)$ is the point $M(\sec\theta, 0)$. As θ increases from 0 to $\frac{\pi}{2}$, $\sec\theta$ assumes all real values greater than or equal to one. When $\theta = \frac{\pi}{2}$, n is parallel to the x-axis; hence, $\sec\theta$ does not exist. As θ increases from $\frac{\pi}{2}$ to π, $\sec\theta$ assumes all real values less than or equal to negative one. Hence, the range of the secant function is $\{y\colon y \leqslant -1 \text{ or } y \geqslant 1\}$. The domain and range of the cotangent and cosecant functions can be determined by asking similar questions about Figures 4 and 5.

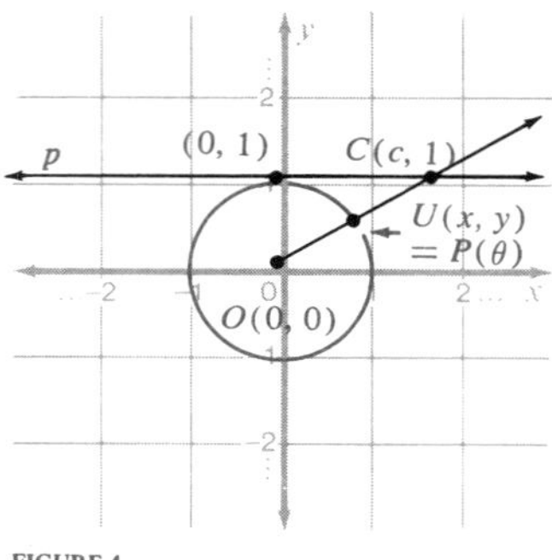

FIGURE 4

For what real numbers θ is the second coordinate of $P(\theta)$ zero? Are there real numbers for which the cotangent function defined by $\cot\theta = \frac{x}{y}$, for $P(\theta) = (x, y)$, is undefined? What is the domain of the cotangent function? In Figure 4, the line p is tangent to the unit circle at (0, 1) and intersects $\overleftrightarrow{OU}$ at $C(c, 1)$ where U is the point $P(\theta)$. Prove that $c = \cot\theta$ and that $C(c, 1)$ is the point $C(\cot\theta, 1)$. If θ increases from 0 to π, what values does $\cot\theta$ assume? What is the range of the cotangent function? Determine the domain and range of the cosecant function by asking yourself similar questions. Use Figure 5.

$P(\pi n) \in \{(1, 0), (-1, 0)\}$; Yes, for $\theta \in \{\pi n\}$; Domain (cotangent): $\{\theta\colon \theta \neq \pi n\}$.

Proof: The slope of line $\overleftrightarrow{OU}$ is $\frac{y}{x} = \frac{1}{c}$. Hence, $c = \frac{y}{x} = \cot\theta$. All real values; Range (cotangent): R, Domain (cosecant): $\{\theta\colon \theta \neq \pi n\}$, Range (cosecant):

$$\{x\colon x \leq -1 \text{ or } x \geq 1\}.$$

The domain and range of each of the six trigonometric functions are shown in the table below.

Function	*Domain*	*Range*
Sine	R	$\{y\colon -1 \leq y \leq 1\}$
Cosine	R	$\{y\colon -1 \leq y \leq 1\}$
Tangent	$\left\{\theta\colon \theta \neq \frac{\pi}{2} + \pi n\right\}$	R
Cotangent	$\{\theta\colon \theta \neq \pi n\}$	R
Secant	$\left\{\theta\colon \theta \neq \frac{\pi}{2} + \pi n\right\}$	$\{y\colon y \geq 1 \text{ or } y \leq -1\}$
Cosecant	$\{\theta\colon \theta \neq \pi n\}$	$\{y\colon y \geq 1 \text{ or } y \leq -1\}$

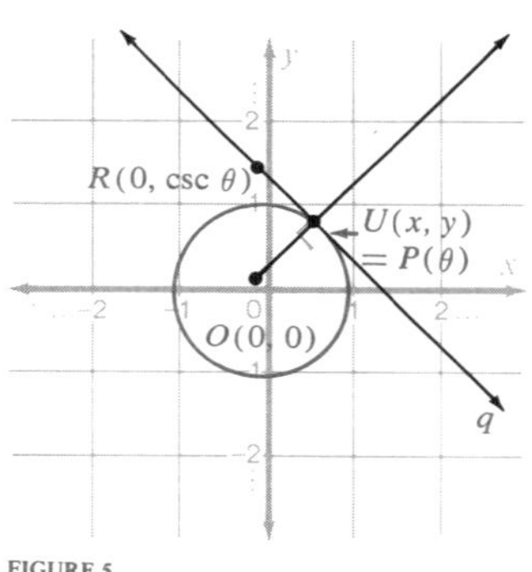

FIGURE 5

Nature exhibits many periodic events, events which cycle through the same states in a regular way. Night and day, the tides, phases of the moon, motions of planets and comets, the orbiting of satellites, and the swing of a pendulum are examples of periodic events. One basic objective of mathematics is to describe the physical world. Accordingly, mathematicians invented the concept of periodic function as one means of representing periodic events. The trigonometric functions are periodic functions.

Definition A function f is a **periodic function** if and only if $f(x) = f(x + k)$ for some nonzero real number k and for all x in the domain of f. The number k is called a **period** of f. The smallest positive period is called the **primitive period** of f.

The graphs of two functions appear in Figures 6 and 7. Which of the functions is periodic, and if periodic, what is the primitive period of the function?

Both functions are periodic with primitive period 1 [Figure 6] and 2π [Figure 7].

Recall that $P(\theta) = P(\theta + 2\pi n)$. Thus, each number in the set $\{2\pi n\colon n \text{ a nonzero integer}\}$ is a period of the wrapping function P. The number 2π is the smallest positive period of the function P; hence, 2π is the primitive period of P. The trigonometric functions are defined in terms of P and consequently inherit the periodicity of P.

Theorem The trigonometric functions are periodic with period 2π.

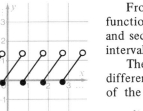

From the discussion of the domain and range of the trigonometric functions, we note that the primitive period of the sine, cosine, cosecant, and secant functions is 2π. That is, an interval of length 2π is the smallest interval possible before the values of these functions are repeated.

The primitive period of the tangent and cotangent functions is different. Note that the ratio of the y-coordinate to the x-coordinate of the point $P(\theta)$ is equal to the ratio of the y-coordinate to the x-coordinate of the point $P(\theta + \pi)$ [Figure 8]. By definition, $\tan \theta = \dfrac{y}{x}$ and $\tan (\theta + \pi) = -\dfrac{y}{-x} = \dfrac{y}{x}$. Thus, the tangent function has primitive period π. Similarly, the primitive period of the cotangent function is also π.

FIGURE 6

No. Because 2π is a period of the tangent and cotangent functions.

Does the fact that π is the primitive period of the tangent and cotangent functions contradict the theorem which states that the trigonometric functions have period 2π? Why?

Exercises

1. With the help of Figure 4 discuss the domain and range of the cotangent function.
2. With the help of Figure 5 discuss the domain and range of the cosecant function.
3. Discuss the table on page 244 taking particular note of the meaning of the following symbols: $\leqslant$, $\geqslant$, R, $\{\pi n\}$, $\left\{\dfrac{\pi}{2} + \pi n\right\}$, $\{\theta\colon \theta \neq \pi n\}$, $\left\{\theta\colon \theta \neq \dfrac{\pi}{2} + \pi n\right\}$, and the word *or*.

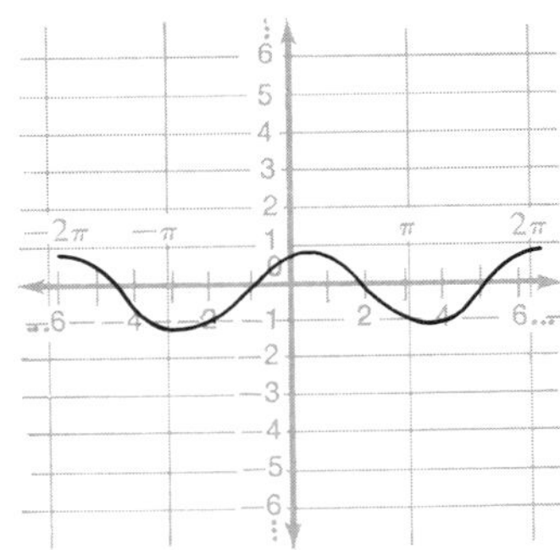

FIGURE 7

4. Name the trigonometric functions (if any) for which each statement below is true.
   a. The domain consists of all real numbers.
   b. Integral multiples of $\dfrac{\pi}{2}$ are in the domain.
   c. Numbers less than one are in the range.
   d. The range does not include 1 and -1.
   e. Integral multiples of π are not in the range.
   f. The domain includes all real numbers greater than π.

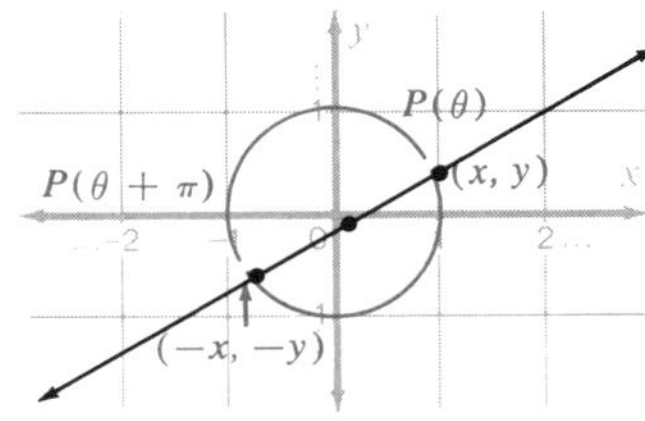

FIGURE 8

5. Name a function which satisfies each statement.
   a. The range is all real numbers.
   b. The domain is all real numbers.
   c. The range is all real numbers greater than or equal to 1.

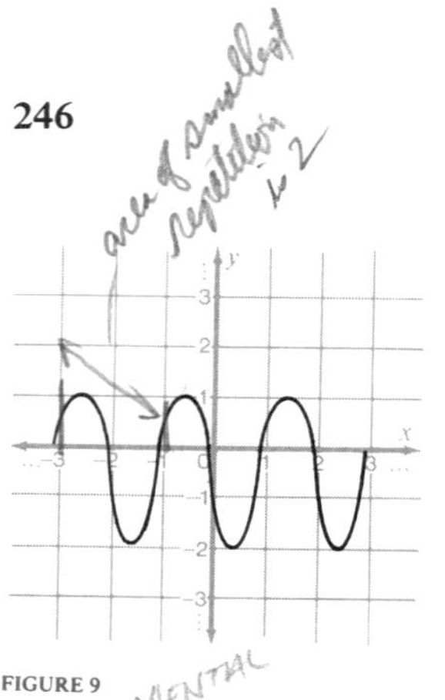

FIGURE 9

d. The domain includes all real numbers except 0.
e. The range is all real numbers less than or equal to 0.
f. The domain does not include −3.
g. The domain includes all real numbers except $\frac{1}{2}$ and −1.
h. The range includes real numbers greater than or equal to 1 and less than or equal to −1.
i. The domain does not include integral multiples of 2π.

6. Determine if each of the functions graphed in Figures 9–17 is periodic, and if periodic, find the primitive period.

7. Sketch the graph of a function which satisfies each condition.
   a. Domain is $\{x\colon -5 \leqslant x \leqslant 3\}$ and primitive period is 2.
   b. Domain is $\{x\colon -1 < x \leqslant 2\}$ and primitive period is 1.
   c. Domain is $\{x\colon 0 \leqslant x \leqslant \pi\}$ and range is $\{y\colon 0 \leqslant y \leqslant \pi\}$.
   d. Domain is R, range is $\{y\colon -3 \leqslant y \leqslant 5\}$, and primitive period is π.
   e. Domain is $\left\{x\colon -\frac{\pi}{2} < x < \frac{\pi}{2}\right\}$ and range is the set of real numbers.
   f. Domain is the set of real numbers, range is $\{y\colon y = 5\}$, and period is 50.

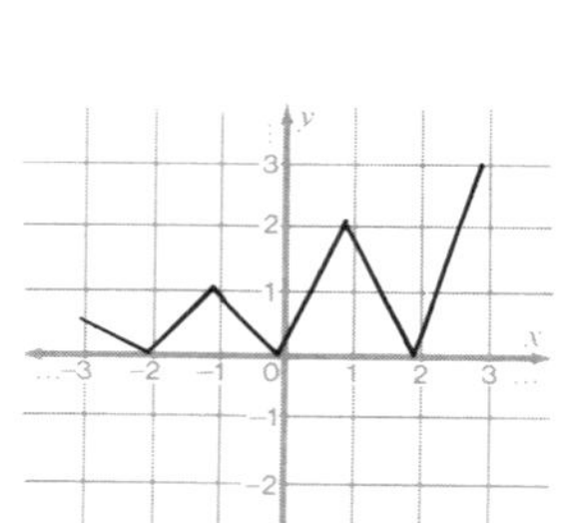

FIGURE 10

8. The function f defined by the equation $f(x) = c$ where $c \in R$ is trivially a periodic function for which any real number is a period. That is, for any real number, say $-\pi$, $f(x + (-\pi)) = f(x)$. What is the primitive period, if any, of f?

9. Assume that the following theorem is true.

   Suppose that f is a periodic function with period p and that g is a periodic function with period q. If h is the function defined by the equation $h(x) = f(x) + g(x)$ and r is an integral multiple of both p and q, then h is a periodic function with period r.

   Use this theorem to determine a period for each function f defined by the equations below.
   a. $f(x) = \sin x + \cos x$
   b. $f(x) = \cos x + \tan x$
   c. $f(x) = \cot x + \sec x$
   d. $f(x) = \sin x + \csc x$
   e. $f(x) = \tan x + \cot x$
   f. $f(x) = \sin x + \cos x + \tan x + \sec x$

FIGURE 11

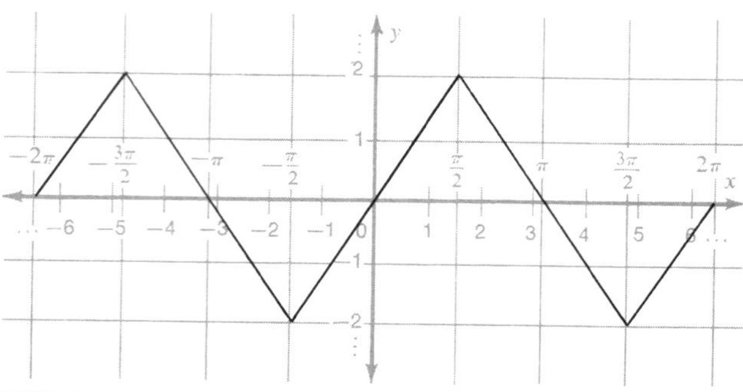

FIGURE 12

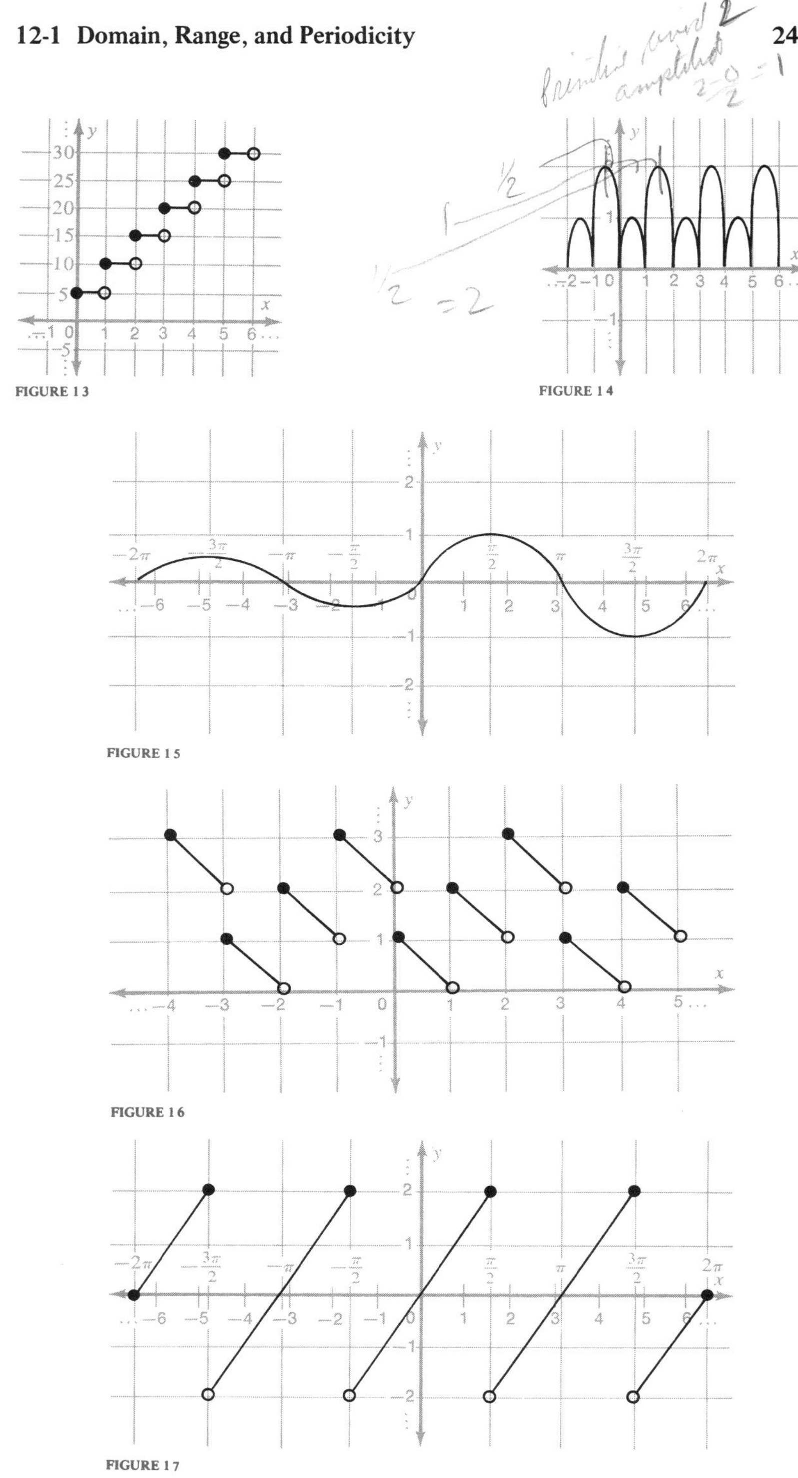

FIGURE 13

FIGURE 14

FIGURE 15

FIGURE 16

FIGURE 17

12-2 GRAPHS OF THE TRIGONOMETRIC FUNCTIONS

Since each trigonometric function is periodic, the branches of the graph of each of these functions over adjacent intervals one primitive period in length, are congruent. Thus, we need only plot the graph of a trigonometric function over an interval one primitive period in length to know what its complete graph looks like.

Since the primitive period of the sine function is 2π, we need only graph the sine function on an interval from 0 to 2π. Begin by locating the point corresponding to 2π (approximately 6.28) on the horizontal θ-axis of a rectangular coordinate system. The points for the special real numbers in the interval from 0 to 2π for which we have computed the values of the trigonometric functions (Column 1, Table 1) are then located on the θ-axis [Figure 1].

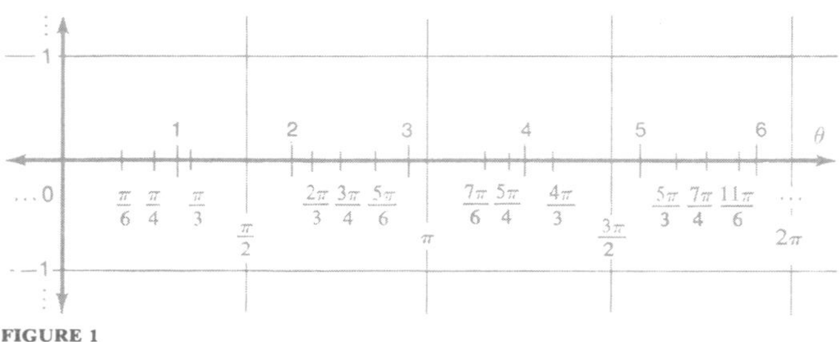

FIGURE 1

The values of the sine function (Column 2, Table 1) are computed using the results of Section 11-4 and the reduction formulas. Finally, the points $(\theta, \sin\theta)$ are plotted and connected with a smooth curve to obtain a fairly accurate indication of the shape of the graph of the sine function for $0 \leqslant \theta \leqslant 2\pi$ [Figure 2].

Sketch the graph of the sine function in the following intervals:

a. $-2\pi \leqslant \theta \leqslant -\pi$
b. $8\pi \leqslant \theta \leqslant 10\pi$
c. $-\pi \leqslant \theta \leqslant \pi$.

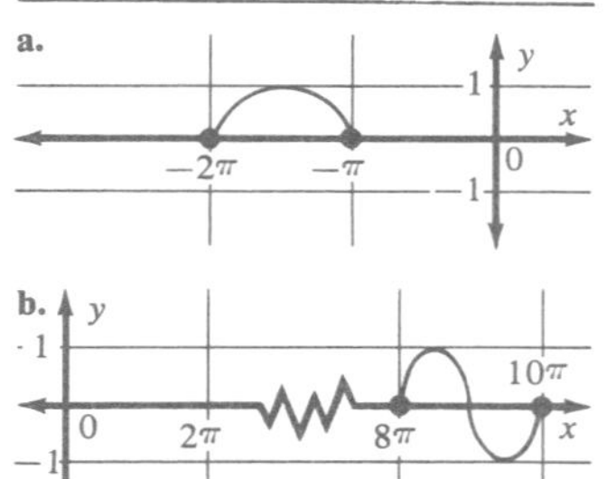

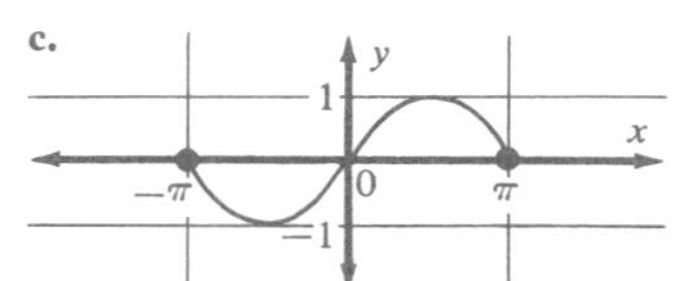

One of the elementary relationships among the trigonometric functions is $\csc\theta = \dfrac{1}{\sin\theta}$ when θ is such that $\sin\theta \neq 0$. Therefore, we can sketch the graph of the cosecant function for θ such that $0 \leqslant \theta \leqslant 2\pi$ by plotting the points $\left(\theta, \dfrac{1}{\sin\theta}\right)$ for any θ for which $\sin\theta$ is not zero [Figure 3]. Note that $\sin\theta$ is zero when $\theta = 0$, $\theta = \pi$, and $\theta = 2\pi$ and that 0, π, and 2π are not elements of the domain of the cosecant function.

TABLE 1

θ	sin θ Exact	sin θ Rational approximation
0	0	0.00
$\frac{\pi}{6}$	$\frac{1}{2}$	0.50
$\frac{\pi}{4}$	$\frac{\sqrt{2}}{2}$	0.71
$\frac{\pi}{3}$	$\frac{\sqrt{3}}{2}$	0.87
$\frac{\pi}{2}$	1	1.00
$\frac{2\pi}{3}$	$\frac{\sqrt{3}}{2}$	0.87
$\frac{3\pi}{4}$	$\frac{\sqrt{2}}{2}$	0.71
$\frac{5\pi}{6}$	$\frac{1}{2}$	0.50
π	0	0.00
$\frac{7\pi}{6}$	$-\frac{1}{2}$	-0.50
$\frac{5\pi}{4}$	$-\frac{\sqrt{2}}{2}$	-0.71
$\frac{4\pi}{3}$	$-\frac{\sqrt{3}}{2}$	-0.87
$\frac{3\pi}{2}$	-1	-1.00
$\frac{5\pi}{3}$	$-\frac{\sqrt{3}}{2}$	-0.87
$\frac{7\pi}{4}$	$-\frac{\sqrt{2}}{2}$	-0.71
$\frac{11\pi}{6}$	$-\frac{1}{2}$	-0.50
2π	0	-0.00

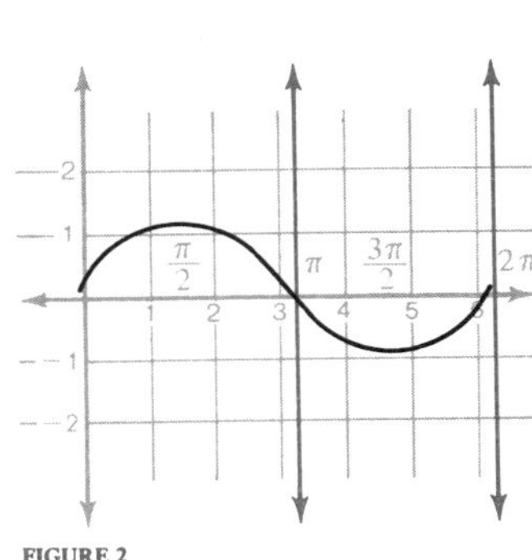

FIGURE 2

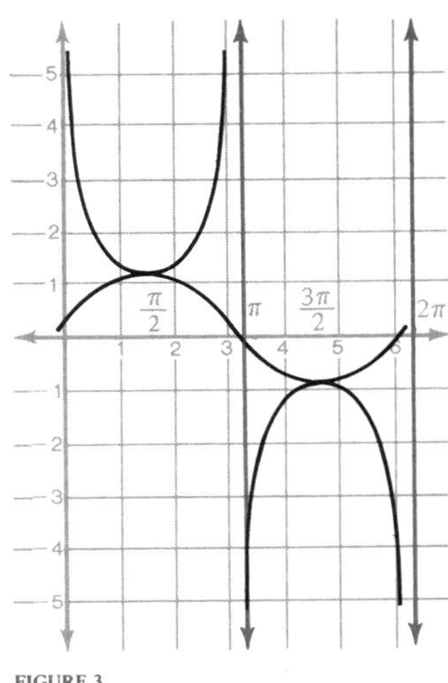

FIGURE 3

Graphs of the cosine and secant functions may be sketched using procedures similar to those employed to obtain the graphs of the sine and cosecant functions. (See Exercises 3 and 4, page 251.) The graph of the tangent function for $0 \leqslant \theta \leqslant \pi$ is shown in Figure 4 and was plotted using Table 2. Recall that tan $\frac{\pi}{2}$ is undefined since $\frac{\pi}{2}$ is not in the domain of the tangent function. Part (a) of the question below calls for the graph of the tangent function for $-\frac{\pi}{2} < \theta < \frac{\pi}{2}$. Notice that the graph of the tangent function on this interval is "continuous"–there are no holes or gaps.

Sketch the graph of the tangent function in the following intervals:

a. $-\frac{\pi}{2} < \theta < \frac{\pi}{2}$ **b.** $\frac{\pi}{2} < \theta < \frac{3\pi}{2}$

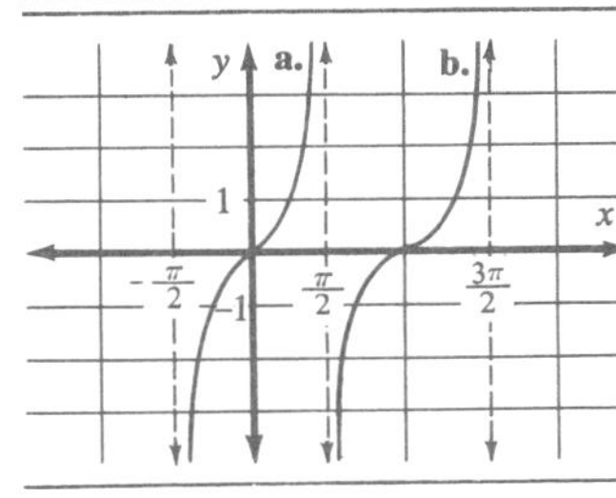

Another elementary relationship among the trigonometric functions is $\cot\theta = \frac{1}{\tan\theta}$ when θ is such that tan θ exists and tan $\theta \neq 0$. Therefore, to graph the cotangent function, we plot points for $\left(\theta, \frac{1}{\tan\theta}\right)$ where θ is in the domain of the tangent function and tan $\theta \neq 0$. Since the tangent function is undefined when $\theta = \frac{\pi}{2}$, we must return to the definition of the cotangent function to find that $\cot\frac{\pi}{2} = 0$. When $\theta = 0$ or $\theta = \pi$, tan $\theta = 0$ and $\cot\theta \neq \frac{1}{\tan\theta}$. In fact, 0 and π are not elements of the domain of the

TABLE 2

θ	tan θ *Exact*	tan θ *Rational approximation*
0	0	0.00
$\frac{\pi}{6}$	$\frac{\sqrt{3}}{3}$	0.58
$\frac{\pi}{4}$	1	1.00
$\frac{\pi}{3}$	$\sqrt{3}$	1.73
$\frac{\pi}{2}$	undefined	undefined
$\frac{2\pi}{3}$	$-\sqrt{3}$	-1.73
$\frac{3\pi}{4}$	-1	-1.00
$\frac{5\pi}{6}$	$-\frac{\sqrt{3}}{3}$	-0.58
π	0	0.00

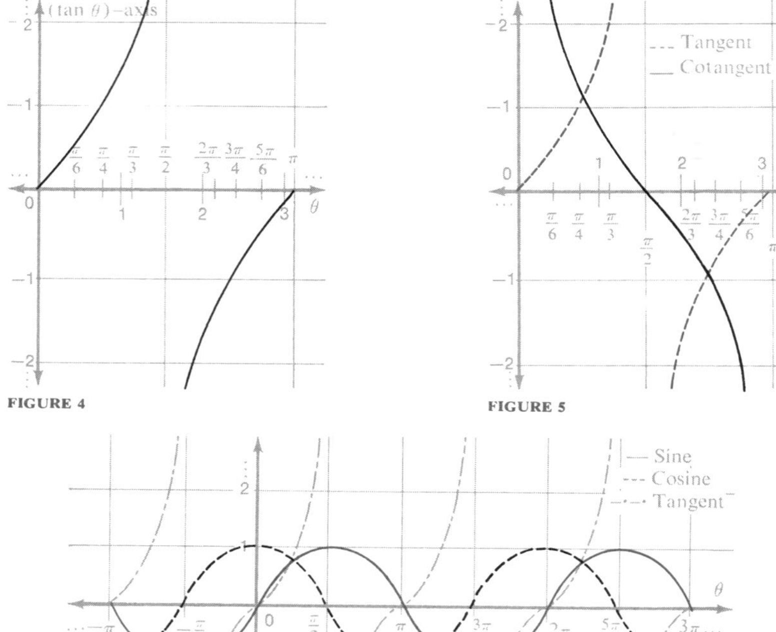

FIGURE 4

FIGURE 5

FIGURE 6

cotangent function. The graphs of the tangent and cotangent functions for $0 \leqslant \theta \leqslant \pi$ are shown in Figure 5.

We have sketched the graphs of several trigonometric functions over one primitive period. Their periodicity allows us to extend the graphs indefinitely on either side of the interval. The graphs of the sine, cosine, and tangent functions for several primitive periods are shown in Figure 6.

Over how many primitive periods is the sine function graphed? The cosine function? The tangent function?

Sine function: 2 primitive periods
Cosine function: 2 primitive periods
Tangent function: 4 primitive periods

Exercises

1. On the horizontal axis of a rectangular coordinate system, locate and label the points that correspond to $0, \pi, -\frac{3\pi}{2}, \frac{7\pi}{2}, \frac{5\pi}{3}, -\frac{5\pi}{6}$, and -3π.

2. On the coordinate system constructed in Exercise 1, locate and label the points $\left(\frac{7\pi}{2}, \cos\frac{7\pi}{2}\right)$, $(\pi, \tan\pi)$, $\left(-\frac{3\pi}{2}, \cot\left(-\frac{3\pi}{2}\right)\right)$, $\left(\frac{5\pi}{3}, \sec\frac{5\pi}{3}\right)$, and $(-3\pi, \sin(-3\pi))$.

3. Make a table of values of the cosine function for the special numbers listed in Table 2 on page 250. Sketch the graph of the function defined by $y = \cos\theta$ for $0 \leqslant \theta \leqslant 2\pi$.

4. Use the graph in Exercise 3 and the elementary relationship between the cosine and secant functions to sketch the graph of the function defined by $y = \sec\theta$ for $0 \leqslant \theta \leqslant 2\pi$ and $\cos\theta \neq 0$.

5. Which trigonometric functions are graphed in Figures 7 and 8? Over how many primitive-periods is each function graphed?

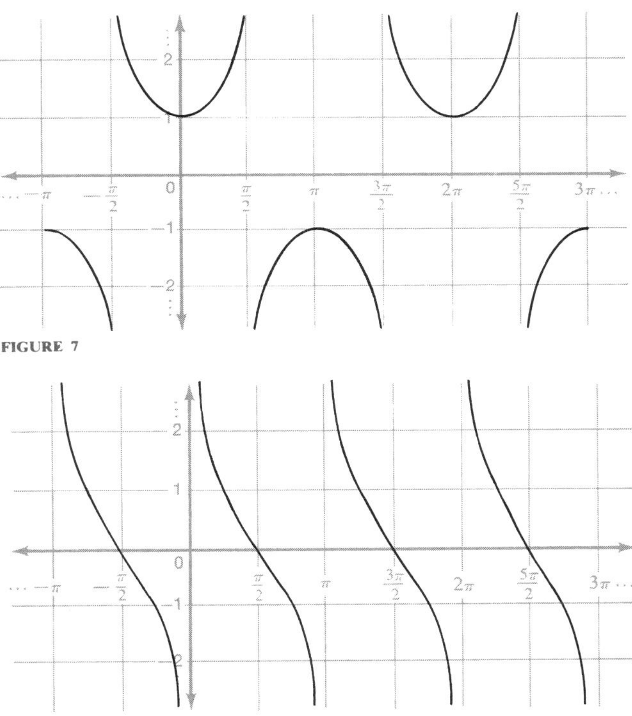

FIGURE 7

FIGURE 8

6. Assume that each of the functions graphed in Figures 9–13 is periodic with primitive period readable from the graph. Use only the labeled points to name what appear to be the endpoints of an interval of length one primitive period.

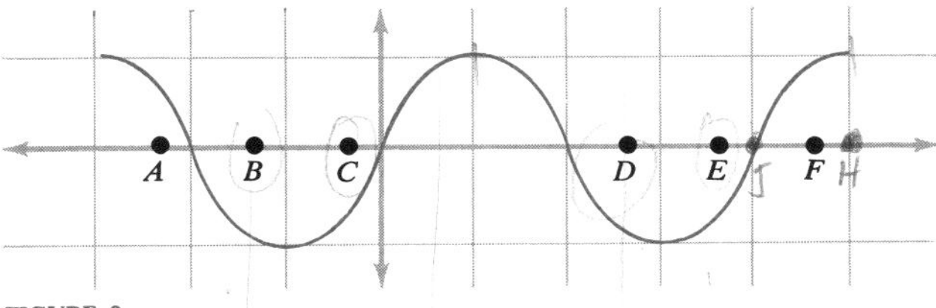

FIGURE 9

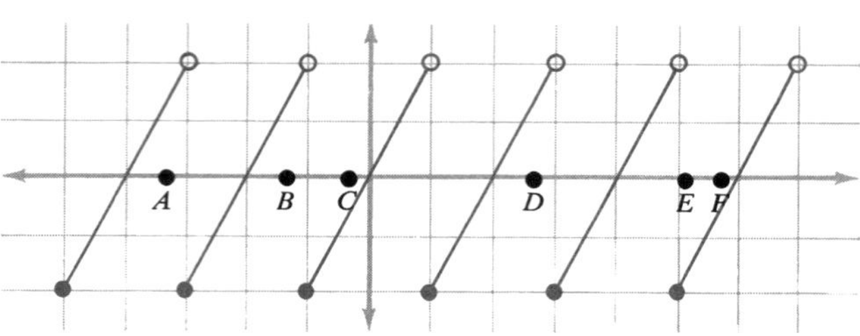

FIGURE 10

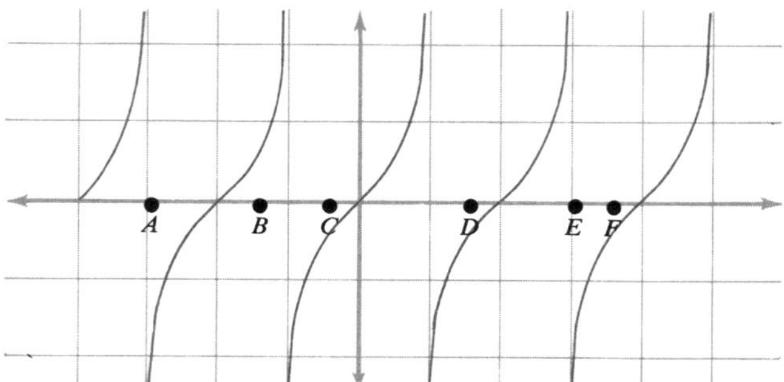

FIGURE 11

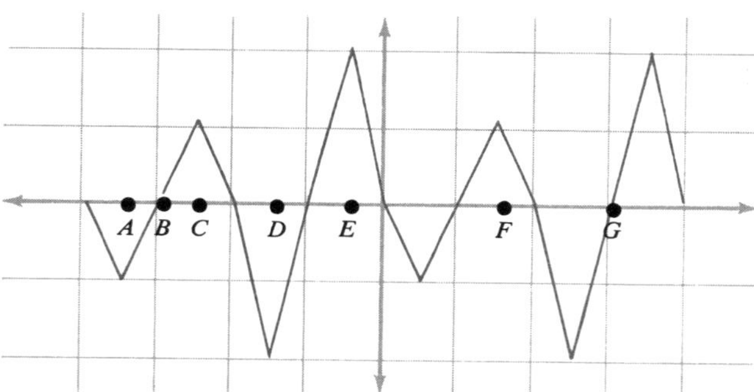

FIGURE 12

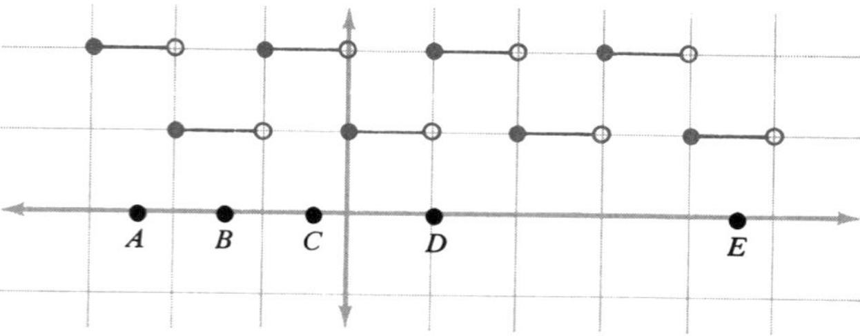

FIGURE 13

7. Sketch the graph of each function on the indicated interval only.

a. $\sec u, \frac{3\pi}{4} \leqslant u \leqslant 2\pi$

b. $\tan \theta, -\frac{3\pi}{2} \leqslant \theta \leqslant -\frac{\pi}{4}$

c. $\sin \varphi, 20\pi \leqslant \varphi \leqslant 23\pi$

d. $\cot z, \frac{5\pi}{4} \leqslant z < 3\pi$

e. $\csc \theta, -3\pi \leqslant \theta \leqslant \frac{\pi}{2}$

f. $\cos x, -7\pi \leqslant x \leqslant -4\pi$

8. Construct a table to assist you in graphing each of the functions defined.

a. $y = 2 \sin x$

b. $y = \cos \frac{1}{2}x$

c. $y = \tan\left(x + \frac{\pi}{2}\right)$

d. $y = -2 \sin 2x$

9. Consider the inverse relation of the sine function. What are the domain and range of the inverse relation of the sine function? Sketch the graph of the inverse sine relation. Is the inverse sine relation a function?

12-3 INVERSE TRIGONOMETRIC RELATIONS

The inverse of a relation f is the set of all ordered pairs (b, a) where $(a, b) \in f$. Thus, the inverse of a relation

$$r = \left\{\left(\frac{\pi}{6}, \frac{1}{2}\right), \left(\frac{\pi}{4}, \frac{\sqrt{2}}{2}\right), \left(\frac{2\pi}{3}, \frac{\sqrt{3}}{2}\right), \left(\frac{3\pi}{4}, \frac{\sqrt{2}}{2}\right)\right\}$$

is

$$r^{-1} = \left\{\left(\frac{1}{2}, \frac{\pi}{6}\right), \left(\frac{\sqrt{2}}{2}, \frac{\pi}{4}\right), \left(\frac{\sqrt{3}}{2}, \frac{2\pi}{3}\right), \left(\frac{\sqrt{2}}{2}, \frac{3\pi}{4}\right)\right\}.$$

Is r a function? What is the domain of r? Range of r? Is r^{-1} a function? What is the domain of r^{-1}? Range of r^{-1}? Is there a one-to-one correspondence between the elements of the domain and the elements of the range of r? If a one-to-one correspondence exists between these two sets, is r^{-1} a function?

Yes, r is a function; Domain of $r = \left\{\frac{\pi}{6}, \frac{\pi}{4}, \frac{2\pi}{3}, \frac{3\pi}{4}\right\}$; Range of $r = \left\{\frac{1}{2}, \frac{\sqrt{2}}{2}, \frac{\sqrt{3}}{2}\right\}$; No, r^{-1} is not a function; Domain of r^{-1} is equal to range of r; range of r^{-1} is equal to domain of r; No, elements of the domain and range of r cannot be matched one-to-one; Yes, if there were a 1-to-1 correspondence, r^{-1} would be a function.

Note that r is a subset of the sine function and r^{-1} is a subset of the inverse of the sine function. Furthermore, r^{-1} is not a function because it contains two ordered pairs, $\left(\frac{\sqrt{2}}{2}, \frac{\pi}{4}\right)$ and $\left(\frac{\sqrt{2}}{2}, \frac{3\pi}{4}\right)$, with equal first coordinates and unequal second coordinates. Consequently, the inverse sine relation is not a function. Similar arguments prove that the other inverse trigonometric relations are not functions.

Two symbols, **$\sin^{-1}$** and **arcsin**, are commonly used for the inverse sine relation. We will use the symbol $\sin^{-1}$, read "the inverse sine relation," because it is similar to other notations that we have used to name inverses, f and f^{-1}. Similarly, $\mathbf{cos}^{-1}$, $\mathbf{cot}^{-1}$, $\mathbf{tan}^{-1}$, $\mathbf{sec}^{-1}$, and $\mathbf{csc}^{-1}$ are used to name the inverse cosine, inverse cotangent, inverse tangent, inverse secant, and inverse cosecant relations, respectively.

Because the trigonometric functions are periodic, an infinite number of elements in the domain of any one of these functions are paired with each element in the range of the function. Hence, each number in the domain of an inverse trigonometric relation is paired with an infinite number of elements in the range of the inverse trigonometric relation. For example, if $t = \sin\theta$ then $t = \sin(\theta + 2\pi n)$ for $n \in I$. Therefore, each of the ordered pairs $(\theta + 2\pi n, t)$ is an element of the sine function. Hence, each of the ordered pairs $(t, \theta + 2\pi n)$ is an element of the inverse sine relation, $\sin^{-1}$, and each element t in the domain of $\sin^{-1}$ is paired with an infinite number of elements $\theta + 2\pi n$ in its range. The symbol $\sin^{-1} t$ denotes any one of the numbers

$$\ldots, \theta - 2\pi, \theta - \pi, \theta, \theta + \pi, \theta + 2\pi, \ldots$$

That is, for each element t in the domain of the inverse sine relation, $\sin^{-1} t$ denotes any number whose sine is t. For example, $\sin^{-1}\frac{1}{2}$ is any element of $\left\{\frac{\pi}{6} + 2\pi n\right\} \cup \left\{\frac{5\pi}{6} + 2\pi n\right\}$.

Caution: Do not equate $\sin^{-1}\theta$ and $\frac{1}{\sin\theta}$! To express $\frac{1}{\sin\theta}$ using negative exponents, we write $(\sin\theta)^{-1}$. *The symbol* $\sin^{-1}\theta$ *is reserved for any number* r *such that* $\sin r = \theta$. In general, however, $\sin^n\theta = (\sin\theta)^n$ for $n \in R$ and $n \neq -1$.

Remarks similar to those made in the preceding two paragraphs apply for each of the other trigonometric functions.

Example Determine the set of all possible y for which $y = \sin^{-1}\frac{\sqrt{2}}{2}$.

Solution: Since $\sin^{-1}\frac{\sqrt{2}}{2}$ represents any number for which the value of the sine function is $\frac{\sqrt{2}}{2}$, $y = \sin^{-1}\frac{\sqrt{2}}{2}$ implies y represents any number such that $\sin y = \frac{\sqrt{2}}{2}$. From Section 11-4 it follows that

$$y \in \left\{\frac{\pi}{4} + 2\pi n\right\} \cup \left\{\frac{3\pi}{4} + 2\pi n\right\}$$

where n is an integer.

What do the graphs of the inverse trigonometric relations look like? We saw in Chapter 9, Section 2, that the graph of a relation and its inverse are symmetric with respect to the line $y = x$. For example, the graphs of the function

$$f = \{(x, y)\colon\ y = x^2\}$$

and its inverse relation

$$f^{-1} = \{(x, y)\colon\ y = \sqrt{x} \text{ or } y = -\sqrt{x}\}$$

are reflections about the line $y = x$ [Figure 1].

	Domain	*Range*
f	R	$\{r\colon r \geqslant 0\}$
f^{-1}	$\{r\colon r \geqslant 0\}$	R

The domain of f is the range of f^{-1} and the range of f is the domain of f^{-1}.

What are the domain and range of f? Of f^{-1}? What relationship exists between the domains and ranges of f and f^{-1}?

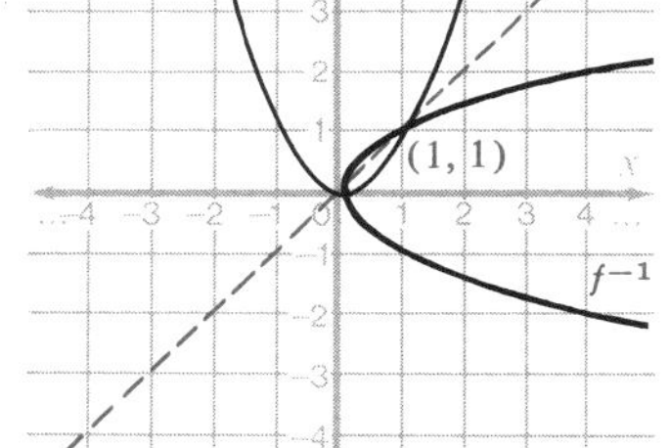

FIGURE 1

The inverse relation f^{-1} is not a function because there are two nonzero elements in the range of f^{-1} paired with each nonzero element in the domain of f^{-1}. Mathematicians frequently distinguish a subset of a relation which is a function by arbitrarily selecting only one of the elements from the range that is paired with a given element in the domain. Such a function is called a **principal value function.** We emphasize that the domain of a principal value function is equal to the domain of the relation from which it is defined. The range of a principal value function is a subset of the range of the parent relation which contains one and only one of the elements from the range for each element in the domain.

We can find one principal value function for

$$f^{-1} = \{(x, y)\colon\ y = \sqrt{x} \text{ or } y = -\sqrt{x}\}$$

by choosing the positive value of y that is paired with each element x in the domain of f^{-1}. Another function is defined by choosing the negative value of y paired with each x. Still another function is defined by choosing the positive value of y when x is a rational number and the negative value of y when x is an irrational number.

Make up other principal value functions for the relation

$$f^{-1} = \{(x, y)\colon\ y = \sqrt{x} \text{ or } y = -\sqrt{x}\}.$$

Answers may vary: function defined by

a. $y = \sqrt{x}$ if $x \leqslant 2$ or $y = -\sqrt{x}$ if $x > 2$

b. $y = \sqrt{x}$ if $x \in I$ or $y = -\sqrt{x}$ if $x \notin I$

A principal value function for f^{-1} is often denoted by F^{-1}. The graph of a principal value function of a relation is called a **principal branch** of the graph of the relation. Figures 2–4 illustrate several principal branches of the graph of f^{-1}.

Exercises

1. For each relation r: (i) Determine the domain and range of r. (ii) Is r a function? Why or why not? (iii) List or describe the elements of r^{-1}. (iv) Determine the domain and range of r^{-1}. (v) Is r^{-1} a function?

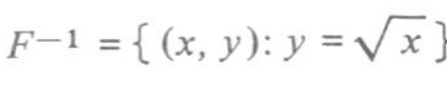

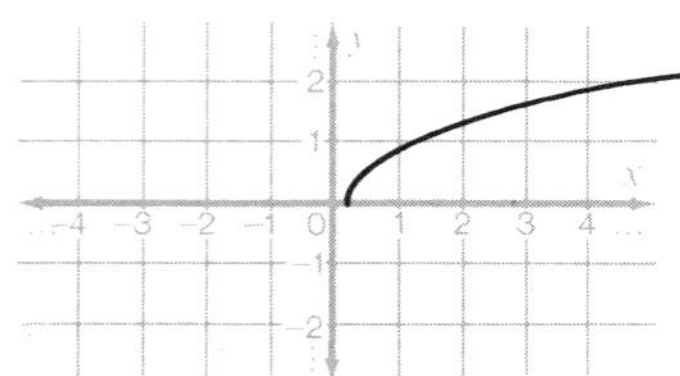

FIGURE 2

(vi) If r^{-1} is a function, what is the relationship between the elements of the domain and range of r?

a. r: To each automobile corresponds its manufacturer
b. r: To each state correspond its inhabitants
c. r: To each child corresponds his father
d. r: To each city correspond its aldermen
e. $r = \{(a, b), (b, c), (a, c)\}$
f. $r = \{(0, 1), (1, 2), (2, 3), (3, 4)\}$
g. $r = \{(x, y): y = x^2\}$

2. Determine the set of all possible replacements for y that make each sentence true.

a. $y = \sin^{-1} \frac{1}{2}$ b. $y = \sec^{-1} \sqrt{2}$ c. $y = \cot^{-1} \sqrt{3}$
d. $y = \left(\sin \frac{\pi}{2}\right)^{-1}$ e. $y = \tan^{-1} \frac{\sqrt{3}}{3}$ f. $y = \csc^{-1} 1$
g. $y = \cos^{-1} \frac{\sqrt{3}}{2}$ h. $y = \tan^2 \frac{\pi}{4}$ i. $y = \cos^{-1} \frac{1}{2}$
j. $y = \tan^{-1} 1$ k. $y = \left(\csc \frac{\pi}{4}\right)^{-1}$ l. $y = \sin^{-1} 0$
m. $y = \sec^{-1} (-1)$ n. $y = \csc^{-1} (-1)$ o. $y = \cot^{-1} 0$
p. $y = (\cos \pi)^{-1}$ q. $y = \cot^4 \frac{13\pi}{4}$

$F^{-1} = \{(x, y): y = \sqrt{x}\}$

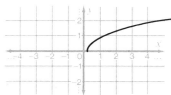

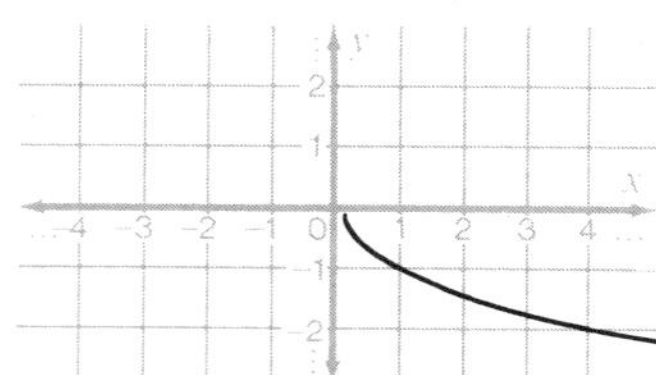

FIGURE 3

3. Respond to parts i–vi of Exercise 1 for the relation r determined by each equation.

a. $y = \frac{1}{x}$ b. $y = \log_{10} x$ c. $y = \frac{1}{\csc x}$ d. $y = \sqrt{x}$

$F^{-1} = \{(x, y): y = \sqrt{x}$ if $0 \leqslant x \leqslant 1$ and $y = -\sqrt{x}$ if $x > 1\}$

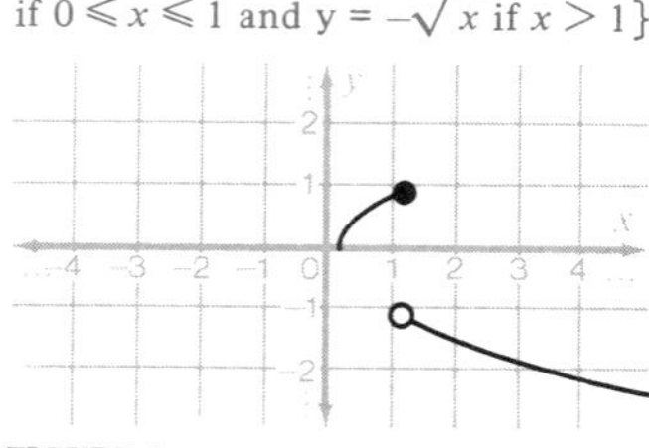

FIGURE 4

4. Make up an example of a relation that is a function and whose inverse is a function.

5. Make up an example of a relation that is a function but whose inverse is not a function.

6. Make up an example of a relation that is not a function but whose inverse is a function.

7. Make up an example of a relation that is not a function and whose inverse is not a function.

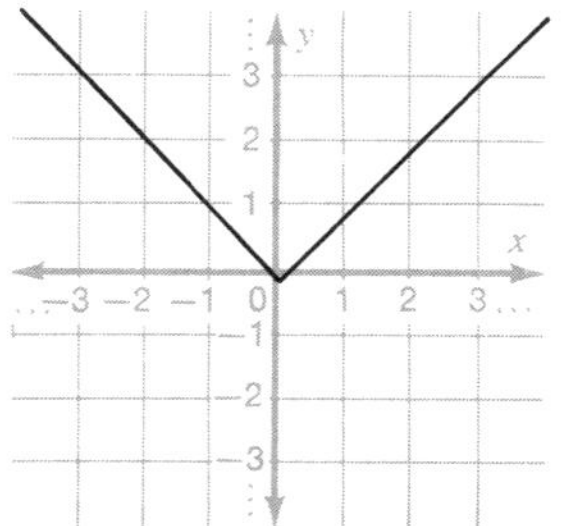

FIGURE 5

8. Sketch the graph of the function defined by the equation $y = -|x|$. Use symmetry to sketch the graph of the inverse of the function. Is the inverse relation a function?

9. Which of the graphs in Figures 5–10 below are the graphs of functions?

10. Sketch the graph of the relation defined by each equation.

a. $y = x + 1$ b. $y = -x - 1$ c. $y = x^2$
d. $y = x^3$ e. $y = |x|$ f. $y = \sqrt{x}$

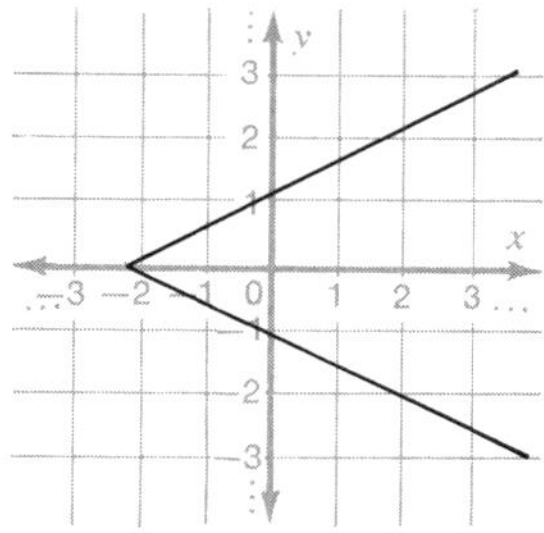
FIGURE 6

FIGURE 7

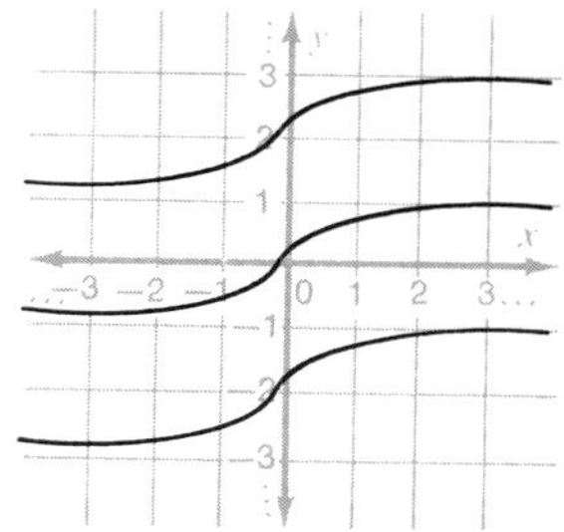
FIGURE 8

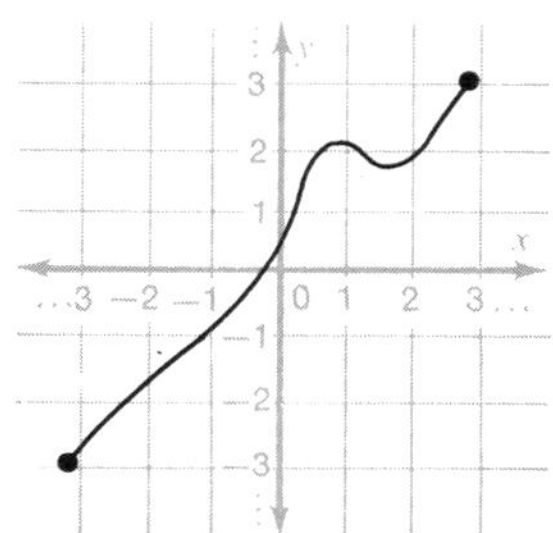
FIGURE 9

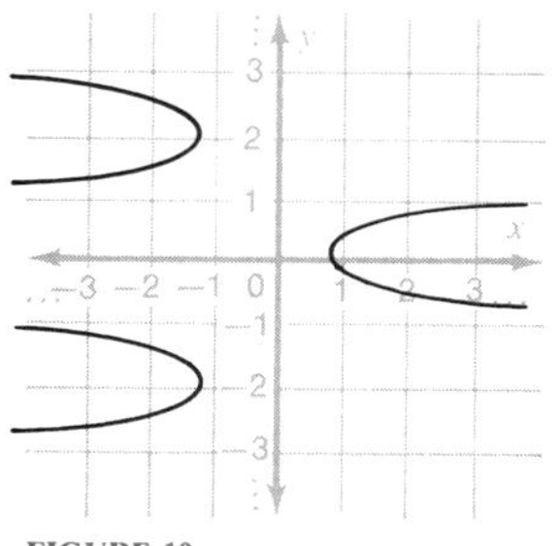
FIGURE 10

g. $y = \dfrac{1}{x}$ h. $y = -x^2 + 2$ i. $y = -\sqrt{x}$

j. $y = \sqrt{x^2 + 1}$

11. Use the concept of symmetry to sketch the graph of the inverse of each relation in Exercise 10.

12. Sketch the graphs of two different principal branches of the graphs in Exercise 11 that are not the graphs of functions.

13. Define a principal value function for the inverse of each relation.

a. r: To each voter corresponds his congressman

b. $r = \{(-5, 4), (\sqrt{2}, \pi), (1, \pi), (\log_{10} 9, 4), (3^{5.6}, 33)\}$

c. $r = \{(x, y)\colon\ y = -x^2 - 1\}$

14. Define a principal value function for the inverse of each relation.

a. r: To each child corresponds his father

b. $r = \left\{(\pi, -1), (0, 3), (4, 1), \left(\dfrac{3}{2}, -1\right)\right\}$

c. $r = \{(x, y)\colon\ y = x^2 + 1\}$

15. Given the relation $\{(x, y)\colon\ y = x\}$, is it necessary to define a principal value function for its inverse? Why or why not?

12-4 PRINCIPAL VALUE FUNCTIONS AND GRAPHS OF THE INVERSE TRIGONOMETRIC RELATIONS

Since $\sin^{-1}$ is the inverse relation of the sine function, the domain of $\sin^{-1}$ is $\{x\colon\ -1 \leqslant x \leqslant 1\}$ (the range of the sine function) and the range of $\sin^{-1}$ is the set of all real numbers (the domain of the sine function). The graph of $\sin^{-1}$ is symmetric to the graph of the sine function about the line $y = x$ [Figure 1].

Since the domain of a principal value function is the domain of the parent relation, the domain of a principal value function for $\sin^{-1}$ is $\{x\colon\ -1 \leqslant x \leqslant 1\}$. The range of a principal value function for $\sin^{-1}$ may be any subset of the range of $\sin^{-1}$ containing one and only one of the numbers paired with each element of $\{x\colon -1 \leqslant x \leqslant 1\}$.

Which of the following sets qualify for the range of a principal value function for $\sin^{-1}$?

a. $\{y\colon 0 \leqslant y \leqslant \pi\}$ **b.** $\left\{y\colon -\frac{\pi}{2} \leqslant y \leqslant \frac{\pi}{2}\right\}$

c. $\left\{y\colon -\frac{3\pi}{2} \leqslant y \leqslant -\frac{\pi}{2}\right\}$ **d.** $\left\{y\colon 0 \leqslant y \leqslant \frac{\pi}{2} \text{ and } \pi \leqslant y \leqslant \frac{3\pi}{2}\right\}$

e. $\left\{y\colon -\frac{\pi}{2} < y < \frac{\pi}{2},\ y = -\frac{7\pi}{2},\ \text{and}\ y = \frac{15\pi}{2}\right\}$

b., c., e.

If we choose the set of real numbers between and including $-\frac{\pi}{2}$ and $\frac{\pi}{2}$ to be the range of a principal value function for $\sin^{-1}$ and this principal value function is denoted by $\mathbf{Sin^{-1}}$ (called the inverse sine function),

$$y = \mathrm{Sin}^{-1}\,x$$

means

$$y = \sin^{-1} x \quad \text{and} \quad -\frac{\pi}{2} \leqslant y \leqslant \frac{\pi}{2}.$$

Therefore, if x is an element of the domain of Sin^{-1}

$$-\frac{\pi}{2} \leqslant \mathrm{Sin}^{-1}\,x \leqslant \frac{\pi}{2}.$$

The graph of Sin^{-1} is shown in Figure 2.

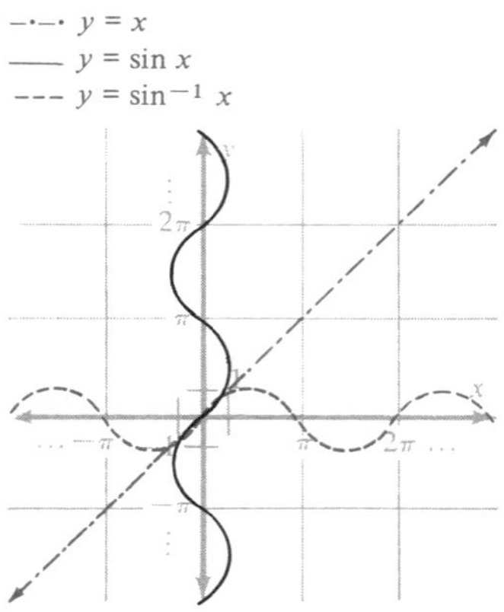

—·—· $y = x$
—— $y = \sin x$
- - - $y = \sin^{-1} x$

FIGURE 1

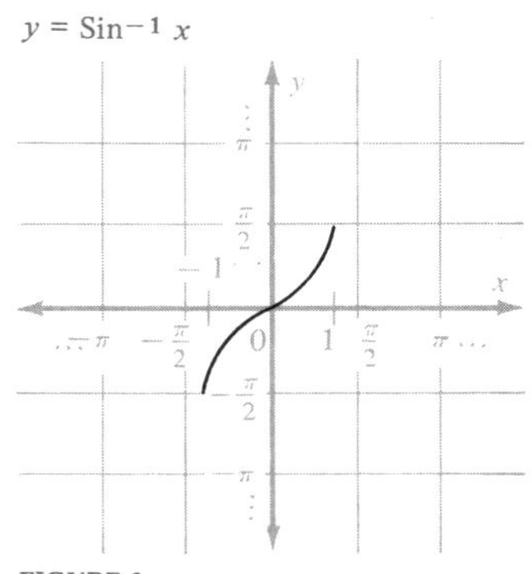

FIGURE 2

Example 1 Evaluate $\text{Sin}^{-1}\,\frac{1}{2}$.

Solution: If $y = \text{Sin}^{-1}\,\frac{1}{2}$, then y is a number such that $\sin y = \frac{1}{2}$ and $-\frac{\pi}{2} \leqslant y \leqslant \frac{\pi}{2}$. The only number y between $-\frac{\pi}{2}$ and $\frac{\pi}{2}$ such that $\sin y = \frac{1}{2}$ is $y = \frac{\pi}{6}$. Thus, $\text{Sin}^{-1}\,\frac{1}{2} = \frac{\pi}{6}$.

y	0	$-\frac{1}{2}$	$\frac{\sqrt{2}}{2}$	1	$-\frac{\sqrt{3}}{2}$
$\text{Sin}^{-1}\,y = x$	0	$-\frac{\pi}{6}$	$\frac{\pi}{4}$	$\frac{\pi}{2}$	$-\frac{\pi}{3}$

What real number x makes each of the following sentences true? $\text{Sin}^{-1}\,0 = x$, $\text{Sin}^{-1}\left(-\frac{1}{2}\right) = x$, $\text{Sin}^{-1}\,\frac{\sqrt{2}}{2} = x$, $\text{Sin}^{-1}\,1 = x$, and $\text{Sin}^{-1}\left(-\frac{\sqrt{3}}{2}\right) = x$.

The graphs of the inverse cosine relation, $\cos^{-1}$, and the inverse cosine function, $\mathbf{Cos}^{-1}$, are shown in Figures 3 and 4, respectively. The domain of $\cos^{-1}$ is $\{x\colon\ -1 \leqslant x \leqslant 1\}$ (the range of the cosine function). The range of $\cos^{-1}$ is the set of real numbers (the domain of the cosine function). The domain of Cos^{-1} must be equal to the domain of $\cos^{-1}$. According to convention, we will choose the set of real numbers between and including 0 and π for the range of Cos^{-1}; that is, $0 \leqslant \text{Cos}^{-1}\,x \leqslant \pi$ for any x in the domain of Cos^{-1}.

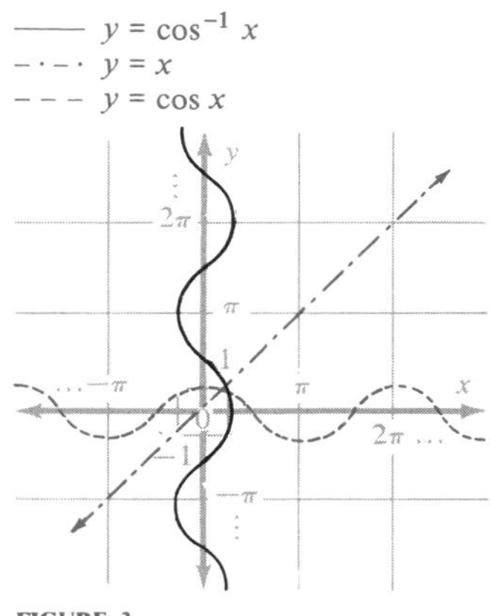

FIGURE 3

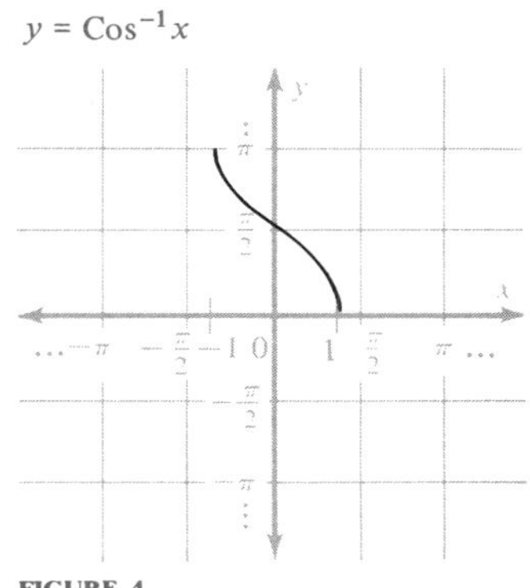

FIGURE 4

Example 2 Evaluate $\sin\left[\text{Cos}^{-1}\left(-\frac{\sqrt{3}}{2}\right)\right]$.

Solution: The sentence $y = \text{Cos}^{-1}\left(-\frac{\sqrt{3}}{2}\right)$ is true if and only if $\cos y = -\frac{\sqrt{3}}{2}$ and $0 \leqslant y \leqslant \pi$. The only number y between and including

0 and π such that $\cos y = -\frac{\sqrt{3}}{2}$ is $y = \frac{5\pi}{6}$. Thus $\text{Cos}^{-1}\left(-\frac{\sqrt{3}}{2}\right) = \frac{5\pi}{6}$ and $\sin\left[\text{Cos}^{-1}\left(-\frac{\sqrt{3}}{2}\right)\right] = \sin\frac{5\pi}{6} = \frac{1}{2}$.

What real numbers x make each of the following sentences true: $\text{Cos}^{-1}\left(-\frac{\sqrt{2}}{2}\right) = x$, $\text{Cos}^{-1}\ 1 = x$, $\text{Cos}^{-1}\ \frac{1}{2} = x$, $\text{Cos}^{-1}\ [\text{Sin}^{-1}\ 0] = x$, and $\text{Cos}^{-1}\left[\tan\left(-\frac{\pi}{4}\right)\right] = x$?

y	$-\frac{\sqrt{2}}{2}$	1	$\frac{1}{2}$	0	-1
$\text{Cos}^{-1}\ y$	$\frac{3\pi}{4}$	0	$\frac{\pi}{3}$	$\frac{\pi}{2}$	π

The domain of the inverse tangent function, **Tan**$^{-1}$, is the set of all real numbers, the domain of $\tan^{-1}$. So that the graph of Tan^{-1} [Figure 6] is a continuous branch of the graph of $\tan^{-1}$ [Figure 5], the range of Tan^{-1} is chosen to be the set of numbers between (but not including) $-\frac{\pi}{2}$ and $\frac{\pi}{2}$; that is, $-\frac{\pi}{2} < \text{Tan}^{-1}\ x < \frac{\pi}{2}$ for any real number x.

If each student is asked to find a real number y such that $y = \tan^{-1}\frac{\sqrt{3}}{3}$, is it possible that each student could have a different number and be correct? Could this happen if each student is asked to find a real number y such that $y = \text{Tan}^{-1}\frac{\sqrt{3}}{3}$?

Yes, if $\tan^{-1}\frac{\sqrt{3}}{3} = x$, then x is in $\left\{\frac{\pi}{6} + \pi n\right\}$; No, $\text{Tan}^{-1}\frac{\sqrt{3}}{3} = \frac{\pi}{6}$.

Exercises

$y = \tan^{-1} x$

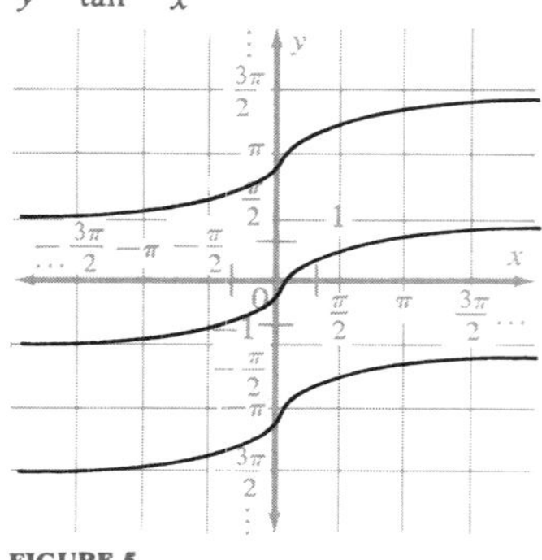

FIGURE 5

1. Determine the set of all real replacements for y that make each statement true.

   a. $\sin^{-1}\left(-\frac{\sqrt{2}}{2}\right) = y$ b. $\tan^{-1}\frac{\sqrt{3}}{3} = y$ c. $\cos^{-1}\left(-\frac{\sqrt{3}}{2}\right) = y$

   d. $\cos^{-1}\left(-\frac{\sqrt{2}}{2}\right) = y$ e. $\cot^{-1}\ 1 = y$ f. $\sec^{-1}\ 2 = y$

   g. $\tan^{-1}\sqrt{3} = y$ h. $\sec^{-1}\left(-\frac{2\sqrt{3}}{3}\right) = y$ i. $\csc^{-1}\sqrt{2} = y$

   j. $\sin^{-1}\frac{1}{2} = y$ k. $\cot^{-1}\frac{\sqrt{3}}{3} = y$ l. $\cot^{-1}\sqrt{3} = y$

2. Evaluate.

   a. $\text{Cos}^{-1}\left(-\frac{\sqrt{2}}{2}\right)$ b. $\text{Sin}^{-1}\frac{\sqrt{2}}{2}$ c. $\text{Tan}^{-1}\frac{\sqrt{3}}{3}$

   d. $\text{Tan}^{-1}\ (-1)$ e. $\text{Sin}^{-1}\left(-\frac{\sqrt{2}}{2}\right)$ f. $\text{Cos}^{-1}\left(\frac{\sqrt{2}}{2}\right)$

   g. $\text{Cos}^{-1}\frac{1}{2}$ h. $\text{Sin}^{-1}\ 1$ i. $\text{Sin}^{-1}\ 0$

   j. $\text{Tan}^{-1}\sqrt{3}$ k. $\text{Tan}^{-1}\ (-\sqrt{3})$ l. $\text{Sin}^{-1}\ (-1)$

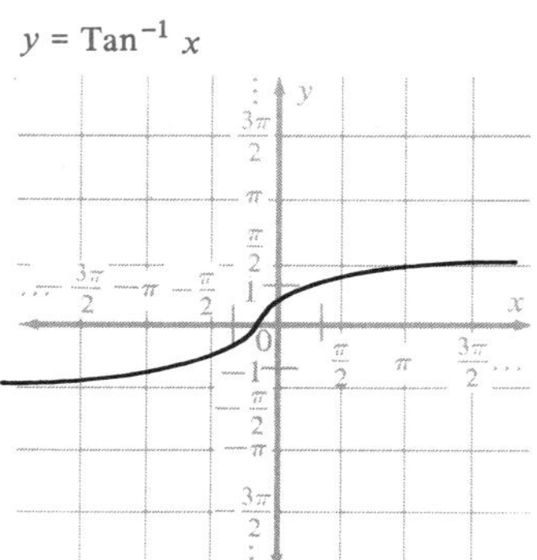

FIGURE 6

m. $\text{Cos}^{-1}\left(-\frac{\sqrt{3}}{2}\right)$ n. $\text{Sin}^{-1}\left(-\frac{\sqrt{3}}{2}\right)$ o. $\text{Tan}^{-1}\left(-\frac{\sqrt{3}}{3}\right)$

p. $\text{Cos}^{-1}\ (-1)$ q. $\text{Cos}^{-1}\ 0$ r. $\text{Tan}^{-1}\ 0$

3. Evaluate, if possible.

a. $\sin\ [\text{Cos}^{-1}\ (-1)]$ b. $\tan\ [\text{Sin}^{-1}\ 1]$ c. $\cos\frac{\pi}{3} - \text{Sin}^{-1}\ (0.5)$

d. $\text{Cos}^{-1}\ 2$ e. $\text{Tan}^{-1}\left[\frac{1}{2}\csc\left(-\frac{5\pi}{3}\right)\right]$ f. $\cos\ (\text{Cos}^{-1}\ x)$

g. $\text{Cos}^{-1}\left(\sec\frac{\pi}{3}\right)$ h. $\text{Sin}^{-1}\left(\sin\frac{\pi}{4}\right)$ i. $\text{Sin}^{-1}\left(\sec\frac{13\pi}{3}\right)^{-1}$

j. $\cos\left(\text{Sin}^{-1}\frac{\sqrt{2}}{2}\right)$ k. $\text{Tan}^{-1}\left(\tan\frac{2\pi}{3}\right)$ l. $\sec\ (\text{Cos}^{-1}\ 1)$

4. Evaluate, if possible.

a. $\sin\left(\text{Cos}^{-1}\frac{1}{2} + \text{Sin}^{-1}\frac{1}{2}\right)$ b. $\text{Cos}^{-1}\left(\frac{2\sin\pi}{6}\right)$

c. $\text{Sin}^{-1}\ (\cos\ [\text{Tan}^{-1}\ (-1)])$ d. $\cot\left(\frac{2\pi}{3}\right) + \text{Tan}^{-1}\left(\frac{\sqrt{3}}{3}\right)$

e. $\text{Tan}^{-1}\left[2\sin\left(\text{Cos}^{-1}\frac{1}{2}\right)\right]$ f. $\tan\ (\text{Sin}^{-1}\ 1)$

g. $\cot\left[\text{Cos}^{-1}\left(-\frac{\sqrt{3}}{2}\right) - \text{Tan}^{-1}\left(-\frac{\sqrt{3}}{3}\right)\right]$

h. $\tan\left(\text{Cos}^{-1}\frac{\sqrt{3}}{2} + \text{Sin}^{-1}\frac{\sqrt{3}}{2}\right)$

5. Given that the range of Cot^{-1} is $\{y\colon\ 0 < y < \pi\}$:
 a. Determine the domain and range of $\cot^{-1}$; of Cot^{-1}.
 b. Sketch the graph of $\cot^{-1}$; of Cot^{-1}.

6. Given that the range of Csc^{-1} is $\{y\colon\ -\frac{\pi}{2} \leqslant y \leqslant \frac{\pi}{2}, y \neq 0\}$:
 a. Determine the domain and range of $\csc^{-1}$; of Csc^{-1}.
 b. Sketch the graph of $\csc^{-1}$; of Csc^{-1}.

7. Given that the range of Sec^{-1} is $\left\{y\colon\ 0 \leqslant y \leqslant \pi \text{ and } y \neq \frac{\pi}{2}\right\}$.
 a. Determine the domain and range of $\sec^{-1}$; of Sec^{-1}.
 b. Sketch the graph of $\sec^{-1}$; of Sec^{-1}.

8. Prove each statement if $a \neq 0$.

a. $\cot^{-1} a = \tan^{-1}\frac{1}{a}$ b. $\sec^{-1} a = \cos^{-1}\frac{1}{a}$ c. $\csc^{-1} a = \sin^{-1}\frac{1}{a}$

12-5 MORE ON PERIODICITY

Many periodic functions can be constructed from known periodic functions. For example, the functions g and h defined by the equations

$$g(x) = a \sin x \qquad \text{and} \qquad h(x) = \sin bx$$

where $a \neq 0$ and $b > 0$ are very much like the sine function. The domain of g is the set of all real numbers and $-|a| \leqslant a \sin x \leqslant |a|$ for all real numbers x. The domain of h is the set of all real numbers and its range is restricted to the same interval as the range of the sine function; that is, $-1 \leqslant \sin bx \leqslant 1$ for all real numbers x. The function g is periodic with primitive period 2π. The function h is periodic, but if $b \neq 1$ then the primitive period of h is not 2π. The following theorems generalize these situations and establish a means for finding the primitive periods of such functions.

TABLE 1

x	$2 \cos x$
0	$2 \cos 0 = 2$
$\frac{\pi}{2}$	$2 \cos \frac{\pi}{2} = 0$
π	$2 \cos \pi = -2$
$\frac{3\pi}{2}$	$2 \cos \frac{3\pi}{2} = 0$
2π	$2 \cos 2\pi = 2$

Theorem 1 If f is a periodic function with primitive period p and g is a function defined by the equation $g(x) = af(x)$ where $a \in R$ and $a \neq 0$, then g is periodic with primitive period p.

Example 1 Find the primitive period of the function g defined by the equation

$$g(x) = 2 \cos x$$

and sketch its graph on an interval the length of its primitive period.

Solution: Since the primitive period of the cosine function is 2π, the primitive period of g is 2π by Theorem 1. To graph g on the interval from 0 to 2π, locate 2π on the x-axis and separate the segment from 0 to 2π into four congruent segments. Plot the points of g corresponding to $0, \frac{\pi}{2}, \pi, \frac{3\pi}{2}$, and 2π [Table 1]. Then sketch a smooth curve similar to the graph of the cosine function [Figure 1].

$g(x) = 2 \cos x$ and $0 \leqslant x \leqslant 2\pi$

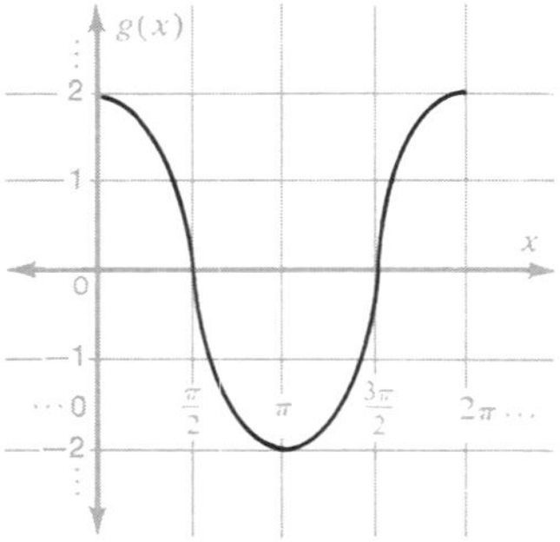

FIGURE 1

Theorem 2 If f is a periodic function with primitive period p, then the function h defined by the equation $h(x) = f(bx)$ where $b > 0$ is periodic with primitive period $\frac{p}{b}$.

Example 2 Find the primitive period of the function defined by the equation

$$h(x) = \sin 2x$$

and sketch the graph of this function on an interval the length of its primitive period.

Solution: The primitive period of the sine function is 2π. Therefore, by Theorem 2 the primitive period of h is $\frac{2\pi}{2}$ or π. To graph the function on the interval from 0 to π, locate π on the x-axis and separate the segment from 0 to π into four congruent segments. Plot the points of the graph corresponding to $0, \frac{\pi}{4}, \frac{\pi}{2}, \frac{3\pi}{4}$, and π [Table 2]. Then sketch a curve which is similar to the graph of the sine function [Figure 2].

TABLE 2

x	$\sin 2x$
0	$\sin 0 = 0$
$\frac{\pi}{4}$	$\sin \frac{\pi}{2} = 1$
$\frac{\pi}{2}$	$\sin \pi = 0$
$\frac{3\pi}{4}$	$\sin \frac{3\pi}{2} = -1$
π	$\sin 2\pi = 0$

$h(x) = \sin 2x$ and $0 \leqslant x \leqslant \pi$

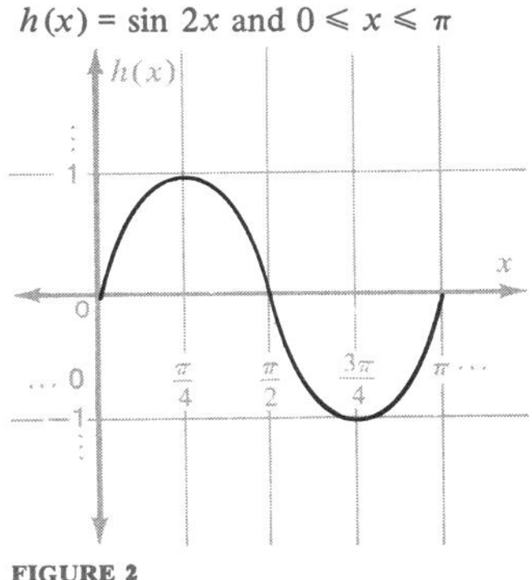

FIGURE 2

TABLE 3

x	$\tan \frac{1}{2}x$
0	$\tan 0 = 0$
$\frac{\pi}{2}$	$\tan \frac{\pi}{4} = 1$
π	$\tan \frac{\pi}{2}$ —undefined
$\frac{3\pi}{2}$	$\tan \frac{3\pi}{4} = -1$
2π	$\tan \pi = 0$

Example 3 Find the primitive period of the function defined by the equation

$$h(x) = \tan \frac{1}{2}x$$

and sketch the graph of this function on an interval the length of its primitive period.

Solution: The primitive period of the tangent function is π; hence, by Theorem 2, the primitive period of h is $\pi \div \frac{1}{2}$ or 2π. To graph h on the interval from 0 to 2π, locate 2π on the x-axis and the points that separate the segment from zero to 2π into four congruent segments. Then plot the points of the graph corresponding to these points $\left(0, \frac{\pi}{2}, \text{ and } \frac{3\pi}{2}\right)$ [Table 3] and sketch a curve similar to the graph of the tangent function [Figure 3].

If f and g are functions and h is the function defined by the equation $h(x) = f(x) + g(x)$ for all x in the domain of f and g, then h is called the **sum** of f and g. For example, the function h_1 defined by

$$h_1(x) = \sin x + \cos x$$

is the sum of the sine and cosine functions. Moreover, the sine and cosine functions are periodic with primitive period 2π. Is h_1 periodic? Is 2π a period of h_1?

$h(x) = \tan \frac{1}{2}x$ and $0 \leqslant x \leqslant 2\pi$

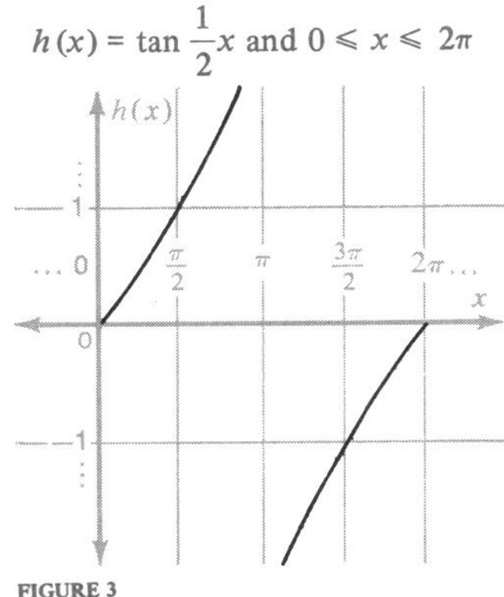

FIGURE 3

Theorem 3 Suppose that f and g are periodic functions and that f and g both have the same period p. If h is the sum of f and g, then h is a periodic function with period p.

Figures a and b are the graphs of two functions f and g, respectively. What is the primitive period of f? Of g? Draw the graph of the function h which is the sum of f and g. What is the period of h? Does h have a primitive period?

Primitive period of f and g is 2 units; • The graph of h is the x-axis; Any nonzero real number is a period of h; • Since there is no smallest positive real number, h has no primitive period.

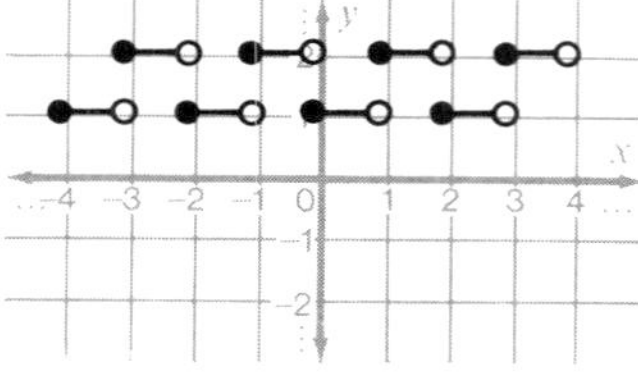

a.

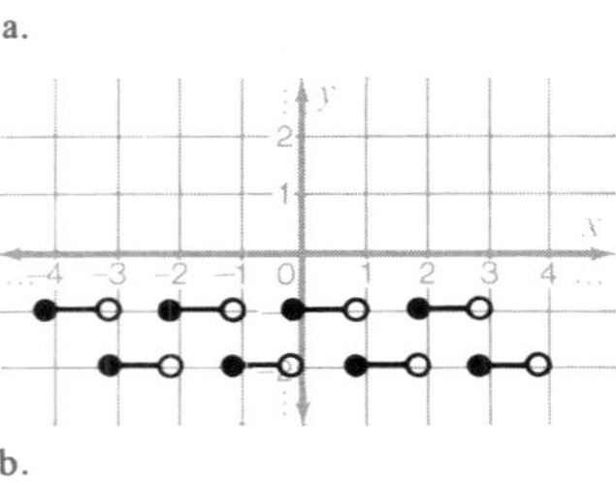

b.

The primitive periods of the functions defined by the equations

$$j(x) = \sin 2x + \cos \frac{x}{2}$$

$$k(x) = \sin \frac{2x}{3} + \cos \frac{x}{2}$$

$$l(x) = \sin \pi x + \cos \frac{x}{2}$$

do not satisfy the equal-period hypothesis of Theorem 3. Does this mean that functions such as j, k, and l are not periodic? A partial answer to this question is given in the statement of Theorem 4.

Theorem 4 Suppose f is a periodic function with period p, g is a periodic function with period q, and h is the function defined by the equation

$$h(x) = f(x) + g(x).$$

If r is a nonzero integral multiple of both p and q, then h is periodic with period r.

Example 4 Find a period of the function defined by the equation

$$h(x) = \sin 2x + \cos x$$

and sketch the graph of h over an interval one period in length.

Solution: The function h is the sum of two periodic functions with primitive periods $\frac{2\pi}{2} = \pi$ and 2π, respectively. The smallest positive integral multiple of π and 2π is 2π, hence 2π is a period of h by Theorem 4. The graph of h on an interval of length 2π can be obtained from the graphs of f defined by $y = \sin 2x$ and g defined by $y = \cos x$ on the interval $0 \leqslant x \leqslant 2\pi$. The length of the ordinates of f and g are added (using compasses or dividers) to obtain the ordinates of h. The graphs of f, g, and h are shown in Figure 4.

FIGURE 4

Exercises

1. Find the primitive period of the function defined by each equation, and sketch the graph of each function on an interval of length one primitive period.

   a. $f(x) = \sin 2x$ b. $f(x) = \tan \dfrac{x}{3}$ c. $g(x) = \csc \dfrac{x}{4}$
   d. $j(x) = \cos 3x$ e. $l(x) = \sec 2\pi x$ f. $f(x) = \tan 3x$
   g. $f(x) = \sec \dfrac{x}{2}$ h. $h(x) = \cot \dfrac{x}{2}$ i. $m(x) = \sin \pi x$
   j. $n(x) = \sec 2x$ k. $k(x) = \tan \pi x$ l. $t(x) = \cos \sqrt{2}\, x$
   m. $f(x) = \sin \pi x$ n. $f(x) = -2 \cos x$ o. $f(x) = \dfrac{1}{2} \sec x$
   p. $f(x) = 3 \tan x$

2. Sketch the graph of the function defined by each equation on an interval of length one primitive period.

   a. $y = 4 \sin x$ b. $y = -4 \sin x$ c. $y = \dfrac{1}{4} \sin x$
   d. $y = -\dfrac{1}{4} \sin x$ e. $y = 2 \cot x$ f. $y = \dfrac{1}{2} \cot x$
   g. $y = -2 \cot x$ h. $y = -\dfrac{1}{2} \cot x$

3. Use the following theorem to sketch the graph of the function defined by each equation on an interval the length of one primitive period.

   If f is a function with primitive period p and m is the function defined by the equation $m(x) = af(bx)$ where $a \neq 0$ and $b > 0$, the m is periodic with primitive period $\dfrac{p}{b}$.

   a. $y = 4 \cos \dfrac{1}{2}x$ b. $y = -4 \cos \dfrac{1}{2}x$ c. $y = \dfrac{1}{4} \cos \dfrac{1}{2}x$
   d. $y = -\dfrac{1}{4} \cos \dfrac{1}{2}x$ e. $y = 4 \sec \dfrac{1}{2}x$ f. $y = -4 \sec \dfrac{1}{2}x$
   g. $-y = \dfrac{1}{4} \sec \dfrac{1}{2}x$ h. $y = \dfrac{1}{4} \sec \dfrac{1}{2}x$

4. Which functions defined by the equations below satisfy the hypotheses of Theorem 4? What is a period of those functions that satisfy the hypotheses of Theorem 4?

   a. $f(x) = \sin \dfrac{x}{2} + \cos \dfrac{x}{2}$ b. $g(x) = \cos \pi x + \tan \dfrac{x}{2}$
   c. $h(x) = \tan 2x + \cot \dfrac{x}{2}$ d. $j(x) = \cot 3x + \sec 6x$
   e. $h(x) = \sec 2x + \csc \dfrac{6x}{3}$ f. $k(x) = \csc \dfrac{\sqrt{3}\, x}{3} + \sin \dfrac{x}{\sqrt{3}}$
   g. $g(x) = \sin x + \cot \dfrac{x}{2}$ h. $j(x) = \sec \dfrac{x}{2} + \tan x$
   i. $h(x) = \csc 4x + \cot 2x$ j. $f(x) = \cos \pi x + \sec \dfrac{x}{\pi}$

5. Find a period for the function defined by each equation.
   a. $f(x) = \sin x + \cos 2x$
   b. $f(x) = \sin \pi x + \tan \pi x$
   c. $f(x) = \sin 3x + \cot \frac{3x}{4}$
   d. $f(x) = \cos \pi x + \sin 2\pi x$
   e. $f(x) = \cos x + \tan \frac{x}{4}$
   f. $f(x) = \tan \frac{\pi x}{3} + \sin \pi x$
   g. $f(x) = \tan \frac{x}{4} + \cos 2x$
   h. $f(x) = \tan x + \cot 8x$
   i. $f(x) = \cot 4x + \sin x$
   j. $f(x) = \cot \pi x + \tan \frac{\pi x}{6}$
6. Why is it impossible to find a period for the function g in Exercise 4b?
7. Sketch the graphs of f in part (a) and part (d) of Exercise 5.
8. Determine the domain of each function in Exercise 5.

12-6 AMPLITUDE AND PHASE SHIFT

A function f whose graph lies between two horizontal lines is called a **bounded function.** The smallest number U such that the graph of f touches or is below the line $y = U$ is called the **least upper bound** of f. Similarly, the largest number L such that the graph of f touches or is above the line $y = L$ is called the **greatest lower bound** of f. Since $-1 \leqslant \sin x \leqslant 1$ and $-1 \leqslant \cos x \leqslant 1$ for every real number x, these functions are bounded, and have least upper bound 1 and greatest lower bound -1. The other trigonometric functions are not bounded. (Why?)

Definition The **amplitude** of a bounded periodic function f is

$$\frac{U - L}{2}$$

where U is the least upper bound and L is the greatest lower bound of f.

Thus, the amplitude of the sine (or cosine) function is

$$\frac{1 - (-1)}{2} = \frac{1 + 1}{2} = 1.$$

What is the amplitude of the function defined by $g(x) = 2 \sin x$? If we let $f(x) = \sin x$ then $g(x) = 2 \sin x = 2 f(x)$. Also, since $-1 \leqslant \sin x \leqslant 1$, we have $-2 \leqslant 2 \sin x \leqslant 2$. Thus the largest and smallest values of g are 2 and -2, and the amplitude of g is

$$\frac{2-(-2)}{2} = \frac{2+2}{2} = 2.$$

In general, if a function f has amplitude m and g is the function defined by $g(x) = af(x)$, then the amplitude of g is $|a| \cdot m$.

Example 1 If g is the function defined by the equation $g(x) = -3 \sin 2x$, find the amplitude and primitive period of g, and sketch the graph of g on an interval whose length is its primitive period.

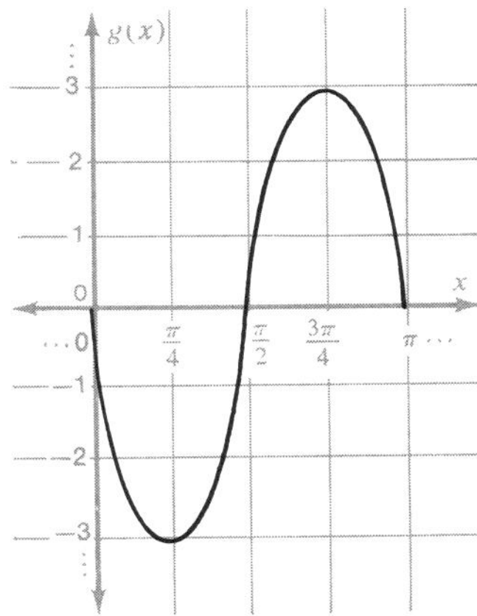

FIGURE 1

Solution: If f is the function defined by the equation $f(x) = \sin 2x$, then f is periodic with primitive period $\frac{2\pi}{2} = \pi$ (Theorem 2, Section 12-5). Moreover, the graph of f is similar to the graph of the sine function, differing only in period. The function g defined by $g(x) = -3 \sin 2x = -3f(x)$, has amplitude $|-3|$ times that of f. Since the amplitude of f is 1, the amplitude of g is $|-3| \cdot 1 = 3$. Figure 1 is the graph of the function g defined by

$$g(x) = -3 \sin 2x \quad \text{for} \quad 0 \leqslant x \leqslant \pi.$$

We have just discussed graphing a function defined by an equation of the form

$$g(x) = a \sin bx, \quad \text{where } b > 0.$$

The graph of g is similar to the graph of the sine function, has period $\frac{2\pi}{b}$, and has amplitude $|a|$. Now consider graphing the function h defined by an equation of the form

$$h(x) = a \sin (bx + c), \quad \text{where } a, b, c \in R, a \neq 0, \text{ and } b > 0.$$

If $u = x + \frac{c}{b}$, then $bu = bx + c$ and $x = u - \frac{c}{b}$. Substituting these new variables into the equation $h(x) = a \sin (bx + c)$, we have

$$h\left(u - \frac{c}{b}\right) = a \sin bu.$$

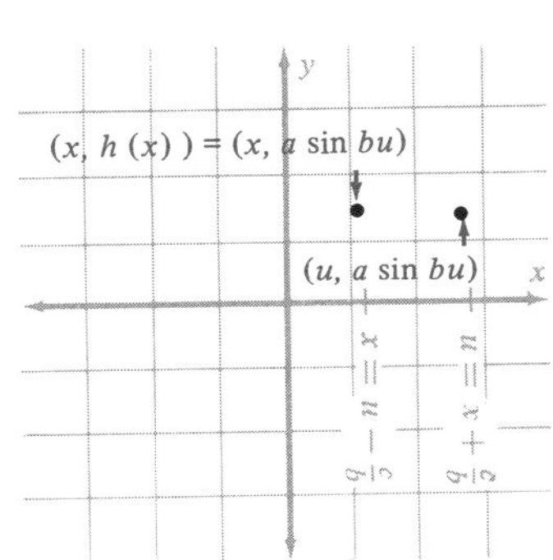

FIGURE 2

Therefore, the graph of h is similar to the graph of the sine function, has period $\frac{2\pi}{b}$, and has amplitude $|a|$. Note that for any x

$$h(x) = h\left(u - \frac{c}{b}\right) = a \sin bu$$

and the point corresponding to $(x, h(x))$ is $\frac{c}{b}$ units to the left of the point for $(u, a \sin bu)$ if $c > 0$ [Figure 2]. Thus, the graph of the function h

defined by

$$h(x) = a \sin (bx + c)$$

is shifted $\frac{c}{b}$ units to the left of the graph of the function g defined by

$$g(x) = a \sin bx.$$

The number $-\frac{c}{b}$ is called the **phase shift** of h.

Let h be the function defined by the equation $h(x) = 3 \sin (4x - \pi)$. Find the period, amplitude, and phase shift of h and then sketch the graph of h over an interval one primitive period in length.

Solution: Comparing the definition of h with the equation

$$h(x) = a \sin (bx + c),$$

we see that $a = 3$, $b = 4$, and $c = -\pi$. The primitive period of h is $\frac{2\pi}{4} = \frac{\pi}{2}$, the amplitude of h is $|a| = |3| = 3$, and the phase shift of h is $-\frac{c}{b} = -\left(-\frac{\pi}{4}\right) = \frac{\pi}{4}$. The graph of the function g defined by the equation

$$g(x) = 3 \sin 4x \quad \text{for} \quad 0 \leqslant x \leqslant \frac{\pi}{2}$$

is then shifted $\frac{\pi}{4}$ units to the right to produce the graph of the function h defined by the equation

$$h(x) = 3 \sin (4x - \pi) \quad \text{for} \quad \frac{\pi}{4} \leqslant x \leqslant \frac{3\pi}{4}.$$

in Figure 3.

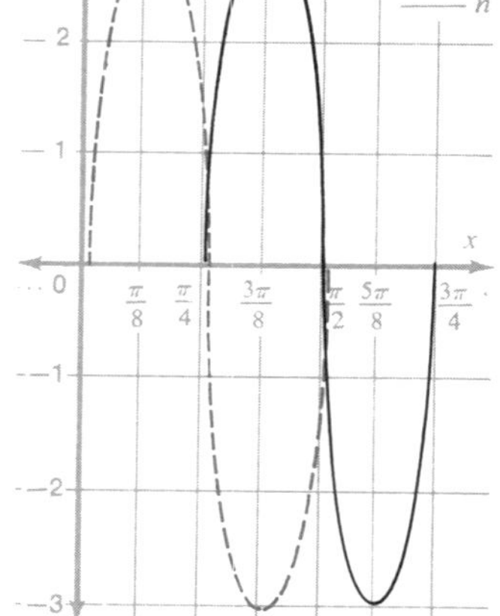

FIGURE 3

Exercises

1. Determine the amplitude (if defined) and primitive period of the function defined by each equation.

   a. $f(x) = \cos x$
   b. $h(x) = \frac{1}{2} \cos 3x$
   c. $f(x) = 2 \sin \frac{x}{2}$
   d. $f(x) = -\frac{1}{3} \cos x$
   e. $f(x) = \pi \sin ax, \quad a > 0$
   f. $f(x) = 6 \sin 3x$
   g. $f(x) = -\frac{1}{2} \cos \pi x$
   h. $f(x) = 2 \tan x$
   i. $g(x) = 5 \sin (x - \pi)$
   j. $j(x) = \frac{4}{3} \sin (6x - \pi)$

2. Determine the amplitude (if defined), primitive period, and phase shift for the function defined by each equation.

a. $f(x) = -5 \sin \left(\frac{1}{2}x - 1\right)$
b. $g(x) = \frac{1}{2} \sin \left(2x + \frac{\pi}{4}\right)$
c. $h(x) = -\cos \left(4x + \frac{3\pi}{4}\right)$
d. $j(x) = 3 \sin (2x + 1)$
e. $f(x) = 2 \tan \left(x - \frac{\pi}{4}\right)$
f. $g(x) = \pi \cos (3x - 9)$
g. $h(x) = 3 \cos (\pi x + 2)$
h. $f(x) = a \sin (bx + c)$
i. $g(x) = 6 \tan (2x - \pi)$
j. $h(x) = \frac{2}{5} \sin (5x - 5)$
k. $f(x) = \frac{1}{2} \cos (\pi x - 2)$
l. $g(x) = -\sqrt{3} \sin \left(\frac{x}{\pi} + 2\right)$
m. $h(x) = \frac{1}{6} \tan (2x + 4)$

3. Sketch the graph of the function defined by each equation.

a. $f(x) = 2 \sin 3x$
b. $h(x) = -\frac{1}{2} \cos \frac{x}{2}$
c. $f(x) = 3 \sin \left(2x + \frac{\pi}{4}\right)$
d. $g(x) = \frac{\sqrt{2}}{2} (\cos x + \sin x)$
e. $g(x) = 2 \tan (x + \pi)$
f. $h(x) = \cos 2x - 2 \sin x$

Review Exercises

1. Determine the domain and range of the function defined by each equation. **12-1**

a. $f(x) = \frac{1}{\tan x}$
b. $f(x) = |x| - x$

2. Define each of the following.
   a. Periodic function
   b. Primitive period of a periodic function

3. Sketch the graph of any periodic function with primitive period $\frac{\pi}{4}$ whose domain is $\{x: -\pi \leqslant x \leqslant 0\}$ and whose range is $\{y: 0 \leqslant y \leqslant 2\}$.

4. Sketch the graph of each function defined by the equations: **12-2**

a. $y = \cos x$ for $\frac{5\pi}{2} \leqslant x \leqslant 7\pi$

b. $y = \tan x$ for $-\frac{\pi}{2} < x < \frac{\pi}{2}$

5. Prove that the inverse tangent relation is not a function. **12-3**

6. Let f be the relation defined by the equation $f(x) = \sec x$.
   a. What is the domain of f?
   b. What is the range of f?
   c. What is f^{-1}?
   d. What is the domain of f^{-1}?

e. What is the range of f^{-1}? f. Is f a function?
g. Is f^{-1} a function?

7. Find the solution set of the equation $\sec^{-1} 2 = y$ if R is the replacement set for y.

8. Consider the relation $r = \{(1, 2), (2, 3), (3, 4), (2, 5), (3, 5)\}$. Define a principal value function for r.

9. Evaluate. 12-4
a. $\left(\text{Sin}^{-1} \frac{\sqrt{3}}{2}\right) \left(\text{Cos}^{-1} \frac{1}{2}\right)$ b. $\left(\sin \frac{\pi}{4}\right)^{-1}$
c. $\tan^2 [\text{Tan}^{-1} (-1)]$

10. Give the domain, range, and sketch the graph of the inverse tangent function.

11. Determine the primitive period of the function defined by each equation. 12-5
a. $g(x) = 3 \cot x$ b. $g(x) = \cot 3x$
c. $g(x) = -\pi \csc x$ d. $g(x) = \csc \pi x$

12. Find a period for each function defined by the equations.
a. $h(x) = \sin 2x + 3 \cos x$ b. $h(x) = \tan \frac{x}{2} + \sec x$

13. Determine the primitive period and amplitude (if defined) for g if g is defined by the equation $g(x) = 3 \tan \pi x$. 12-6

14. Sketch the graph of the function defined by the equation $g(x) = 3 \tan \pi x$.

15. Determine the amplitude, period, and phase shift for h if h is defined by the equation $h(x) = -2 \cos \left(2x + \frac{\pi}{2}\right)$.

16. Sketch the graph of the function h defined by the equation $h(x) = -2 \cos \left(2x + \frac{\pi}{2}\right)$

13. TRIGONOMETRIC IDENTITIES AND EQUATIONS

13-1 IDENTITIES AND THEIR PROOF

If either member of an equation involves a trigonometric function, then the equation is called a **trigonometric equation**. For example,

$$x^2 + \tan x = \frac{x^3 - 6}{x(x + 1)}$$

is a trigonometric equation in a single variable x.

In the remainder of this chapter, the set of real numbers will be the replacement set for each variable. The **domain of definition** for an equation in one variable will be the largest subset of the set of real numbers for which both members of the equation denote real numbers. For example, the left member of

$$\frac{x}{x + 5} = x^2\sqrt{x - 1}$$

is undefined when $x = -5$ and the right member does not denote a real number when $x < 1$. Therefore, the domain of definition for the equation is

$$\{x\colon x \geqslant 1\}.$$

What is the domain of definition for the trigonometric equation

$$x^2 + \tan x = (x^3 - 6) \div x(x + 1)?$$

$\left\{x\colon x \neq \frac{\pi}{2} + n\pi \text{ or } x \neq -1 \text{ or } x \neq 0\right\}$

An equation may not yield true statements for any element of its domain definition. In this case the solution set is the empty set. For example, the sentence $\cos x = -7$ is a trigonometric equation whose domain of definition is the set of all real numbers, but there is no real number such that $\cos x = -7$. (Why?) The domain of definition for the equation

$x^2 + 5x = -6$ is the set of all real numbers and the solution set is $\{-2, -3\}$. The set of all real numbers is both the domain of definition and the solution set of the equation $x^2 - 4 = (x + 2)(x - 2)$. Thus the solution set of an equation can be the empty set, a proper subset of, or equal to, the domain of definition for the equation.

Definition An equation for which the domain of definition and solution set are equal is called an **identity**. If an equation is not an identity it is called a **conditional equation.**

Stated in another way, an equation for which every element of the domain of definition is a solution is called an identity. If there is at least one element of the domain of definition which is not a solution, then the equation is called a conditional equation.

The definition of an identity is based on the concept of the domain of definition for an equation. Therefore, we must know the domain of definition for an equation when trying to prove whether or not the equation is an identity.

To determine the domain of definition for an equation, first find those numbers for which (1) a denominator is zero, or (2) there is an even root of a negative number, or (3) any trigonometric function in the equation is undefined. The domain of definition for the equation is the set of all real numbers except those determined by (1), (2), or (3). For example, the domain of definition for the equation

$$\tan\theta = \frac{1}{\theta^2 - 4} + \sqrt{\theta^2 - 1}$$

is the set of all real numbers θ other than:

(1) $\theta = 2 \quad \text{or} \quad \theta = -2$

since the denominator $\theta^2 - 4$ is equal to 0 for these numbers;

(2) $-1 < \theta < 1$

because $\theta^2 - 1$ is negative for these numbers; and

(3) $\theta = \frac{\pi}{2} + \pi n$, when n an integer

because $\tan\theta$ is undefined for these numbers. This is a conditional equation, for if $\theta = \pi$, we obtain the false statement

$$\tan\pi = \frac{1}{\pi^2 - 4} + \sqrt{\pi^2 - 1} \quad \text{or} \quad 0 = \frac{1}{\pi^2 - 4} + \sqrt{\pi^2 - 1}.$$

Thus the number π is not a solution of the equation and the equation is not an identity.

Determine the domain of definition for each of the following equations: $\sqrt{1-x}=4$, $\dfrac{x^2-2x-6}{x-2}=\dfrac{1}{x^2-4}$, $\sin x = \tan x$. Prove that each sentence is a conditional equation.

Domain of definition: $\{x\colon x \leqslant 1\}$. $\{x\colon x \neq 2 \text{ or } x \neq -2\}$, $\left\{x\colon x \neq \frac{\pi}{2} + n\pi\right\}$; ●

If $x = 0$, then $\sqrt{1-x} = \sqrt{1} = 1 \neq 4$ and $\sqrt{1-x} = 4$ is a conditional equation. Similarly, $x = 3$ and $x = \frac{\pi}{4}$ yield false statements for the last two equations, respectively.

One method of proving that an equation is an identity involves showing that the equation is equivalent to another identity with the same domain of definition. For example, to prove that

$$\frac{x^2+2x+1}{x+1} = x+1$$

is an identity, we begin with the equation

$$(1) \qquad \frac{x^2+2x+1}{x+1} = \frac{x^2+2x+1}{x+1}$$

which is clearly an identity. Equation (1) is equivalent to each of the following equations:

$$(2) \qquad \begin{aligned} \frac{x^2+2x+1}{x+1} &= \frac{(x+1)(x+1)}{x+1} \\ \frac{x^2+2x+1}{x+1} &= x+1. \end{aligned}$$

Therefore, Equation (1) and Equation (2) are equivalent; that is, they have the same solution set. Since Equation (1) is an identity, its solution set is its domain of definition $\{x\colon x \neq -1\}$. Thus, the solution set of Equation (2) is $\{x\colon x \neq -1\}$, which is its domain of definition, and so Equation (2) is an identity.

A short-cut technique that is generally used for the above procedure is:

1. Determine the domain of definition for the original equation;
2. Transform one member of the equation, usually the more complicated, so that it is identical to the other member using:
   a. known identities;
   b. properties of equality;
   c. valid algebraic operations.

Example 1 Show whether or not this equation is an identity:

$$\frac{8}{x^2-16} + \frac{1}{x+4} = \frac{1}{x-4}.$$

Solution: The domain of definition for the equation is

$$\{x\colon x \neq 4 \quad \text{or} \quad x \neq -4\}.$$

(Why?) Since we are unable to think of any number in the domain of definition which is not a solution of the equation, we try to transform one member of the equation so that it is identical to the other member of the equation. Why is each of the following transformations possible?

$$\frac{8}{x^2 - 16} + \frac{1}{x+4} = \frac{8}{(x-4)(x+4)} + \frac{1}{x+4}$$

$$= \frac{8 + (x-4)}{(x+4)(x-4)}$$

$$= \frac{x+4}{(x+4)(x-4)}$$

$$= \frac{1}{x-4}.$$

Thus the given equation is an identity for $\{x\colon\ x \neq 4 \quad \text{or} \quad x \neq -4\}$.

Example 2 Determine whether or not this equation is an identity.

$$\frac{1}{x+1} + \frac{1}{x+2} = \frac{1}{x+3}.$$

Solution: The domain of definition for the equation is the set

$$\{x\colon\ x \neq -1 \quad \text{or} \quad x \neq -2 \quad \text{or} \quad x \neq -3\}.$$

(Why?) If $x = 0$, we obtain the statement

$$\frac{1}{1} + \frac{1}{2} = \frac{1}{3}.$$

This statement is false; hence, 0 is not a solution and the equation is not an identity.

Show whether or not the equation $\frac{18}{x^4 - 81} + \frac{1}{x^2+9} = \frac{1}{x^2 - 9}$ is an identity. What valid conclusion can we reach if we are not able to find an element of the domain of definition which is not a solution of the equation and, in addition, not able to transform one member of the equation so that it is identical to the other member?

Domain of definition is

$$\{x\colon\ x \neq 3 \text{ or } x \neq -3\}$$

and the right member of the equation can be transformed as follows:

$$\frac{18}{(x^2-9)(x^2+9)} + \frac{1}{x^2+9}$$

$$= \frac{18 + (x^2 - 9)}{(x^2-9)(x^2+9)}$$

$$= \frac{x^2+9}{(x^2-9)(x^2+9)}$$

$$= \frac{1}{x^2 - 9}$$

By transitivity of the equality relation, the equation is an identity; ● Conclusion: we still do not know whether the equation is an identity or a conditional equation.

Exercises

1. Determine the domain of definition for each equation.
   a. $x + 9 = 15$ b. $5x^2 + 3 = x - 7$
   c. $\csc\theta - 1 = \cot^2\theta$ d. $\sqrt{x-1} = \sqrt{x} - 1$

e. $\sin x + \cos x = 1$

f. $\cos^2 \theta + \sec^2 \theta = 5$

g. $5 - \tan \theta = \cot \theta$

h. $\sin^2 \theta + \cos^2 \theta = 1$

i. $1 + \tan^2 x = \sec^2 x$

j. $\tan 2\theta = \dfrac{2 \tan \theta}{1 - \tan^2 \theta}$

k. $\csc^2 x - 1 = \cot^2 x$

l. $\sin 2\theta = 2 \sin \theta \cos \theta$

m. $\left|\sin \dfrac{\alpha}{2}\right| = \sqrt{\dfrac{1 - \cos \alpha}{2}}$

2. Find the domain of definition and solution set of each equation.

a. $x + 4 = 3$

b. $2x^2 + 7x = 15$

c. $2x^2 + 7x - 15 = (2x - 3)(x + 5)$

d. $x^3 + 8 = 0$

e. $5x^2 + 5x = -5$

f. $\dfrac{x^2 - 4}{x - 2} = 0$

g. $\sin x = 1$

h. $\tan x = 1$

3. Determine the domain of definition for each equation.

a. $\cos 2\theta = 1 - 2 \sin^2 \theta$

b. $\left|\tan \dfrac{\alpha}{2}\right| = \sqrt{\dfrac{1 - \cos \alpha}{1 + \cos \alpha}}$

c. $\cos 2\theta = 2 \cos \theta - 1$

d. $\left|\cos \dfrac{\alpha}{2}\right| = \sqrt{\dfrac{1 + \cos \alpha}{2}}$

4. Determine the domain of definition and the solution set of each equation.

a. $\sin^2 x - \sin x - 2 = 0$

b. $2 \cos^2 \theta + 7 \cos \theta = 15$

c. $|x - 5| = 6$

d. $\tan \theta = \dfrac{\sin \theta}{\cos \theta}$

e. $3x^2 - 4x + 5 = 0$

5. Review the definition of a conditional equation. Find a condition that, if satisfied, proves that an equation is a conditional equation. Use the condition to prove that the equation $\sin x = -\dfrac{\sqrt{2}}{2}$ is a conditional equation.

6. Prove that each equation is a conditional equation.

a. $x - 6 = 3$

b. $x^2 + 5x + 6 = (x + 2)(x - 3)$

c. $(x - 2)^2 + (x^2 + 2)^2 = 16$

d. $\dfrac{1}{2 - x} + \dfrac{1}{x - 2} = \dfrac{1}{(x - 2)^2}$

e. $\sqrt{\dfrac{x}{3}} - \sqrt{\dfrac{x}{4}} = \sqrt{\dfrac{x}{12}}$

f. $\sqrt{(x - 1)^2} = x - 1$

g. $|x + 6| = |x| + |6|$

h. $\cos^2 \varphi = \dfrac{1}{2}$

7. Determine which equations are identities.

a. $(x + 2)(3x - 6) = 3x^2 - 12$

b. $\sqrt{\dfrac{3}{x}} - \sqrt{\dfrac{5}{x}} = -\sqrt{\dfrac{2}{x}}$

c. $\sqrt{x^2} = x$

d. $|y + 10| = |y| - |10|$

e. $\dfrac{1}{x-1} + \dfrac{1}{(x-1)^2} = \dfrac{x}{(x-1)^2}$

f. $x^2 - 10x + 25 = (x + 5)^2$

g. $\dfrac{2}{x} + \dfrac{5}{x} = \dfrac{7}{x}$

h. $\sin \theta + \cos \theta = 1$

i. $8x^3 - 27 = (2x - 3)(4x^2 + 6x + 9)$

8. Read the discussion and definition of the wrapping function P in Section 11-3.
   a. For any real number θ, what is $P(\theta)$?
   b. For any real number θ, what equation do the coordinates $P(\theta)$ always satisfy? (Hint: Use the fact that the point $P(\theta)$ is on the unit circle and the Pythagorean theorem.)
   c. The trigonometric functions are defined in terms of the coordinates of $P(\theta)$. How are $\sin \theta$ and $\cos \theta$ defined?
   d. Use the results of parts (b) and (c) to prove that the equation $\sin^2 \theta + \cos^2 \theta = 1$ is an identity.
9. Prove that each equation is an identity. (Hint: Start with the identity $\sin^2 \theta + \cos^2 \theta = 1$.)
   a. $\sin^2 \theta = 1 - \cos^2 \theta$
   b. $\cos^2 \theta = 1 - \sin^2 \theta$

13-2 FUNDAMENTAL TRIGONOMETRIC IDENTITIES

A trigonometric equation that is an identity is called a **trigonometric identity**. We observed several trigonometric identities from the definitions of the trigonometric functions in Section 11-4:

(1) $\csc \theta = \dfrac{1}{\sin \theta}$ or $\sin \theta \cdot \csc \theta = 1$

(2) $\sec \theta = \dfrac{1}{\cos \theta}$ or $\cos \theta \cdot \sec \theta = 1$

(3) $\cot \theta = \dfrac{1}{\tan \theta}$ or $\tan \theta \cdot \cot \theta = 1$

(4) $\tan \theta = \dfrac{\sin \theta}{\cos \theta}$

(5) $\cot \theta = \dfrac{\cos \theta}{\sin \theta}$.

In order to discover another important trigonometric identity, we recall the wrapping function P and the definitions of the sine and cosine

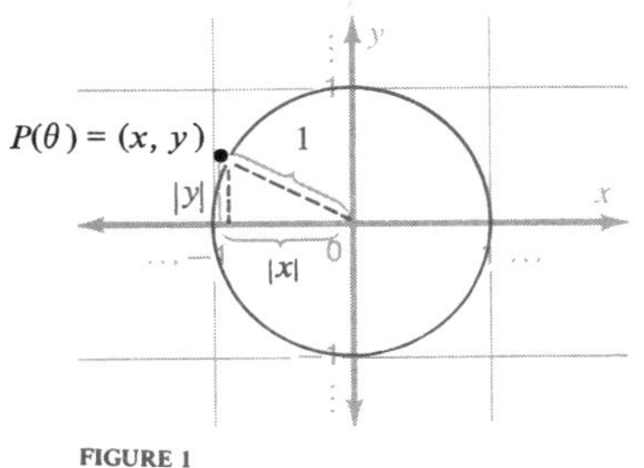

FIGURE 1

functions. The wrapping function P associates each real number θ with the coordinates of a point on the unit circle. Furthermore, if $P(\theta) = (x, y)$, then by definition

$$x = \cos\theta \quad \text{and} \quad y = \sin\theta.$$

As a consequence of the Pythagorean theorem [Figure 1], the coordinates of every point on the unit circle satisfy the equation

$$x^2 + y^2 = 1.$$

Substituting $\cos\theta$ for x and $\sin\theta$ for y, we obtain the equivalent equation

(6) $$\cos^2\theta + \sin^2\theta = 1$$

What is the domain of definition for Equation (6)? What is the solution set of the equation? Is Equation (6) an identity?

Since $\sin\theta$ and $\cos\theta$ are defined for every real number θ, the domain of definition for Equation (6) is R. For each real number θ, the point for $(\cos\theta, \sin\theta)$ is on the unit circle; hence, every real number is a solution of Equation (6). The domain of definition and the solution set are equal. Thus, Equation (6) is an identity.

Since Equation (6) is an identity, the equations below are equivalent:

$$\frac{\sin^2\theta + \cos^2\theta}{\cos^2\theta} = \frac{1}{\cos^2\theta}$$

$$\frac{\sin^2\theta}{\cos^2\theta} + \frac{\cos^2\theta}{\cos^2\theta} = \frac{1}{\cos^2\theta}$$

$$\left(\frac{\sin\theta}{\cos\theta}\right)^2 + 1 = \frac{1}{\cos^2\theta}$$

$$\tan^2\theta + 1 = \sec^2\theta.$$

The domain of definition and solution set of each equation is

$$\{\theta\colon\ \cos\theta \neq 0\} = \left\{\theta\colon\ \theta \neq \frac{\pi}{2} + \pi n\right\}.$$

Hence, the equation

(7) $$1 + \tan^2\theta = \sec^2\theta$$

is a trigonometric identity. If we divide each member of $\sin^2\theta + \cos^2\theta = 1$ by $\sin^2\theta$ instead of $\cos^2\theta$ we obtain another trigonometric identity

(8) $$1 + \cot^2\theta = \csc^2\theta.$$

The domain of definition and solution set of this identity is

$$\{\theta\colon\ \sin\theta \neq 0\} = \{\theta\colon\ \theta \neq \pi n\}.$$

The eight trigonometric identities listed above are called the **fundamental trigonometric identities.** It is helpful to use these identities when determining if other trigonometric equations are identities. When you do not see a method for proving that a given equation is an identity, it is frequently helpful to express all of the trigonometric functions in the member of the equation that is to be transformed in terms of the sine and cosine functions.

Example 1 Prove that the equation $\cos^4 x - \sin^4 x = 1 - 2\sin^2 x$ is an identity.

Solution: The domain of definition for the equation is R. It is not obvious that some real number fails to be a solution of the equation. Therefore, we attempt to transform one member of the equation so that it is identical to the other member. We choose to work with $\cos^4 x - \sin^4 x$ since this expression is the difference of two squares. Each equation in the sequence of equations below is an identity with domain of definition R.

$$\begin{aligned}\cos^4 x - \sin^4 x &= (\cos^2 x - \sin^2 x)(\cos^2 x + \sin^2 x)\\ &= (\cos^2 x - \sin^2 x)(1)\\ &= (1 - \sin^2 x) - \sin^2 x\\ &= 1 - 2\sin^2 x.\end{aligned}$$

Thus, $\cos^4 x - \sin^4 x = 1 - 2\sin^2 x$ for all real numbers x.

Example 2 Prove that the equation $\sec^2 x + \csc^2 x = \sec^2 x \cdot \csc^2 x$ is an identity.

Solution: The domain of definition for the equation is the set of all real numbers other than those for which either $\sec x$ or $\csc x$ is undefined. This is the set of all real numbers other than the integral multiples of $\frac{\pi}{2}$.

$$\begin{aligned}\sec^2 x + \csc^2 x &= \frac{1}{\cos^2 x} + \frac{1}{\sin^2 x}\\ &= \frac{\sin^2 x + \cos^2 x}{\cos^2 x \sin^2 x}\\ &= \frac{1}{\cos^2 x \sin^2 x}\\ &= \left(\frac{1}{\cos^2 x}\right)\left(\frac{1}{\sin^2 x}\right)\\ &= \sec^2 x \csc^2 x.\end{aligned}$$

Thus the given equation is an identity.

Exercises

1. Express each of the following in terms of the sine and cosine functions and simplify.
   a. $\sec^2 \theta + \csc^2 \theta$
   b. $\sin^2 \theta (1 + \tan^2 \theta)$
   c. $\dfrac{\sec x + \tan x}{\dfrac{1}{1 - \sin x}}$
   d. $\tan^2 x + \cot^2 x$
   e. $\dfrac{\tan x + \cot x}{\sec x \csc x}$
   f. $(\tan x + \cot x)^2$

2. Express each of the following in terms of a single trigonometric function.
   a. $\tan \theta \csc \theta \sec \theta$
   b. $\sin^2 \varphi + \cos^2 \varphi + \cot^2 \varphi$
   c. $\tan^2 \varphi + \cos^2 \varphi + \sin^2 \varphi$
   d. $\sec^2 \varphi - \tan^2 \varphi - \cos^2 \varphi$
   e. $\dfrac{\sin^2 \theta \cot^2 \theta}{\sec \theta}$
   f. $\cot \alpha \csc \alpha \cos \alpha$
   g. $\sec x \csc x + \tan^2 x$
   h. $\dfrac{\sqrt{1 + \tan^2 \beta}}{\sqrt{1 - \sin^2 \beta}}$
   i. $\dfrac{\sqrt{\sec^2 \varphi - 1}}{\sqrt{1 - \cos^2 \varphi}}$
   j. $\sqrt{\dfrac{1 - \cos^2 y}{1 + \cot^2 y}}$

3. Prove that each equation is an identity.
   a. $\cos^2 \theta = 1 - \sin^2 \theta$
   b. $\sin^2 \theta = 1 - \cos^2 \theta$
   c. $1 + \cot^2 \theta = \csc^2 \theta$
   d. $\sec y = \tan y \sin y + \cos y$
   e. $\tan^4 x - \sec^4 x = 1 - 2 \sec^2 x$

4. Prove that the equations below are identities.
   a. $1 + \tan^2 \theta = \sec^2 \theta$
   b. $\tan x = \sin x \sec x$
   c. $\dfrac{\cos y}{\sin y \cot y} = 1$
   d. $1 - 2 \sin^2 x = 2 \cos^2 x - 1$
   e. $(\sin x + \cos x)^2 = 1 + 2 \sin x \cos x$
   f. $\tan x + \cot x = \csc x \sec x$
   g. $(\tan x + \cot x)^2 = \sec^2 x + \csc^2 x$
   h. $1 - \cos A \sin A \cot A = \sin^2 A$
   i. $\sin^4 \varphi - \cos^4 \varphi = 2 \sin^2 \varphi - 1$
   j. $\dfrac{1 - \sin \theta}{1 + \sin \theta} = \sec^2 \theta - 2 \sec \theta \tan \theta + \tan^2 \theta$
   k. $\csc u - \cot u = \dfrac{\sin u}{1 + \cos u}$

5. Prove that each equation is an identity.
   a. $\dfrac{1 - \cos x}{1 + \cos x} = (\cot x - \csc x)^2$
   b. $\dfrac{1 + \sin x}{\cos x} = \dfrac{\cos x}{1 - \sin x}$
   c. $\dfrac{1}{\sec y - \tan y} = \sec y + \tan y$
   d. $\sec x + \tan x = \dfrac{\cos x}{1 - \sin x}$

e. $\dfrac{\sin \omega + \tan \omega}{1 + \cos \omega} = \tan \omega$

f. $\dfrac{\sec \theta + \tan \theta}{\dfrac{1}{1 - \sin \theta}} = \cos \theta$

g. $\sin^2 t \sec^2 t + \sin^2 t \csc^2 t = \sec^2 t$

h. $\dfrac{\sin u \cos u}{1 + \cos u} - \dfrac{\sin u}{1 - \cos u} = -\csc u - \csc u \cos^2 u$

i. $(1 - \cos^2 \theta)(1 + \cos^2 \theta) = 2 \sin^2 \theta - \sin^4 \theta$

j. $\dfrac{1 - \cos^2 z}{\sin^2 z} = \sin^2 z + \cos^2 z$

6. Prove that the following equations are identities.

a. $\dfrac{\sin \theta}{1 + \cos(-\theta)} - \cot(-\theta) = \csc(\pi - \theta)$

b. $\dfrac{\cos^2(\pi - x) - \sin(-x) \sin x}{\csc x} = \sin(\pi - x)$

c. $\dfrac{1 - \cos^6(-\varphi)}{\sin^2(\pi - \varphi)} = 3 - 3 \sin^2(\pi - \varphi) + \sin^4(\pi - \varphi)$

7. Prove that each equation is an identity.

a. $\dfrac{1}{\sec(-\theta) - \tan(\pi + \theta)} = \sec \theta + \tan \theta$

b. $\dfrac{\tan^2 x + 1}{2 \cos(-x) - \sin\left(\dfrac{\pi}{2} - x\right)} = \sec^3 x$

c. $\dfrac{\sec(-n)}{1 - \cos(-n)} = \dfrac{\sec n + 1}{\sin^2(-n)}$

13-3 Cos $(\theta \pm \varphi)$; Sin $(\theta \pm \varphi)$; Tan $(\theta \pm \varphi)$

The elements of the domain of definition for an equation in two variables are ordered pairs of numbers (x, y) for which each member of the equation is defined. A solution for such an equation is a pair of numbers for which the equation is a true sentence. If the equation yields a true statement for every pair in the domain of definition then it is an identity. For example,

$$x^2 - y^2 = (x - y)(x + y)$$

is an identity since it is true for every $(x, y) \in R \times R$, the domain of definition for the equation. In this section we will develop identities in two variables with which we will be able to evaluate the sine, cosine, and

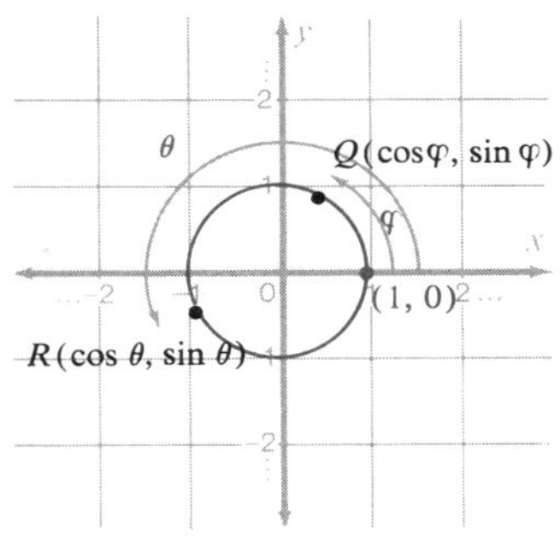

FIGURE 1

tangent of the sum and difference of two real numbers, θ and φ, if $\sin\theta$, $\cos\theta$, $\sin\varphi$, and $\cos\varphi$ are known. For example, we will be able to evaluate $\cos\frac{\pi}{12}$ because $\frac{\pi}{12} = \frac{\pi}{4} - \frac{\pi}{6}$ and $\sin\frac{\pi}{4}$, $\cos\frac{\pi}{4}$, $\sin\frac{\pi}{6}$, and $\cos\frac{\pi}{6}$ are known.

The real numbers θ and φ correspond to $(\cos\theta, \sin\theta)$ and $(\cos\varphi, \sin\varphi)$ under the wrapping function P. Let R and Q be the points $P(\theta)$ and $P(\varphi)$ [Figure 1]. The square of the distance between R and Q is equal to the length of $\overline{RQ}$ squared:

$$\begin{aligned}[m(\overline{RQ})]^2 &= (\cos\theta - \cos\varphi)^2 + (\sin\theta - \sin\varphi)^2 \\ &= 2 - 2\cos\theta\cos\varphi - 2\sin\theta\sin\varphi.\end{aligned}$$

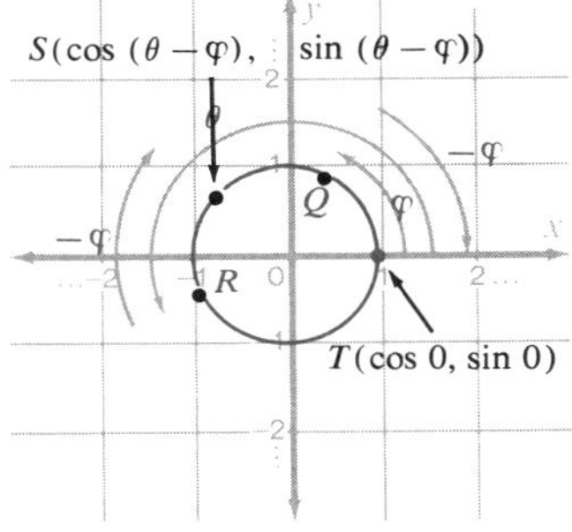

FIGURE 2

Next consider the points S and T which are $(-\varphi)$ units along the unit circle from R and Q. These are the points $P(\theta - \varphi) = (\cos(\theta - \varphi), \sin(\theta - \varphi))$ and $P(\varphi - \varphi) = P(0) = (\cos 0, \sin 0) = (1, 0)$ [Figure 2]. Observe that $P(\varphi - \varphi) = P(0) = (1, 0)$ and that the length of arc RSQ is $\theta - \varphi$. Moreover, since $(\theta - \varphi) - (\varphi - \varphi) = \theta - \varphi$, the length of arc SQT is equal to the length of arc RSQ. Since arcs with equal lengths subtend chords with equal measures, we have

$$m(\overline{RQ}) = m(\overline{ST}).$$

Furthermore, since

$$\begin{aligned}[m(\overline{ST})]^2 &= [\cos(\theta - \varphi) - 1]^2 + [\sin(\theta - \varphi) - 0]^2 \\ &= 2 - 2\cos(\theta - \varphi)\end{aligned}$$

and $[m(\overline{RQ})]^2 = [m(\overline{ST})]^2$, we have

$$2 - 2\cos\theta\cos\varphi - 2\sin\theta\sin\varphi = 2 - 2\cos(\theta - \varphi).$$

This equation is equivalent to

$$\cos(\theta - \varphi) = \cos\theta\cos\varphi + \sin\theta\sin\varphi.$$

The discussion above outlines a proof for the following theorem.

Theorem 1 The equation $\cos(\theta - \varphi) = \cos\theta\cos\varphi + \sin\theta\sin\varphi$ is an identity.

Earlier in this section, we introduced the problem of computing $\cos\frac{\pi}{12}$. It is now possible to make this computation since $\frac{\pi}{12} = \frac{\pi}{4} - \frac{\pi}{6}$.

$$\cos\frac{\pi}{12} = \cos\left(\frac{\pi}{4} - \frac{\pi}{6}\right) = \cos\frac{\pi}{4}\cos\frac{\pi}{6} + \sin\frac{\pi}{4}\sin\frac{\pi}{6}$$

$$= \frac{\sqrt{2}}{2} \cdot \frac{\sqrt{3}}{2} + \frac{\sqrt{2}}{2} \cdot \frac{1}{2}$$

$$= \frac{\sqrt{6} + \sqrt{2}}{4}.$$

Theorem 1 indicates that the equation

$$\cos(\theta - \varphi) = \cos\theta \cos\varphi + \sin\theta \sin\varphi$$

is true for any two real numbers. In particular, the equation is true for θ and $-\varphi$. If φ is replaced by $-\varphi$ in the identity, we obtain

$$\cos[\theta - (-\varphi)] = \cos\theta \cos(-\varphi) + \sin\theta \sin(-\varphi).$$

We can use the reduction formulas (Theorem 3, Section 11-5)

$$\cos(-\theta) = \cos\theta$$

$$\sin(-\theta) = -\sin\theta$$

and the identity $\theta - (-\varphi) = \theta + \varphi$ to verify that

$$\cos(\theta + \varphi) = \cos\theta \cos\varphi - \sin\theta \sin\varphi.$$

Since every real number has an additive inverse, this equation is true for all pairs of real numbers and, hence, is an identity. The preceding discussion proves Theorem 2.

Theorem 2 The equation $\cos(\theta + \varphi) = \cos\theta \cos\varphi - \sin\theta \sin\varphi$ is an identity.

The reduction formulas

$$\sin x = \cos\left(\frac{\pi}{2} - x\right) \quad \text{and} \quad \cos x = \sin\left(\frac{\pi}{2} - x\right)$$

were introduced in Section 11-5. If we substitute $x = \theta - \varphi$ in the formula $\sin x = \cos\left(\frac{\pi}{2} - x\right)$, we obtain the following equivalent equations:

$$\begin{aligned}\sin(\theta - \varphi) &= \cos\left[\left(\frac{\pi}{2} - \theta\right) + \varphi\right] \\ &= \cos\left(\frac{\pi}{2} - \theta\right)\cos\varphi - \sin\left(\frac{\pi}{2} - \theta\right)\sin\varphi \\ &= \sin\theta \cos\varphi - \cos\theta \sin\varphi.\end{aligned}$$

The argument outlined in this paragraph proves Theorem 3.

Theorem 3 The equation $\sin(\theta - \varphi) = \sin\theta \cos\varphi - \cos\theta \sin\varphi$ is an identity.

Theorem 3 is true for any two real numbers θ and φ; hence, it is true for the real numbers θ and $-\varphi$. Substituting $-\varphi$ for φ in the equation in Theorem 3, we obtain Theorem 4.

Theorem 4 The equation $\sin(\theta + \varphi) = \sin\theta \cos\varphi + \cos\theta \sin\varphi$ is an identity.

Using the identity $\tan(\theta + \varphi) = \dfrac{\sin(\theta + \varphi)}{\cos(\theta + \varphi)}$, we can prove Theorem 5.

Theorem 5 The equation $\tan(\theta + \varphi) = \dfrac{\tan\theta + \tan\varphi}{1 - \tan\theta \tan\varphi}$ is an identity.

What is the domain of definition for the equation in Theorem 5? What is the solution set of the equation?

The domain of definition is the set of all real numbers θ and φ except those for which θ, φ, and $\theta + \varphi$ belong to $\left\{\frac{\pi}{2} + \pi n\right\}$ and $\tan\theta \tan\varphi = 1$; The solution set is equal to the domain of definition.

Substituting $-\varphi$ for φ in the equation in Theorem 5, we obtain Theorem 6.

Theorem 6 The equation $\tan(\theta - \varphi) = \dfrac{\tan\theta - \tan\varphi}{1 + \tan\theta \tan\varphi}$ is an identity.

Exercises

1. Evaluate each trigonometric functional value if $\sin\theta = \frac{3}{5}$, $\cos\theta = \frac{4}{5}$, $\sin\varphi = \frac{5}{13}$, $\cos\varphi = \frac{12}{13}$.
   a. $\cos(\theta - \varphi)$ b. $\cos(\theta + \varphi)$ c. $\sin(\theta - \varphi)$
   d. $\sin(\theta + \varphi)$ e. $\tan(\theta + \varphi)$ f. $\tan(\theta - \varphi)$
2. If $\sin\alpha = 0.96$, $\cos\alpha = 0.27$, $\sin\beta = 0.25$, and $\cos\beta = 0.97$, evaluate each trigonometric functional value.
   a. $\cos(\alpha - \beta)$ b. $\cos(\alpha + \beta)$ c. $\sin(\alpha - \beta)$
   d. $\sin(\alpha + \beta)$ e. $\tan(\alpha + \beta)$ f. $\tan(\alpha - \beta)$
3. If $\alpha = \frac{\pi}{3}$ and $\beta = \frac{5\pi}{6}$, evaluate each expression.
   a. $\cos(\alpha - \beta)$ b. $\cos(\alpha + \beta)$ c. $\sin(\alpha - \beta)$
   d. $\sin(\alpha + \beta)$ e. $\tan(\alpha + \beta)$ f. $\tan(\alpha - \beta)$
4. If $\theta = \frac{3\pi}{4}$ and $\varphi = \frac{\pi}{6}$, evaluate each expression.
   a. $\cos(\theta - \varphi)$ b. $\cos(\theta + \varphi)$ c. $\sin(\theta - \varphi)$
   d. $\sin(\theta + \varphi)$ e. $\tan(\theta + \varphi)$ f. $\tan(\theta - \varphi)$
5. Use Theorems 1-6 to evaluate each trigonometric functional value.
   a. $\tan\frac{\pi}{12}$ b. $\sin\frac{13\pi}{12}$ c. $\cos\left(\frac{-7\pi}{12}\right)$

d. $\cos \frac{5\pi}{12}$ e. $\tan \frac{5\pi}{6}$ f. $\tan \frac{7\pi}{12}$
g. $\sin \frac{2\pi}{3}$ h. $\sin \left(\frac{-5\pi}{12}\right)$ i. $\cos \frac{19\pi}{12}$
j. $\sin \frac{\pi}{12}$ k. $\tan \frac{2\pi}{3}$ l. $\cos \frac{5\pi}{6}$
m. $\sin \left(\frac{-7\pi}{12}\right)$

6. Use Theorem 3 to show that $\sin \left(\frac{3\pi}{2} - u\right) = -\cos u$ is an identity.
7. Use Theorem 5 to show that $\tan (\pi + u) = \tan u$ is an identity.
8. Use Theorem 4 to show that $\sin (\pi + y) = -\sin y$ is an identity.
9. Simplify each expression.
   a. $\cos (\theta + \varphi) \cos \varphi + \sin (\theta + \varphi) \sin \varphi$
   b. $\sin (\alpha - \beta) \cos \beta + \cos (\alpha - \beta) \sin \beta$
10. a. Use Theorem 1 to prove that $\cos \left(\frac{\pi}{2} - y\right) = \sin y$ is an identity.
    b. Substitute $\frac{\pi}{2} - x$ for y in the equation $\cos \left(\frac{\pi}{2} - y\right) = \sin y$ and prove that $\cos x = \sin \left(\frac{\pi}{2} - x\right)$ is an identity.
11. Use Theorems 1-6 to prove the reduction formulas in Theorems 1-4 of Section 11-5 are identities.

13-4 THE SINE, COSINE, AND TANGENT OF 2θ AND $\frac{\theta}{2}$

In this section we will discover identities which equate the $\sin 2\theta$, $\cos 2\theta$, and $\tan 2\theta$ or $\sin \frac{\theta}{2}$, $\cos \frac{\theta}{2}$, and $\tan \frac{\theta}{2}$ with expressions involving the sine, cosine, or tangent of the number θ. With these identities it will be possible to compute the sine, cosine, and tangent of twice (and half of) most numbers for which we already know or can compute the trigonometric functional values.

Theorem 4, Section 13-3, holds if $\varphi = \theta$. Thus

$$\sin (\theta + \theta) = \sin \theta \cos \theta + \cos \theta \sin \theta$$

which is equivalent to the equation in Theorem 1.

Theorem 1 The equation $\sin 2\theta = 2 \sin \theta \cos \theta$ is an identity.

The equation of Theorem 2, Section 13-3, is also true for $\varphi = \theta$:

$$\cos(\theta + \theta) = \cos \theta \cos \theta - \sin \theta \sin \theta.$$

This equation is equivalent to

$$\cos 2\theta = \cos^2 \theta - \sin^2 \theta$$

and when the right member of the equation is transformed by substituting $1 - \cos^2 \theta$ for $\sin^2 \theta$ or $1 - \sin^2 \theta$ for $\cos^2 \theta$, we obtain the two equivalent equations listed in Theorem 2.

Theorem 2 Each of the following equivalent equations is an identity:

$$\cos 2\theta = \cos^2 \theta - \sin^2 \theta$$

$$\cos 2\theta = 1 - 2 \sin^2 \theta$$

$$\cos 2\theta = 2 \cos^2 \theta - 1.$$

Applying arguments similar to those used in the previous two paragraphs to the identity

$$\tan(\theta + \varphi) = \frac{\tan \theta + \tan \varphi}{1 - \tan \theta \tan \varphi}$$

we obtain Theorem 3.

Theorem 3 The equation $\tan 2\theta = \dfrac{2 \tan \theta}{1 - \tan^2 \theta}$ is an identity.

What is the domain of definition for the above equation? What is the solution set of this equation? Is this equation an identity?

Domain of definition: The set of all real numbers θ *except* those for which

$$\theta \in \left\{\frac{\pi}{2} + n\pi\right\}$$

$$2\theta \in \left\{\frac{\pi}{2} + n\pi\right\};$$

● The solution set is equal to the domain of definition; ● Yes, the equation is an identity.

From Theorem 2 we know that

$$\cos 2\theta = 2 \cos^2 \theta - 1$$

for all real numbers θ. Equivalently, we have

$$\cos^2 \theta = \frac{1 + \cos 2\theta}{2}$$

for all real numbers θ. If we let $\theta = \frac{\alpha}{2}$, then

$$\cos^2 \frac{\alpha}{2} = \frac{1 + \cos \alpha}{2}$$

or

$$\cos \frac{\alpha}{2} = \pm \sqrt{\frac{1 + \cos \alpha}{2}}$$

for all real numbers α. The appropriate sign depends on the quadrant of the point $P\left(\frac{\alpha}{2}\right)$.

What relationship exists between the value of $\cos \frac{\alpha}{2}$ and the quadrant of the point $P\left(\frac{\alpha}{2}\right)$? What is the domain of definition for the equation $\cos \frac{\alpha}{2} = \pm \sqrt{\frac{1 + \cos \alpha}{2}}$? What is the solution set of the equation?

See the statement of Theorem 4; Domain of definition: R; Solution set: R

Theorem 4 The equation $\cos \frac{\alpha}{2} = \pm \sqrt{\frac{1 + \cos \alpha}{2}}$ is an identity; $\cos \frac{\alpha}{2}$ is positive if the point $P\left(\frac{\alpha}{2}\right)$ is in the first or fourth quadrant and negative if the point $P\left(\frac{\alpha}{2}\right)$ is in the second or third quadrant.

Use the identity $\cos 2\theta = 1 - 2 \sin^2 \theta$ to show that

$$\sin \frac{\alpha}{2} = \pm \sqrt{\frac{1 - \cos \alpha}{2}}$$

is an identity. For what values of $\frac{\alpha}{2}$ is $\sin \frac{\alpha}{2}$ positive? Negative?

$$\cos 2\theta = 1 - 2 \sin^2 \theta$$

$$2 \sin^2 \theta = 1 - \cos 2\theta$$

$$\sin^2 \theta = \frac{1 - \cos 2\theta}{2}$$

$$\sin \theta = \pm \sqrt{\frac{1 - \cos 2\theta}{2}}$$

If $\theta = \frac{\alpha}{2}$, then

$$\sin \frac{\alpha}{2} = \pm \sqrt{\frac{1 - \cos \alpha}{2}}.$$

The domain of definition and solution set is R; ● See the statement of Theorem 5

Theorem 5 The equation $\sin \frac{\alpha}{2} = \pm \sqrt{\frac{1 - \cos \alpha}{2}}$ is an identity; $\sin \frac{\alpha}{2}$ is positive if the point $P\left(\frac{\alpha}{2}\right)$ is in the first or second quadrant and negative if the point $P\left(\frac{\alpha}{2}\right)$ is in the third or fourth quadrant.

From the identity $\tan \theta = \dfrac{\sin \theta}{\cos \theta}$ it is not difficult to show that the equation

$$\tan^2 \frac{\alpha}{2} = \frac{\sin^2 \frac{\alpha}{2}}{\cos^2 \frac{\alpha}{2}}$$

is also an identity. Since $\sin^2 \dfrac{\alpha}{2} = \dfrac{1 - \cos \alpha}{2}$ and $\cos^2 \dfrac{\alpha}{2} = \dfrac{1 + \cos \alpha}{2}$,

$$\tan^2 \frac{\alpha}{2} = \frac{1 - \cos \alpha}{1 + \cos \alpha}$$

for α such that $\cos \alpha \neq -1$ and $\tan \dfrac{\alpha}{2}$ is defined.

What are the domain of definition and the solution set for the equation $\tan^2 \dfrac{\alpha}{2} = \dfrac{1 - \cos \alpha}{1 + \cos \alpha}$? Is the equation an identity?

Domain of definition and solution set: $\{\alpha\colon\ \alpha \neq \pi + 2\pi n\}$; Yes, the equation is an identity.

Theorem 6 The equation $\tan \dfrac{\alpha}{2} = \pm \sqrt{\dfrac{1 - \cos \alpha}{1 + \cos \alpha}}$ is an identity; $\tan \dfrac{\alpha}{2}$ is positive if the point $P\left(\dfrac{\alpha}{2}\right)$ is in the first or third quadrant and negative if the point $P\left(\dfrac{\alpha}{2}\right)$ is in the second or fourth quadrant.

Exercises

1. Find $\sin 2\alpha$ and $\tan 2\alpha$, if $\sin \alpha = \dfrac{5}{12}$ and $\cos \alpha = \dfrac{12}{13}$.
2. Find $\cos 2\beta$ if $\sin \beta = 0.5227$ and $\cos \beta = 0.8525$.
3. If $\sin \alpha = \dfrac{1}{2}$ and $\cos \alpha = -\dfrac{\sqrt{3}}{2}$, find $\sin \dfrac{\alpha}{2}$, $\cos \dfrac{\alpha}{2}$, and $\tan \dfrac{\alpha}{2}$.
4. Express $\cos \dfrac{\pi}{8}$ and $\sin \dfrac{\pi}{16}$ in radical form.
5. Express $\cos \dfrac{\pi}{12}$ and $\sin \dfrac{\pi}{24}$ in radical form.
6. Prove that the equation $\dfrac{\sin 2\theta}{2 \sin^2 \theta} = \cot \theta$ is an identity.
7. Prove that the equation $\left|\tan \dfrac{\theta}{2}\right| = |\csc \theta - \cot \theta|$ is an identity.

8. Simplify each of the following expressions.
   a. $(\sin x + \cos x)^2 - \sin 2x$
   b. $\left|\tan \frac{\alpha}{2}\right| \left|\cos \frac{\alpha}{2}\right|$
   c. $\sin^2 2x - \cos^2 2x$
   d. $\cos 2\theta + \sin^2 2\theta$
   e. $\left|\sin \frac{\alpha}{2}\right| \left|\cos \frac{\alpha}{2}\right|$
   f. $\sin^2 2t + \cos^2 2t$
9. Derive an identity with $\cot 2\theta$ for one member of the equation.
10. Derive an identity with $\cot \frac{\theta}{2}$ for one member of the equation.
11. Prove that each equation is an identity.
   a. $\cos 2\alpha = \cos^4 \alpha - \sin^4 \alpha$
   b. $\cos x = \dfrac{1 - \tan^2 \dfrac{x}{2}}{1 + \tan^2 \dfrac{x}{2}}$

13-5 TRIGONOMETRIC EQUATIONS

Because the forms of trigonometric equations are so varied, procedures to follow when solving them are few and very general. It is often helpful to express all of the trigonometric expressions in the equation in terms of a single trigonometric function; then, apply ordinary algebraic techniques when possible.

Example 1 Find the solution set for the equation

$$\cos^2 \theta + \sin \theta = -1.$$

Solution: We substitute $1 - \sin^2 \theta$ for $\cos^2 \theta$. The resulting equation involves only a single trigonometric function:

$$\cos^2 \theta + \sin \theta = -1$$

$$(1 - \sin^2 \theta) + \sin \theta = -1$$

$$\sin^2 \theta - \sin \theta - 2 = 0.$$

Factoring yields

$$(\sin \theta - 2)(\sin \theta + 1) = 0$$

which is equivalent to

$$\sin \theta - 2 = 0 \quad \text{or} \quad \sin \theta + 1 = 0$$

and $\sin\theta = 2$ or $\sin\theta = -1$.

Since the range of the sine function is the set of real numbers between and including -1 and 1, the solution set of $\sin\theta = 2$ is the empty set. The only number θ between 0 and 2π for which $\sin\theta = -1$ is $\frac{3\pi}{2}$. Hence, the solution set of the equation $\cos^2\theta + \sin\theta = -1$ is $\left\{\frac{3\pi}{2} + 2\pi n\right\}$.

Example 2 Solve the equation $\sin\frac{x}{2} + 1 = 0$.

Solution: This equation is already in terms of a single trigonometric function. Moreover, the only algebraic procedure that seems to be helpful is to add -1 to both members of the equation to obtain the equivalent equation

$$\sin\frac{x}{2} = -1.$$

As we observed in the previous example, the only number between 0 and 2π for which the value of the sine function is -1 is $\frac{3\pi}{2}$. Thus, if n is an integer

$$\frac{x}{2} = \frac{3\pi}{2} + 2\pi n$$

or equivalently,

$$x = 3\pi + 4\pi n.$$

The solution set for $\sin\frac{x}{2} + 1 = 0$ is $\{3\pi + 4\pi n\}$.

Example 3 Find the solution set of $\csc x + \cot x = 2$.

Solution: Recall that the cosecant and cotangent functions are related by the identity

$$1 + \cot^2\theta = \csc^2\theta$$

and consider the following development:

$$\csc x + \cot x = 2$$

$$\csc x = 2 - \cot x$$

$$\csc^2 x = 4 - 4\cot x + \cot^2 x$$

$$1 + \cot^2 x = 4 - 4\cot x + \cot^2 x$$

$$4 \cot x = 3$$

$$\cot x = \frac{3}{4}.$$

Since the range of the cotangent function is R, we know there exists x such that $0 \leqslant x \leqslant 2\pi$ and $\cot x = \frac{3}{4}$. But we have never discovered any such numbers. A way to approximate the values of x will be presented in the next section. The best we can do now is to say that the solution set of the equation is a subset of $\left\{\cot^{-1} \frac{3}{4}\right\}$.

Since both members of the equation were squared in the process of solving the equation, each element of $\left\{\cot^{-1} \frac{3}{4}\right\}$ must be checked in the original equation. (Why?) Also, there are two numbers, x_1 and x_2, between 0 and 2π such that $0 < x_1 < \frac{\pi}{2}$ and $\pi < x_2 < \frac{3\pi}{2}$ and for which the value of the cotangent function is $\frac{3}{4}$. Since the primitive period of the cotangent function is π, $x_2 = x_1 + \pi$ and all numbers differing from x_1 by an integral multiple of π form the set $\left\{\cot^{-1} \frac{3}{4}\right\}$. It is therefore sufficient to check only the number x_1 in the original equation.

Example 4 Find the solution set of the equation $\sin^2 x = 1$.

Solution: We see that $\sin^2 x = 1$ if and only if

$$\sin x = 1 \quad \text{or} \quad \sin x = -1.$$

For what x is $\sin x = 1$? $\sin x = -1$? What is the solution set of the equation $\sin^2 x = 1$?

Equation	*Solution set*
$\sin x = 1$	$\left\{\frac{\pi}{2} + 2\pi n\right\}$
$\sin x = -1$	$\left\{\frac{3\pi}{2} + 2\pi n\right\}$
$\sin^2 x = 1$	$\left\{\frac{\pi}{2} + n\pi\right\}$

Example 5 Find the solution set of the equation $\sin x + \cos x = 1$.

Solution: The identity $\sin^2 \theta + \cos^2 \theta = 1$ indicates the relationship between the sine and cosine functions. Guided by this observation, we square both members of the equation $\sin x + \cos x = 1$ to obtain

$$\sin^2 x + 2 \sin x \cos x + \cos^2 x = 1$$

$$2 \sin x \cos x = 0$$

$$\sin 2x = 0.$$

What is the solution set of $\sin 2x = 0$? The solution set of $\sin x + \cos x = 1$ must be a subset of this solution set. What is the solution set of the equation $\sin x + \cos x = 1$?

Equation	*Solution set*
$\sin 2x = 0$	$\left\{\frac{\pi}{2} n\right\}$
$\sin x + \cos x = 1$	$\{0 + 2\pi n\} \cup \left\{\frac{\pi}{2} + 2\pi n\right\}$

Exercises

1. Find the solution set of each equation.
   a. $2 \sin x + \csc x = 3$
   b. $\sin^2 \theta + \cos \theta = 1$
   c. $\sin x + \cos x = 1$
   d. $\cos \frac{\beta}{2} - 1 = 2$
   e. $\tan 2x + 1 = 0$
   f. $2 \sin^2 y + \cos^2 y = 2$
   g. $\sin z = \cos z$
   h. $\cos^2 2x = -1$
   i. $2 \sec^2 x + \tan^2 x = 2$
   j. $\sin 2y = 1$

2. Find the solution set of each equation.
   a. $\sin^2 x - 4 \sin x + 3 = 0$
   b. $\tan^2 \frac{t}{2} + 1 = 0$
   c. $\cos u - 2 \cos u \sin u = 0$
   d. $2 \cos t - \sqrt{3} \sin t = 0$
   e. $\cot \frac{x}{2} = \sqrt{3}$
   f. $2 \sin^2 \frac{x}{2} = 3 \cos^2 \frac{x}{2}$
   g. $\tan \frac{y}{4} = \sqrt{3}$
   h. $\sin 2t + \cos t = 0$
   i. $\cos 2y = 1$
   j. $\sin 2t + \sin t = 0$

3. Find the solution set of each equation.
   a. $2 \cos^2 2x - 4 \cos 2x = -\frac{3}{2}$
   b. $8 \cos^2 y - 16 \cos y + 6 = 0$
   c. $\sin (\pi - \theta) = \cos (\pi + \theta)$
   d. $\sec \frac{x}{2} + \cos \frac{x}{2} = -2$
   e. $2 \cos^2 \frac{x}{2} = \cos^2 x$
   f. $\tan^2 (\pi - \theta) + \cos^{-2} (-\theta) = 1$
   g. $\csc (-\theta) + \cot (\pi - \theta) = -2$
   h. $\sin^2 (\pi + \varphi) + \cos (-\varphi) = 1$

13-6 TRIGONOMETRIC TABLES

To find many numbers for which we could compute the values of the trigonometric functions by the means we have been using is tedious and not very fruitful. Instead we accept the results of methods beyond the scope of this book. Tables of approximations for the values of the trigonometric functions for many numbers are available. Table II, pages 374–377, is one such table of approximations for real numbers θ such that $0 \leqslant \theta \leqslant 1.60$. The procedure for reading Table II is to first find the number for which you want the value of some trigonometric function in the left-hand column (θ-column) of the table. Then read the value of the trigonometric function for the number from the row containing the number and the column containing the appropriate trigonometric function. The table at the top of the next page shows the format of Table II.

θ	$\sin\theta$	$\cos\theta$	$\tan\theta$	$\cot\theta$	$\sec\theta$	$\csc\theta$
0.00	0.0000	1.0000	0.0000	.....	1.0000	.....
⋮	⋮	⋮	⋮	⋮	⋮	⋮
0.60	0.5646	0.8253	0.6841	1.462	1.212	1.771
⋮	⋮	⋮	⋮	⋮	⋮	⋮
1.60	0.9996	−0.0292	−34.233	−0.0292	−34.25	1.0000

In the table above, $\tan 0.60 \approx 0.6841$ and $\sec 0.60 \approx 1.212$. Recall that the symbol "≈" is used to remind us that table values are usually approximations.

What are sin 0.60, cot 1.60, and csc 0.00?

$\sin 0.60 \approx 0.5646$
$\cot 1.60 \approx -0.0292$
csc 0.00 is undefined

The process of linear interpolation described on pages 177–179 is used to approximate the values of the trigonometric functions for any real number θ between 0 and 1.60 that is not listed in Table II.

Example 1 Find sin 0.385

Solution: To find sin 0.385 we use $\sin 0.38 \approx 0.3709$ and $\sin 0.39 \approx 0.3802$ from Table II.

		x	$\sin x$		
0.01	0.005	0.38	0.3709	d	0.0093
		0.385	y		
		0.39	0.3802		

Referring to the discussion on linear interpolation we see that the appropriate slopes are

$$\frac{d}{0.005} \approx \frac{0.0093}{0.01}$$

where $d = y - 0.3709$. After computing, $d \approx 0.00465$. Thus

$$y = 0.3709 + d \approx 0.3709 + 0.00465 \approx 0.3756$$

and $\sin 0.385 \approx 0.3756$.

Example 2 Find cos 0.385.

Solution: To find cos 0.385, we use Table II to find $\cos 0.38 \approx 0.9287$ and $\cos 0.39 \approx 0.9249$, and the form on the facing page.

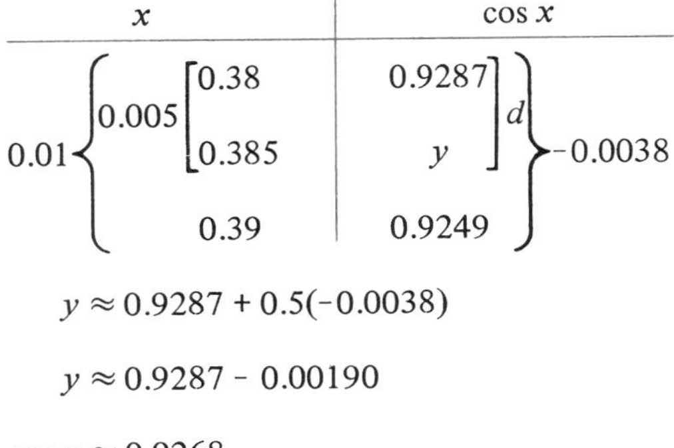

	x	$\cos x$	
0.01 {0.005 [	0.38	0.9287] d	} −0.0038
	0.385	y	
	0.39	0.9249	

$$y \approx 0.9287 + 0.5(-0.0038)$$

$$y \approx 0.9287 - 0.00190$$

$$\cos x \approx 0.9268$$

Since 0.385 is 0.5 of the way from 0.38 to 0.39, the assumption underlying linear interpolation is that cos 0.385 is 0.5 of the way from cos 0.38 to cos 0.39. Thus,

$$\cos 0.385 \approx 0.9287 + 0.5(0.9249 - 0.9287)$$

$$= 0.9287 - 0.0019$$

$$= 0.9268.$$

Notice that since $0 \leqslant 0.38 < 0.39 \leqslant \frac{\pi}{2}$, cos 0.38 > cos 0.39. Hence, 0.0019 was subtracted from 0.0287 rather than added.

$P(0.385) \approx (0.9268, 0.3756)$

What are approximations for the coordinates $P(0.385)$?

In Section 11-5, reduction formulas were used to find the values of the trigonometric functions for any real number in terms of the corresponding trigonometric functional values of a reference number. Since $\frac{\pi}{2}$ is approximately 1.571, the set of numbers $0 \leqslant \theta \leqslant 1.60$ in the θ-column of Table II contains the set of reference numbers. Therefore, using the reduction formulas and Table II, it is possible to approximate the values of the trigonometric functions for any real number. When working the problems in this section, use 6.283 to approximate 2π, 4.712 to approximate $\frac{3\pi}{2}$, 3.142 to approximate π, and 1.571 to approximate $\frac{\pi}{2}$.

Example 3 Find sec 16.283.

Solution: Since 16.283 is a positive number, the smallest positive number θ such that $P(\theta) = P(16.283)$ is obtained by subtracting multiples of 2π from 16.283: $16.283 - 2(6.283) = 3.717$. Now, since $\pi \approx 3.142$ and $\frac{3\pi}{2} \approx 4.712$, the point $P(3.717)$ is in the third quadrant. Thus, if we set $\pi + \varphi = 3.717$, $\varphi = 3.717 - 3.142 = 0.575$ is the reference number for 16.283. Using linear interpolation, $\sec 0.575 = |\sec 16.283| \approx 1.192$. Since the values of the secant function are negative for numbers which the wrapping function P assigns points in the third quadrant, $\sec 16.283 \approx -1.192$.

Example 3, Section 13-5, asks for the solution set of the equation $\csc x + \cot x = 2$. At that time it was noted that the best description of the solution set of the equation which we could provide was a subset of $\left\{\cot^{-1} \frac{3}{4}\right\}$. We may now approximate the number z such that $0 \leqslant z \leqslant 1.60$ and $\cot z = \frac{3}{4}$.

The number 0.7500 does not appear in the cot θ-column but lies between 0.7615 and 0.7458. These numbers correspond to the numbers 0.92 and 0.93, respectively, in the θ-column. Using the tabular form introduced earlier we have:

		θ	$\cot \theta$		
0.01	a	0.92	0.7615	0.0115	0.0157
		z	0.7500		
		0.93	0.7458		

$$\frac{0.0115}{a} \approx \frac{0.0157}{0.01}$$

$$a \approx \frac{0.0115}{0.0157}(0.01)$$

$$z - 0.92 \approx (0.73)(0.01)$$

$$z \approx 0.9273$$

$$\cot 0.9273 \approx 0.7500.$$

The numbers for which the value of the cotangent function is 0.7500 and between 0 and 2π are the reference number z and $z + \pi$. Thus

$$\{0.9273 + \pi n\}$$

is the solution set of the equation $\csc x + \cot x = 2$.

Find the solution set of $\cos x = \frac{9}{10}$.

Approximate solutions are elements of $\{0.4507 + 2\pi n\} \cup \{2\pi n - 0.4507\}$ where $n \in I$.

Exercises

1. Use Table II and (if necessary) linear interpolation to approximate each of the following.
   a. sin 0.75 b. cos 0.98 c. tan 1.52 d. csc 0.27
   e. sin 0.02 f. tan 0.37 g. sec 0.115 h. cot 0.212
   i. cos 1.281 j. tan 0.005 k. csc 0.999 l. cot 1.453
   m. sin 0.983 n. sec 1.238 o. tan 0.451 p. cos 0.786
2. Use Table II, the reduction formulas, and (if necessary) linear interpolation to approximate each of the following.
   a. sec (-1.00) b. cot (-1.32) c. cos (-1.07)
   d. csc (-1.175) e. sin (-0.666) f. sec (-1.565)
   g. sin (-0.628) h. sec (-1.52) i. $\tan z = -0.6000$
   j. $\sin z = 0.7200$ k. $\cos z = -0.700$ l. $\tan z = 0.4000$
   m. $\sec z = 0.9000$ n. $\csc z = 1.163$ o. $\cot x = -60.000$
   p. $\sin y = 1.732$ q. $\sec u = -2.425$
3. Find the solution set of each equation.
   a. $2 \tan v \sin v - \tan v = 0$ b. $5 \sec \theta - \frac{5\pi}{2} = 0$
   c. $\sin^2 y - \sin y - 2 = 0$ d. $\cos^2 x = 0.09$
   e. $\sin^2 2x + \cos^2 2x = 1$ f. $\tan 2\varphi - 4.6 = 0$
   g. $\cos^2 \frac{x}{2} = 0.325$ h. $2 \tan^2 \frac{x}{2} - 5 \tan \frac{x}{2} - 25 = 0$
4. Find sin (0.73 + 0.27)
   a. using the identity for the sine of the sum of two numbers.
   b. by first adding the numbers.
5. Find the solution set of each equation.
   a. $\sin^2 \frac{\theta}{2} \cos^2 \frac{\theta}{2} = \frac{1}{4}$ b. $\tan \theta = \cot \theta$
6. Find the solution set of each equation.
   a. $\cos 2u - 2 \cos u = 0$ b. $\cot^2 x = 9 \sin^2 x + \cos^2 x$
   c. $\sqrt{1 - \sin^2 \theta} = 1$ d. $2 \sin x - \csc x = 1$
   e. $\cos 2t + \sin t = 0$ f. $\cos\left(\frac{\pi}{4} - 2\beta\right) = \frac{\sqrt{2}}{2}$
7. Find the solution set of the equation $\tan^2 \theta + 4 \tan \theta + 4 = 0$.

Review Exercises

1. What is the domain of definition for the equation 13-1

$$\sqrt{1 - x^2} = \tan x + \frac{3}{x^2 - 9}?$$

2. Show whether or not each equation is an identity.
   a. $\frac{10}{4x^2 - 25} - \frac{1}{2x - 5} = \frac{-1}{2x + 5}$
   b. $x^2 + x = x(x - 1)$
3. Prove that the equation $\sin^2 \theta + \cos^2 \theta = 1$ is an identity. 13-2

4. Prove that each trigonometric equation is an identity.
   a. $1 + \sin 2x = (\sin x + \cos x)^2$
   b. $\csc x - \sin x = \cot x \cos x$

5. Using only the identities involving the trigonometric functions of the sum or difference of two numbers, find each trigonometric functional value. **13-3**

   a. $\cos \frac{\pi}{12}$ b. $\tan \frac{2\pi}{3}$

6. Suppose $\sin \alpha = 0.52$ and $\cos \alpha = 0.85$. Find $\sin 2\alpha$. **13-4**

7. Suppose $\cos \beta = \frac{4}{5}$ and the point $P(\beta)$ is in the first quadrant. Find $\sin \beta$ and $\cos \frac{\beta}{2}$.

8. Find the solution set of each trigonometric equation. **13-5**

   a. $\sin^2 t - 1 = 0$ b. $2 \sin t - \cos t = 0$ c. $\cot \frac{x}{2} + 1 = 0$

9. Use Table II to approximate **13-6**
   a. $\tan(-11.693)$ b. $\mathrm{Tan}^{-1}\ 1.063$

14. TRIGONOMETRY OF TRIANGLES

14-1 TRIGONOMETRIC FUNCTIONS OF ANGLES

In this chapter, we will study trigonometry from the viewpoint of "triangle measurement." To do this, we must first see how the trigonometric functions are related to a set of trigonometric functions whose domain is the set of all angles.

Consider an angle, the coordinate system of the plane of the angle for which it is in standard position [Figure 1], and the unit circle centered at the origin of this coordinate system. The intersection of the initial side of the angle and the unit circle is the point $R(1, 0)$. Let Q be the point at which the unit circle and the terminal side of the angle intersect. Since circle O is a unit circle, the "length" of the arc subtended by the angle is equal to the radian measure θ of $\angle ROQ$. Thus, Q is the point $P(\theta)$ where P is the wrapping function and Q has coordinates $(\cos\theta, \sin\theta)$ where $\theta = m_r \angle ROQ$.

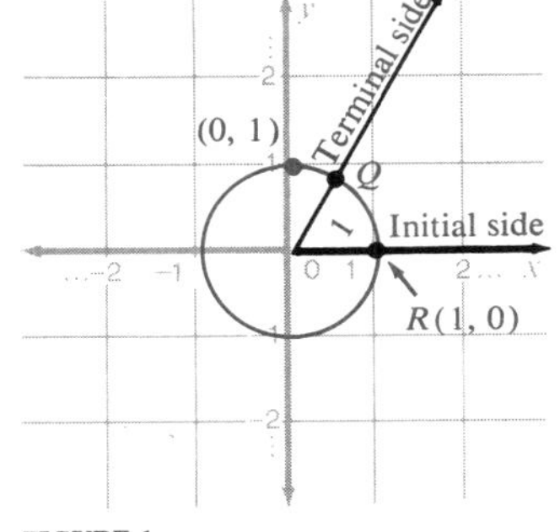

FIGURE 1

Let $\angle ROS$ be the angle in standard position in the figure on the next page. Suppose $m_r \angle ROS = \varphi$. What are the coordinates of S?

These observations motivate the following definition.

Definition If $\angle A$ is an angle with radian measure α, then

$$\begin{array}{ll} \sin\angle A = \sin\alpha & \csc\angle A = \csc\alpha \\ \cos\angle A = \cos\alpha & \sec\angle A = \sec\alpha \\ \tan\angle A = \tan\alpha & \cot\angle A = \cot\alpha \end{array}$$

That is, the value of a trigonometric function of an angle is the trigonometric functional value of the radian measure of the angle. Thus, if the radian measure of the angle is known, approximate values of the trigonometric functions of the angle can be found by using Table II. We will usually consider an angle as being in standard position.

Example 1 Approximate $\sin \angle A$, if $m_r \angle A = 0.785$

Solution: By definition, $\sin \angle A = \sin (0.785)$. Using Table II and linear interpolation, $\sin \angle A = \sin (0.785) \approx 0.7068$.

Example 2 Approximate $\sec \angle A$, if $m_r \angle A = 16.283$.

Solution: By definition, $\sec \angle A = \sec 16.283$. The procedure and the calculations for determining that $\sec 16.283 \approx -1.192$ may be found in the solution of Example 3, Section 13-6. A different interpretation of the first step of the procedure is useful. In the discussion above we observed that if $\angle A$ is in standard position, the point on the terminal side of $\angle A$ one unit from the origin is the point $(\cos 16.283, \sin 16.283) = P(16.283)$. Thus, locating the quadrant of the point $P(16.283)$ can be interpreted as locating the quadrant containing the terminal side of $\angle A$ when $\angle A$ is in standard position.

Example 3 Find $\cos \angle A$ if $m_d \angle A = 45$.

Solution: By definition, $\cos \angle A = \cos \alpha$ where α is the radian measure of $\angle A$. Therefore, we must express the degree measure of $\angle A$, 45, in its radian measure equivalent, $\frac{\pi}{4}$. Thus, $\cos \angle A = \cos \frac{\pi}{4} = \frac{\sqrt{2}}{2}$.

S is the point $(\cos \varphi, \sin \varphi)$.

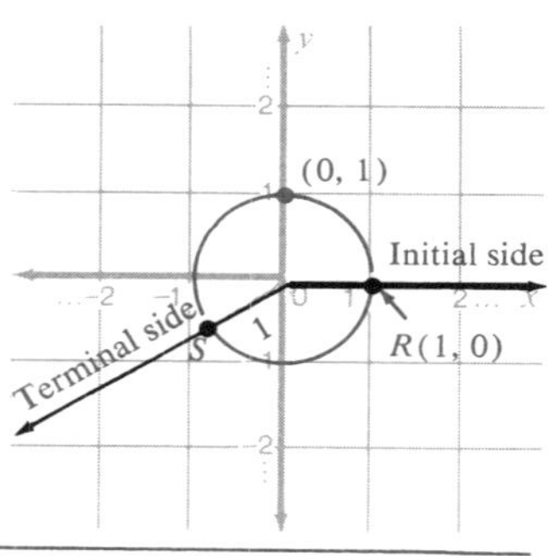

In the last example we were asked to find the value of the cosine of an angle for which we knew the degree measure. Since the trigonometric functions of an angle are defined in terms of the radian measure of the angle, it was necessary to change the degree measure of the angle to the equivalent radian measure. For convenience, Table III, pages 378–382, gives the values of the trigonometric functions of angles measured in degrees. In the remainder of this chapter, expressions such as "tan $\angle A$ if $\angle A$ has radian measure 0.785" will be abbreviated by "tan 0.785." Similarly, "tan $\angle A$ if $\angle A$ has degree measure 45" is abbreviated by "tan 45°."

Only the values of the trigonometric functions for angles with measures between 0° and 90° are listed in Table III, as the reduction formulas on page 236 make it possible to construct Table III so that it is half as long as you might expect. The left-hand column and the column headings at the top of the page are used to find the values of the trigonometric functions for angles with measures between 0° and 45°. If the measure of an angle is between 45° and 90°, the right-hand column and the column labels at the bottom of the page are used.

Find sin 20° and cos 70°; tan 13° and cot 77°; sec 43° and csc 47°. From these few examples, make a generalization which would enable you to determine cot 26° (without using the table) given that tan 64° ≈ 2.0503. This generalization makes the brief form of Table III possible.

$\sin 20° \approx 0.3420$, $\cos 70° \approx 0.3420$
$\tan 13° \approx 0.2309$, $\cot 77° \approx 0.2309$
$\sec 43° \approx 1.367$, $\csc 47° \approx 1.367$

For the generalization replace $\frac{\pi}{2}$ by 90° in the reduction formulas on page 236.

Linear interpolation can be used to approximate the values of trigonometric functions for angles with degree measures not listed in Table III.

Example 4 Find $\cos 13^\circ 45'$.

Solution: Cos $13^\circ 45'$ is between Table III entries for $\cos 13^\circ 40'$ and $\cos 13^\circ 50'$.

		θ	$\cos\theta$	
$10'$	$5'$	$13^\circ 40'$	0.9717	-0.0007
		$13^\circ 45'$	$\cos 13^\circ 45'$	
		$13^\circ 50'$	0.9710	

Since $13^\circ 45'$ is $\frac{5}{10}$ of the way from $13^\circ 40'$ to $13^\circ 50'$, $\cos 13^\circ 45'$ is $\frac{5}{10}$ of the way from 0.9717 to 0.9710. Therefore,

$$\cos 13^\circ 45' \approx 0.9717 + \frac{5}{10}(0.9710 - 0.9717) \approx 0.9714.$$

Notice that $\cos 13^\circ 40' > \cos 13^\circ 50'$ and that 0.00035 is subtracted from 0.9717 rather than added.

Given the approximate value of a trigonometric function of an angle, we can use Table III to approximate the degree measure of the angle to the nearest minute.

Example 5 Find the degree measure of an acute angle, $\angle A$, if $\cot \angle A = 0.3040$.

Solution: Looking in the *Cot*-columns, we see that 0.3040 is between 0.3057 ($\cot 73^\circ 00'$) and 0.3026 ($\cot 73^\circ 10'$).

	θ	$\cot\theta$		
$10'$	$73^\circ 00'$	0.3057	-0.0017	-0.0031
	y	0.3040		
	$73^\circ 10'$	0.3026		

Since 0.3040 is $\frac{17}{31}$ of the way from 0.3057 to 0.3026, y is $\frac{17}{31}$ of the way from $73^\circ 00'$ to $73^\circ 10'$:

$$y \approx 73^\circ 00' + \frac{17}{31}(73^\circ 10' - 73^\circ 00') \approx 73^\circ 05'.$$

Similarly, if $\cot \angle A = 0.3040$, the radian measure of $\angle A$ can be approximated by using Table II and linear interpolation.

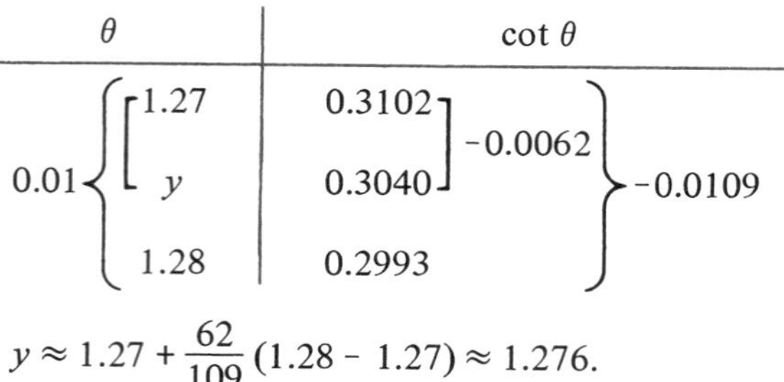

θ	$\cot\theta$
1.27	0.3102
y	0.3040
1.28	0.2993

0.01; −0.0062; −0.0109

$$y \approx 1.27 + \frac{62}{109}(1.28 - 1.27) \approx 1.276.$$

$73°05' = 73\frac{1}{12}° \approx 73.08333°$
$\approx 73.08333(0.01745)$ radians
≈ 1.275

● "≈" is used in computations that involve table entries which are approximations.

Transform 73°05′ into radians and compare the result with 1.276. Discuss the use of the symbol ≈ in each example.

Exercises

1. Find the trigonometric functional value of each angle.
   a. $\sin \angle A$, if $m_r \angle A = \frac{\pi}{3}$
   b. $\cos \angle C$, if $m_r \angle C = \frac{5\pi}{6}$
   c. $\tan \angle D$, if $m_r \angle D = 3.283$
   d. $\cot \angle B$, if $m_r \angle B = \frac{\pi}{2}$
   e. $\sec \angle A$, if $m_r \angle A = \frac{7\pi}{4}$
   f. $\csc \angle B$, if $m_r \angle B = \frac{\pi}{2}$
   g. $\sin \angle C$, if $m_r \angle C = 12\pi$
   h. $\cos \angle D$, if $m_r \angle D = -\frac{17\pi}{4}$
   i. $\tan \angle A$, if $m_r \angle A = \frac{25\pi}{6}$
   j. $\cot \angle D$, if $m_r \angle D = 7.283$
   k. $\sec \angle C$, if $m_r \angle C = -9.45$
   l. $\csc \angle B$, if $m_r \angle B = 3.654$
2. Find the value of the trigonometric function for each angle.
   a. sin 57°15′ b. sec 25°40′ c. cot 30°45′
   d. csc 89°32′ e. tan 9°9′ f. cos 12°15′
   g. csc 75°16′ h. sin 23°46′ i. cos 18°20′
   j. tan 80°30′ k. sec 67°37′ l. cot 46°27′
   m. sin 0.732 n. tan 1.323 o. sec 1.000
   p. csc 0.980 q. cot 1.460 r. cos 0.583
3. Find the degree measure of an acute angle that is a solution of each equation.
   a. $\sin \angle A = 0.9980$ b. $\sec \angle A = 1.392$ c. $\cos \angle A = 0.7000$
   d. $\cot \angle A = 0.7177$ e. $\tan \angle A = 1.150$ f. $\cot \angle A = 0.1565$
   g. $\cos \angle A = 0.9670$ h. $\csc \angle A = 1.083$ i. $\sin \angle A = 0.4699$
   j. $\tan \angle A = 2.000$
4. Find the radian measure of an acute angle that is a solution of each equation in Exercise 3.
5. Let $\angle A$ be an angle in standard position with radian measure α. Let P be the point of intersection of the terminal side of $\angle A$ and the unit circle centered at the origin. What are the coordinates of P? Why?
6. Write a detailed discussion of the rationale for the short-cut construction of Table III.

7. Devise a procedure for using Table III to find the values of the trigonometric functions of angles with degree measures less than 0 and greater than 90. $\left(\right.$Hint: The procedure is analogous to that for using Table II to find the values of the trigonometric functions for angles with radian measures less than 0 and greater than $\left.\frac{\pi}{2}\right)$.

8. Find $\cos(-15)^\circ$ and $\tan 135^\circ$ by the procedure that you devised in Exercise 7.

14-2 VALUES OF THE TRIGONOMETRIC FUNCTIONS FOR ANGLES OF ANY DEGREE MEASURE

Thus far, Table III has only been used to find the values of the trigonometric functions for angles with degree measure between 0 and 90. Here we discuss a procedure for using Table III to find the values of the trigonometric functions for angles with degree measure less than 0 or greater than 90. The procedure is analogous to that for using Table II.

The values of the trigonometric functions for each number in the set $\{\theta + 2\pi n\}$ are equal to the values for every other number in the set because each number corresponds to the same point on the unit circle under the wrapping function P. This point is the intersection of the unit circle and the terminal side of an angle in standard position with any number in the set $\{\theta + 2\pi n\}$ as its radian measure. Thus, we see that all of the angles in standard position with $\theta + 2\pi n$ as their radian measures have the same terminal side. Angles in standard position with the same terminal side are called **coterminal angles.**

Which of the angles ($\angle A$, $\angle B$, $\angle C$, $\angle D$, $\angle E$) in standard position are coterminal, if $m_r \angle A = \frac{\pi}{2}$, $m_r \angle B = \frac{3\pi}{2}$, $m_r \angle C = \left(\frac{\pi}{2} + 2\pi\right)$, $m_r \angle D = -\frac{3\pi}{2}$, $m_r \angle E = \frac{9\pi}{2}$? Give the degree measures of six angles in standard position that are coterminal. Why are the values of a trigonometric function for each angle in a set of coterminal angles equal?

Coterminal angles: $\angle A$, $\angle C$, $\angle D$, $\angle E$; ● Examples: 180°, −180°, 540°, −540°, 900°, −900°; ● Terminal sides of coterminal angles intersect the unit circle at the same point and the trigonometric functions are defined in terms of this point of intersection.

Let φ be the degree equivalent of the radian measure θ and recall that 2π radians = 360°. Then for any integer n

$$(\theta + 2\pi n) \text{ radians} = (\varphi + 360n)^\circ$$

and all of the angles in standard position with the numbers $\varphi + 360n$ as their degree measures are coterminal. The smallest nonnegative number in $\{\varphi + 360n\}$ is less than 360.

Example 1 Find the smallest nonnegative number that is the degree measure of an angle coterminal to the angle with measure 843°.

Solution: Since 843 is a degree measure which is positive and greater than 360, we add a negative multiple of 360 to 843:

$$843 - 2(360) = 843 - 720 = 123.$$

Thus, the angle in standard position with measure 123° is the required angle.

Example 2 Find the smallest nonnegative number that is the degree measure of an angle coterminal to the angle with measure $(-1056)^\circ$.

Solution: Since (-1056) is a degree measure which is negative, we add a positive multiple of 360 to (-1056) to find the number:

$$-1056 + 3(360) = -1056 + 1080 = 24.$$

Thus, 24 is the degree measure of an angle coterminal to the angle with measure (-1056°).

Since the values for a trigonometric function for coterminal angles are equal and since one of these coterminal angles has nonnegative measure less than 360°, we may consider only the trigonometric functions for angles with this measure. Then the values of the trigonometric functions of angles coterminal to these angles are determined.

Let $\angle AOB$ be an angle such that $m_r \angle AOB = \theta$ and $0 < \theta < \frac{\pi}{2}$. Then the terminal side of $\angle AOB$ is in the first quadrant [Figure 1] and intersects the unit circle at the point

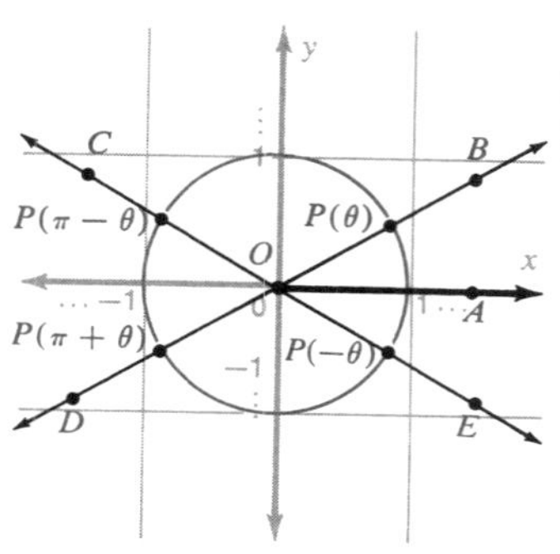

FIGURE 1

$$P(\theta) = (\cos\theta, \sin\theta) = (\cos\angle AOB, \sin\angle AOB).$$

If $m_d \angle AOB = \varphi$, then $0 < \varphi < 90$ and

$$(\cos\theta, \sin\theta) = (\cos\varphi^\circ, \sin\varphi^\circ)$$

since φ is the degree measure equivalent to θ.

In the discussion of the reduction formulas, we noted that the point $P(\theta)$ is related to three other points on the unit circle, $P(\pi - \theta)$, $P(\pi + \theta)$, and $P(-\theta)$. The radian measures of $\angle AOC$, $\angle AOD$, and $\angle AOE$, the angles in standard position with terminal sides containing these points are $\pi - \theta$, $\pi + \theta$, and $-\theta$, respectively. The equivalent degree measure of $\angle AOC$, $\angle AOD$, and $\angle AOE$ are $180 - \varphi$, $180 + \varphi$, and $-\varphi$, respectively. That is, the measure of an angle in standard position can be expressed in terms of an angle with degree measure between 0 and 90.

Definition An angle in standard position with degree measure φ, where $0 \leqslant \varphi \leqslant 90$, is called a **reference angle.**

To find the value of the trigonometric functions for $\angle AOC$ where $m_d \angle AOC = 180 - \varphi$, we recall that

$$P(\theta) = (\cos \theta, \sin \theta) = (\cos \varphi^\circ, \sin \varphi^\circ)$$

and that

$$P(\pi - \theta) = (-\cos \theta, \sin \theta)$$

according to the reduction formulas. Therefore

$$P(\pi - \theta) = (-\cos \theta, \sin \theta) = (-\cos \varphi^\circ, \sin \varphi^\circ).$$

Furthermore, observe that

$$P(\pi - \theta) = (\cos (\pi - \theta), \sin (\pi - \theta))$$
$$= (\cos (180 - \varphi)^\circ, \sin (180 - \varphi)^\circ)$$

since $(180 - \varphi)$ is the degree equivalent of $(\pi - \theta)$. Hence

$$(-\cos \varphi^\circ, \sin \varphi^\circ) = (\cos (180 - \varphi)^\circ, \sin (180 - \varphi)^\circ)$$

$$\cos (180 - \varphi)^\circ = -\cos \varphi^\circ, \quad \text{and} \quad \sin (180 - \varphi)^\circ = \sin \varphi^\circ.$$

These are precisely the reduction formulas found in Section 11-5 with $(\pi - \theta)$ replaced by its degree measure equivalent, $180 - \varphi$. It can be shown that the degree equivalents of all the reduction formulas hold.

$\tan (180 - \varphi)^\circ = -\tan \varphi^\circ$
$\cot (180 - \varphi)^\circ = -\cot \varphi^\circ$
$\sec (180 - \varphi)^\circ = -\sec \varphi^\circ$
$\csc (180 - \varphi)^\circ = \csc \varphi^\circ$

What are the reduction formulas for $\tan (180 - \varphi)^\circ$; $\cot (180 - \varphi)^\circ$; $\sec (180 - \varphi)^\circ$; and $\csc (180 - \varphi)^\circ$?

Example 3 Find $\cos 858^\circ$.

Solution: The smallest nonnegative number that is the degree measure of an angle which is coterminal to $\angle A$ is $858 - 2(360) = 138$. Since 138 is the degree measure of an angle with terminal side in the second quadrant

$$138 = 180 - \varphi$$

where φ is the degree measure of the reference angle, and

$$\varphi = 42.$$

Therefore, $\cos 858^\circ = \cos 138^\circ = -\cos 42^\circ$. (Why?) Since

$$\cos 42^\circ \approx 0.7431,$$

we have

$$\cos 858^\circ \approx -0.7431.$$

In summary, we use the procedure below to find the values of the trigonometric functions for angles with degree measure less than 0 or greater than 90.

1. Find the smallest nonnegative number ψ that is the degree measure of an angle coterminal to the given angle by adding integral multiples of 360 to the degree measure of the original angle.
2. Use the number ψ to determine the quadrant containing the terminal side of the given angle.
3. Determine the degree measure φ of the reference angle. If the terminal side of the angle with degree measure ψ is in the
   (a) first quadrant, then $\psi = \varphi$;
   (b) second quadrant, then $\psi = 180 - \varphi$ and $\varphi = 180 - \psi$;
   (c) third quadrant, then $\psi = 180 + \varphi$ and $\varphi = \psi - 180$;
   (d) fourth quadrant, then $\psi = 360 - \varphi$ and $\varphi = 360 - \psi$.
4. Use Table III to find the absolute value of the value of the trigonometric function for the given angle.
5. Determine whether or not the value of the trigonometric function for the given angle is positive or negative.

Example 4 Find $\tan \angle B$ if $m_d \angle B = -435$; that is, find $\tan (-435)^\circ$.

Solution: The smallest nonnegative number that is the degree measure of an angle which is coterminal to $\angle B$ is

$$-435 + 2(360) = 285.$$

Since the terminal side of an angle with degree measure 285 is in the fourth quadrant, the degree measure of the reference angle is

$$\varphi = 360 - 285 = 75.$$

Therefore, $|\tan (-435)^\circ| = \tan 75^\circ \approx 3.732$. Since the tangent function is negative in the fourth quadrant,

$$\tan (-435)^\circ \approx -3.732.$$

The degree-equivalent forms of all of the identities which we have proved are also valid. For example, the equation

$$\cos (\alpha - \beta) = \cos \alpha \cos \beta + \sin \alpha \sin \beta$$

is an identity when α and β are degree measures of angles.

Example 5 Without using the tables, find sin 15°

Solution: We first observe that $\sin 15° = \sin (60 - 45)°$. Using the identity $\sin (\theta - \varphi) = \sin \theta \cos \varphi - \cos \theta \sin \varphi$, we have

$$\sin 15° = \sin (60 - 45)° = \sin 60° \cos 45° - \cos 60° \sin 45°$$

$$= \frac{\sqrt{3}}{2} \cdot \frac{\sqrt{2}}{2} - \frac{1}{2} \cdot \frac{\sqrt{2}}{2}$$

$$= \frac{\sqrt{6} - \sqrt{2}}{4}.$$

Exercises

1. Find the smallest nonnegative number that is the degree measure of an angle which is coterminal to the angle with the given measure.
   a. 798° b. (-250)° c. 444°
   d. 1375° e. (-19)° f. 98°
   g. (-532)° h. 988° i. 397°
   j. (-495)° k. 1238° l. (-797)°
2. Approximate each trigonometric functional value.
   a. sin 135° b. tan 227° c. sec 315°
   d. csc 405° e. cot (-78)° f. cos (-832)°
   g. csc 495°20′ h. sin (-554)° i. tan 947°30′
   j. cot 675°50′ k. cos 1395° l. sec (-197)°
   m. sin 560° n. tan 167°30′ o. sec (-459°24′)
3. Use an appropriate identity but not the tables to evaluate each trigonometric functional value.
   a. sin 75° b. tan 300° c. cos 105°
   d. cos 150° e. tan (-15)° f. sin 120°
   g. cos (-135)° h. tan 210° i. sin 330°
   j. sin 135° k. cos (-225)°
4. Approximate each trigonometric functional value.
   a. csc (-12°14′) b. cot 147°37′ c. sin 523°18′
   d. cos 621°56′ e. tan (-98°54′) f. sec 115°43′
5. Evaluate by simplifying each to an expression involving a single trigonometric function.
   a. $\cos 73° \cos 41° + \sin 73° \sin 41°$
   b. $\sin 3.14 \cos 4.45 - \cos 3.14 \sin 4.45$
   c. $\cos 18° \cos 27° - \sin 18° \sin 27°$
   d. $\sin 123° \cos 87° + \cos 123° \sin 87°$
   e. $\dfrac{\tan 21° + \tan 159°}{1 - \tan 21° \tan 159°}$ f. $2 \sin \dfrac{\pi}{12} \cos \dfrac{\pi}{12}$
   g. $\dfrac{2 \tan 37°}{1 - \tan^2 37°}$ h. $\cos^2 15° - \sin^2 15°$
6. Consider $\Delta OQQ'$ and $\Delta OPP'$ in Figure 2. If $\angle OQ'Q$ and $\angle OP'P$ are right angles, prove that the triangles are similar. State a relationship that exists among $\overline{OQ}$, $\overline{QQ'}$, $\overline{OP}$, and $\overline{PP'}$.

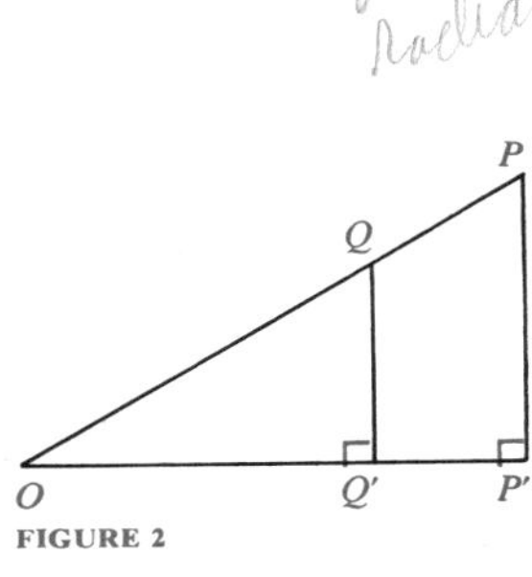

FIGURE 2

14-3 THE RIGHT-TRIANGLE DEFINITIONS OF THE TRIGONOMETRIC FUNCTIONS

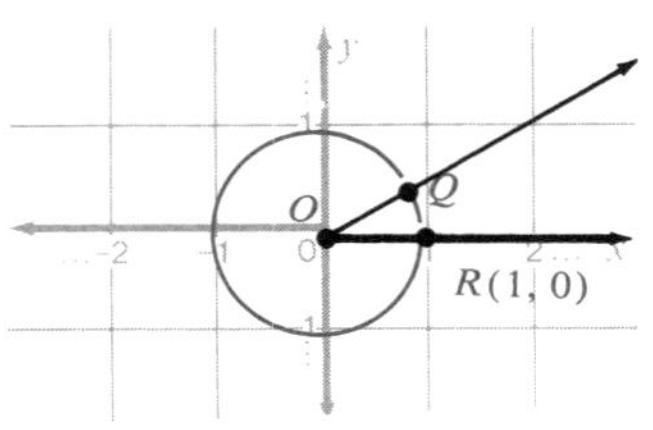

FIGURE 1

Let $\angle ROQ$ be an angle in standard position, where R is the point of intersection of the unit circle O and the initial side, and Q is the point of intersection of the unit circle O and the terminal side [Figure 1]. Then the length of $\overline{OQ}$ is 1 and if the radian measure of $\angle ROQ$ is θ, the coordinates of Q are $(\cos\theta, \sin\theta)$.

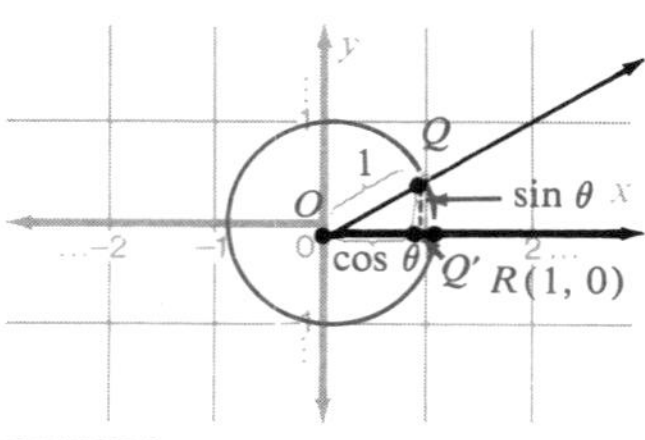

FIGURE 2

Let Q' be the point of intersection of the line which is perpendicular to the x-axis and contains Q [Figure 2]. Then the coordinates of Q' are $(\cos\theta, 0)$ and the length of $\overline{QQ'}$ is $\sin\theta$.

Let $P(x, y)$ be a point on the terminal side of $\angle ROQ$ which is r units from the origin where $r > 0$ [Figure 3]. If $\overline{PP'}$ is perpendicular to the x-axis, then the coordinates of P' are $(x, 0)$. The triangles $QQ'O$ and $PP'O$ are right triangles with corresponding angles equal in measure. (Why?) Hence, $\triangle QQ'O$ and $\triangle PP'O$ are similar triangles, and the lengths of the corresponding sides are proportional. That is

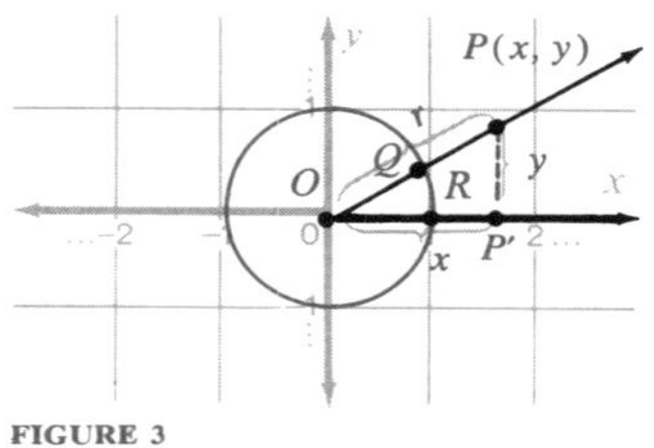

FIGURE 3

$$\frac{m(\overline{QQ'})}{m(\overline{OQ})} = \frac{y}{r}; \quad \frac{m(\overline{OQ'})}{m(\overline{OQ})} = \frac{x}{r}; \quad \text{and} \quad \frac{m(\overline{QQ'})}{m(\overline{OQ'})} = \frac{y}{x}.$$

But $m(\overline{QQ'}) = \sin\theta$, $m(\overline{OQ'}) = \cos\theta$, and $m(\overline{OQ}) = 1$. Hence

$$\frac{\sin\theta}{1} = \frac{y}{r}, \quad \frac{\cos\theta}{1} = \frac{x}{r}, \quad \text{and} \quad \frac{\sin\theta}{\cos\theta} = \frac{y}{x} \quad \text{or} \quad \tan\theta = \frac{y}{x}.$$

$\sec\theta = \frac{r}{x}$, $\csc\theta = \frac{r}{y}$, $\cot\theta = \frac{x}{y}$; $x^2 + y^2 = r^2$.

Write similar statements for $\sec\theta$, $\csc\theta$, and $\cot\theta$. What relationship exists among x, y, and r?

The preceding discussion outlines a proof for the following theorem.

Theorem If $P(x, y)$ is a point on the terminal side of an angle, $\angle A$, in standard position and $P(x, y)$ is r units from the origin where $r > 0$, then

$$\sin\angle A = \frac{y}{r} \qquad \csc\angle A = \frac{r}{y}$$

$$\cos\angle A = \frac{x}{r} \qquad \sec\angle A = \frac{r}{x}$$

$$\tan\angle A = \frac{y}{x} \qquad \cot\angle A = \frac{x}{y}.$$

Example 1 Find $\cos\angle A$ where $\angle A$ is an angle in standard position with the point $(5, 12)$ on its terminal side.

Solution: The numbers x, y, and r $(r > 0)$ in the Theorem are the measures of the sides of a right triangle with r the measure of the hypotenuse. Hence $x = 5, y = 12$, and

$$r = \sqrt{x^2 + y^2} = \sqrt{5^2 + 12^2} = 13.$$

Finally, applying the Theorem, we have $\cos \angle A = \frac{5}{13}$.

Example 2 Given that $\cos \angle A = \frac{3}{5}$, and that $\angle A$ is an angle in standard position, what is $\tan \angle A$?

Solution: From the Theorem, $\cos \angle A = \frac{x}{r}$. Since $\cos \angle A = \frac{3}{5}$ and $r > 0$, we can conclude that $x = 3$ and $r = 5$. Also, $\tan \angle A = \frac{y}{x}$. Therefore, we must find the second coordinate of any point which has first coordinate 3 and which is 5 units from the origin [Figure 4]. Substituting $x = 3$ and $r = 5$ into $x^2 + y^2 = r^2$ we have

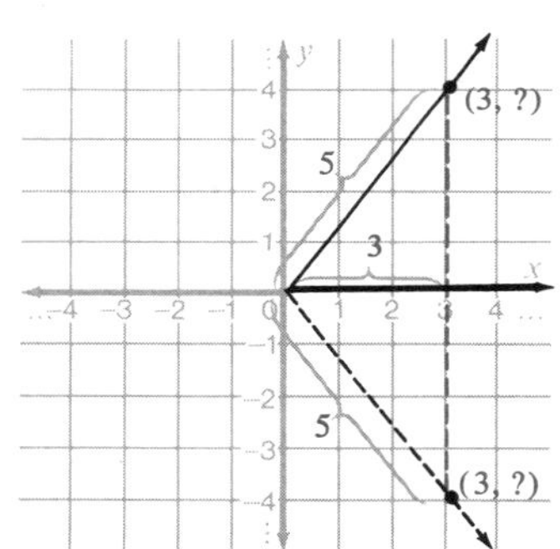

FIGURE 4

$$3^2 + y^2 = 5^2, \quad y^2 = 25 - 9 = 16, \quad y = 4 \text{ or } y = -4.$$

Therefore

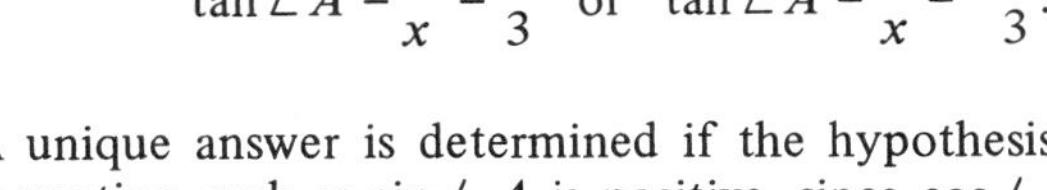

$$\tan \angle A = \frac{y}{x} = \frac{4}{3} \text{ or } \tan \angle A = \frac{y}{x} = -\frac{4}{3}.$$

A unique answer is determined if the hypothesis includes additional information such as $\sin \angle A$ is positive, since $\cos \angle A$ and $\sin \angle A$ are both positive only in quadrant I.

In what quadrant is the point (x, y) if $\cos \angle A = \frac{3}{5}$ and $\csc \angle A = -\frac{5}{4}$?

Quadrant IV

Consider a right triangle, ΔABC, where $\angle C$ is the right angle and $\angle A$ is in standard position. Denote the lengths of the sides opposite $\angle A$, $\angle B$, and $\angle C$ by a, b, and c, respectively. Then the length of the hypotenuse is c and the coordinates of B are (b, a) [Figure 5]. Applying the Theorem above, we obtain the following equations which comprise the **right-triangle definitions** of the trigonometric functions.

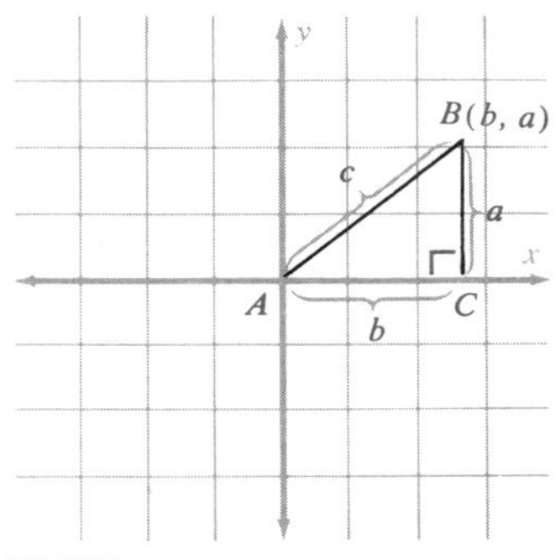

FIGURE 5

$$\sin \angle A = \frac{a}{c} \qquad \csc \angle A = \frac{c}{a}$$

$$\cos \angle A = \frac{b}{c} \qquad \sec \angle A = \frac{c}{b}$$

$$\tan \angle A = \frac{a}{b} \qquad \cot \angle A = \frac{b}{a}.$$

The above equations may be interpreted as follows:

$$\sin \angle A = \frac{\text{length of the side opposite } \angle A}{\text{length of the hypotenuse}}$$

$$\cos \angle A = \frac{\text{length of the side adjacent to } \angle A}{\text{length of the hypotenuse}}$$

$$\tan \angle A = \frac{\text{length of the side opposite } \angle A}{\text{length of the side adjacent to } \angle A}.$$

Write equations relating csc $\angle A$, sec $\angle A$, and cot $\angle A$ to the lengths of the various sides of ΔABC. How is the side opposite $\angle A$ related to the side adjacent to $\angle B$? The side adjacent to $\angle A$ related to the side opposite $\angle B$?

$$\csc \angle A = \frac{\text{length of the hypotenuse}}{\text{length of the side opposite } \angle A}$$

$$\sec \angle A = \frac{\text{length of the hypotenuse}}{\text{length of the side adjacent to } \angle A}$$

$$\cot \angle A = \frac{\text{length of the side adjacent to } \angle A}{\text{length of the side opposite } \angle A}$$

The side opposite $\angle A$ is the side adjacent to $\angle B$; the side adjacent to $\angle A$ is the side opposite $\angle B$.

Note that the side opposite $\angle A$ is the side adjacent to $\angle B$ and vice versa. So

$$\sin \angle A = \frac{\text{length of the side opposite } \angle A}{\text{length of the hypotenuse}}$$

$$= \frac{\text{length of the side adjacent to } \angle B}{\text{length of the hypotenuse}}$$

$$= \cos \angle B.$$

Compare sec $\angle A$ and csc $\angle B$; tan $\angle A$ and cot $\angle B$. How do these observations relate to the reduction formulas on page 236?

---

sec $\angle A$ = csc $\angle B$;

● tan $\angle A$ = cot $\angle B$; ● Since ΔABC is a right triangle, $\angle A$ and $\angle B$ are complementary; that is

$$m_r\angle A = \frac{\pi}{2} - m_r\angle B$$

If $m_r\angle B = \theta$, then $m_r\angle A = \frac{\pi}{2} - \theta$, and sec $\angle A$ = csc $\angle B$ and tan $\angle A$ = cot $\angle B$ imply

$$\sec\left(\frac{\pi}{2} - \theta\right) = \csc\theta$$

$$\tan\left(\frac{\pi}{2} - \theta\right) = \cot\theta$$

which are two of the reduction formulas.

Historically, the trigonometric functions were defined in terms of the lengths of the sides of a right triangle. These definitions reflect the nature of the applied problems in which they were used. The definitions in Chapter 11 reflect applications of trigonometry today; however, the right-triangle definitions are still very useful.

Exercises

1. Each ordered pair corresponds to a point on the terminal side of an angle in standard position. Find the values of all six trigonometric functions of each angle.

a. (2, 2)	**b.** (1, -6)	**c.** (4, 1)
d. (-3, 5)	**e.** (7, -2)	**f.** (-7, -5)
g. (5, 3)	**h.** (-6, 0)	**i.** (-2, -4)
j. (3, -4)	**k.** (0. 7)	**l.** (5, -13)
m. (-4, -6)	**n.** (-7, 3)	**o.** (5, 0)

2. Replace m and n to make each sentence true if $\triangle ABC$ is a right triangle with right angle, $\angle C$.

   a. $\sin \angle A = \dfrac{m}{\text{length of the hypotenuse}} = \dfrac{a}{n}$.

   b. $\cot \angle A = \dfrac{m}{\text{length of the side opposite } \angle A} = \dfrac{n}{a}$.

   c. $\tan \angle B = \dfrac{\text{length of the side opposite } \angle B}{m} = \dfrac{n}{a}$.

   d. $\sec \angle B = \dfrac{m}{\text{length of the side adjacent to } \angle B} = \dfrac{n}{a}$.

   e. $\csc \angle A = \dfrac{\text{length of the hypotenuse}}{m} = \dfrac{c}{n}$.

   f. $\cos \angle B = \dfrac{\text{length of the side adjacent to } \angle B}{m} = \dfrac{n}{c}$.

3. For those sets of three numbers below that are the measures of the sides of a right triangle, find the values of the trigonometric functions for the acute angles.

   a. $a = 3,\ b = 4,\ c = 5$
   b. $a = 2,\ b = 3,\ c = \sqrt{13}$
   c. $a = 3,\ b = \sqrt{3},\ c = 2\sqrt{3}$
   d. $a = \sqrt{5},\ b = \sqrt{11},\ c = 4$
   e. $a = \sqrt{2},\ b = \sqrt{10},\ c = 2\sqrt{3}$
   f. $a = \sqrt{6},\ b = 4,\ c = \sqrt{10}$
   g. $a = 6,\ b = 6,\ c = 6\sqrt{2}$
   h. $a = 2,\ b = \sqrt{7},\ c = \sqrt{3}$
   i. $a = 7\sqrt{2},\ b = 7,\ c = 7$
   j. $a = 13,\ b = 11,\ c = 4\sqrt{3}$

4. Let $\angle A$ be an angle in standard position with terminal side in the second quadrant. If $\tan \angle A = \dfrac{-5}{12}$, find $\sec \angle A$.

5. Find $\cos \angle A$ given that $\tan \angle A = \dfrac{5}{7}$ and $\sin \angle A$ is negative.

6. Given that $\sec \angle B = -\dfrac{2}{\sqrt{3}}$ and $\tan \angle B = -\dfrac{\sqrt{3}}{3}$, find $\sin \angle B$.

7. Suppose that $\angle A$ is an angle in standard position with terminal side in the first quadrant. If $\cos \angle A = \dfrac{5}{13}$, find $\cot \angle A$.

8. If $\angle C$ is an angle in standard position with terminal side in the second quadrant and $\tan \angle C = -\dfrac{1}{\sqrt{3}}$, find $\csc \angle C$.

9. Given that $\sec \theta = 2$ where θ is the radian measure of an angle in standard position with terminal side in the fourth quadrant, what is $\sin \theta$?

10. Given that $\sin \angle C = -\frac{5}{13}$ and $\cot \angle C = \frac{12}{5}$, find $\cos \angle C$.

11. If $\cos \angle A = -\frac{6}{7}$ and $\tan \angle A = -\frac{\sqrt{13}}{6}$, find $\sec \angle A$.

12. Given that $\tan \angle A = -\frac{5}{4}$ and $\sec \angle A$ is negative, find $\sin \angle A$.

13. If $\cot \angle B = \frac{3}{4}$ and $\csc \angle B$ is positive, find $\cos \angle B$.

14. Given that $\cos \angle A = -\frac{3}{5}$ and that $\angle A$ is in standard position, what is $\tan \angle A$?

15. Given that $\sin \angle B = \frac{5}{13}$ and that $\angle B$ is in standard position, find $\cos \angle B$.

16. Let the ray for each set be the terminal side of an angle in standard position. Find the values of the sine, cosine, and tangent functions for each angle.
   a. $\{(x, y)\colon 2x - 3y = 0,\ x \geqslant 0\}$
   b. $\{(x, y)\colon 5x - 8y = 0,\ x \geqslant 0\}$
   c. $\{(x, y)\colon 2x - y = 0,\ x \geqslant 0\}$
   d. $\{(x, y)\colon 8x - 7y = 0,\ x \geqslant 0\}$
   e. $\{(x, y)\colon 7x - 5y = 0,\ x \geqslant 0\}$
   f. $\{(x, y)\colon 5x - y = 0,\ x \geqslant 0\}$
   g. $\{(x, y)\colon x - 3y = 0,\ x \geqslant 0\}$
   h. $\{(x, y)\colon 5x - 3y = 0,\ x \geqslant 0\}$
   i. $\{(x, y)\colon 2x - y = 0,\ x \leqslant 0\}$
   j. $\{(x, y)\colon 2x - 3y = 0,\ x \leqslant 0\}$
   k. $\{(x, y)\colon 8x - 7y = 0,\ x \leqslant 0\}$

17. Let $\angle C$ be the right angle in $\triangle ABC$ and $\sin \angle A = \frac{\sqrt{10}}{10}$. Without any calculations, we can immediately conclude that $\cos \angle B = \frac{\sqrt{10}}{10}$. Why is this so? Base your arguments on the right-triangle definitions of the trigonometric functions.

18. Express the area of right triangle, $\triangle ABC$, with right angle, $\angle C$, in terms of $\sin \angle A$.

14-4 SOLVING RIGHT TRIANGLES

We can apply the right-triangle definitions of the trigonometric functions to find values of the trigonometric functions of angles with radian (degree) measures $\frac{\pi}{6}$ (30°), $\frac{\pi}{3}$ (60°), and $\frac{\pi}{4}$ (45°). If $\triangle ABC$ is a right triangle, $\angle C$ is the right angle, and $m_r \angle A = \frac{\pi}{6}$ $[m_d \angle A = 30]$, then

$m_r \angle B = \frac{\pi}{3}$ $[m_d \angle B = 60]$. The values of all the trigonometric functions for $\frac{\pi}{3}$ and $\frac{\pi}{6}$ can be deduced using the right triangle definitions of the trigonometric functions and remembering that $\sin \frac{\pi}{6} = \frac{1}{2}$. Since $m_r \angle A = \frac{\pi}{6}$, we have

$$\sin \frac{\pi}{6} = \frac{\text{length of the side opposite } \angle A}{\text{length of the hypotenuse}} = \frac{1}{2}$$

and construct the triangle in Figure 1. Thus, by using the right triangle definition of the tangent function, $\tan \angle A = \tan \frac{\pi}{6} = \frac{1}{\sqrt{3}} = \frac{\sqrt{3}}{3}$ and $\sin \angle B = \sin \frac{\pi}{3} = \frac{\sqrt{3}}{2}$.

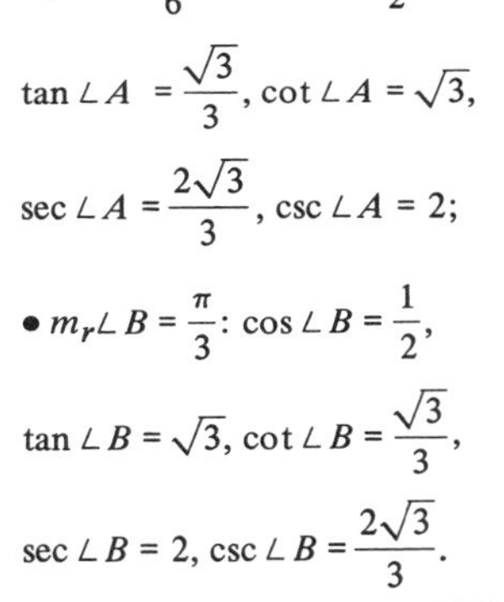

$m_r \angle A = \frac{\pi}{6}$: $\cos \angle A = \frac{\sqrt{3}}{2}$,

$\tan \angle A = \frac{\sqrt{3}}{3}$, $\cot \angle A = \sqrt{3}$,

$\sec \angle A = \frac{2\sqrt{3}}{3}$, $\csc \angle A = 2$;

• $m_r \angle B = \frac{\pi}{3}$: $\cos \angle B = \frac{1}{2}$,

$\tan \angle B = \sqrt{3}$, $\cot \angle B = \frac{\sqrt{3}}{3}$,

$\sec \angle B = 2$, $\csc \angle B = \frac{2\sqrt{3}}{3}$.

What are the values of the other trigonometric functions for $\angle A$? For $\angle B$?

The numerals $\frac{\pi}{6}$ and $\frac{\pi}{3}$ in Figure 1 name the measures of $\angle A$ and $\angle B$, respectively. Henceforth, whenever a symbol appears in the interior of the picture of an angle, it will represent the measure of the angle. If the symbol represents the degree measure of an angle, it will carry the degree superscript as in Figure 2. Otherwise, it will denote the radian measure of the angle.

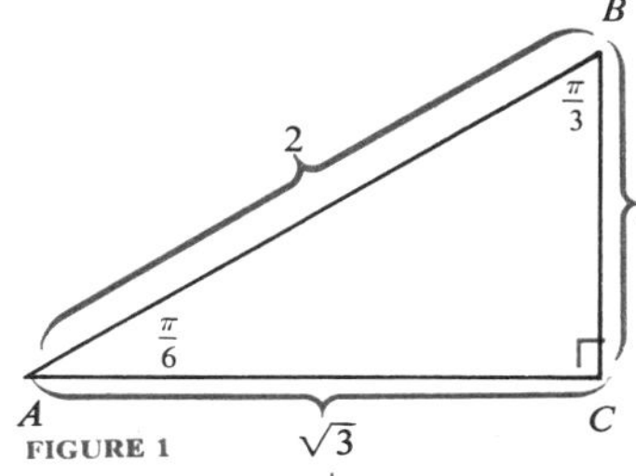

FIGURE 1

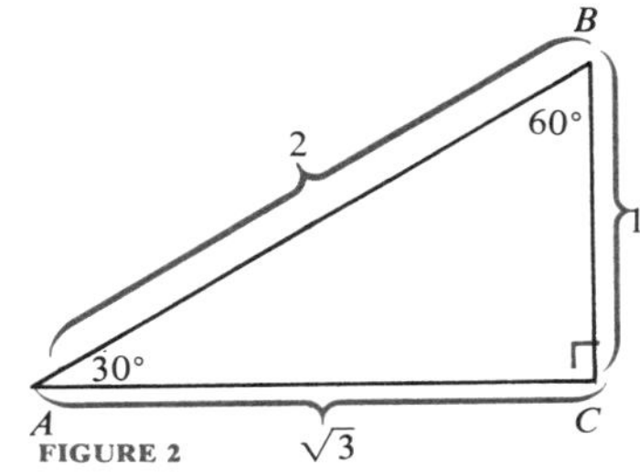

FIGURE 2

Construct a right triangle with hypotenuse of length $\sqrt{2}$ and each leg of length 1. Thus

$$\sin \frac{\pi}{4} = \frac{\sqrt{2}}{2} = \cos \frac{\pi}{4}$$

$$\tan \frac{\pi}{4} = 1 = \cot \frac{\pi}{4}$$

$$\sec \frac{\pi}{4} = \sqrt{2} = \csc \frac{\pi}{4}.$$

Use the right-triangle definitions of the trigonometric functions and the fact that $\sin \frac{\pi}{4} = \frac{1}{\sqrt{2}} = \frac{\sqrt{2}}{2}$ to construct a triangle from which you can obtain the values of all the trigonometric functions for $\frac{\pi}{4}$ $(45°)$.

There are six positive numbers associated with any triangle—the measure of its three sides and three angles. To **solve a triangle** means to find these six numbers for the triangle. A right triangle may be solved if two of these measures are known, at least one of which is the measure of a side.

Example Solve the right triangle, $\triangle ABC$, if $\angle C$ is the right angle, $m_d \angle A = 47$, and $a = 5.75$.

Solution: Since the sum of the degree measures of the acute angles of a right triangle is 90

$$m_d \angle B = 90 - m_d \angle A = 90 - 47 = 43.$$

Also, since

$$\sin \angle A = \frac{\text{length of the side opposite } \angle A}{\text{length of the hypotenuse}},$$

$a = 5.75$, $m_d \angle A = 47$, and c is the length of the hypotenuse [Figure 3], we have

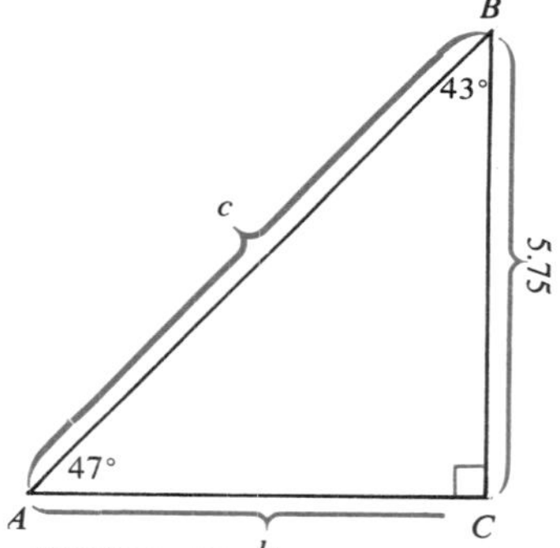

FIGURE 3

$$\sin 47° = \frac{a}{c} = \frac{5.75}{c}$$

$$c = \frac{5.75}{\sin 47°} \approx \frac{5.75}{0.7314}.$$

Such computations are easier if logarithms are used:

$$\log c \approx \log 5.75 - \log 0.7314 \approx 0.8956$$

and

$$c \approx 7.863.$$

Using the right-triangle definition of the cotangent function of an angle, we have

$$\cot 47° = \frac{b}{a} = \frac{b}{5.75},$$

and

$$b = (5.75)(\cot 47°) \approx (5.75)(1.072)$$

$$b \approx 6.164.$$

Exercises

1. Without using tables, solve the right triangle, $\triangle ABC$, ($\angle C$ the right angle) for each set of data.

   a. $m_r \angle A = \frac{\pi}{6}$, $c = 6$ b. $m_d \angle B = 45$, $a = 2$

   c. $m_d \angle A = 60$, $a = 2\sqrt{3}$ d. $a = 3$, $b = 3\sqrt{3}$

   e. $a = 2$, $c = 2\sqrt{2}$ f. $m_d \angle A = 30$, $b = \sqrt{3}$

   g. $b = 2\sqrt{3}$, $c = 4$ h. $m_r \angle B = \frac{\pi}{3}$, $c = 6$

   i. $m_r \angle B = \frac{\pi}{4}$, $b = 3$ j. $b = 3$, $c = 3\sqrt{2}$

2. Each set of data pertains to a right triangle, $\triangle ABC$, with $\angle C$ the right angle. Find x in each case.

   a. $a = 9.7$, $b = 6$, $c = x$
   b. $a = 9$, $b = 3$, $m_r \angle A = x$
   c. $a = 5$, $c = 7$, $m_d \angle B = x$
   d. $m_d \angle A = 12$, $a = 20$, $b = x$
   e. $m_r \angle B = 0.9$, $b = 12$, $c = x$
   f. $b = 9$, $c = 11$, $m_r \angle B = x$
   g. $m_r \angle A = 1.53$, $b = 15$, $m_d \angle B = x$
   h. $m_r \angle A = 1.2$, $c = 11$, $m_d \angle B = x$
   i. $m_d \angle B = 78$, $c = 10$, $a = x$
   j. $m \angle A = 42°50'$, $a = 11$, $c = x$
   k. $b = 2$, $c = 6$, $m_d \angle A = x$
   l. $m_d \angle B = 75$, $b = 9$, $c = x$
   m. $a = 1$, $b = 2$, $m_r \angle B = x$
   n. $m_r \angle B = 0.7$, $b = 3$, $m_d \angle A = x$
   o. $b = 12$, $c = 13$, $a = x$
   p. $m \angle A = 62°26'$, $b = 17$, $m_r \angle B = x$
   q. $a = 23$, $c = 25$, $m_d \angle B = x$
   r. $m \angle B = 20°60'$, $a = 7$, $m_d \angle A = x$
   s. $m_r \angle A = 1$, $c = 3$, $b = x$
   t. $a = 8$, $c = 7$, $b = x$

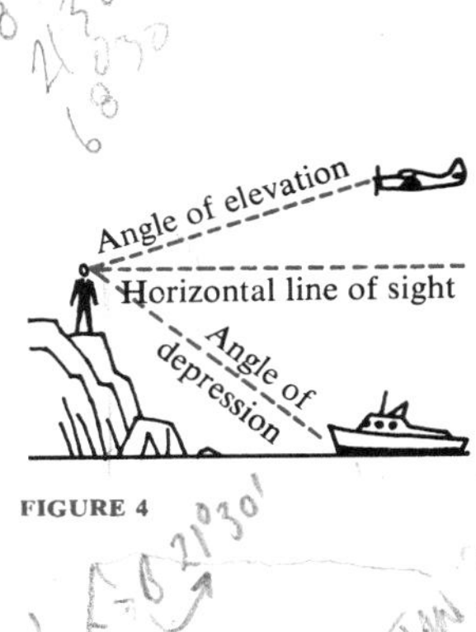

FIGURE 4

Angles are often sighted by a horizontal line of sight and a line of sight to an object above or below the horizontal line of sight. Such angles are often used in applications and have been named the **angle of elevation** and the **angle of depression**, respectively [Figure 4].

3. A man 6-ft tall is standing on a cliff which is at the edge of a lake and 24 feet above the water. The measure of the angle of depression to a boat on the lake is $21°30'$. What is the distance from the foot of the cliff to the boat?

4. A flag-pole sitter observes that the measure of the angle of depression from the top of the flag pole to a small bare spot on the lawn surrounding the flag pole is $67°13'$. After tiring of his perch, he descends, measures the distance from the base of the flag pole to the bare spot, and finds that the distance is approximately 32 feet. What is the height of the flag pole?

5. Two airplanes depart Baer Field, Fort Wayne, at 10:00 A.M. and 12:00 noon, respectively. The earlier plane heads north at 400 miles per hour. The second plane heads east at 500 miles per hour. How far apart are the planes one hour after the departure of the second plane?

6. A young man bought three-hundred feet of kite string. Six feet of string was needed to make the kite and to attach the string to the kite. After letting out all the string, he held the end of the string three feet from the ground and observed that the measure of the angle of elevation from that point to the kite was 37°. How high was the kite? How far was it from where he stood to the point directly beneath the kite?

7. Company rules prohibit a painter from working on a ladder that forms an angle with the ground of measure greater than 60°. If a painter is assigned to paint a sign 17 feet above ground level and the only ladder he has is 20 feet long, can this ladder be used or must he get a longer ladder?

8. At the instant when an observer notes that the measure of the angle of elevation to an approaching airplane is 18°36′, a second observer, 12 miles away, reports that the airplane is directly overhead. How high was the plane flying at that instant?

9. Two helicopters head west and south, respectively, at the same time. Both helicopters travel at the same constant rate of speed. An hour after takeoff, the helicopter flying south finds a ship in need of help. The helicopter traveling west is radioed for help the instant the ship is found. How long will it take the helicopter heading west to arrive with help for the ship? How many degrees must the west-bound helicopter change its course to be headed in the direction of the ship?

10. What tacit assumption did you make when solving Exercise 5?

11. A range finder is used to determine the distance from the instrument to an object. To use a coincidence range finder, the operator peers through an eyepiece and turns one prism (P_2) until the image seen through it coincides with the image seen through a second fixed prism (P_1). The angle through which the prism is turned and the distance between the prisms are used to calculate the distance from the range finder to the object. (See Figure 5.) If the distance between prisms P_1 and P_2 is 20 feet, the angle at P_1 is a right angle, and the measure of the angle at P_2 is 89°, what is the distance from the range finder to the object?

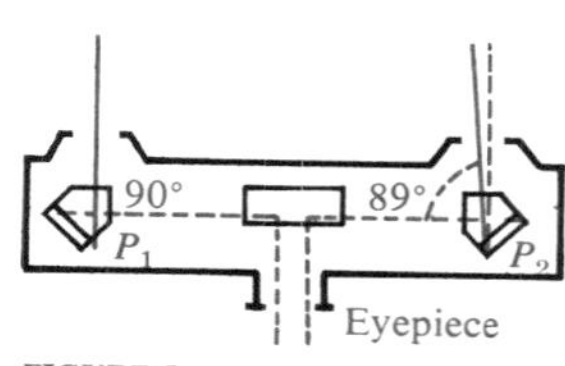

FIGURE 5

12. What is the distance from the range finder mentioned in Exercise 11 to an object if the measure of the angle at P_2 is 85°?

14-5 LOGARITHMS OF THE VALUES OF THE TRIGONOMETRIC FUNCTIONS

In the example in the previous section it was necessary to first find sin 47° (0.7314) and then find the common logarithm of 0.7314. Table IV (pages 383–387), a table of the logarithms of the values of the trigonometric functions for angles, eliminates the step of finding the value of the trigonometric function before finding the logarithm of that value.

Table IV is constructed and read in much the same way as the tables of values of trigonometric functions. There is one notable difference: The complete logarithms of the values of the trigonometric functions are not listed. Each entry in the body of the table has been increased by 10; hence, 10 should be subtracted from each value when computing with it.

To find the logarithm of the tangent of an angle measured in degrees, first find the degree measure in the left-most or right-most column of Table IV. If the number is in the left-most column, the logarithm of the tangent of the angle is in the column headed *L Tan* at the *top*. If the number is in the right-most column, the logarithm of the tangent of the angle is in the column with *L Tan* at the *bottom*. For example,

$$\log(\tan 16°) \approx 9.4575 - 10 \quad \text{and} \quad \log(\tan 66°) \approx 10.3514 - 10.$$

If the angle measure is given in radians, the conversion formula must be used to find its degree equivalent.

Linear interpolation is used to find the logarithms of values of the trigonometric functions of angles whose measures are not in Table IV. Table IV contains a feature—two columns headed *d* and one column headed *cd*, to facilitate interpolations. Each of the *d*-columns is to the right of an *L Sin*- or *L Cos*-column. Each entry in a *d*-column is the difference between consecutive entries in the *L Sin*- or *L Cos*-column to the left of it. The *cd*-column is the difference between consecutive entries in the *L Tan*- and *L Cot*-column. The letters "*d*" and "*cd*" stand for "difference" and "common difference."

Example 1 Use Table IV to find log (sin 19°18′).

Solution: We must interpolate to find log (sin 19°18′).

		$\theta°$	L Sin $\theta°$	
10′	8′	19°10′	9.5163 − 10	
		19°18′	log (sin 19°18′)	0.0036
		19°20′	9.5199 − 10	

Since $19°18'$ is $\frac{8}{10}$ of the way from $19°10'$ to $19°20'$, $\log(\sin 19°18')$ is $\frac{8}{10}$ of the way from

$$\log(\sin 19°10') \approx 9.5163 - 10$$

to

$$\log(\sin 19°20') \approx 9.5199 - 10.$$

Thus

$$\log(\sin 19°18') \approx 9.5163 - 10 + \frac{8}{10}(0.0036)$$

$$\approx 9.5192 - 10.$$

Note: The difference between log (sin 19°20′) and log (sin 19°10′) is found in the *d*-column to the right of the *L Sin*-column. The entry in the *d*-column through which a horizontal line passes between log (sin 19°10′) and log (sin 19°20′) in the table is the difference between these values.

Example 2 Use Table IV to find log (cos 29°13′).

Solution: Since 29°13′ is between 29°10′ and 29°20′, we interpolate:

		$\theta°$	L Cos $\theta°$	
10′	3′	29°10′	9.9411 - 10	-0.0007
		29°13′	log (cos 29°13′)	
		29°20′	9.9404 - 10	

$$\log(\cos 29°13') \approx 9.9411 - 10 - \left(\frac{3}{10}\right)(0.0007)$$
$$\approx 9.9409 - 10$$

Note: In Example 2, $\frac{3}{10}$ of the difference between log (cos 29°10′) and log (cos 29°20′) was subtracted from log (cos 29°10′). It should be easy to convince yourself that this procedure is correct by observing that as $\theta°$ increases from 29°10′ to 29°20′, log (cos $\theta°$) decreases.

log (cos 19° 18′) ≈ 9.9749 – 10
log (sin 29° 13′) ≈ 9.6885 – 10

Find log (cos 19°18′) and log (sin 29°13′) using Table IV.

Example 3 Find the degree measure of the acute angle, $\angle A$, if

$$\sin \angle A = \frac{\sin 33°20'}{17.2}.$$

Solution:

$$\log(\sin \angle A) = \log\left(\frac{\sin 33°20'}{17.2}\right)$$

$$= \log(\sin 33°20') - \log 17.2.$$

The subtraction is easier to perform when it is expressed in vertical form:

$\log(\sin 33°20')$	$\approx 9.7400 - 10$	(Table IV)
$(-)\log 17.2$	≈ 1.2355	(Table I)
$\log(\sin \angle A)$	$\approx 8.5045 - 10$	

Thus far, we have obtained an approximation for $\log(\sin \angle A)$. The degree measure of $\angle A$ is found by using linear interpolation:

	$\theta°$	L Sin $\theta°$		
10′	$1°40'$	$8.4637 - 10$	0.0408	0.0413
	$m_d \angle A$	$8.5045 - 10$		
	$1°50'$	$8.5050 - 10$		

$$m\angle A \approx 1°40' + \frac{408}{413}(10')$$

$$\approx 1°40' + 7' = 1°47'.$$

Exercises

1. Evaluate each of the following using Tables IV and I.
   a. $17.9 \sin 23°10'$
   b. $\sin 23° \cos 14°$
   c. $\sec 57°20' \tan 17°40'$
   d. $\dfrac{\tan 37°30'}{\cot 52°30'}$
   e. $18^{1/3} \csc 9°31'$
   f. $\dfrac{231 \sec 43°20'}{\cos 7°50'}$
   g. $\sqrt[3]{\tan 17°90'}$
   h. $342\sqrt{\sin 52°49'}$
   i. $\sqrt{\tan 18° \csc 18°}$
   j. $\dfrac{\sin 13°10' \cos 47°30'}{\sec 35°20'}$
   k. $\sqrt{192} \div \sqrt[3]{\cos 44°8'}$
   l. $\sqrt[4]{\dfrac{\sec 37°19' \sin 47°20'}{\tan 23°12'}}$

2. Find the degree measure of each acute angle, $\angle A$.
   a. $\sin \angle A = \dfrac{18.7}{21.1}$
   b. $\cos \angle A = \dfrac{31.9}{64.7}$
   c. $\tan \angle A = \dfrac{\sin 33°10'}{\cos 24°20'}$
   d. $\cot \angle A = \sin 18° \sec 18°$

e. $\sec \angle A = \dfrac{24.6}{\cot 5°10'}$ f. $\csc \angle A = \dfrac{997}{\sec 88°50'}$

g. $\sin \angle A = \sin 79°13' \cos 15°47'$ h. $\cos \angle A = \dfrac{\tan 81°32'}{\cot 6°53'}$

i. $\tan \angle A = \dfrac{22.5}{\cot 11°28'}$ j. $\cot \angle A = \dfrac{\cot 28°54'}{\sin 30°22'}$

k. $\sec \angle A = \dfrac{\tan 85°60'}{7.25}$ l. $\csc \angle A = \cos 41°55' \csc 45°36'$

3. If $\log (\sec \angle A) = 0.1046$, use Table IV to find $m_d \angle A$ if $\angle A$ is an acute angle.

4. Use Tables IV and I to evaluate the product and quotient below.
   a. $\csc 32°10' \cot 78°30'$ b. $\cos 67°49' \div \cot 27°57'$

5. Find the degree measure of each acute angle, $\angle A$.
   a. $\sin A = \dfrac{\cot 42°10'}{2.75}$ b. $\sec \angle A = \sec 63°14' \cot 38°42'$

6. Use the fact that $\csc 32°10' = \dfrac{1}{\sin 32°10'}$ to verify that

$$\log (\csc 32°10') = -\log (\sin 32°10').$$

7. Use the fact that $\sec 63°14' = \dfrac{1}{\cos 63°14'}$ to verify that

$$\log (\sec 63°14') = -\log (\cos 63°14').$$

8. Prove that $\log (\sec \angle A) = -\log (\cos \angle A)$ is an identity.

9. Prove that $\log (\csc \angle A) = -\log (\sin \angle A)$ is an identity.

10. Prove that $\log (\cot \angle A) = -\log (\tan \angle A)$ is an identity.

14-6 SOLVING OBLIQUE TRIANGLES

In this section we will investigate techniques for solving **oblique triangles**—triangles which do not have a right angle. Consider $\triangle ABC$ [Figure 1] with $\angle A$ in standard position. If the lengths of the sides opposite $\angle A$, $\angle B$, and $\angle C$ are denoted by a, b, and c, respectively, and the point of intersection of the perpendicular to the x-axis which contains B is denoted by D, then $\triangle BDA$ is a right triangle. From the right-triangle definitions of the trigonometric functions we see that B has coordinates $(c \cos \angle A, c \sin \angle A)$. Since C corresponds to $(b, 0)$, $m(\overline{CD}) = c \cos \angle A - b$ and a^2, the square of the length of the side opposite $\angle A$ in $\triangle ABC$ and the square of the length of the hypotenuse in $\triangle BCD$ [Figure 2], can be found using the Pythagorean theorem:

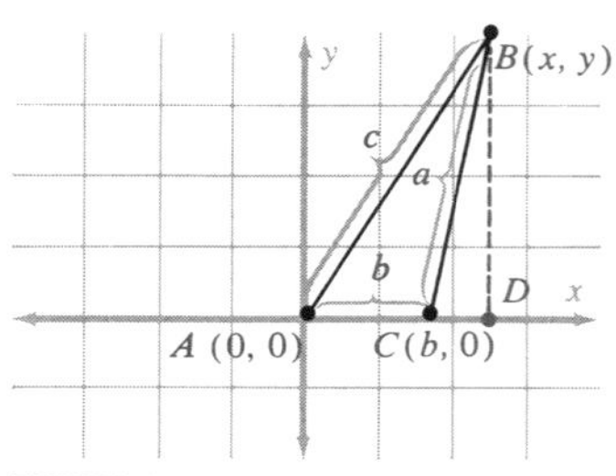

FIGURE 1

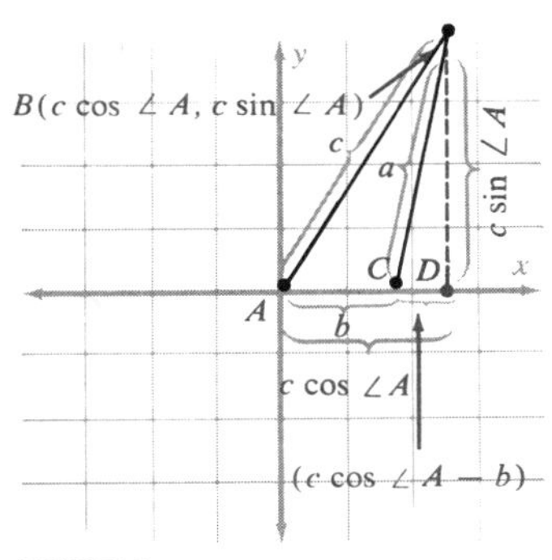

FIGURE 2

$$a^2 = (c \sin \angle A)^2 + (c \cos \angle A - b)^2$$

$$a^2 = c^2 \sin^2 \angle A + c^2 \cos^2 \angle A - 2bc \cos \angle A + b^2$$

$$a^2 = c^2(\sin^2 \angle A + \cos^2 \angle A) + b^2 - 2bc \cos \angle A$$

$$a^2 = b^2 + c^2 - 2bc \cos \angle A.$$

By considering $\angle B$ and $\angle C$ of $\triangle ABC$ in standard position, we derive the other two equations in Theorem 1.

Theorem 1 (*Law of cosines*) In any triangle, $\triangle ABC$, if a, b, and c are the measures of the sides opposite $\angle A$, $\angle B$, and $\angle C$, respectively, then

$$a^2 = b^2 + c^2 - 2bc \cos \angle A$$

$$b^2 = a^2 + c^2 - 2ac \cos \angle B$$

$$c^2 = a^2 + b^2 - 2ab \cos \angle C.$$

If the measures of the three sides of a triangle are known, which formula would you use to find $m_d \angle A$? $m_d \angle B$? $m_d \angle C$? If b, c, and $m_d \angle A$ are known, which formula would you use to find a?

Use $a^2 = b^2 + c^2 - 2bc \cos \angle A$;
- use $b^2 = a^2 + c^2 - 2bc \cos \angle B$;
- use $c^2 = a^2 + b^2 - 2ab \cos \angle C$;
- Use $a^2 = b^2 + c^2 - 2ab \cos \angle A$.

Example 1 Given that $m_d \angle C = 13$, $b = 4$, and $a = 5$, find c.

Solution: Since $\angle C$ is the angle containing the sides $\overline{AC}$ and $\overline{BC}$, the third equation of the theorem applies:

$$c^2 = a^2 + b^2 - 2ab \cos \angle C$$

$$c^2 \approx 5^2 + 4^2 - 2(5)(4)(0.9744)$$

$$c \approx 2.024.$$

The law of cosines applies when solving triangles for which the measures of three sides or two sides and the included angle are known. A unique triangle is also determined if the measures of two angles and any side are known. What restrictions must there be on the measures of the two angles in order for them to be angles of a triangle? The measures of two sides and an angle opposite one of them may determine a unique triangle. Triangles for which measures of two angles and any side or two sides and any angle opposite one of them are known are solved by applying the law of sines.

Theorem 2 (*Law of sines*) In any triangle, $\triangle ABC$, if a, b, and c are the measures of the sides opposite $\angle A$, $\angle B$, and $\angle C$, respectively, then

$$\frac{\sin \angle A}{a} = \frac{\sin \angle B}{b} = \frac{\sin \angle C}{c}.$$

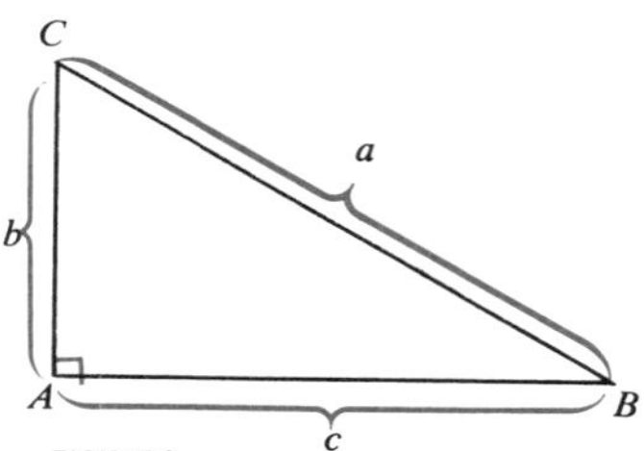

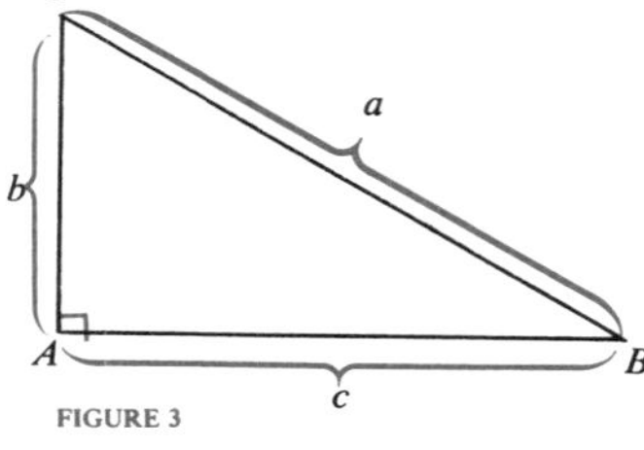

FIGURE 3

Proof: There are two cases to consider: (1) the triangle is a right triangle and (2) the triangle is not a right triangle.

(Case 1): Suppose $\triangle ABC$ is a right triangle and $\angle A$ is the right angle [Figure 3]. Then from the right-triangle definitions of the trigonometric functions, we have

$$\sin \angle B = \frac{b}{a} \quad \text{and} \quad \sin \angle C = \frac{c}{a}$$

or

$$\frac{\sin \angle B}{b} = \frac{1}{a} \quad \text{and} \quad \frac{\sin \angle C}{c} = \frac{1}{a}.$$

Since $\angle A$ is a right angle, we also have

$$\sin \angle A = 1 = \frac{a}{a} \quad \text{or} \quad \frac{\sin \angle A}{a} = \frac{1}{a}.$$

So if $\angle A$ is a right angle

$$\frac{\sin \angle A}{a} = \frac{\sin \angle B}{b} = \frac{\sin \angle C}{c}.$$

$\angle A$ is obtuse

$\angle A$ is acute

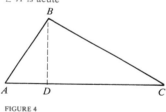

FIGURE 4

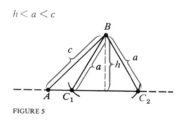

Using similar reasoning, the same conclusion is reached if either $\angle B$ or $\angle C$ is the right angle.

$h < a < c$

FIGURE 5

(Case 2): If $\triangle ABC$ is not a right triangle, then the altitude containing any of the vertices does not intersect the opposite side of $\triangle ABC$ at another vertex of $\triangle ABC$. For this reason, the intersection, D, of the altitude containing vertex B and the line AC is neither point A nor point C [Figure 4]. Let h denote the length of the altitude $\overline{BD}$. Since $\overline{BD}$ is perpendicular to $\overline{AC}$ (or to $\overline{AC}$ extended if $\angle A$ is an obtuse angle), the two triangles, $\triangle ABD$ and $\triangle CBD$, are right triangles. (Why?) From the right-triangle definition of the sine function, we have

$h < c < a$

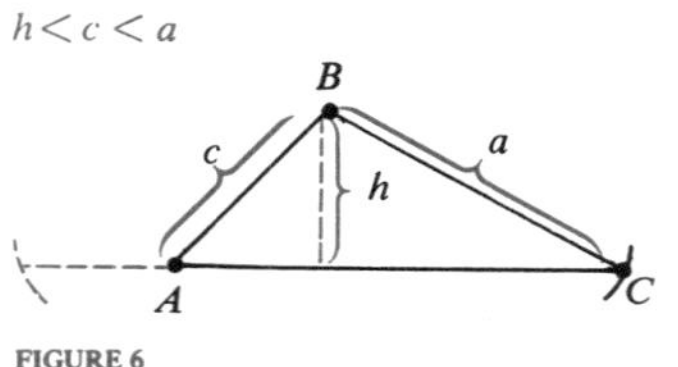

FIGURE 6

$$\sin \angle A = \frac{h}{c} \quad \text{and} \quad \sin \angle C = \frac{h}{a}$$

or

$$h = c \sin \angle A \quad \text{and} \quad h = a \sin \angle C.$$

Thus, $\quad c \sin \angle A = a \sin \angle C$

and $$\frac{\sin \angle A}{a} = \frac{\sin \angle C}{c}$$

which is part of the statement that was to be proved. The remainder of the proof, showing that

$$\frac{\sin \angle A}{a} = \frac{\sin \angle B}{b} \quad \text{or} \quad \frac{\sin \angle C}{c} = \frac{\sin \angle B}{b}$$

is left as Exercise 15, page 323.

$a < h < c$

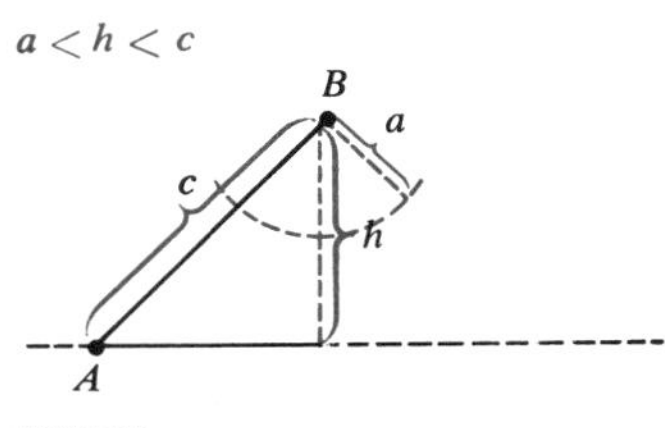

FIGURE 7

When the measures of two sides and an angle opposite one of them are given, it may be that two triangles [Figure 5], one triangle [Figure 6], or no triangles [Figure 7] are determined. Suppose $\angle A$, a, and c are given and h is the measure of the altitude of the triangle containing vertex B. In Figure 5, we see that $h < a < c$ and that two triangles are determined, $\triangle ABC_1$ and $\triangle ABC_2$. The case when $h < a < c$ and two triangles are determined is frequently called the **ambiguous case**. What relationship exists among h, a, and c if a right triangle is determined [Figure 8]? If a unique triangle that is not a right triangle is determined? If no triangle is determined?

Given $\angle A$, c, and a

if $\sin \angle A < \frac{a}{c} < 1$, then Figure 5;

if $\sin \angle A < 1 < \frac{a}{c}$, then Figure 6;

if $\frac{a}{c} < \sin \angle A < 1$, then Figure 7;

if $\sin \angle A = \frac{a}{c}$ and $\frac{a}{c} < 1$, then Figure 8.

Devise a procedure for testing the measures of two sides and an angle opposite one of them to see if the data describes Figure 5, 6, 7, or 8. (Hint: Use the right-triangle definition of the sine function.)

Example 2 Given that $m_r \angle A = \frac{\pi}{6}$, $a = 10$, and $c = 30$, solve $\triangle ABC$.

Solution: Applying the law of sines,

$$\frac{\sin \frac{\pi}{6}}{10} = \frac{\sin \angle C}{30}$$

$$\frac{\frac{1}{2}}{10} = \frac{\sin \angle C}{30}.$$

$h = a$ and $a < c$

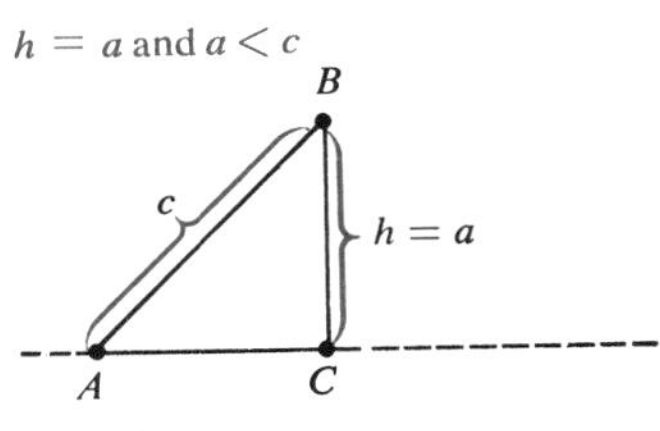

FIGURE 8

Hence, $\sin \angle C = 1.5$ which is impossible. (Why?) Thus, if $m_r \angle A = \frac{\pi}{6}$, $a = 10$, and $c = 30$, then no triangle is determined.

Exercises

1. Find the indicated part of $\triangle ABC$.
   a. $m \angle A = 30°10'$, $b = 4$, $c = 7$; a
   b. $m \angle C = 47°40'$, $a = 6$, $b = 4$; c
   c. $m \angle B = 80°35'$, $a = 12$; $c = 6$; b
   d. $m \angle A = 32°27'$, $b = 3$, $c = 5$; a
   e. $a = 2$, $b = 3$, $c = 4$; $m_d \angle A$

f. $m_r \angle C = \frac{\pi}{2}$, $a = 5$, $b = 12$; $m_r \angle B$
g. $m \angle B = 75°58'$, $a = 4$, $c = 8$; $m_d \angle C$
h. $m_r \angle A = \pi$, $b = 9$, $c = 13$; a
i. $m \angle A = 75°10'$, $c = 3140$, $m \angle B = 58°20'$; a
j. $m \angle A = 37°20'$, $m \angle C = 14°40'$, $a = 6.25$; b
k. $a = 3.24$, $b = 6.48$, $m \angle B = 57°20'$; $m_d \angle A$
l. $a = 9.78$, $c = 18.24$, $m \angle A = 24°30'$; $m_d \angle C$
m. $m \angle B = 107°13'$, $m \angle C = 19°47'$, $a = 32.93$; b
n. $m \angle A = 60°34'$, $m \angle C = 39°24'$, $b = 832.4$; c
o. $a = 5.321$, $b = 17.76$, $m \angle A = 12°43'$; c

2. In navigation, the **bearing** (compass direction) of a ship or airplane is the degree measure of the angle measured clockwise from the north. Thus, if the bearing of a ship is 090, then the ship is traveling east.
   a. What is the bearing of a ship traveling south? West?
   b. In what direction is the ship traveling if its bearing is 000?

3. An airplane with a cruising speed of 100 miles per hour leaves Smith Field, Fort Wayne, on a bearing of 225. An hour later, a second airplane leaves Smith Field flying on a bearing of 035 at 150 miles per hour. If you assume that both airplanes fly at a constant speed equal to their cruising speeds, how far apart are the airplanes 3 hours after the first plane left the field?

4. A boat's radio can communicate with a station on the dock if the boat is within 20 miles of the dock. The boat travels east for 15 miles and then changes its bearing to 045. If the boat travels at 3 miles per hour, how long can it maintain this bearing before it is out of communication range with the dock?

5. A ship leaves port on a bearing of 045, travels in that direction for 6 hours, and then changes its course so that it is on a bearing of 160. If the ship travels at 18 knots and maintains the 160 bearing, how long will it take the ship to reach a point due east of the port?

6. The good ship HMS Pinafore leaves San Francisco at 12 noon and travels due west at 20 knots (nautical miles per hour). A second ship leaves San Francisco at 2 PM and travels northwest at 10 knots. How far apart are they at 6 PM of the same day?

7. Find the measures of the angles of a triangle, if the vertices of the triangle are the points $F(1, 1)$, $G(3, 0)$, and $H(-2, -3)$.

8. If the vertices of a triangle are the points $C(-2, -5)$, $D(-3, -5)$, and $E(2, -3)$, find the measures of the angles of the triangle.

9. A lighthouse attendant observes that the measure of the angle formed by his line of sight to a ship and a weather station is 36°. He is simultaneously informed by phone from the weather station that the measure of the angle formed by the line of sight to the ship and the lighthouse is 47°. If the lighthouse and weather station are 3 miles apart, how far from the lighthouse is the ship?

10. What is the area of a parallelogram, if the length of a diagonal is 12 units and the diagonal meets two adjacent sides of the parallelogram at angles with measures 32° and 45°?

11. An astronomer observes that the measure of the angle formed by his line of sight to the stars Canopus and Betelgeuse is 35°17′. If he knows that the distances to Canopus and Betelgeuse are 200 and 300 light years, respectively, how far apart are these two stars?

12. Two ships, one traveling at 30 knots and the other at 15 knots, leave port at the same time. If the ships are 100 nautical miles apart after three hours and one of the ships is known to be on a bearing of 075, what is the bearing of the other ship?

13. A ship leaves port A on a bearing of 075. After traveling at 20 knots for 10 hours and 30 minutes, the ship is ordered to proceed to port B. The navigator determines that the new bearing of the ship should be 300. If it takes 7 hours for the ship traveling at an adjusted speed of 18 knots to reach port B, what is the distance between ports A and B?

14. Assume the law of cosines and prove the Pythagorean theorem for any right triangle, ΔABC.

15. Prove the part of the law of sines (page 321) that is not given in the text.

16. Forest rangers in two lookout towers which are 4.3 miles apart observe a forest fire on bearings 135 and 225 respectively. If the second tower is on a bearing of 090 from the first tower, which tower is closer to the fire and how far is it from this tower to the fire?

17. Two airplanes, each traveling at a ground speed of 500 miles per hour, leave Chicago's O'Hare airport at the same time. One of the airplanes flies on a bearing of 185, destination New Orleans. The airplane arrives in New Orleans one hour and fifty-one minutes after takeoff. The second airplane flies on a bearing of 087, destination New York. The time to New York is one hour and forty-two minutes. How far apart are New York and New Orleans?

18. Find a formula for the area of any triangle, ΔABC, in terms of the measure of two sides, a and b, and the included angle, $\angle C$. What happens to the area of ΔABC if a or b is doubled?

19. In Exercise 18, a formula was found for the area of any triangle, ΔABC, in terms of a, b, and $\angle C$. Find a similar formula for the area of the same triangle using the measures of each of the other possible pairs of sides and included angle. Use these formulas and the fact that the area of a triangle is constant to prove the law of sines.

20. An engineering firm has contracted to design a highway bridge which will span a deep gorge. A fundamental piece of information needed to fulfill the contract is the distance across the gorge (length of the bridge). Keeping the law of sines in mind, how might the engineers obtain an indirect measurement for this distance?

14-7 VECTORS

Many quantities, called **scalar quantities**, are described by their **magnitude**; that is, a number and a unit of measure. Length, area, speed, temperature, height, weight, and age are examples of scalar quantities. Other quantities, called **vector quantities**, require the measure of two or more properties of the quantity for a complete description. For example, to describe a force it is necessary to indicate both its magnitude and direction. Other examples of vector quantities are velocity, acceleration, and the displacement of an object.

Geometrically, a scalar quantity can be represented by a line segment. A vector quantity can be represented geometrically by a directed line segment. Directed line segments that represent vector quantities are called **vectors**. A vector representing a force is indicated in Figure 1. The length of the vector represents the magnitude and the direction of the vector indicates the direction of the force. Vectors are often denoted by a half-headed arrow above the letters for the endpoints of the directed line segment. The symbol $\overrightarrow{AB}$ denotes the vector in Figure 1.

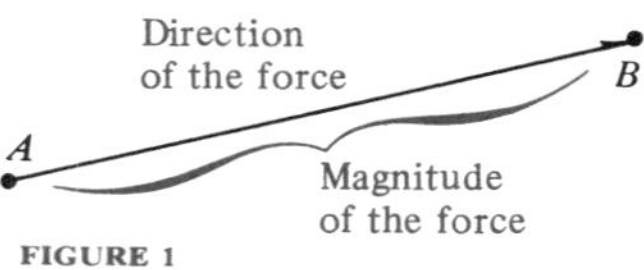

FIGURE 1

If a man starts at point A, walks 5 miles east to point B, and then 3 miles northeast to point C, we can illustrate these displacements by vectors $\overrightarrow{AB}$ and $\overrightarrow{BC}$ in Figure 2. The vector $\overrightarrow{AC}$, representing the net displacement, is called the **sum** of vectors $\overrightarrow{AB}$ and $\overrightarrow{BC}$.

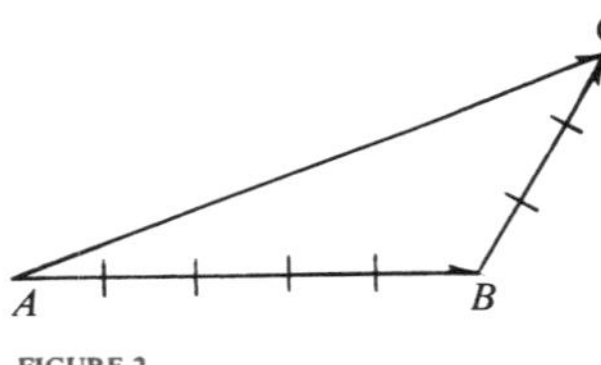

FIGURE 2

In the above example, one displacement followed another and the second displacement started at the end of the first displacement. A different situation is presented when several forces are simultaneously applied at a point. In this case, there is a single force that creates the same effect. The single force, called the **resultant** of the forces, is obtained by the parallelogram law of physics. Two forces, F_1 and F_2, simultaneously applied to a point and the resultant of F_1 and F_2 are illustrated in Figure 3. By the parallelogram law, $\overrightarrow{AD}$, the vector representing the resultant R, is the diagonal of the parallelogram determined by $\overrightarrow{AB}$ and $\overrightarrow{AC}$, the vectors representing the two forces F_1 and F_2. By noting that $\overrightarrow{CD}$ has the same magnitude and direction as $\overrightarrow{AB}$ and begins at the "tip" of $\overrightarrow{AC}$ [Figure 3], the resultant vector is the sum of $\overrightarrow{AB}$ and $\overrightarrow{AC}$ in the sense of the first example.

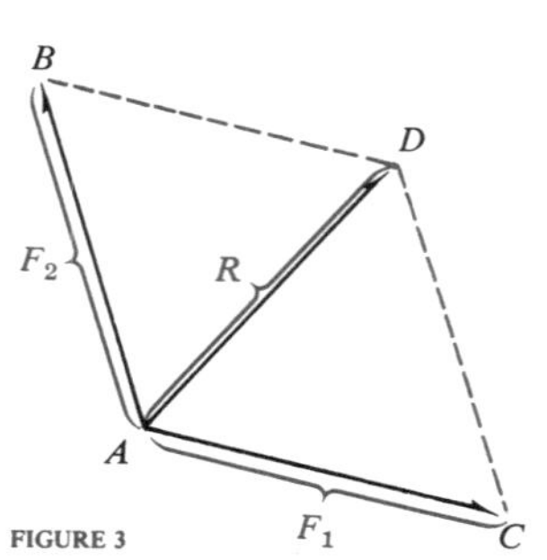

FIGURE 3

For any vector $\overrightarrow{AB}$, any two vectors whose sum is $\overrightarrow{AB}$ are called components of $\overrightarrow{AB}$. The process of finding components for a vector is called **resolving** the vector. Pairs of components ($\overrightarrow{AC}$ and $\overrightarrow{AD}$) for the vector $\overrightarrow{AB}$ are illustrated in Figures 4–6.

We can solve a variety of problems by combining the idea of a vector with the Pythagorean theorem and the trigonometry of triangles.

Example 1 A man rows his boat at 4 mph directly south across a river that flows west

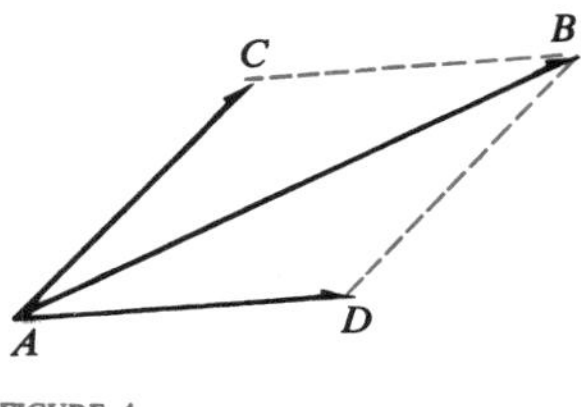

FIGURE 4

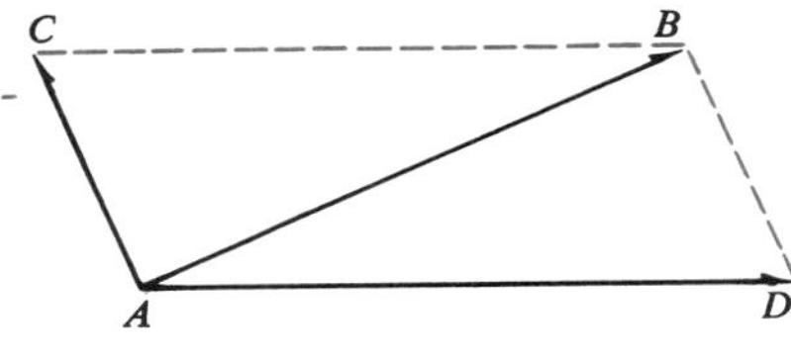

FIGURE 5

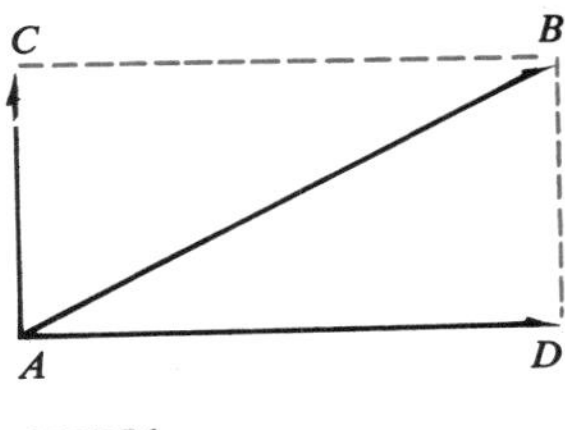

FIGURE 6

at 3 mph. What is the velocity of the boat?

Solution: Velocity is a vector quantity having two dimensions, speed (magnitude) and direction. The situation in this problem is illustrated by $\overrightarrow{AC}$ and $\overrightarrow{CB}$ in Figure 7. The velocity of the boat is the effect of two component velocities, the velocity of the rowing and the velocity of the river's flow. Since these vectors are at right angles to each other, we can obtain the magnitude of the velocity vector, $\overrightarrow{AB}$, using the Pythagorean theorem. Thus, the magnitude of $\overrightarrow{AB}$ is $\sqrt{3^2 + 4^2} = 5$. The direction $\overrightarrow{AB}$ can be found by determining the measure of $\angle BAC$ and mentioning that the direction is west of south. By the trigonometry of right triangles,

A
4 mph
B 3 mph C

FIGURE 7

$$\tan \angle BAC = \frac{3}{4} = 0.75.$$

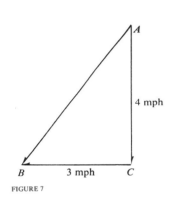

Using Table III, the measure of $\angle BAC$ is approximately $37°52'$. Therefore, the velocity of the boat is 5 mph in the direction $37°52'$ west of south.

Example 2 A weight of 100 pounds is stationary on an inclined plane that makes an angle of 30° with the horizontal. Find the components of the weight that are perpendicular and parallel to the inclined plane.

Solution: The force F exerted by the weight has a magnitude of 100 pounds and a direction which is down [Figure 8]. Let the components of F that are perpendicular and parallel to the inclined plane be F_1 and F_2, respectively. Since F is perpendicular to the horizontal and F_1 is perpendicular to the surface of the inclined plane, the degree measure of the angle determined by F and F_1 is 30. Using the trigonometry of right triangles, we have

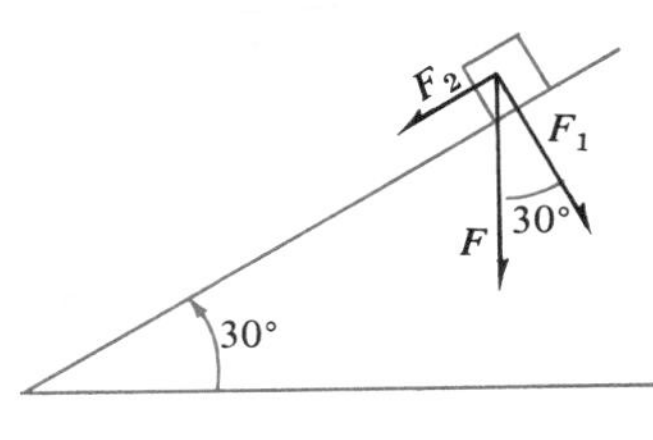

FIGURE 8

$$\cos 30° = \frac{\text{magnitude of } F_1}{\text{magnitude of } F} = \frac{\text{magnitude of } F_1}{100}.$$

Thus, the magnitude of F_1 is 86.6 pounds. The direction of F_1, with respect to F is 30°. Since the degree measure of the angle determined by vectors F and F_1 is 30 and F_1 is perpendicular to F_2, the degree measure of the angle determined by F and F_2 is 60. Again from the trigonometry of triangles, we have

$$\cos 60^\circ = \frac{\text{magnitude of } F_2}{\text{magnitude of } F} = \frac{\text{magnitude of } F_2}{100}.$$

Therefore, the magnitude of F_2 is 50 pounds and the direction of F_2 with respect to F is 60°.

Example 3 Consider two forces, F_1 and F_2, simultaneously applied at a point. If F_1 has magnitude 6 tons, F_2 has magnitude 8 tons, and the angle between the vectors representing these forces has measure 53°, what is the magnitude and direction of the resultant of F_1 and F_2?

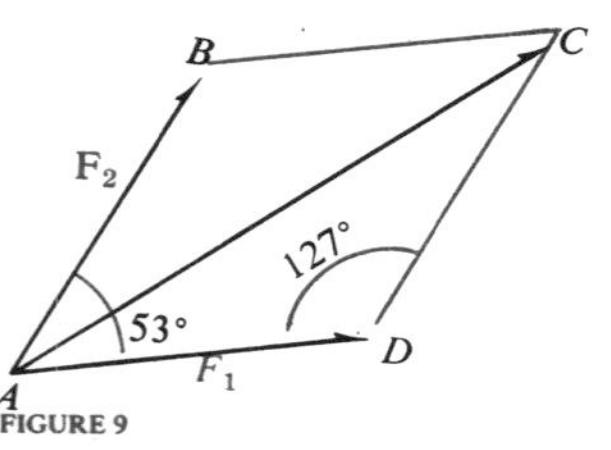

FIGURE 9

Solution: Let force F_1 be represented by $\overrightarrow{AD}$, force F_2 be represented by $\overrightarrow{AB}$ and the resultant be represented by $\overrightarrow{AC}$ [Figure 9]. Since adjacent angles of a parallelogram are supplementary and the opposite sides and angles of a parallelogram are congruent, $\angle ADC$ has measure 127° and and $m(\overline{DC}) = m(\overline{AB})$. Hence, the measures of $\overline{AD}$ and $\overline{DC}$ are 6 and 8, respectively. We now know the measure of two sides, $\overline{AD}$ and $\overline{DC}$, and the included angle, $\angle ADC$, of $\triangle ADC$. Hence, we can apply the law of cosines to obtain the magnitude of $\overline{AC}$ which is $m(\overline{AC})$:

$$[m(\overline{AC})]^2 = [m(\overline{AD})]^2 + [m(\overline{DC})]^2 - 2m(\overline{AC})m(\overline{DC}) \cos \angle D$$

$$= 6^2 + 8^2 - 2(6)(8) \cos 127^\circ$$

$$\approx 36 + 64 - 96(-0.6018) = 157.77.$$

Therefore, $m(\overline{AC}) \approx 12.56$ and the magnitude of the resultant vector, $\overrightarrow{AC}$, is 12.56 tons. The direction of $\overrightarrow{AC}$ can be expressed by the measure of the angle determined by the resultant $\overrightarrow{AC}$ and each component $\overrightarrow{AD}$ and $\overrightarrow{AB}$. The law of sines can be used to find $m \angle CAD$, the measure of the angle determined by the resultant $\overrightarrow{AC}$ and the component $\overrightarrow{AD}$. By the law of sines

$$\frac{\sin \angle D}{m(\overline{AC})} = \frac{\sin \angle CAD}{m(\overline{CD})}$$

$$\sin \angle CAD = \frac{8 \sin 127^\circ}{12.56}$$

$$\approx \frac{8(0.7986)}{12.56} \approx 0.5087.$$

Therefore, $m \angle CAD \approx 30^\circ 30' 30''$. Finally

$$m \angle BAC \approx 53^\circ - 30^\circ 30' 30'' = 22^\circ 29' 30''$$

since $m \angle BAC + m \angle CAD = 53^\circ$.

Exercises

1. Classify each of the following as a vector quantity or a scalar quantity: **(a)** the number of players on a football team; **(b)** the velocity of a train; **(c)** the temperature; **(d)** the length of a rectangular lot; **(e)** a trip of 100 miles.
2. Illustrate a displacement of 350 miles at 10° west of south by means of a vector $\overrightarrow{CD}$.
3. Illustrate a force of 300 pounds downward by means of a vector $\overrightarrow{EF}$.
4. Illustrate a wind velocity of 3 miles per hour from the east by $\overrightarrow{GH}$.
5. Draw two noncollinear vectors, so that the second begins at the "tip" of the first. Construct the sum of the two vectors.
6. Suppose a man rows his boat directly across a river at 4 miles per hour and that the current of the river carries him downstream at the rate of 3 miles per hour. Draw a vector diagram illustrating these two displacements and the net displacement after one hour.
7. Draw any two noncollinear vectors from a point. Construct the resultant of the two vectors.
8. A man pulling a wagon exerts a tug of 35 pounds on the handle. If the handle makes an angle of 15° with the horizontal and if friction is ignored, what is the force pulling the wagon forward?
9. A pilot is instructed to fly due east. If the meteorologist's report indicates a north wind of 30 mph and if the speed of the plane is 200 mph, what heading should he set in order to fly due east?
10. How much force must be exerted to prevent a 4000 pound automobile from moving down a street inclined at 10°?
11. Two forces F_1 and F_2 are simultaneously applied at a point. What is the measure of the angle between the forces that gives a resultant with maximum magnitude? What is the measure of the angle between the forces that gives a resultant with minimum magnitude?
12. Let F_1 and F_2 be two forces that act on the same point. If the magnitude of F_1 is 50 pounds, the magnitude of F_2 is unknown, the magnitude of the resultant is 70 pounds, and the angle determined by the two forces has measure 135°, what is the magnitude of F_2?

Review Exercises

1. Define the cosine of an angle. **14-1**
2. Use Table II and linear interpolation (when necessary) to find the trigonometric functional value of each angle.
   a. $\sin \angle A$ when $m_r \angle A = 0.95$
   b. $\sin \angle B$ when $m_r \angle B = 7.892$
   c. $\sin \angle C$ when $m_d \angle C = 78$
   d. $\sin \angle D$ when $m_d \angle D = -495$
3. What is $\csc 53^\circ$ if $\sec 37^\circ = 1.252$?

4. Use Table III and linear interpolation (when necessary) to find the trigonometric functional value of each angle.
   a. $\sin 47°20'$ **b.** $\tan 19°45'$ **c.** $\sec 87°12'$

5. Find the degree measure of an acute angle that is a solution of $\cot \angle A = 0.7940$.

6. Approximate the value of $\csc(-347°15')$. **14-2**

7. Evaluate $\cos(-18)° \cos 27° - \sin(-18)° \sin 27°$ by simplifying the phrase to an expression involving a single trigonometric function.

8. Suppose the point $(3, -2)$ is on the terminal side of an angle in standard position. What are the values of all six trigonometric functions of the angle? **14-3**

9. Given that $\sin \angle C = -\dfrac{3}{\sqrt{10}}$ and $\tan \angle C = 3$, find $\sec \angle C$.

10. Replace m and n to make each sentence true for right triangle, $\triangle XYZ$, with right angle, $\angle Y$.
    a. $\tan \angle X = \dfrac{\text{length of the side opposite } \angle X}{m} = \dfrac{n}{z}$.
    b. $\sec \angle Z = \dfrac{\text{length of the hypotenuse}}{m} = \dfrac{n}{x}$.

11. Verify that the three numbers $a = \sqrt{2}$, $b = 2\sqrt{3}$, and $c = \sqrt{10}$ could be the measures of the sides of a right triangle and determine the values of the six trigonometric functions for $\angle A$ and for $\angle C$.

12. Complete the table below with the aid of the right-triangle definitions of the trigonometric functions. **14-4**

θ	$\sin\theta$	$\cos\theta$	$\tan\theta$	$\cot\theta$	$\sec\theta$	$\csc\theta$
$30°$						
$\frac{\pi}{3}$						
$45°$						

13. If $\triangle ABC$ is a right triangle with right angle, $\angle C$, find b if $m\angle A = 38°15'$ and $c = 13.25$. Tables may be used to solve this problem.

14. At a point 75 feet from a building, the measure of the angle of inclination from the ground level to the top of the building is $63°44'$. What is the height of the building?

15. Use Tables I and IV to evaluate $(27.5)^{1/2}(\sec 83°42')$. **14-5**

16. If in a given triangle, $\triangle ABC$, $a = 8$, $b = 12$, and $m \angle C = 44°20'$, find c. 14-6

17. Given that the measures of the sides of $\triangle ABC$ are $a = 10$, $b = 9$, and $c = 6$, find the measure of the largest angle.

18. If in a given triangle, $\triangle ABC$, $m \angle A = 52°30'$, $m \angle C = 30°10'$, and $b = 9$, find a.

19. Given that the measures of parts of $\triangle ABC$ are $b = 7.35$, $c = 3.7$, and $m \angle C = 67°$, find $m_d \angle B$.

20. An airplane is directed on an eastward course at 150 miles an hour. A north wind is blowing at 30 miles an hour. What is the velocity (direction and speed) of the airplane? 14-7

21. Two forces, F_1 and F_2 are simultaneously applied at a point. If F_1 has magnitude 12 pounds, F_2 has magnitude 8 pounds, and the angle between the vectors representing these forces has measure $\frac{\pi}{4}$ radians, what is the magnitude and direction of the resultant of F_1 and F_2?

15. POWERS AND ROOTS OF COMPLEX NUMBERS

15-1 POLAR COORDINATES

In the rectangular-coordinate plane, there is a one-to-one correspondence between the set of points in the plane and the set of all ordered pairs of real numbers. The correspondence between these two sets provides the rationale for graphing and is the corner stone of analytic or coordinate geometry.

In this section, we will study a coordinate system in which a correspondence between ordered pairs of real numbers and the points in a plane is established using the concepts of distance and angle measure. Consider a ray, $\overrightarrow{OX}$, which we will call the **polar axis**, and its endpoint O, which we will call the **pole** or **origin**. Let P be any point in the plane and assign to P the ordered pair of real numbers (r, θ), where r is the distance from O to P (hence, $r \geqslant 0$) and θ is the measure of $\angle XOP$ [Figure 1]. The numbers r and θ are called the **polar coordinates** of P. The segment, $\overline{OP}$, is called the **radius vector** of point P.

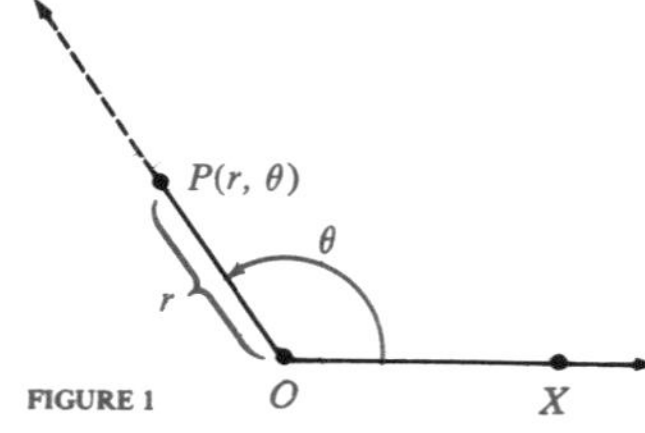

FIGURE 1

This correspondence between ordered pairs of real numbers (r, θ) with $r \geqslant 0$ and points in the plane is not one-to-one. For, if (r, θ) corresponds to a point P and θ is expressed in radians, then each ordered pair in the set $\{(r, \theta + 2\pi k)$: k is an integer$\}$ corresponds to P. (Why?) We can make the correspondence one-to-one by requiring the second coordinate to be the smallest nonnegative number of the set $\{\theta + 2\pi k\}$. If the angle is measured in degrees, the second coordinate is required to be the smallest nonnegative number that is the degree measure of $\angle XOP$. In this chapter, when the polar coordinates of a point are not given, we will consider an element of

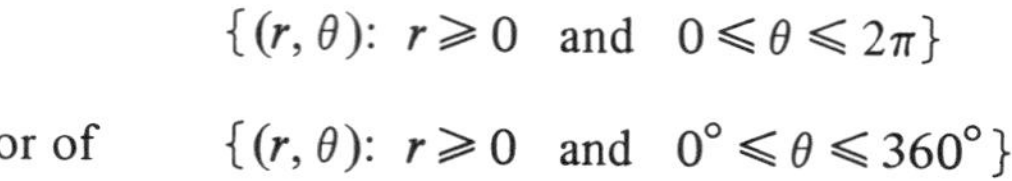

$$\{(r, \theta)\colon\ r \geqslant 0 \ \text{ and } \ 0 \leqslant \theta \leqslant 2\pi\}$$

or of $$\{(r, \theta)\colon\ r \geqslant 0 \ \text{ and } \ 0^\circ \leqslant \theta \leqslant 360^\circ\}$$

as specifying its polar coordinates. The coordinate system that we have just described is called the **polar coordinate system.**

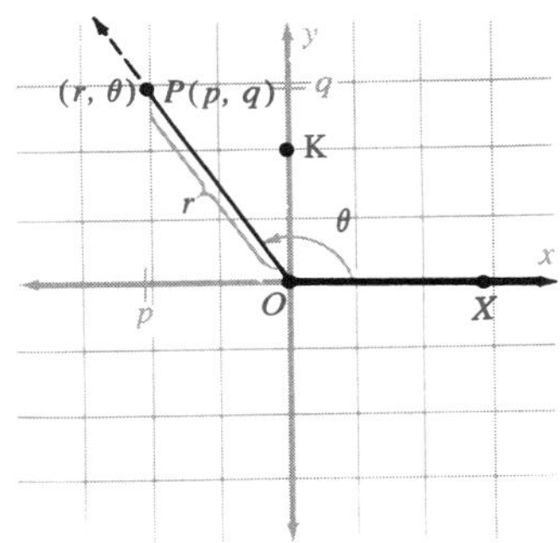

FIGURE 2

A natural relationship exists between the rectangular and polar coordinate systems for the plane. Let the polar axis, $\overrightarrow{OX}$, be the ray of the x-axis that corresponds to the nonnegative real numbers and let the pole, O, be the origin of the rectangular-coordinate system [Figure 2]. Then any point P in the plane has both rectangular and polar coordinates indicated by (p, q) and (r, θ), respectively. For example, the rectangular and polar coordinates of K in Figure 3 are given by $(0, 2)$ and $(2, 90°)$ or $\left(2, \frac{\pi}{2}\right)$, respectively.

To express the rectangular coordinates of P in terms of its polar coordinates, recall that the values of the trigonometric functions of an angle in standard position are related to the coordinates of any point on its terminal side. Using the Theorem from Section 14-3 we have

$$\cos\theta = \frac{p}{r} \quad \text{and} \quad \sin\theta = \frac{q}{r}$$

from which we obtain expressions for p and q in terms of r and θ:

$$p = r\cos\theta \quad \text{and} \quad q = r\sin\theta.$$

Give the rectangular coordinates for each of the following points whose coordinates are given in polar form: $P(3, 19°)$, $Q\left(1, \frac{2\pi}{3}\right)$, $R(14, 225°)$, $S\left(\sqrt{2}, \frac{5\pi}{3}\right)$.

$P(2.8365, 0.9768)$, $Q\left(-\frac{1}{2}, \frac{\sqrt{3}}{2}\right)$

$R(-7\sqrt{2}, -7\sqrt{2})$, $S\left(\frac{\sqrt{2}}{2}, -\frac{\sqrt{6}}{2}\right)$

To write expressions for the polar coordinates of P in terms of its rectangular coordinates, recall that r is the distance from the origin to the point P and observe that $\overline{OP}$ is the hypotenuse of a right triangle, $\triangle OPQ$, [Figure 3] with legs of length $|p|$ and $|q|$. Applying the Pythagorean theorem to $\triangle OPQ$, we see that

$$r = \sqrt{p^2 + q^2}.$$

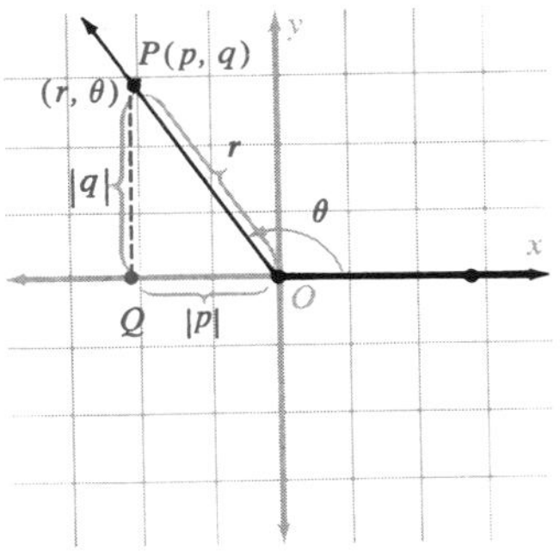

FIGURE 3

To find θ first locate the quadrant containing the terminal side of the angle with measure θ by noting whether p and q are positive or negative. Then use the values of p and q and

$$\tan\theta = \frac{q}{p}$$

to uniquely determine θ.

Example 1 Find the polar coordinates r and θ, where $0 \leqslant \theta < 2\pi$, of the point with rectangular coordinates $(-\sqrt{3}, 1)$.

Solution: From the discussion above we know that

$$r = \sqrt{p^2 + q^2}$$

and

$$\tan\theta = \frac{q}{p} \quad \text{or} \quad \theta = \tan^{-1}\frac{q}{p}$$

where the quadrant of the point (p, q) determines a unique value for θ. Since $p = -\sqrt{3}$ and $q = 1$, we have

$$r = \sqrt{3 + 1} = 2$$

and

$$\theta = \tan^{-1}\left(\frac{1}{-\sqrt{3}}\right) = \tan^{-1}\left(-\frac{\sqrt{3}}{3}\right)$$

where $0 \leqslant \theta < 2\pi$ and θ is the measure of an angle in standard position with terminal side in the second quadrant. Therefore, $\theta = \frac{5\pi}{6}$ and the polar coordinates of the point are $\left(2, \frac{5\pi}{6}\right)$.

Example 2 Find the polar coordinates r and θ, where $0^\circ \leqslant \theta < 360^\circ$, of the point with rectangular coordinates $(7, -3)$.

Solution: Since $r = \sqrt{p^2 + q^2}$

$$r = \sqrt{49 + 9} = \sqrt{58}.$$

Furthermore, θ is the measure of an angle in the fourth quadrant such that

$$\theta = \tan^{-1}\left(-\frac{3}{7}\right) \approx \tan^{-1}(-0.4286).$$

Using Table III, the measure of the associated reference angle is approximately $23^\circ 12'$. The measure (between 0° and 360°) of the angle in quadrant IV which is associated with this reference angle is $\theta = 336^\circ 48'$. Therefore, the polar coordinates of the point are given by $(\sqrt{58}, 336^\circ 48')$.

Exercises

1. Determine the quadrant in which each point lies. (Degree measures indicate sets of polar coordinates.)

a. $(-3, 5)$ b. $(6, 1)$ c. $(2, 30^\circ)$
d. $(-1, -2)$ e. $(1, 420^\circ)$ f. $(5, -30^\circ)$
g. $(4, -1)$ h. $(6, -300^\circ)$ i. $(2, 147^\circ)$
j. $(10, 5620^\circ)$

2. Give two other sets of polar coordinates for each point.

a. $(5, 343°)$ b. $(2, -17°)$ c. $\left(\sqrt{5}, \frac{\pi}{6}\right)$

d. $\left(\sqrt{2}, \frac{10\pi}{4}\right)$ e. $(1, 265°)$ f. $(3, -1045°)$

g. $(10, 2\pi)$ h. $\left(\sqrt{7}, -\frac{13\pi}{2}\right)$ i. $(3, -30°)$

j. $\left(\sqrt{2}, \frac{5\pi}{2}\right)$

3. Find the polar coordinates for each point whose coordinates are given in rectangular form.

a. $(2, 2)$ b. $(-1, 3)$ c. $(-2, -3)$ d. $(4, -10)$

e. $(6, 3)$ f. $\left(-5, \frac{3}{2}\right)$ g. $(2, 6)$ h. $(-1, 4)$

4. Find the rectangular coordinates for each point whose coordinates are given in polar form.

a. $\left(1, \frac{\pi}{4}\right)$ b. $\left(5, -\frac{\pi}{6}\right)$ c. $(2, 186°)$

d. $(4, -216°)$ e. $\left(1, \frac{13\pi}{4}\right)$ f. $(105, 10°)$

g. $\left(3, \frac{\pi}{3}\right)$ h. $(10, -100°)$

5. What are the polar coordinates of the pole (origin)?

6. Use polar coordinates to write an equation for which the polar axis is the solution set.

7. Write the polar form of each equation

a. $x^2 + y^2 = 4$ b. $2x + y = 4$ c. $y = x^2$

8. Write the rectangular form of each equation.

a. $r = 4 \cos \theta$ b. $\theta = \frac{\pi}{4}$ c. $r = 7$

If the set of nonnegative real numbers is the replacement set for r and $0 \leqslant \theta < 2\pi$, graph the solution set of each sentence below on polar coordinate paper.

9. $r = 3$

10. $\theta = \pi$

11. $r = 1$ or $r = 3$, and $\theta = \pi$

12. $r = 3$, and $\theta = \pi$ or $\theta = \frac{\pi}{2}$

13. $r \sin \theta = 3$

14. $r \cos \theta = 3$

15. $r \sin \theta = 3$ and $r \cos \theta = 3$

16. $r = \theta$

17. $r = 2 \cos \theta$

18. $r = 3 \sin \theta$

15-2 TRIGONOMETRIC FORM OF A COMPLEX NUMBER

The complex number system was developed in Chapter 7. At that time, we noted that any complex number $p + qi$ can be represented by the ordered pair (p, q). With the ordered-pair notation for complex numbers, it was natural to represent the complex number $p + qi$ geometrically as the point $P(p, q)$ in the plane [Figure 1]. The $p + qi$ form is called the **rectangular form** of the complex number. As was observed in the preceding section, if the polar coordinates of P are r and θ [Figure 2], the relationships between p and q, and r and θ are

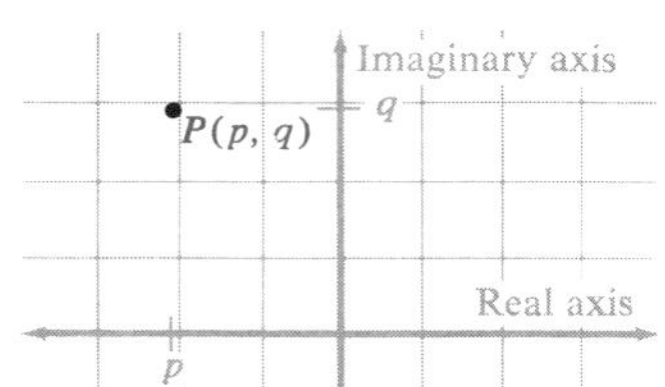

FIGURE 1

$$p = r \cos \theta \qquad q = r \sin \theta \qquad r^2 = p^2 + q^2.$$

Thus, the complex number $p + qi$ can be denoted by

$$p + qi = r \cos \theta + (r \sin \theta)i = r(\cos \theta + i \sin \theta).$$

Definition The expression $r(\cos \theta + i \sin \theta)$ is called the **polar** or **trigonometric form** of the complex number $p + qi$. The number θ is called the **argument** or **amplitude** of the complex number. The number $r = \sqrt{p^2 + q^2}$ is called the **absolute value** or **modulus** of $p + qi$ and is denoted by $|p + qi|$.

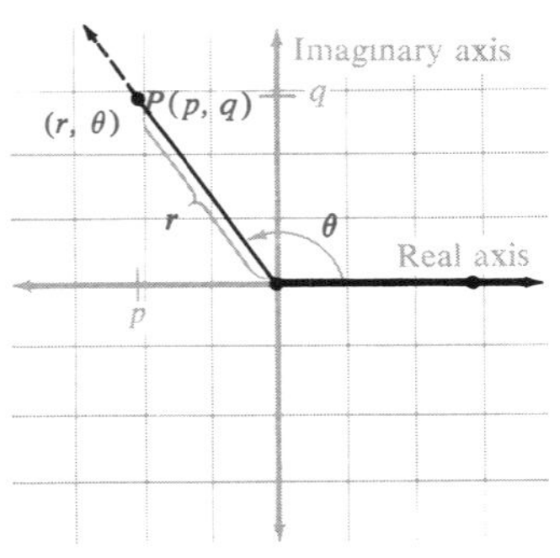

FIGURE 2

Note that $|p + qi|$, the distance from the pole to the point corresponding to the complex number, is a nonnegative real number.

The expression "$r(\cos \theta + i \sin \theta)$" is frequently abbreviated by "r cis θ," where "cis" stands for the first letter of "cosine," the imaginary "i," and the first letter of "sine." Thus

$$\text{cis}\,\frac{\pi}{6} = \cos\frac{\pi}{6} + i \sin\frac{\pi}{6}$$

$$\cos\frac{5\pi}{6} + i \sin\frac{5\pi}{6} = \text{cis}\,\frac{5\pi}{6}$$

$$r\,\text{cis}\,\frac{3\pi}{2} = r\left(\cos\frac{3\pi}{2} + i \sin\frac{3\pi}{2}\right).$$

Since $\cos(\theta + 2\pi k) = \cos \theta$ and $\sin(\theta + 2\pi k) = \sin \theta$

$$r\,\text{cis}\,\theta = r\,\text{cis}\,(\theta + 2\pi k)$$

for any integer k.

Example 1 Find the trigonometric form of the complex number $(-1 - \sqrt{3}i)$.

Solution: The absolute value of the complex number is

$$r = |-1 - \sqrt{3}i| = \sqrt{(-1)^2 + (-\sqrt{3})^2} = 2.$$

Absolute value	4	3	1
Argument	π	$\frac{\pi}{2}$	$\frac{7\pi}{4}$

• $1 + 0i = 1 \text{ cis } 0$
$= 1 (\cos 0 + i \sin 0)$

$0 + 1i = 1 \text{ cis } \frac{\pi}{2}$

$= 1 \left(\cos \frac{\pi}{2} + i \sin \frac{\pi}{2}\right)$

$-1 = 1 \text{ cis } \pi = 1 (\cos \pi + i \sin \pi)$

$-i = 1 \text{ cis } \frac{3\pi}{2}$

$= 1 \left(\cos \frac{3\pi}{2} + i \sin \frac{3\pi}{2}\right)$

To find the argument of a complex number consider the relationship

$$\tan \theta = \frac{q}{p} \quad \text{or} \quad \theta = \tan^{-1} \frac{q}{p}.$$

Sixty degrees and 240° are the only two values of θ in the interval 0° to 360° for which $\theta = \tan^{-1}\left(\frac{-\sqrt{3}}{-1}\right)$. Since $(-1, -\sqrt{3})$ is in the third quadrant, $\theta = 240°$. Hence the trigonometric form of the complex number $-1 - \sqrt{3}i$ is 2 cis 240°.

Find the absolute value and argument of each of the following complex numbers: -4, $3i$, $\frac{\sqrt{2}}{2} - \frac{\sqrt{2}}{2}i$. Write the trigonometric form of the complex numbers $1 + 0i$, $0 + i$, -1, and $-i$.

Example 2 Find the rectangular form of the complex number $5 \text{ cis } \frac{\pi}{3}$.

Solution:

$$5 \text{ cis } \frac{\pi}{3} = 5 \left(\cos \frac{\pi}{3} + i \sin \frac{\pi}{3}\right)$$

$$= 5 \left[\frac{1}{2} + i\left(\frac{\sqrt{3}}{2}\right)\right]$$

$$= \frac{5}{2} + \frac{5\sqrt{3}}{2} i$$

Thus, the rectangular form of the complex number $5 \text{ cis } \frac{\pi}{3}$ is $\frac{5}{2} + \frac{5\sqrt{3}}{2} i$.

Exercises

1. Locate the point in the plane that is the geometric representation of each complex number.
   a. $3 - 5i$ b. $4 + 9i$ c. $-7 + 2i$
   d. $-9 - 6i$ e. $-2 - i$ f. $-i$
   g. -1 h. $0 + i$ i. $1 - 0i$
   j. $0 + 0i$ k. $3 \text{ cis } \frac{\pi}{2}$ l. $2 \text{ cis } \frac{7\pi}{6}$
   m. $3 \text{ cis } (-310)°$ n. $7 \text{ cis } (-33\pi)$ o. $5 \text{ cis } 1572°$
2. Write the trigonometric form of each complex number.
   a. $1 - 2i$ b. $1 + \sqrt{3}i$ c. $\sqrt{3} - i$
   d. $-6 - 6i$ e. $\sqrt{2} - \sqrt{2}i$ f. $-2 + \sqrt{3}i$
   g. $-5 - 7i$ h. $5 - 2i$ i. $-\sqrt{3} + 2i$
   j. $13 + 12i$ k. $-\sqrt{3} + i$ l. $3 + 4i$

3. Write the rectangular form of each complex number.

a. $3 \text{ cis } \frac{\pi}{3}$ b. $2 \text{ cis } \frac{\pi}{6}$ c. $7 \text{ cis } \pi$

d. $5 \text{ cis } \frac{\pi}{4}$ e. $4 \text{ cis } 75°$ f. $6 \text{ cis } (0.768)$

g. $\text{cis } 187°$ h. $0 \text{ cis } (5.75)$ i. $5 \text{ cis } 13°47'$

j. $5 \text{ cis } 166°13'$ k. $2 \text{ cis } (-\pi)$ l. $7 \text{ cis } (-769)°$

4. Find the trigonometric form of $0 + i$ and $1 + \sqrt{3}i$, and of the product $(0 + i)(1 + \sqrt{3}i)$. What relationship exists between the trigonometric forms of the factors and the product?

5. Find the trigonometric form for the conjugate of each complex number in Exercise 2.

6. In Chapter 7, two complex numbers $a + bi$ and $c + di$ were defined to be equal if and only if $a = c$ and $b = d$. When are two complex numbers $r_1 \text{ cis } \theta_1$ and $r_2 \text{ cis } \theta_2$ equal? Prove that your conjecture is true.

7. Consider the two complex numbers

$$x + yi \quad \text{and the product} \quad i^2(x + yi).$$

Prove that the arguments of the two complex numbers differ by π radians but that they have the same absolute value. Describe the relationship between the geometric representations of the two complex numbers.

15-3 MULTIPLYING COMPLEX NUMBERS EXPRESSED IN TRIGONOMETRIC FORM

In Chapter 7 the product of two complex numbers $a + bi$ and $c + di$, expressed in rectangular form, was defined to be

$$(a + bi)(c + di) = (ac - bd) + (ad + bc)i$$

Let us now consider the problem of multiplying two complex numbers represented in trigonometric form. If

$$r_1(\cos\theta_1 + i\sin\theta_1) \quad \text{and} \quad r_2(\cos\theta_2 + i\sin\theta_2)$$

are the trigonometric forms of two complex numbers, then

$$[r_1(\cos\theta_1 + i\sin\theta_1)][r_2(\cos\theta_2 + i\sin\theta_2)]$$

$$(1) \qquad = r_1 r_2[(\cos\theta_1 + i\sin\theta_1)(\cos\theta_2 + i\sin\theta_2)]$$

(1) Commutative and associative properties for multiplication
(2) Definition of complex-number multiplication
(3) Theorems 2 and 4, pages 282 and 283.

$\cos(\theta + \varphi) = \cos\theta \cos\varphi - \sin\theta \sin\varphi$
$\sin(\theta + \varphi) = \sin\theta \cos\varphi + \cos\theta \sin\varphi$

(2) $$= r_1 r_2 \left[(\cos\theta_1 \cos\theta_2 - \sin\theta_1 \sin\theta_2) + i(\cos\theta_1 \sin\theta_2 + \sin\theta_1 \cos\theta_2)\right]$$

(3) $$= r_1 r_2 \left[\cos(\theta_1 + \theta_2) + i \sin(\theta_1 + \theta_2)\right]$$

Supply a reason for each step in the above sequence of equations.

The above computations prove Theorem 1.

Theorem 1 For any complex numbers $r_1 \text{ cis } \theta_1$ and $r_2 \text{ cis } \theta_2$,

$$(r_1 \text{ cis } \theta_1)(r_2 \text{ cis } \theta_2) = r_1 r_2 \text{ cis } (\theta_1 + \theta_2)$$

An interesting observation can be made concerning the relationship between the geometric representations of $r_1 \text{ cis } \theta_1$ and $r_2 \text{ cis } \theta_2$, and their product [Figure 1]. The absolute value of the product of two complex numbers is the product of the absolute values of the two numbers and the argument of the product is the sum of the arguments of the numbers.

Example 1 Find $\left(2 \text{ cis } \frac{\pi}{2}\right)\left(3 \text{ cis } \frac{\pi}{6}\right)$.

Solution: The product is obtained by applying Theorem 1:

$$\left(2 \text{ cis } \frac{\pi}{2}\right)\left(3 \text{ cis } \frac{\pi}{6}\right) = (2)(3) \text{ cis } \left(\frac{\pi}{2} + \frac{\pi}{6}\right) = 6 \text{ cis } \frac{2\pi}{3}$$

Example 2 Find $\dfrac{r_1 \text{ cis } \theta_1}{r_2 \text{ cis } \theta_2}$, where $r_2 \neq 0$.

Solution: Since

$$\frac{r_1 \text{ cis } \theta_1}{r_2 \text{ cis } \theta_2} = \frac{r_1(\cos\theta_1 + i \sin\theta_1)}{r_2(\cos\theta_2 + i \sin\theta_2)} = \frac{r_1}{r_2} \cdot \frac{(\cos\theta_1 + i \sin\theta_1)}{(\cos\theta_2 + i \sin\theta_2)}$$

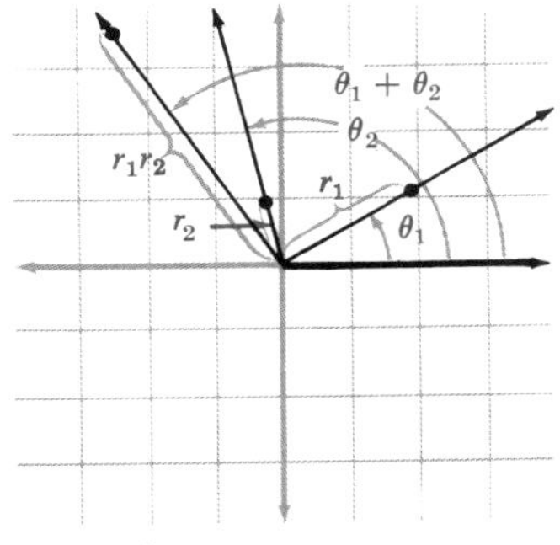

FIGURE 1

and $\cos\theta_1$, $\sin\theta_1$, $\cos\theta_2$, and $\sin\theta_2$ are real numbers (Why?), we can proceed as though we were considering the quotient of two complex numbers in rectangular form. Recall from Chapter 7, Section 4 that to divide complex numbers we multiply by 1 which we express as the conjugate of the denominator divided by itself:

(1) $$\frac{r_1 \text{ cis } \theta_1}{r_2 \text{ cis } \theta_2} = \frac{r_1}{r_2} \cdot \frac{\cos\theta_1 + i \sin\theta_1}{\cos\theta_2 + i \sin\theta_2} \cdot \frac{\cos\theta_2 - i \sin\theta_2}{\cos\theta_2 - i \sin\theta_2}$$

(2) $$= \frac{r_1}{r_2} \cdot \frac{(\cos\theta_1 \cos\theta_2 + \sin\theta_1 \sin\theta_2) + i(\sin\theta_1 \cos\theta_2 - \cos\theta_1 \sin\theta_2)}{\cos^2\theta_2 + \sin^2\theta_2}$$

(1) Definition of cis θ and of multiplication with fractions
(2) Definition of multiplication with fractions and of complex-number multiplication
(3) Theorems 1 and 3, pages 281 and 283.
(4) Fundamental identity (6), page 277 and definition of cis θ

(3) $= \frac{r_1}{r_2} \cdot \frac{\cos(\theta_1 - \theta_2) + i \sin(\theta_1 - \theta_2)}{1}$

(4) $= \frac{r_1}{r_2} \text{ cis } (\theta_1 - \theta_2).$

Supply a reason for each step in the above sequence of equations.

The solution of this example is a proof of Theorem 2.

Theorem 2 For any complex numbers r_1 cis θ_1 and r_2 cis θ_2 such that

$$r_2 \neq 0, \quad \frac{r_1 \text{ cis } \theta_1}{r_2 \text{ cis } \theta_2} = \frac{r_1}{r_2} \text{ cis } (\theta_1 - \theta_2).$$

Example 3 Find $\left(2 \text{ cis } \frac{\pi}{3}\right)^4$.

Solution: Using the definition of a positive integral exponent, associativity, and Theorem 1, we have:

$$\left(2 \text{ cis } \frac{\pi}{3}\right)^4 = \left[\left(2 \text{ cis } \frac{\pi}{3}\right)\left(2 \text{ cis } \frac{\pi}{3}\right)\right]\left[\left(2 \text{ cis } \frac{\pi}{3}\right)\left(2 \text{ cis } \frac{\pi}{3}\right)\right]$$

$$= \left[4 \text{ cis } \left(\frac{\pi}{3} + \frac{\pi}{3}\right)\right]\left[4 \text{ cis } \left(\frac{\pi}{3} + \frac{\pi}{3}\right)\right]$$

$$= 16 \text{ cis } \left(\frac{2\pi}{3} + \frac{2\pi}{3}\right)$$

$$= 16 \text{ cis } \frac{4\pi}{3}.$$

$(\sqrt[3]{2} \text{ cis } 90^\circ)^3$
$= (\sqrt[3]{2} \text{ cis } 90^\circ)(\sqrt[3]{2} \text{ cis } 90^\circ) \cdot (\sqrt[3]{2} \text{ cis } 90^\circ)$
$= (\sqrt[3]{2} \cdot \sqrt[3]{2} \cdot \sqrt[3]{2})\, [(\text{cis } 90^\circ \text{ cis } 90^\circ) \cdot \text{cis } 90^\circ]$
$= 2 \text{ cis } (90^\circ + 90^\circ) \cdot \text{cis } 90^\circ$
$= 2 \text{ cis } [(90^\circ + 90^\circ) + 90^\circ]$
$= 2 \text{ cis } 270^\circ$
$= 2 (\cos 270^\circ + i \sin 270^\circ)$
$= 2\, [0 + i(-1)] = -2i$

Evaluate $(\sqrt[3]{2} \text{ cis } 90^\circ)^3$ using $(\sqrt[3]{2} \text{ cis } 90^\circ)$ as a factor three times.

The pattern from Example 3 is generalized in the statement of De Moivre's theorem.

Theorem 3 (*De Moivre's theorem*) For any complex number r cis θ and any positive integer n, $(r \text{ cis } \theta)^n = r^n \text{ cis } n\theta$.

To prove that a proposition is true for every positive integer (natural number) requires the use of the postulate of mathematical induction which is discussed in Chapter 16. The proof of De Moivre's theorem is deferred and left as an exercise in Section 16-6, page 368.

Example 4 Find $(-1+\sqrt{3}i)^5$.

Solution: The trigonometric form of $-1+\sqrt{3}i$ is $2 \text{ cis } \frac{2\pi}{3}$. Applying De Moivre's theorem, we get

$$(-1+\sqrt{3}\,i)^5 = \left(2 \text{ cis } \frac{2\pi}{3}\right)^5 = 2^5 \text{ cis } \left[5\left(\frac{2\pi}{3}\right)\right]$$

$$= 32 \text{ cis } \frac{10\pi}{3} = 32\left(\cos\frac{10\pi}{3} + i \sin\frac{10\pi}{3}\right)$$

$$= 32\left(-\frac{\sqrt{3}}{2} - \frac{1}{2}i\right) = -16\sqrt{3} - 16i.$$

Thus, $(-1+\sqrt{3}i)^5 = -16\sqrt{3} - 16i$.

Exercises

1. Write the trigonometric and rectangular forms of each product.
   a. $\left(2 \text{ cis } \frac{\pi}{3}\right)\left(3 \text{ cis } \frac{5\pi}{6}\right)$
   b. $\left(\text{cis } \frac{7\pi}{3}\right)(2 \text{ cis } 2\pi)$
   c. $(5 \text{ cis } 47^\circ)(10 \text{ cis } 108^\circ)$
   d. $(6 \text{ cis } 98^\circ)\left(\text{cis } \frac{\pi}{2}\right)$
   e. $[2 \text{ cis } (-30)^\circ]\ (\text{cis } 225^\circ)$
   f. $(2+3i)(1-i)$
   g. $(-3+8i)(10+3i)$
   h. $(1+i)(1-i)(i)$
   i. $(7-10i)(-4+3i)$
   j. $(1+2i)(2-i)(-1-i)$

2. Find each quotient.
   a. $\dfrac{3 \text{ cis } \frac{\pi}{4}}{2 \text{ cis } \frac{\pi}{3}}$
   b. $\dfrac{4 \text{ cis } 32^\circ}{6 \text{ cis } 1062^\circ}$
   c. $\dfrac{4 \text{ cis } \frac{5\pi}{3}}{5 \text{ cis } \frac{7\pi}{2}}$
   d. $\dfrac{2 \text{ cis } 70^\circ}{3 \text{ cis } 30^\circ}$
   e. $\dfrac{10 \text{ cis } 100^\circ}{5 \text{ cis } 50^\circ}$
   f. $\dfrac{7+i}{8+3i}$
   g. $\dfrac{1-i}{2-3i}$
   h. $\dfrac{4+3i}{-1-2i}$
   i. $\dfrac{3-2i}{-4+i}$
   j. $\dfrac{6+i}{6-i}$

3. Evaluate each power.
   a. $\left(2 \text{ cis } \frac{\pi}{4}\right)^2$
   b. $(1+\sqrt{3}i)^3$
   c. $\left(\frac{1}{2} \text{ cis } \frac{\pi}{6}\right)^4$
   d. $(2-i)^4$
   e. $(1+i)^3$
   f. $\left(3 \text{ cis } \frac{7\pi}{2}\right)^2$
   g. $(-3+4i)^3$
   h. $\left(\text{cis } \frac{5\pi}{4}\right)^3$

4. Evaluate each power.

a. $\left(\text{cis}\,\frac{5\pi}{3}\right)^8$ b. $(1-i)^5$ c. $(\text{cis}\,42°)^6$ d. $(\sqrt{2}\,\text{cis}\,18°)^4$

5. If the set of complex numbers is the replacement set for each variable, find the solution of each equation.

a. $2iu - 3 = 0$ b. $z^{1/2} - \sqrt{3} = i$

c. $\left(5\,\text{cis}\,\frac{7\pi}{3}\right)z - i = 0$ d. $(\sqrt{2} - \sqrt{2}i)u = 5\,\text{cis}\,\frac{\pi}{4}$

15-4 ROOTS OF A COMPLEX NUMBER

A complex number z is called an nth root of the complex number u if

$$z^n = u.$$

Verify that $\frac{1}{2} + \frac{\sqrt{3}}{2}i$, cis π, and $\frac{1}{2} - \frac{\sqrt{3}}{2}i$ are cube roots of -1.

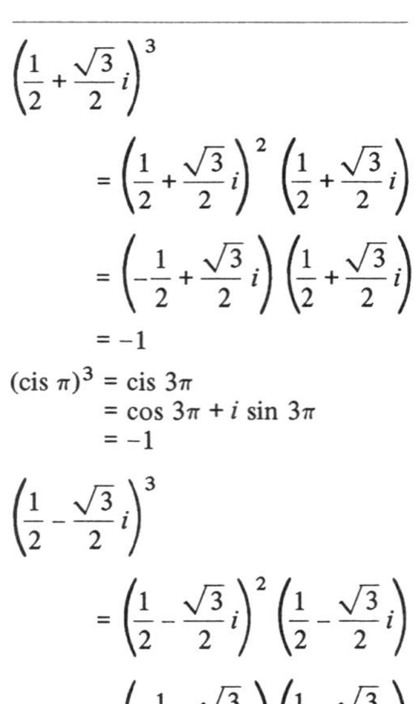

$$\left(\frac{1}{2}+\frac{\sqrt{3}}{2}i\right)^3 = \left(\frac{1}{2}+\frac{\sqrt{3}}{2}i\right)^2\left(\frac{1}{2}+\frac{\sqrt{3}}{2}i\right) = \left(-\frac{1}{2}+\frac{\sqrt{3}}{2}i\right)\left(\frac{1}{2}+\frac{\sqrt{3}}{2}i\right) = -1$$

$$(\text{cis}\,\pi)^3 = \text{cis}\,3\pi = \cos 3\pi + i\sin 3\pi = -1$$

$$\left(\frac{1}{2}-\frac{\sqrt{3}}{2}i\right)^3 = \left(\frac{1}{2}-\frac{\sqrt{3}}{2}i\right)^2\left(\frac{1}{2}-\frac{\sqrt{3}}{2}i\right) = \left(-\frac{1}{2}-\frac{\sqrt{3}}{2}i\right)\left(\frac{1}{2}-\frac{\sqrt{3}}{2}i\right) = -1$$

Consider the complex number r cis θ, where $0 \leqslant \theta < 2\pi$ and recall that

$$r\,\text{cis}\,\theta = r\,\text{cis}\,(\theta + 2\pi k)$$

for any integer k. This fact together with De Moivre's theorem provides the rationale for the statement that

$$\left[r^{1/n}\,\text{cis}\,\frac{\theta + 2\pi k}{n}\right]^n = r\,\text{cis}\,(\theta + 2\pi k) = r\,\text{cis}\,\theta$$

is a true sentence for any integers k and $n \neq 0$. Thus, any of the complex numbers

$$r^{1/n}\,\text{cis}\,\frac{\theta + 2\pi k}{n}$$

where k is any integer, is an nth root of r cis θ. It can be shown that all of the nth roots of r cis θ are obtained by substituting the numbers $0, 1, 2, \ldots, n-1$ for k. Thus, there are exactly n, nth roots of a complex number r cis θ. These remarks are summarized in the statement of the next theorem.

Theorem For any complex number $r \text{ cis } \theta$, the complex numbers

$$r^{1/n} \text{ cis } \frac{\theta + 2\pi k}{n}, \text{ where } k \in \{0, 1, 2, \ldots, n-1\}$$

are the n, nth roots of $r \text{ cis } \theta$.

Example 1 Find the trigonometric and rectangular forms of the four 4th roots of 1.

Solution: The trigonometric form of the complex number $1 + 0i$ is $1 \text{ cis } 0$. Thus, according to the Theorem, the four 4th roots of 1 are

$$1^{1/4} \text{ cis } \frac{0 + 2\pi k}{4} \quad \text{where } k \in \{0, 1, 2, 3\}.$$

That is, when

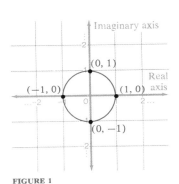

$$k = 0: \quad 1^{1/4} \text{ cis } \frac{0 + 2\pi(0)}{4} = 1 \text{ cis } 0 = 1\,(\cos 0 + i \sin 0) = 1 + 0i$$

$$k = 1: \quad 1^{1/4} \text{ cis } \frac{0 + 2\pi(1)}{4} = 1 \text{ cis } \frac{\pi}{2} = 1\left(\cos \frac{\pi}{2} + i \sin \frac{\pi}{2}\right) = 0 + i$$

$$k = 2: \quad 1^{1/4} \text{ cis } \frac{0 + 2\pi(2)}{4} = 1 \text{ cis } \pi = 1\,(\cos \pi + i \sin \pi) = -1 + 0i$$

$$k = 3: \quad 1^{1/4} \text{ cis } \frac{0 + 2\pi(3)}{4} = 1 \text{ cis } \frac{3\pi}{2} = 1\left(\cos \frac{3\pi}{2} + i \sin \frac{3\pi}{2}\right) = 0 - i.$$

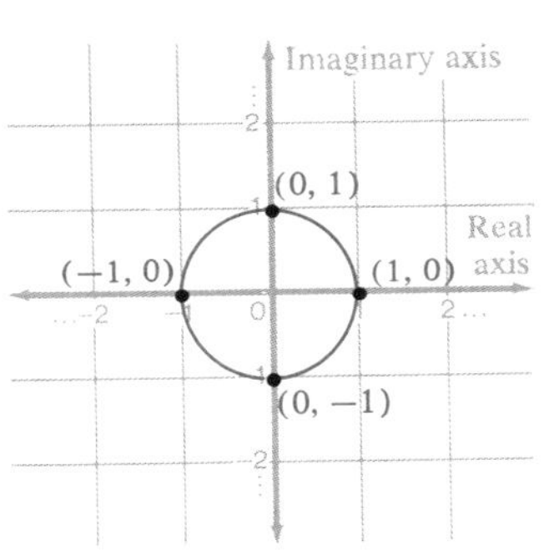

FIGURE 1

Geometrically, the 4th roots of 1 correspond to four equally spaced points on a circle with center at the origin and radius of length 1 [Figure 1].

Example 2 Find the trigonometric and rectangular forms of the four 4th roots of $-2 - 2\sqrt{3}\,i$.

Solution: The trigonometric form of $-2 - 2\sqrt{3}i$ is $4 \text{ cis } \frac{4\pi}{3}$. From the above Theorem, the four 4th roots of $-2 - 2\sqrt{3}i$ are

$$4^{1/4} \text{ cis } \frac{\frac{4\pi}{3} + 2\pi k}{4} \quad \text{where } k \in \{0, 1, 2, 3\}.$$

Since $4^{1/4} = \sqrt{2}$, the four 4th roots of $-2 - 2\sqrt{3}i$ are as follows:

$$k = 0: \quad \sqrt{2} \text{ cis } \frac{4\pi}{12} = \sqrt{2}\left(\cos \frac{\pi}{3} + i \sin \frac{\pi}{3}\right) = \frac{\sqrt{2}}{2} + \frac{\sqrt{6}}{2} i$$

$$k = 1: \quad \sqrt{2} \text{ cis } \frac{10\pi}{12} = \sqrt{2}\left(\cos \frac{5\pi}{6} + i \sin \frac{5\pi}{6}\right) = -\frac{\sqrt{6}}{2} + \frac{\sqrt{2}}{2} i$$

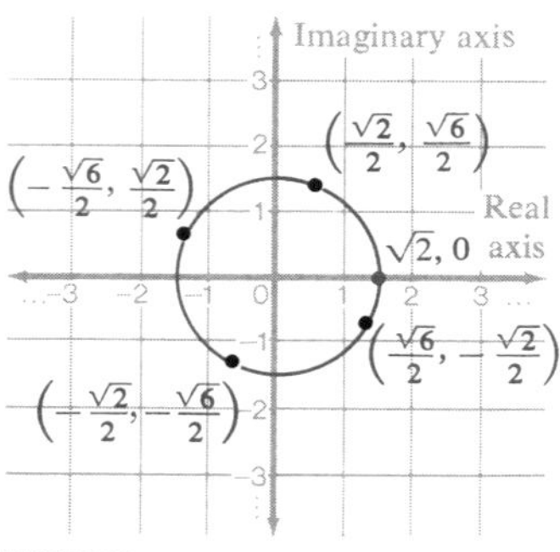

FIGURE 2

$$k = 2: \quad \sqrt{2} \text{ cis } \frac{16\pi}{12} = \sqrt{2}\left(\cos\frac{4\pi}{3} + i \sin\frac{4\pi}{3}\right) = -\frac{\sqrt{2}}{2} - \frac{\sqrt{6}}{2} i$$

$$k = 3: \quad \sqrt{2} \text{ cis } \frac{22\pi}{12} = \sqrt{2}\left(\cos\frac{11\pi}{6} + i \sin\frac{11\pi}{6}\right) = \frac{\sqrt{6}}{2} - \frac{\sqrt{2}}{2} i$$

Observe that the points corresponding to the four 4th roots of $-2 - 2\sqrt{3}i$ [Figure 2] are equally spaced on a circle with center at the origin and radius of length

$$|-2 - 2\sqrt{3}i|^{1/4} = \sqrt{2}$$

Find the trigonometric forms of the four 4th roots of $2 + 2\sqrt{3}i$; of $-2 + 2\sqrt{3}i$; of $2 - 2\sqrt{3}i$.

Fourth roots of $2 + 2\sqrt{3}\, i = 4 \text{ cis } \frac{\pi}{3}$:

$\sqrt{2} \text{ cis } \frac{\pi}{12}$ $\quad \sqrt{2} \text{ cis } \frac{13\pi}{12}$

$\sqrt{2} \text{ cis } \frac{7\pi}{12}$ $\quad \sqrt{2} \text{ cis } \frac{19\pi}{12}$

- Fourth roots of $-2 + 2\sqrt{3}\, i = 4 \text{ cis } \frac{2\pi}{3}$:

$\sqrt{2} \text{ cis } \frac{\pi}{6}$ $\quad \sqrt{2} \text{ cis } \frac{7\pi}{6}$

$\sqrt{2} \text{ cis } \frac{2\pi}{3}$ $\quad \sqrt{2} \text{ cis } \frac{5\pi}{3}$

- Fourth roots of $2 - 2\sqrt{3}\, i = 4 \text{ cis } \frac{5\pi}{3}$:

$\sqrt{2} \text{ cis } \frac{5\pi}{12}$ $\quad \sqrt{2} \text{ cis } \frac{17\pi}{12}$

$\sqrt{2} \text{ cis } \frac{11\pi}{12}$ $\quad \sqrt{2} \text{ cis } \frac{23\pi}{12}$

In general, the geometric representation of the nth roots of a complex number r cis θ is a set of n equally spaced points on a circle with radius of length $r^{1/n}$. The points also lie on the terminal sides of angles in standard position with measures

$$\frac{\theta}{n}, \frac{\theta}{n} + \frac{2\pi}{n}, \frac{\theta}{n} + 2\left(\frac{2\pi}{n}\right), \ldots, \frac{\theta}{n} + (n-1)\left(\frac{2\pi}{n}\right), \text{ respectively.}$$

Exercises

1. Find the n, nth roots of each complex number and graph the corresponding points.
   a. Cube roots of 1
   b. Square roots of $-7 + 24i$
   c. Fourth roots of i
   d. Cube roots of $-1 - i$
   e. Square roots of -1
   f. Cube roots of $2 - 2i$
   g. Fourth roots of 16
   h. Cube roots of 27 cis π
   i. Sixth roots of 1
   j. Fifth roots of $-1 - 3i$
2. If $z \in C$, find the solution set of each equation.
   a. $z^4 - 1 - \sqrt{3}\, i = 0$
   b. $z^2 + i = 0$
3. If $z \in C$, find the solution set of each equation.
   a. $z^3 = -1 + \sqrt{3}\, i$
   b. $z^2 - 1 = i$
   c. $z^2 - i = 0$

Review Exercises

1. Give two sets of polar coordinates for the point with rectangular coordinates indicated by $(2, -2)$. **15-1**

2. Find the rectangular coordinates for $\left(2, -\frac{\pi}{3}\right)$ and $(3, 210^\circ)$. **15-2**

3. Write the trigonometric form of $-5 + 5\sqrt{3}\, i$.

4. Change $4 \text{ cis } 72^\circ 10'$ and $5 \text{ cis } 2$ to rectangular form.

5. Write $(-1 - i)(2 \text{ cis } 3\pi)$ in trigonometric and rectangular form. **15-3**

6. Find the rectangular form of $2 \text{ cis } \left(\frac{9\pi}{2}\right) \div (-2 - 2\sqrt{2}\, i)$.

7. State De Moivre's theorem for $a \text{ cis } \alpha$ raised to the bth power.

8. Evaluate $\left(6 \text{ cis } \frac{11\pi}{6}\right)^3$ and $(-2\sqrt{3} + 2i)^6$.

9. If $z \in C$, what is the solution set of $\left(2 \text{ cis } \frac{13\pi}{6}\right) z - 4\sqrt{3}\, i = 4$?

10. Find the four 4th roots of $-i$ and graph the corresponding points. **15-4**

11. If $z \in C$, find the solution set of each equation.
    a. $z^{1/4} + i = 0$
    b. $z^3 - 27 \text{ cis } 90^\circ = 0$

16. SEQUENCES AND SERIES

16-1 FINITE SEQUENCES

In this chapter, we consider a special class of functions where the domain of each is the set of positive integers or the first n positive integers.

Consider the functions

$$f = \{(1, 3), (2, 5), (3, 7), (4, 9), (5, 11), (6, 13), (7, 15)\}$$

$$g = \{(1, 3), (2, 6), (3, 12), (4, 24), (5, 48)\}$$

$$h = \left\{\left(1, \frac{1}{2}\right), \left(2, -\frac{1}{4}\right), \left(3, \frac{1}{8}\right), \left(4, -\frac{1}{16}\right), \left(5, \frac{1}{32}\right), \left(6, -\frac{1}{64}\right)\right\}.$$

Their domains are the positive integers from 1 to n, where n is 7, 5, and 6, respectively. One method of denoting these functions involves listing the order of the elements in the domain:

$$f\text{: } 3, 5, 7, 9, 11, 13, 15$$

$$g\text{: } 3, 6, 12, 24, 48$$

$$h\text{: } \frac{1}{2}, -\frac{1}{4}, \frac{1}{8}, -\frac{1}{16}, \frac{1}{32}, -\frac{1}{64}.$$

The functions f, g, and h are called finite sequences.

Definition A function is a **finite sequence** if and only if its domain is a set of positive integers from 1 to n.

Use ordered-pair notation to denote the finite sequence

$$j\text{: } 4, 9, 14, 19, 24, 29, 34, 39.$$

$j = \{(1, 4), (2, 9), (3, 14), (4, 19)$ $(5, 24), (6, 29), (7, 34), (8, 39)\}$

If a is a finite sequence, we often show only the elements of the range:

$$a = \{a(1),\ a(2),\ a(3), \ldots,\ a(n)\}$$

It is customary to use a special notation for sequences. In this notation, the ith term is indexed by the subscript i. The sequence a above would appear as

$$a_1, a_2, a_3, \ldots, a_n$$

It is also customary for a_1 to be called the **first term**; a_2 the **second term**; and a_i, the ***i*th term** of the finite sequence.

Finite sequence	*Term* 1st	3rd	5th
f	3	7	11
g	3	12	48
h	$\frac{1}{2}$	$\frac{1}{8}$	$\frac{1}{32}$
j	4	14	24

What are the 1st, 3rd, and 5th terms of f, g, h, and j?

The finite sequence f: 3, 5, 7, 9, 11, 13, 15 possesses a simple pattern. Beginning with the number 3, we add 2 to a term to derive the next term of f. Thus, the difference of any two successive terms is 2.

Definition A finite sequence is an **arithmetic progression** (A.P.) if and only if the difference of two successive terms is the same number in all cases. This number is called the **common difference.**

If the finite sequence a: $a_1, a_2, a_3, \ldots, a_n$ is an arithmetic progression with common difference d, then

$$a_1 = a_1,\ a_2 = a_1 + d,\ a_3 = a_2 + d, \ldots, a_n = a_{n-1} + d$$

$$a_1 = a_1,\ a_2 = a_1 + d,\ a_3 = a_1 + 2d, \ldots,\ a_n = a_1 + (n - 1)\,d.$$

All the members of this sequence can be expressed in terms of a_1 and d.

Which of the finite sequences f, g, and h are arithmetic progressions? In the A.P. with first term 4, common difference 3, and 45 terms, what is the 2nd term? 10th term? 35th term? Last (45th) term?

The finite sequence f is an A.P.;
- $a_2 = 4 + 3 = 7$;
- $a_{10} = 4 + 9(3) = 31$;
- $a_{35} = 4 + 34(3) = 106$;
- $a_{45} = 4 + 44(3) = 136$.

The finite sequence g: 3, 6, 12, 24, 48 possesses a pattern but it is not an arithmetic progression. Beginning with the number 3, we double a term to derive the next term of g. Thus, the quotient of any two successive terms is 2.

Definition A finite sequence is a **geometric progression** (G.P.) if and only if the quotient of two successive terms is the same number in all cases. This number is called the **common ratio.**

If the finite sequence a: $a_1, a_2, a_3, \ldots, a_n$ is a geometric progression with common ratio r, then

$$a_1 = a_1,\ a_2 = a_1 r,\ a_3 = a_2 r, \ldots,\ a_n = a_{n-1} r$$

Substituting, we get

$$a_1 = a_1,\ a_2 = a_1 r,\ a_3 = a_1 r^2, \ldots,\ a_n = a_1 r^{n-1}$$

which expresses the sequence only in terms of a and r.

The finite sequence h on page 344 is a geometric progression. What is the common ratio?

The common ratio for h is

$$\frac{-\frac{1}{4}}{\frac{1}{2}} = -\frac{1}{2}.$$

The sum of a finite sequence is simply the sum of its terms. For example, the sums of f, g, and h below are 63, 93, and $\frac{21}{64}$, respectively.

f: 3, 5, 7, 9, 11, 13, 15

g: 3, 6, 12, 24, 48

h: $\frac{1}{2}, -\frac{1}{4}, \frac{1}{8}, -\frac{1}{16}, \frac{1}{32}, -\frac{1}{64}$

The sum of the finite sequence a: $a_1, a_2, a_3, \ldots, a_n$ is denoted by

$$\sum_{i=1}^{n} a_i$$

where $\sum$ (read *sigma*) denotes addition and i takes on all integral values from 1 to n. The finite sequence f can be represented by

$$3 + 0 \cdot 2,\ 3 + 1 \cdot 2,\ 3 + 2 \cdot 2,\ 3 + 3 \cdot 2,$$

$$3 + 4 \cdot 2,\ 3 + 5 \cdot 2,\ 3 + 6 \cdot 2$$

and its sum by

$$\sum_{i=1}^{7} [3 + (i - 1) \cdot 2]$$

Similarly, the sums of the finite sequences g and h can be denoted by

$$\sum_{i=1}^{5} 3 \cdot 2^{i-1} \quad \text{and} \quad \sum_{i=1}^{6} \frac{1}{2}\left(-\frac{1}{2}\right)^{i-1}$$

respectively.

Sums of finite sequences which are either arithmetic progressions or geometric progressions can be computed by means other than listing all of the terms and computing the sum. For example, if S is the sum of the arithmetic progression f, we find that

$$S = 3 + 5 + 7 + 9 + 11 + 13 + 15.$$

Now consider the same sequence with the terms listed in reverse order:

$$S = 15 + 13 + 11 + 9 + 7 + 5 + 3.$$

Adding the two sequences, we obtain

$$2S = 7 \cdot (3 + 15)$$

$$S = \frac{7}{2} \cdot (3 + 15)$$

That is, S is equal to the product of half the number of terms and the sum of the first and last terms. Theorem 1 generalizes this result.

Theorem 1 If $a_1, a_2, \ldots, a_n$ is an arithmetic progression with sum S, then

$$S = \sum_{i=1}^{n} a_i = \frac{n}{2}(a_1 + a_n).$$

What is the sum of the first 50 positive integers? Find the sum of the even integers between 100 and 300, inclusive.

$\sum_{i=1}^{50} i = \frac{50}{2}(1 + 50) = 1275;$ ● If

$$300 = 100 + (n - 1)2$$

then $n = 101$ and

$$\sum_{i=50}^{150} 2i = \frac{101}{2}(100 + 300)$$

$$= 20{,}200.$$

For the sum S of a geometric progression with n terms, common ratio r, and first term a_1, we note that

$$S = a_1 + a_1 r + a_1 r^2 + \cdots + a_1 r^{n-1}$$

Multiplying on the right by r and subtracting, we obtain

$$S - rS = (a_1 + a_1 r + \cdots + a_1 r^{n-1})$$

$$- (a_1 r + a_1 r^2 + \cdots + a_1 r^{n-1} + a_1 r^n)$$

$$= a_1 - a_1 r^n$$

as the difference between the first term and r times the last term. Thus

$$S(1 - r) = a_1 - a_1 r^n$$

If $r \neq 1$, we can solve for S. We state this result as Theorem 2.

Theorem 2 If $a_1, a_2, \ldots, a_n$ is a geometric progression with common ratio $r \neq 1$ and sum S, then

$$S = \sum_{i=1}^{n} a_i = \frac{a_1 - a_1 r^n}{1 - r} = \frac{a_1 - a_n r}{1 - r}.$$

Applying Theorem 2 to the geometric progression h, we find that its sum is

$$S = \frac{\frac{1}{2} - \frac{1}{2}\left(-\frac{1}{2}\right)^6}{1 - \left(-\frac{1}{2}\right)} = \frac{21}{64}.$$

$S = \dfrac{3 - 48 \cdot 2}{1 - 2} = 93$

Use Theorem 2 to find the sum of the geometric progression g on page 346.

Exercises

Refer to the following finite sequences for Exercises 1-4.

$f_1 = (1, 1), (2, 1), (3, 1), (4, 1), (5, 1)$ f_2: 3, 1, $\frac{1}{3}, \frac{1}{9}, \frac{1}{27}, \frac{1}{81}, \frac{1}{243}$

f_3: 10, 33, 56, 79, 102, 125 f_4: 2, 5, 4, 12, 1, 1

1. State the domain for each function.
2. Give the 1st, 2nd, 4th, and 5th terms of each finite sequence.
3. Which functions are arithmetic progressions? What is the common difference in each case?
4. Which functions are geometric progressions? What is the common ratio in each case?

In Exercises 5-17, the notation a_i, $i \in N$, refers to a term of the finite sequence a: $a_1, a_2, a_3, \ldots, a_n$.

5. Find the 8th term of the A.P., where $a_1 = 3$ and $a_2 = 10$.
6. Find the 50th term of the A.P., where $a_1 = \pi$ and $a_2 = -\frac{\pi}{2}$.
7. Find the 1st and 10th terms of the A.P., where $a_4 = \frac{1}{4}$ and $a_5 = \frac{1}{3}$.
8. Find the first four terms of the A.P., if $a_5 = 3\sqrt{3}$ and $a_7 = 6\sqrt{3}$.
9. Find the 7th term of the G.P., where $a_1 = \sqrt{2}$ and $a_2 = -\sqrt{6}$.
10. Find the 10th term of the G.P., where $a_1 = \sin\frac{\pi}{6}$ and $a_2 = \sin\frac{\pi}{2}$.
11. Find the 1st and 6th terms of the G.P., where $a_3 = \frac{1}{\pi}$ and $a_4 = \frac{2}{\pi}$.

12. Find the first three terms of the G.P., where $a_4 = 3$ and $a_7 = -24$.

13. In an A.P., if $a_1 = 3$ and $a_{50} = -11$, what is the common difference d?

14. In a G.P., if $a_1 = 16$ and $a_4 = -2$, what is the common ratio r?

15. Compute the sum of each finite sequence.
   a. The A.P. having 8 terms with first three terms 3, 10, 17.
   b. The A.P. having 20 terms with first three terms 3, 7, 11.
   c. The A.P. having 31 terms with first three terms 99, 90, 81.
   d. The A.P. having 50 terms with first three terms π, $-\frac{\pi}{2}$, -2π.
   e. The A.P. having 10 terms with fourth term $\frac{1}{4}$ and fifth term $\frac{1}{3}$.
   f. The A.P. having 20 terms with fifth term $3\sqrt{3}$, seventh term $6\sqrt{3}$.
   g. The G.P. having 4 terms with first two terms $\sqrt{2}$ and $-\sqrt{6}$.
   h. The G.P. having 10 terms with first two terms $\sin\frac{\pi}{6}$ and $\sin\frac{\pi}{2}$.
   i. The G.P. having 10 terms with first three terms 2, 4, 8.
   j. The G.P. having 8 terms with first three terms 2, $-\frac{2}{3}$, $\frac{2}{9}$.
   k. The G.P. having 6 terms with third term $\frac{1}{\pi}$ and fourth term $\frac{2}{\pi}$.
   l. The G.P. having 8 terms with fourth term 3 and seventh term -24.

16. In this exercise, a: $a_1, a_2, \ldots, a_n$ is an arithmetic progression with sum S and common difference d. For each arithmetic progression, find the missing two numbers from the list a_1, a_n, d, S, and n.
   a. $a_1 = 4$, $d = 5$, $n = 16$
   b. $a_1 = -3$, $n = 11$, $a_n = 67$
   c. $d = -\frac{1}{3}$, $n = 13$, $S = 4$
   d. $a_1 = 12$, $d = -\frac{1}{2}$, $a_n = \frac{3}{2}$
   e. $a_1 = 10$, $d = -3$, $S = 7$
   f. $a_1 = 5$, $d = 2$, $S = 165$

17. In this exercise, a: a: $a_1, a_2, \ldots, a_n$ is a geometric progression with sum S and common ratio r. For each geometric progression find the missing two numbers from the list a_1, a_n, r, S, and n.
   a. $a_1 = 6$, $r = -3$, $S = -120$
   b. $a_1 = 3$, $r = -2$, $a_n = 3072$
   c. $a_1 = \frac{1}{2}$, $r = 3$, $n = 6$
   d. $a_1 = 1$, $n = 6$, $a_n = -1$
   e. $r = \frac{1}{2}$, $n = 5$, $S = 62$
   f. $a_1 = 2$, $n = 3$, $S = \frac{3}{2}$

18. Compute each sum.
   a. $\sum_{i=1}^{12} i$
   b. $\sum_{i=1}^{12} i^2$
   c. $\sum_{i=1}^{12} 2i$
   d. $\sum_{i=1}^{12} (i + i^2)$
   e. $\sum_{i=1}^{12} 2$
   f. $\sum_{k=2}^{6} k(k-1)$

19. If a: a_1, a_2, a_3 is an A.P. with three terms, then a_2 is called the **arithmetic mean between a_1 and a_3.**
    a. Find the arithmetic mean between 87 and 91; $-\frac{3}{2}$ and $\frac{7}{3}$.
    b. Show that $a_2 = (a_1 + a_3) \div 2$.
20. If a: $a_1, a_2, a_3, \ldots, a_{n-1}, a_n$ is an A.P., then the terms

    $$a_2, a_3, \ldots, a_{n-1}$$

    are called **arithmetic means between a_1 and a_n.**
    a. Insert three arithmetic means between 1 and 9.
    b. Insert four arithmetic means between 4 and -7.
21. If a: a_1, a_2, a_3 is a G.P. with three terms, then a_2 is called **the geometric mean between a_1 and a_3.**
    a. Show that $a_2 = \sqrt{a_1 a_3}$ if a_1, a_2, and a_3 are positive real numbers.
    b. Find the geometric mean between 9 and 25; between 1 and 5.
22. If a motorcycle depreciates 25% of its value every year and the original value is \$600.00, compute an 8-year depreciation schedule.
23. A building lot purchased for \$5,000 appreciates 8% of its current value each year for 10 years. At that point what is its value?
24. If \$100.00 is invested at 5% compounded annually, what is the total amount at the end of 6 years?
25. If \$200.00 is invested at 4% compounded quarterly, what is the total amount at the end of 4 years?

16-2 SEQUENCES

We now extend our notion of finite sequence to include all the positive integers in the domain.

Definition A function is a **sequence** if and only if the domain of the function is the set of positive integers.

Thus, a sequence has infinitely many terms. An example of a sequence is the function b below:

$$b = \{(i, 2i): \ i \text{ a positive integer}\}$$

As is the case for finite sequences, there are several methods for denoting sequences. For example, b can be denoted by

$$2, 4, 6, \ldots, 2m, \ldots$$

or: $b_1, b_2, b_3, \ldots, b_m, \ldots$ where $b_m = 2m$

Since there are infinitely many terms in a sequence, a general term must be stated explicitly in order that the functional value assigned to each positive integer is clearly understood.

The tenth terms of the sequences

$$u\colon\ 1, \frac{1}{2}, \frac{1}{3}, \frac{1}{4}, \ldots, \frac{1}{m}, \ldots$$

$$v\colon\ 1, \frac{1}{2}, \frac{1}{4}, \frac{1}{8}, \ldots, \frac{1}{2^{m-1}}, \ldots$$

$$w\colon\ -1, 1, -1, 1, \ldots, (-1)^m, \ldots$$

are $\frac{1}{10}$, $\frac{1}{2^{10-1}} = \frac{1}{512}$, and $(-1)^{10} = 1$, respectively.

	Term		
Sequence	*5th*	*7th*	*12th*
u	$\frac{1}{5}$	$\frac{1}{7}$	$\frac{1}{12}$
v	$\frac{1}{16}$	$\frac{1}{64}$	$\frac{1}{2048}$
w	-1	-1	1

Find the 5th, 7th, and 12th terms of the sequences u, v, and w.

If we examine the sequence u, we find that as the number of the term increases, the terms get closer and closer to the number 0. The distance between the origin on the real number line and the points for the terms of the sequence becomes very small as m increases. For example, the distance between $\frac{1}{m}$, where $m > 100$ and 0 is less than 0.01:

$$\left|\frac{1}{m} - 0\right| < 0.01 \qquad (m > 100)$$

For every term $\frac{1}{m}$, where $m > 1{,}000{,}000$, we have

$$\left|\frac{1}{m} - 0\right| < 0.000001 \qquad (m > 10^6)$$

The graph of function u [Figure 1] indicates that the functional values approximate 0 for large integers in the domain.

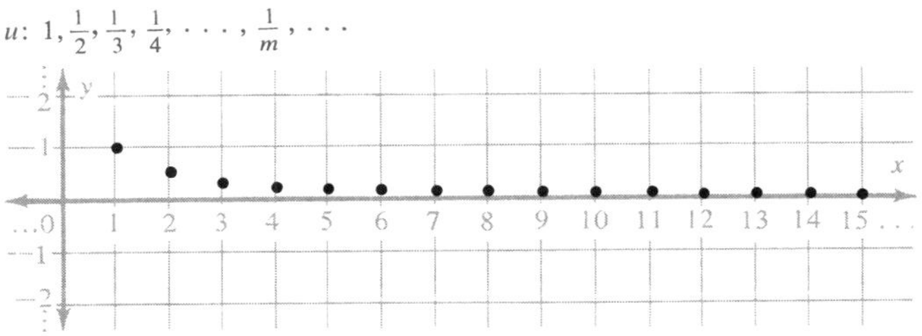

FIGURE 1

Suppose we want to find a term of u that will be within a certain distance d of the number 0. Which term will meet this requirement? We only need to let m be greater than $\frac{1}{d}$, since $m > \frac{1}{d}$ implies $d > \frac{1}{m}$ and

$$\left|\frac{1}{m} - 0\right| = \frac{1}{m} < d.$$

So all terms $\frac{1}{m}$ where $m > d$ meet the requirement. If m is large enough, we can get as close as we like to 0. Since we are able to approximate 0 with terms of the sequence in this manner, we say that the sequence has a limit and its limit is 0.

Definition A sequence a: $a_1, a_2, a_3, \ldots, a_m, \ldots$ **has a limit r** if and only if for every positive number d (**degree of accuracy**), there is a real number M such that

$$|a_m - r| < d$$

for every integer $m > M$.

If the number $|a_m - r|$ is less than d, then the distance between the points for a_m and r is less than d.

The dots in Figure 2 illustrate a sequence with limit r. For a specified positive real number d, all points of the graph lie between the lines for $y = r + d$ and $y = r - d$ when integers in the domain are larger than M.

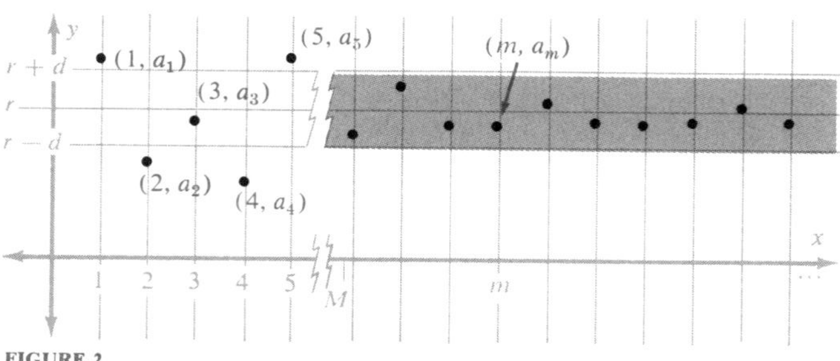

FIGURE 2

Some sequences do not have limits. The sequences

$$a\text{: } 2, 4, 6, 8, \ldots, 2m, \ldots$$

has the property that for every real number r which you could propose as a limit, if $m > \frac{r}{2}$ then $2m > r$, and the distance $|2m - r|$ becomes quite large for integers m considerably larger than $\frac{r}{2}$.

Exercises

1. For each sequence, find the 5th, 8th, 10th, jth, $(j-1)$st, and $(j+1)$st terms.
   a. $3, 6, 9, 12, \ldots, 3m, \ldots$
   b. $\frac{1}{3}, -\frac{1}{6}, \frac{1}{9}, -\frac{1}{12}, \ldots, \frac{(-1)^m}{3m}, \ldots$
   c. $1, 1 \cdot 2, 1 \cdot 2 \cdot 3, 1 \cdot 2 \cdot 3 \cdot 4, \ldots, m!, \ldots$ (where $m!$ is read "m factorial" and $m! = 1 \cdot 2 \cdot 3 \cdots (m-1) \cdot m$.)
2. Which sequences in Exercise 1 do you think have limits?
3. For each sequence find the 5th, 7th, 10th, kth, $(k-1)$st, and $(k+1)$st terms.
   a. $\frac{1}{2}, 1, \frac{3}{2}, 2, \ldots, \frac{m}{2}, \ldots$
   b. $0, \log_{10} 2, \log_{10} 3, \log_{10} 4, \ldots, \log_{10} m, \ldots$
   c. $1, 4, 9, 16, \ldots, m^2, \ldots$
   d. $-1, \frac{1}{4}, -\frac{1}{9}, \frac{1}{16}, \ldots, \frac{(-1)^m}{m^2}, \ldots$
   e. $2, 0, 2, 0, \ldots, [1 + (-1)^{m+1}], \ldots$
   f. $\frac{1}{2}, \frac{2}{3}, \frac{3}{4}, \frac{4}{5}, \ldots, \frac{m}{m+1}, \ldots$
   g. $1, 0, -1, 0, \ldots, \sin \frac{m\pi}{2}, \ldots$
   h. $0.3, 0.33, 0.333, 0.3333, \ldots \sum_{i=1}^{m} \frac{3}{10^i}, \ldots$
   i. $a_1, a_2, a_3, a_4, \ldots, a_m, \ldots$ where $a_m = \pi$ for every positive integer m
   j. $1, \frac{1}{2}, \frac{1}{6}, \frac{1}{24}, \ldots, \frac{1}{m!}, \ldots$
   k. $1, \sqrt{2}, \sqrt{3}, 2, \ldots, \sqrt{m}, \ldots$
4. How many elements are in the range of each sequence in Exercise 3?
5. Which terms of the sequence in Exercise 3a are larger than 100? Than 1000? Than 1,000,000? Than the real number M?
6. Perform Exercise 5 with the sequence in Exercise 3b.
7. Perform Exercise 5 with the sequence in Exercise 3c.
8. Perform Exercise 5 with the sequence in Exercise 3k.
9. Which sequences in Exercise 3 seem to have limits?

16-3 INFINITE GEOMETRIC SERIES

The number $\frac{1}{3}$ can be approximated by each of the following:

$$s_1 = 0.3 = \frac{3}{10}$$

$$s_2 = 0.33 = \frac{3}{10} + \frac{3}{100}$$

$$s_3 = 0.333 = \frac{3}{10} + \frac{3}{100} + \frac{3}{1000}$$

$$\cdots$$

$$s_m = \frac{3}{10} + \frac{3}{100} + \frac{3}{1000} + \cdots + \frac{3}{10^m} = \sum_{i=1}^{m} \frac{3}{10^i}$$

Each approximation is a term in the sequence

$$s\colon\ s_1, s_2, s_3, \ldots, s_m, \ldots$$

and this sequence has $\frac{1}{3}$ as its limit. Notice that each term of this sequence is the sum of a geometric progression. The term s_m is the sum of the finite sequence

$$\frac{3}{10}, \frac{3}{10}\left(\frac{1}{10}\right), \frac{3}{10}\left(\frac{1}{10}\right)^2, \ldots, \frac{3}{10}\left(\frac{1}{10}\right)^m$$

which is a geometric progression having first term $\frac{3}{10}$ and common ratio $\frac{1}{10}$. Thus, we have a sequence of sums of geometric progressions

$$\sum_{i=1}^{1} \frac{3}{10}\left(\frac{1}{10}\right)^{i-1}, \quad \sum_{i=1}^{2}\left(\frac{3}{10}\ \frac{1}{10}\right)^{i-1},$$

$$\sum_{i=1}^{3} \frac{3}{10}\left(\frac{1}{10}\right)^{i-1}, \ldots, \quad \sum_{i=1}^{m} \frac{3}{10}\left(\frac{1}{10}\right)^{i-1}, \ldots$$

which has a limit $\frac{1}{3}$. The sequence s illustrates the concept of an infinite geometric series.

Definition An **infinite geometric series** is a sequence of sums of geometric progressions with the identical first term and common ratio such that each successive geometric progression has one more term than the preceding geometric progression and the first geometric progression has one term.

Let $s_1, s_2, s_3, s_4, \ldots, s_m, \ldots$ below be the terms of an infinite geometric series (that is, each term is the sum of a geometric progression) where a is the first term and r is the common ratio of each geometric progression. $\left(\text{In our previous example, } a = \frac{3}{10} \text{ and } r = \frac{1}{10}.\right)$

$$s_1 = a$$

$$s_2 = a + ar$$

$$s_3 = a + ar + ar^2$$

$$s_m = a + ar + ar^2 + \cdots + ar^{m-1}$$

If the infinite geometric series $s_1, s_2, s_3, \ldots, s_m, \ldots$ has a limit, we denote this limit by $\sum_{i=1}^{\infty} ar^{i-1}$ or

$$a + ar + ar^2 + \cdots + ar^{m-1} + \cdots .$$

Hence

$$\frac{3}{10} + \frac{3}{100} + \frac{3}{1000} + \cdots + \frac{3}{10^m} + \cdots$$

is the number $\frac{1}{3}$; or in its usual form, 0.333 . . . is the limit $\frac{1}{3}$. Notice that the notation $a + ar + ar^2 + \cdots + ar^{m-1} + \cdots$ denotes a number which is defined as the limit of the sequence

$$a,\ a + ar,\ a + ar + ar^2, \ldots,\ a + ar + ar^2 + \cdots + \cdots + ar^{m-1}, \ldots$$

Not all infinite geometric series have limits. Using the above notation for the geometric series, we notice that the mth geometric progression has a sum

$$a + ar + ar^2 + \cdots + ar^{m-1} = \frac{a - ar^m}{1 - r} = \frac{a}{1 - r} - \frac{ar^m}{1 - r}.$$

If the common ratio r is restricted so that $-1 < r < 1$, then for large integers m, r^m is an approximation for 0. Thus, if $-1 < r < 1$

$$\frac{a}{1 - r} - \frac{ar^m}{1 - r} \quad \text{is an approximation for} \quad \frac{a}{1 - r}$$

with error

$$\left|\frac{a}{1 - r} - \frac{ar^m}{1 - r} - \frac{a}{1 - r}\right| = \left|\frac{ar^m}{1 - r}\right|$$

For large integers m, this error is "near" 0. Thus an infinite geometric series

$$a,\ a + ar,\ a + ar + ar^2, \ldots,\ a + ar + ar^2 + \cdots + ar^{m-1}, \ldots$$

has a limit $\frac{a}{1-r}$ if $-1 < r < 1$. We write $\sum_{i=1}^{\infty} ar^{i-1} = \frac{a}{1-r}$.

r: $\frac{1}{7}, -\frac{1}{10}$; r^2: $\frac{1}{49}, \frac{1}{100}$; r^3: $\frac{1}{343}, -\frac{1}{1000}$; r^4: $\frac{1}{2401}, \frac{1}{10{,}000}$; r^5: $\frac{1}{16{,}807}, -\frac{1}{100{,}000}$.

Find r^2, r^3, r^4, and r^5 if $r = \frac{1}{7}$; if $r = -\frac{1}{10}$.

It is also true that if the common ratio r for the infinite geometric series satisfies $r \geqslant 1$ or $r \leqslant -1$, then this infinite geometric series does not have a limit, provided $a \neq 0$.

a. This sequence is the sequence of partial sums for the infinite geometric series with $a = 1$ and $r = \frac{1}{2}$. Since $-1 < \frac{1}{2} < 1$, the sequence has a limit which is

$$\frac{a}{1-r} = \frac{1}{1-\frac{1}{2}} = 2.$$

b. This sequence is the sequence of partial sums for the infinite geometric series with $a = 1$ and $r = 2$. Since $|r| \geqslant 1$, the sequence does not have a limit.

Show that the sequence in a has a limit but that the sequence in b does not.

a. $1,\ 1 + \frac{1}{2},\ 1 + \frac{1}{2} + \frac{1}{4}, \ldots,\ 1 + \frac{1}{2} + \frac{1}{4} + \cdots + \frac{1}{2^{m-1}}, \ldots$

b. $1,\ 1 + 2,\ 1 + 2 + 4, \ldots,\ 1 + 2 + 4 + \cdots + 2^{m-1}, \ldots$

The number 0.123123123123123123 . . . refers to the limit of the infinite geometric series

$$\frac{123}{1000},\ \frac{123}{1000} + \frac{123}{1000}\left(\frac{1}{1000}\right), \ldots,\ \sum_{i=1}^{m} \frac{123}{1000}\left(\frac{1}{1000}\right)^{i-1}, \ldots$$

which has first term $\frac{123}{1000}$ and common ratio $\frac{1}{1000}$. Its limit is

$$\frac{a}{1-r} = \frac{\frac{123}{1000}}{1 - \frac{1}{1000}} = \frac{123}{999} = \frac{41}{333}.$$

Hence, $0.123123123123 \cdots = \frac{41}{333}$.

Infinite geometric series are special cases from the class of functions called infinite series. The **infinite series** often denoted by

$$a_1 + a_2 + a_3 + \cdots + a_m + \cdots \quad \text{or} \quad \sum_{i=1}^{\infty} a_i$$

is simply the infinite sequence of partial sums displayed below:

$$s_1 = a_1$$

$$s_2 = a_1 + a_2$$

$$s_3 = a_1 + a_2 + a_3$$

$$\cdots$$

$$s_m = a_1 + a_2 + a_3 + \cdots + a_m = \sum_{i=1}^{m} a_i$$

$$\cdots$$

If this sequence $s_1, s_2, s_3, \ldots, s_m, \ldots$ has a limit S, we say that the infinite series **converges**, that this limit is the "sum" (not a sum of arithmetic) of the series, and that this "sum" is denoted by

$$\sum_{i=1}^{\infty} a_i = S.$$

Our examples have been limited to infinite geometric series. In the exercises, limits for these series are also called "sums."

Exercises

Show that each sequence in Exercises 1-11 is an infinite geometric series. Find the first terms and common ratios. Which sequences have limits? What are these limits or "sums"?

1. $0.4,\ 0.44,\ 0.444,\ 0.4444, \ldots, \sum_{i=1}^{m} (0.4)(0.1)^{i-1}, \ldots$

2. $0.6,\ 0.66,\ 0.666,\ 0.6666, \ldots, \sum_{i=1}^{m} (0.6)(0.1)^{i-1}, \ldots$

3. $1,\ 1+3,\ 1+3+9,\ 1+3+9+27, \ldots, \sum_{i=1}^{m} 3^{i-1}, \ldots$

4. $0.23,\ 0.2323,\ 0.232323,\ 0.23232323, \ldots, \sum_{i=1}^{m} (0.23)(0.01)^{i-1}, \ldots$

5. $1,\ 1+0.1,\ 1+0.1+0.01, \ldots \sum_{i=1}^{m} (0.1)^{i-1}, \ldots$

6. $1, 2, 3, 4, 5, \ldots, \sum_{i=1}^{m} 1(1)^{i-1}, \ldots$

7. $\frac{1}{4}, \frac{3}{4}, 1\frac{3}{4}, 3\frac{3}{4}, \ldots, \sum_{i=1}^{m} \frac{1}{4}(2^{i-1}), \ldots$

8. $100,\ 150,\ 175,\ 187.5, \ldots, \sum_{i=1}^{m} 100(0.5)^{i-1}, \ldots$

9. $3.1,\ 3.131,\ 3.13131,\ 3.1313131, \ldots, \sum_{i=1}^{m} (3.1)(0.01)^{i-1}, \ldots$

10. $1, -1, 3, -5, \ldots, \sum_{i=1}^{m} (-2)^{i-1}, \ldots$

11. $1, \frac{2}{3}, \frac{7}{9}, \frac{20}{27}, \ldots, \sum_{i=1}^{m} \left(-\frac{1}{3}\right)^{i-1}, \ldots$

12. If $a + ar + ar^2 + \cdots + ar^{m-1} + \cdots$ is a convergent infinite geometric series with "sum" S and if c is a real number, show that the series $ca + car + car^2 + \cdots + car^{m-1} + \cdots$ is an infinite geometric series and that its "sum" is cS.

13. Use the result in Exercise 12 and the fact that $0.111 \cdots = \frac{1}{9}$ in deriving each "sum."

    a. $0.222 \cdots = \frac{2}{9}$
    b. $0.333 \cdots = \frac{1}{3}$
    c. $0.444 \cdots = \frac{4}{9}$
    d. $0.666 \cdots = \frac{2}{3}$

14. A ball rebounds two-thirds the distance it is dropped. Suppose it is dropped from a height of 15 meters. How far does it travel before coming to rest? (Assume infinitely many bounces, a convenient idealization of reality, which gives a good approximation of the actual distance.)

15. Each swing of a pendulum bob is 75% as long as the preceding swing. If the first swing is 20 feet long, how far does the bob travel before coming to rest?

16. In another application the pendulum bob's first swing measured 30 inches. If each successive swing were $\frac{3}{5}$ as long as its predecessor, how far would the bob have traveled before coming to rest?

17. Suppose a ball rebounds 0.5 the distance it drops. When dropped from a height of 24 yards, how far does it travel before coming to rest?

16-4 BINOMIAL EXPANSION

The coefficients of the polynomials $(x + y)^n$ where n is a positive integer have several interesting properties which lead to the so-called binomial theorem. First, we will examine a few cases where n is small and observe several consistent patterns. In these examples, n takes on the values 1, 2, 3, 4, and 5 in turn.

$$(x + y)^1 = x + y$$

$$(x + y)^2 = x^2 + 2xy + y^2$$

$$= x^2 + \frac{2}{1}xy + \frac{2 \cdot 1}{1 \cdot 2}y^2$$

$$(x + y)^3 = x^3 + 3x^2y + 3xy^2 + y^3$$

$$= x^3 + \frac{3}{1}x^2y + \frac{3 \cdot 2}{1 \cdot 2}xy^2 + \frac{3 \cdot 2 \cdot 1}{1 \cdot 2 \cdot 3}y^3$$

$$(x + y)^4 = x^4 + 4x^3y + 6x^2y^2 + 4xy^3 + y^4$$

$$= x^4 + \frac{4}{1}x^3y + \frac{4 \cdot 3}{1 \cdot 2}x^2y^2 + \frac{4 \cdot 3 \cdot 2}{1 \cdot 2 \cdot 3}xy^3$$

$$+ \frac{4 \cdot 3 \cdot 2 \cdot 1}{1 \cdot 2 \cdot 3 \cdot 4}y^4$$

$$(x + y)^5 = x^5 + \frac{\overset{+1}{5}}{1}x^4y^1 + \frac{5 \cdot \overset{+1}{4}}{1 \cdot 2}x^3y^2 + \frac{5 \cdot 4 \cdot \overset{+1}{3}}{1 \cdot 2 \cdot 3}x^2y^3$$

$$+ \frac{5 \cdot 4 \cdot 3 \cdot \overset{+1}{2}}{1 \cdot 2 \cdot 3 \cdot 4}x^1y^4 + \frac{5 \cdot 4 \cdot 3 \cdot 2 \cdot 1}{1 \cdot 2 \cdot 3 \cdot 4 \cdot 5}y^5$$

In each of these cases:

1. the polynomial has $n + 1$ terms;
2. the first term is $x^n = x^ny^0$;
3. the exponent of x decreases and the exponent of y increases as the number of the term increases;
4. the sum of the exponents for x and y in any term is n (i.e., x^ny^0, $x^{n-1}y^1, \ldots, x^0y^n$);

5. the first term and $(n+1)$st term have the same coefficient, the second terr. and nth term have the same coefficient, and for $m \leqslant \frac{n}{2}$, the mth term and $[(n+2)-m]$th term have the same coefficient;
6. the denominator of the y^m term can be expressed as $1 \cdot 2 \cdot 3 \cdot \ldots \cdot m$;
7. the numerator of the y^m term can be expressed as $n(n-1)\cdots(n-m+1)$.

The denominator of the coefficients has a form that occurs so often that there is a special name and notation for it.

Definition For any natural number m, we define ***m* factorial** (denoted $\boldsymbol{m!}$) to be the product

$$1 \cdot 2 \cdot 3 \cdot \ldots \cdot m$$

In addition, we define

$$0! = 1$$

These properties for the general polynomial $(x+y)^n$ and a representative $(m+1)$st term, where $0 \leqslant m \leqslant n$, have the following form.

Theorem (*The binomial theorem*) In the expansion of the binomial $(x+y)$ to the nth power, where n is a positive integer

$$(x+y)^n = x^n + nx^{n-1}y + \cdots + nxy^{n-1} + y^m$$

and the $(m+1)$st term, where $0 \leqslant m \leqslant n$, is

$$\frac{n(n-1)\cdots(n-m+1)}{1 \cdot 2 \cdots m} x^{n-m}y^m$$

Since

$$\frac{n(n-1)\cdots(n-m+1)}{1 \cdot 2 \cdots m}$$

$$= \frac{n(n-1)\cdots(n-m+1)(n-m)(n-m-1)\cdots 2 \cdot 1}{1 \cdot 2 \cdots m \qquad (n-m)(n-m-1)\cdots 2 \cdot 1}$$

$$= \frac{n!}{m!(n-m)!}$$

the $(m+1)$st terms of the binomial expansion is also

$$\frac{n!}{m!(n-m)!} x^{n-m}y^m.$$

In case $m = 0$, $(m + 1)$ refers to the first term, x^n. Thus

$$\frac{n!}{0!(n-0)!} = 1$$

since we defined $0! = 1$.

Example 1 Use the binomial theorem to find the polynomial form for $(x + y)^6$.

Solution:

$$(x+y)^6 = x^6 + \frac{6}{1}x^5y + \frac{6 \cdot 5}{1 \cdot 2}x^4y^2 + \frac{6 \cdot 5 \cdot 4}{1 \cdot 2 \cdot 3}x^3y^3$$

$$+ \frac{6 \cdot 5 \cdot 4 \cdot 3}{1 \cdot 2 \cdot 3 \cdot 4}x^2y^4 + 6xy^5 + y^6$$

$$= x^6 + 6x^5y + 15x^4y^2 + 20x^3y^3 + 15x^2y^4 + 6xy^5 + y^6.$$

Example 2 Expand $(a + 3b)^7$ by the binomial theorem.

Solution: Replacing x by a and y by $3b$ in the binomial theorem, we have

$$(a+3b)^7 = a^7 + 7a^6(3b) + \frac{7 \cdot 6}{1 \cdot 2}a^5(3b)^2 + \frac{7 \cdot 6 \cdot 5}{1 \cdot 2 \cdot 3}a^4(3b)^3$$

$$+ \frac{7 \cdot 6 \cdot 5 \cdot 4}{1 \cdot 2 \cdot 3 \cdot 4}a^3(3b)^4 + 21a^2(3b)^5 + 7a(3b)^6 + (3b)^7$$

$$= a^7 + 21a^6b + 189a^5b^2 + 945a^4b^3 + 2835a^3b^4$$

$$+ 5103a^2b^5 + 5103ab^6 + 2187b^7.$$

Expand $(2a + b)^8$ by the binomial theorem.

$$(2a+b)^8 = (2a)^8 + 8(2a)^7b$$

$$+ \frac{8 \cdot 7}{1 \cdot 2}(2a)^6b^2$$

$$+ \frac{8 \cdot 7 \cdot 6}{1 \cdot 2 \cdot 3}(2a)^5b^3$$

$$+ \frac{8 \cdot 7 \cdot 6 \cdot 5}{1 \cdot 2 \cdot 3 \cdot 4}(2a)^4b^4$$

$$+ \frac{8 \cdot 7 \cdot 6}{1 \cdot 2 \cdot 3}(2a)^3b^5$$

$$+ \frac{8 \cdot 7}{1 \cdot 2}(2a)^2b^6$$

$$+ 8(2a)b^7 + b^8$$

$$= 256a^8 + 1024a^7b$$

$$+ 1792a^6b + 1792a^5b^3$$

$$+ 1120a^4b^4 + 448a^3b^5$$

$$+ 112a^2b^6 + 16ab^7$$

$$+ b^8$$

Example 3 Find the first four terms in the expansion of $(2a - 3b)^{10}$.

Solution: Replacing x by $2a$ and y by $-3b$ in the binomial theorem, we have the first four terms

$$(2a)^{10},\ \frac{10}{1}(2a)^9(-3b),\ \frac{10\cdot 9}{1\cdot 2}(2a)^8(-3b)^2,$$

$$\frac{10\cdot 9\cdot 8}{1\cdot 2\cdot 3}(2a)^7(-3b)^3$$

or

$$1024a^{10};\ -15{,}360a^9b;\ 103{,}680a^8b^2;\ -414{,}720a^7b^3.$$

Example 4 Find the 5th and 8th terms in the expansion of $\left(\frac{1}{a} - 2b\right)^{12}$.

Solution: Replacing x by $\frac{1}{a}$, y by $-2b$, m by 4 for the 5th term, and m by 7 for the 8th term in the binomial theorem, we have

$$\text{5th term: } \frac{12!}{4!8!}\left(\frac{1}{a}\right)^8(-2b)^4 = 7920b^4a^{-8}$$

$$\text{8th term: } \frac{12!}{7!5!}\left(\frac{1}{a}\right)^5(-2b)^7 = -101{,}376b^7a^{-5}.$$

Exercises

Use the binomial theorem and expand each binomial in Exercises 1-11.

1. $(x+y)^7$
2. $(x+y)^9$
3. $(2x-y)^5$
4. $(2x-3y)^4$
5. $(3x-2y)^4$
6. $\left(\frac{1}{2}x+2y\right)^6$
7. $(-x-2y)^7$
8. $\left(\frac{x}{2}-\frac{y}{3}\right)^5$
9. $\left(\frac{x}{3}-\frac{y}{2}\right)^5$
10. $\left(\frac{1}{x}+\frac{1}{y}\right)^4$
11. $\left(\frac{2}{x}-\frac{1}{y}\right)^5$
12. Use the binomial theorem to find the first four terms and the indicated term of each expansion.
    a. $(x+y)^8$; 8th term
    b. $(x-y)^9$; 6th term

Use the binomial theorem and expand each binomial in Exercises 13-16.

13. $\left(\frac{1}{2x}+\frac{1}{y}\right)^4$
14. $\left(\frac{2}{x}-\frac{3}{y}\right)^5$
15. $\left(\frac{2}{3x}+\frac{3}{y}\right)^5$
16. $\left(\frac{1}{3x}-\frac{y}{2}\right)^6$

In Exercises 17-22, use the binomial theorem to find the first four terms and the indicated term of each expansion.

17. $(x+y)^{13}$; 8th term
18. $(x-y)^{14}$; 7th term
19. $(1+x)^{13}$; 10th term
20. $(2x-y)^{18}$; 6th term
21. $(x-2y)^{10}$; term containing y^6
22. $(x^2+y^2)^7$; term containing y^8

16-5 BINOMIAL SERIES

In the binomial theorem, if we replace $(x + y)^n$ by $(1 + a)^n$, the expansion is

$$(1 + a)^n = 1 + \frac{n}{1!} a + \frac{n(n-1)}{2} a^2 + \cdots$$

$$+ \frac{n(n-1)\cdots(n-r+1)}{r!} a^r + \cdots + a^n.$$

For example

$$(1 + 0.1)^4 = 1 + 4(0.1) + \frac{4 \cdot 3}{1 \cdot 2}(0.1)^2 + \frac{4 \cdot 3 \cdot 2}{1 \cdot 2 \cdot 3}(0.1)^3 + (0.1)^4$$

$$= 1 + 0.4 + 0.06 + 0.004 + 0.0001 = 1.4641.$$

Even if n is not a positive integer we may apply the pattern suggested by the binomial theorem. However, in this case, there are not $(n + 1)$ terms in the expansion; in fact, the procedure does not terminate.

For example, if $n = \frac{1}{2}$, then

$$(1 + a)^{1/2} = 1 + \frac{1}{2} a + \frac{\frac{1}{2}\left(\frac{1}{2} - 1\right)}{2!} a^2 + \cdots$$

$$+ \frac{\frac{1}{2}\left(\frac{1}{2} - 1\right)\left(\frac{1}{2} - 2\right)\cdots\left(\frac{1}{2} - r + 1\right)}{r!} a^r + \cdots.$$

Since there is a term in this expansion for each positive integer we have a series, called a **binomial series.** If $a = 0.2$, then the binomial series expansion would be

$$1 + \frac{1}{2}(0.2) + \frac{\frac{1}{2}\left(-\frac{1}{2}\right)}{2!}(0.2)^2 + \cdots$$

$$+ \frac{\frac{1}{2}\left(-\frac{1}{2}\right)\cdots\left(\frac{1}{2} - r + 1\right)}{r!}(0.2)^r + \cdots$$

where the sequence of partial sums for this series has

$$s_1 = 1$$

$$s_2 = 1 + \frac{1}{2}(0.2) = 1.1$$

$$s_3 = 1 + \frac{1}{2}(0.2) + \frac{\frac{1}{2}\left(-\frac{1}{2}\right)}{2!}(0.2)^2 = 1.095.$$

These partial sums are approximations for $(1 + 0.2)^{1/2} = \sqrt{1.2}$ and this binomial series converges to $\sqrt{1.2}$.

Find the first four terms in the binomial series expansion of $(1 + a)^{1/2}$ where $a = 0.1$. Compute the first four terms in the sequence of partial sums for this series.

$$(1 + 0.1)^{1/2} = 1 + \frac{1}{2}(0.1) + \frac{\frac{1}{2}\left(-\frac{1}{2}\right)}{2}(0.1)^2 + \frac{\left(\frac{1}{2}\right)\left(-\frac{1}{2}\right)\left(-\frac{3}{2}\right)}{2 \cdot 3} \cdot (0.1)^3 + \cdots$$

- $s_1 = 1$

$$s_2 = 1 + \frac{1}{2}(0.1) = 1.05$$

$$s_3 = 1 + \frac{1}{2}(0.1) + \frac{\left(\frac{1}{2}\right)\left(-\frac{1}{2}\right)}{2} \cdot (0.1) = 1.04875$$

$$s_4 = 1 + \frac{1}{2}(0.1) + \frac{\left(\frac{1}{2}\right)\left(-\frac{1}{2}\right)}{2} \cdot (0.1)^2 + \frac{\left(\frac{1}{2}\right)\left(-\frac{1}{2}\right)\left(-\frac{3}{2}\right)}{2 \cdot 3} \cdot (0.1)^3 = 1.0488125$$

If $n = \frac{1}{2}$ and $a = -0.1$, then the binomial series expansion becomes

$$[1 + (-0.1)]^{1/2} = 1 + \frac{1}{2}(-0.1) + \frac{\frac{1}{2}\left(-\frac{1}{2}\right)}{2!}(-0.1)^2 + \cdots + \frac{\frac{1}{2}\left(-\frac{1}{2}\right)\cdots\left(\frac{1}{2} - r + 1\right)}{r!}(-0.1)^r + \cdots$$

and the sequence (of partial sums) for this series has

$$1$$

$$1 + \frac{1}{2}(-0.1) = 0.95$$

$$1 + \frac{1}{2}(-0.1) + \frac{\frac{1}{2}\left(-\frac{1}{2}\right)}{2!}(-0.1)^2 = 0.94875$$

as the first three terms.

The binomial series for $(1 + a)^n$ which is

$$1 + na + \frac{n(n-1)}{2!}a^2 + \cdots + \frac{n(n-1)\cdots(n-r+1)}{r!}a^r + \cdots$$

where n is not a positive integer and $|a| < 1$ is a convergent series. This result is usually proved in a calculus course.

Exercises

1. Find the first four terms in each binomial series expansion.
   a. $\left(1 + \frac{1}{10}\right)^{-1}$ b. $\left(1 - \frac{1}{10}\right)^{-1}$ c. $\left(1 + \frac{1}{4}\right)^{-1}$
   d. $\left(1 + \frac{1}{5}\right)^{-2}$ e. $\left(1 - \frac{1}{5}\right)^{-2}$ f. $\left(1 + \frac{1}{100}\right)^{-1}$
2. Find the first four terms of the sequence of partial sums for each series in Exercise 1.
3. For each series in Exercise 1, find the error of approximation if the fourth partial sum is used to estimate the "sum" of the series.
4. Find the first four terms in each binomial series expansion and the first four terms in the sequence of partial sums for each series.
   a. $(1 + 0.1)^{1/3}$ b. $(1 - 0.1)^{1/3}$ c. $\left(1 + \frac{1}{4}\right)^{1/2}$
   d. $\left(1 - \frac{1}{4}\right)^{1/2}$ e. $\left(1 + \frac{1}{3}\right)^{1/2}$ f. $\left(1 - \frac{1}{3}\right)^{1/2}$
5. Approximate each number using a binomial series.
   a. $\sqrt{1.02}$ b. $\sqrt{0.98}$
   c. $\sqrt[3]{1.1}$ d. $\sqrt[3]{0.9}$

16-6 MATHEMATICAL INDUCTION

The binomial theorem discussed in Section 16-4 is proved by mathematical induction. Before displaying this proof in Section 16-7 we develop the concept of mathematical induction.

Consider the following question: If a subset S of the set of N natural numbers contains the number 1 and if for every number in the subset, the corresponding number which is 1 greater is also in the subset, what natural numbers belong to the subset?

$$1 \in S$$

Since $1 \in S$ then $1 + 1 \in S$ or $2 \in S$.

Since $2 \in S$ then $2 + 1 \in S$ or $3 \in S$.

Since $3 \in S$ then $3 + 1 \in S$ or $4 \in S$.

. . .

Continuing this process it seems plausible that $S = N$. We will accept this property for the natural numbers.

Property (*Mathematical induction*) If S is a subset of N such that (i) $1 \in S$ and (ii) $n + 1 \in S$ whenever $n \in S$ then $S = N$.

A proof by mathematical induction requires showing that (i) the assertion is true for $n = 1$, and (ii) if the assertion is true for any n, then it is also true for $n + 1$. The above method is used to prove that an infinite number of statements are true. For example, the first four statements in the following listing are true, and a general pattern seems to be established for an arbitrary natural number n:

$$1 = \frac{1 \cdot 2}{2}$$

$$1 + 2 = \frac{2 \cdot 3}{2}$$

$$1 + 2 + 3 = \frac{3 \cdot 4}{2}$$

$$1 + 2 + 3 + 4 = \frac{4 \cdot 5}{2}$$

$$\cdots$$

$$(1) \qquad 1 + 2 + 3 + 4 + \cdots + (n - 1) + n = \frac{n(n + 1)}{2}$$

$$\cdots$$

Thus, we have a statement for each natural number. We cannot test the statement for each n as there are infinitely many cases to consider. But we can apply the property of mathematical induction to prove (1).

As the description of the method implies, the proof is in two parts.

Part 1: Since

$$1 = \frac{1 \cdot 2}{2}$$

the assertion holds for $n = 1$.

Part 2: We now assume that (1) holds for n. (This is called the inductive assumption.) Our task is to prove that if (1) holds for n, then (1) holds for $n + 1$:

$$(2) \qquad 1 + 2 + \cdots + (n + 1) = \frac{(n + 1)[(n + 1) + 1]}{2}$$

We rewrite the left member of (2) as

(3) $$1 + 2 + \cdots + (n+1) = (1 + 2 + \cdots + n) + (n+1).$$

We can now apply our inductive assumption to the first n terms in the right member of (3):

$$(1 + 2 + \cdots + n) + (n+1) = \left(\frac{n(n+1)}{2}\right) + (n+1)$$

$$= \left(\frac{(n+1)}{2}\right)(n+2)$$

$$= \frac{(n+1)[(n+1)+1]}{2}$$

This last expression is identical to the right member of (2), and we have proved the assertion.

What is the sum of the first 300 natural numbers? The first 1000 natural numbers? (Use the result that we have proved.)

$\frac{300 \cdot 301}{2} = 45{,}150$; ● $\frac{1000 \cdot 1001}{2} = 500{,}500$

Exercises

1. Which part of the property of mathematical induction is not satisfied by the following subsets of N?
   a. $\{1, 2, 3, 4, 5, 6, 7, 8, 9, 10\}$ **b.** $\{2, 3, 4, \ldots\}$
2. Find real numbers to replace the question marks and a general pattern for each of the following:
   a. $1 = 1^2,\ 1 + 3 = 2^2,\ 1 + 3 + 5 = ?^2, \ldots$
   b. $\frac{1}{1 \cdot 2} = \frac{1}{2},\ \frac{1}{1 \cdot 2} + \frac{1}{2 \cdot 3} = \frac{2}{3},\ \frac{1}{1 \cdot 2} + \frac{1}{2 \cdot 3} + \frac{1}{3 \cdot 4} = \frac{3}{4},$
   $\frac{1}{1 \cdot 2} + \frac{1}{2 \cdot 3} + \frac{1}{3 \cdot 4} + \frac{1}{4 \cdot 5} = ?, \ldots$
3. In Exercise 2a, let P_1 denote the first statement, P_2 the second, P_3 the third, and if n is a natural number, let P_n denote the nth statement. What is P_n?
4. Apply the instructions in Exercise 3 to the statements in Exercise 2b.
5. Use mathematical induction to prove that P_n is true for every natural number n in Exercise 3.
6. In Exercise 4, prove that P_n is true for every natural number n.
7. If n is a natural number and P_n is the statement

   $$2 + 4 + 6 + \cdots + 2n = n(n+1)$$

   prove that P_n is true for every natural number n.

8. If n is a natural number and P_n is the statement

$$2^{n-1} \leqslant 1 \cdot 2 \cdot 3 \cdots (n-1) \cdot n$$

prove that P_n is true for every natural number n.

9. Use mathematical induction to prove DeMoivre's theorem, page 338.

16-7 PROOF OF THE BINOMIAL THEOREM

Recall from the binomial theorem (page 360) that the $(m+1)$st term of $(x+y)^n$, where n is a positive integer and $0 \leqslant m \leqslant n$, is

$$\frac{n!}{m!(n-m)!} x^{n-m} y^m .$$

Before giving a proof, let us apply the theorem for $n = 3$ and $n = 4$:

$$(x+y)^3 = \frac{3!}{0!3!} x^{3-0} y^0 + \frac{3!}{1!2!} x^{3-1} y^1 + \frac{3!}{2!1!} x^{3-2} y^2 + \frac{3!}{3!0!} x^{3-3} y^3$$

$$(x+y)^4 = \frac{4!}{0!4!} x^{4-0} y^0 + \frac{4!}{1!3!} x^{4-1} y^1 + \frac{4!}{2!2!} x^{4-2} y^2 + \frac{4!}{3!1!} x^{4-3} y^3 + \frac{4!}{4!0!} x^{4-4} y^4$$

Using the distributive property

$$(x+y)^4 = (x+y)(x+y)^3 = x(x+y)^3 + y(x+y)^3 .$$

If we express $(x+y)^3$ in factorial notation which was derived from the binomial theorem, we have the expansion of $(x+y)^4$.

$$(x+y)^4 = \left(\frac{3!}{0!3!} x^3 y^0 + \frac{3!}{1!2!} x^2 y^1 + \frac{3!}{2!1!} x^1 y^2 + \frac{3!}{3!0!} x^0 y^3\right)$$
$$+ y\left(\frac{3!}{0!3!} x^3 y^0 + \frac{3!}{1!2!} x^2 y^1 + \frac{3!}{2!1!} x^1 y^2 + \frac{3!}{3!0!} x^0 y^3\right)$$

$$= \frac{3!}{0!3!}x^4y^0 + \left(\frac{3!}{1!2!} + \frac{3!}{0!3!}\right)x^3y^1 + \left(\frac{3!}{2!1!} + \frac{3!}{1!2!}\right)x^2y^2$$

$$+ \left(\frac{3!}{3!0!} + \frac{3!}{2!1!}\right)x^1y^3 + \frac{3!}{2!0!}x^0y^4$$

$$= x^4 + \frac{3!(3+1)}{1!3!}x^3y^1 + \frac{3!(2+2)}{2!2!}x^2y^2$$

$$+ \frac{3!(1+3)}{3!1!}x^1y^3 + y^4$$

$$= \frac{4!}{0!4!}x^4y^0 + \frac{4!}{1!3!}x^3y^1 + \frac{4!}{2!2!}x^2y^2 + \frac{4!}{3!1!}x^1y^3$$

$$+ \frac{4!}{4!0!}x^0y^4$$

The proof of the binomial theorem by mathematical induction is similar to the above discussion. We assume that the binomial theorem applies to n and prove that it also applies to $(n + 1)$.

Proof: Since $(x + y)^1 = \frac{1!}{0!1!}x^1y^0 + \frac{1!}{1!0!}x^0y^1$, the theorem holds for $n = 1$. By hypothesis, we assume that the theorem holds for the positive integer n. Since $(x + y)^{n+1} = (x + y)(x + y)^n = x(x + y)^n + y(x + y)^n$, the $(m + 1)$st term in the expansion of $(x + y)^{n+1}$ is the sum

$$x\left[\frac{n!}{m!(n-m)!}x^{n-m}y^m\right] + y\left[\frac{n!}{(m-1)!(n-m+1)!}x^{n-m+1}y^{m-1}\right].$$

We want to show that this term is the result we obtain when we apply the theorem for $(n + 1)$:

$$\frac{(n+1)!}{m!(n+1-m)!}x^{n+1-m}y^m.$$

This conclusion follows from the computations

$$\frac{n!}{m!(m-n)!} + \frac{n!}{(m-1)!(n-m+1)!}$$

$$= \frac{n!(n-m+1)}{m!(n-m+1)!} + \frac{n!m}{m!(n-m+1)!}$$

$$= \frac{n!(n - m + 1 + m)}{m!(n - m + 1)!}$$

$$= \frac{(n + 1)!}{m!(n + 1 - m)!}$$

By the property of mathematical induction, the theorem is proved.

Review Exercises

1. Find the first four terms of an A.P. with $a_5 = \frac{7}{2}$ and $a_8 = \frac{23}{4}$. **16-1**

2. Find the first and seventh terms of a G.P. where $a_4 = -\frac{56}{27}$ and $a_5 = \frac{112}{81}$.

3. Find the sum of the A.P. having 20 terms with fourth term -5.7 and seventh term 0.6.

4. Find the sum of the G.P. having 9 terms with third term 0.9 and fourth term -0.27.

5. Find the 5th, 8th, 11th, kth, $(k - 1)$st, and $(k + 1)$st terms of **16-2**
$$0.21,\ 0.2121,\ 0.212121, \ldots, \sum_{i=1}^{m} \frac{21}{100}\left(\frac{1}{100}\right)^{i-1}, \ldots.$$

6. What are the first four terms in the sequence of partial sums for **16-3** the series $\sum_{i=1}^{\infty} \frac{21}{100}\left(\frac{1}{100}\right)^{i-1}$? What is the "sum" of this series?

7. Find the first four terms and the 7th term in the binomial expan- **16-4** sion for $(x - 2y)^9$.

8. Find the first four terms in the binomial series expansion for **16-5** $(1 + a)^{-(2/3)}$.

9. State the postulate of mathematical induction. **16-6**

TABLES

Table I

Mantissas for common logarithms of numbers (four places)

N	0	1	2	3	4	5	6	7	8	9
55	7404	7412	7419	7427	7435	7443	7451	7459	7466	7474
56	7482	7490	7497	7505	7513	7520	7528	7536	7543	7551
57	7559	7566	7574	7582	7589	7597	7604	7612	7619	7627
58	7634	7642	7649	7657	7664	7672	7679	7686	7694	7701
59	7709	7716	7723	7731	7738	7745	7752	7760	7767	7774
60	7782	7789	7796	7803	7810	7818	7825	7832	7839	7846
61	7853	7860	7868	7875	7882	7889	7896	7903	7910	7917
62	7924	7931	7938	7945	7952	7959	7966	7973	7980	7987
63	7993	8000	8007	8014	8021	8028	8035	8041	8048	8055
64	8062	8069	8075	8082	8089	8096	8102	8109	8116	8122
65	8129	8136	8142	8149	8156	8162	8169	8176	8182	8189
66	8195	8202	8209	8215	8222	8228	8235	8241	8248	8254
67	8261	8267	8274	8280	8287	8293	8299	8306	8312	8319
68	8325	8331	8338	8344	8351	8357	8363	8370	8376	8382
69	8388	8395	8401	8407	8414	8420	8426	8432	8439	8445
70	8451	8457	8463	8470	8476	8482	8488	8494	8500	8506
71	8513	8519	8525	8531	8537	8543	8549	8555	8561	8567
72	8573	8579	8585	8591	8597	8603	8609	8615	8621	8627
73	8633	8639	8645	8651	8657	8663	8669	8675	8681	8686
74	8692	8698	8704	8710	8716	8722	8727	8733	8739	8745
75	8751	8756	8762	8768	8774	8779	8785	8791	8797	8802
76	8808	8814	8820	8825	8831	8837	8842	8848	8854	8859
77	8865	8871	8876	8882	8887	8893	8899	8904	8910	8915
78	8921	8927	8932	8938	8943	8949	8954	8960	8965	8971
79	8976	8982	8987	8993	8998	9004	9009	9015	9020	9025
80	9031	9036	9042	9047	9053	9058	9063	9069	9074	9079
81	9085	9090	9096	9101	9106	9112	9117	9122	9128	9133
82	9138	9143	9149	9154	9159	9165	9170	9175	9180	9186
83	9191	9196	9201	9206	9212	9217	9222	9227	9232	9238
84	9243	9248	9253	9258	9263	9269	9274	9279	9284	9289
85	9294	9299	9304	9309	9315	9320	9325	9330	9335	9340
86	9345	9350	9355	9360	9365	9370	9375	9380	9385	9390
87	9395	9400	9405	9410	9415	9420	9425	9430	9435	9440
88	9445	9450	9455	9460	9465	9469	9474	9479	9484	9489
89	9494	9499	9504	9509	9513	9518	9523	9528	9533	9538
90	9542	9547	9552	9557	9562	9566	9571	9576	9581	9586
91	9590	9595	9600	9605	9609	9614	9619	9624	9628	9633
92	9638	9643	9647	9652	9657	9661	9666	9671	9675	9680
93	9685	9689	9694	9699	9703	9708	9713	9717	9722	9727
94	9731	9736	9741	9745	9750	9754	9759	9763	9768	9773
95	9777	9782	9786	9791	9795	9800	9805	9809	9814	9818
96	9823	9827	9832	9836	9841	9845	9850	9854	9859	9863
97	9868	9872	9877	9881	9886	9890	9894	9899	9903	9908
98	9912	9917	9921	9926	9930	9934	9939	9943	9948	9952
99	9956	9961	9965	9969	9974	9978	9983	9987	9991	9996

By permission from WILLIAM E. KLINE, ROBERT A. OESTERLE, and LEROY M. WILLSON, *Foundations of Advanced Mathematics*, American Book Company, New York, 1965.

Table I

Mantissas for common logarithms of numbers (four places)

N	0	1	2	3	4	5	6	7	8	9
10	0000	0043	0086	0128	0170	0212	0253	0294	0334	0374
11	0414	0453	0492	0531	0569	0607	0645	0682	0719	0755
12	0792	0828	0864	0899	0934	0969	1004	1038	1072	1106
13	1139	1173	1206	1239	1271	1303	1335	1367	1399	1430
14	1461	1492	1523	1553	1584	1614	1644	1673	1703	1732
15	1761	1790	1818	1847	1875	1903	1931	1959	1987	2014
16	2041	2068	2095	2122	2148	2175	2201	2227	2253	2279
17	2304	2330	2355	2380	2405	2430	2455	2480	2504	2529
18	2553	2577	2601	2625	2648	2672	2695	2718	2742	2765
19	2788	2810	2833	2856	2878	2900	2923	2945	2967	2989
20	3010	3032	3054	3075	3096	3118	3139	3160	3181	3201
21	3222	3243	3263	3284	3304	3324	3345	3365	3385	3404
22	3424	3444	3464	3483	3502	3522	3541	3560	3579	3598
23	3617	3636	3655	3674	3692	3711	3729	3747	3766	3784
24	3802	3820	3838	3856	3874	3892	3909	3927	3945	3962
25	3979	3997	4014	4031	4048	4065	4082	4099	4116	4133
26	4150	4166	4183	4200	4216	4232	4249	4265	4281	4298
27	4314	4330	4346	4362	4378	4393	4409	4425	4440	4456
28	4472	4487	4502	4518	4533	4548	4564	4579	4594	4609
29	4624	4639	4654	4669	4683	4698	4713	4728	4742	4757
30	4771	4786	4800	4814	4829	4843	4857	4871	4886	4900
31	4914	4928	4942	4955	4969	4983	4997	5011	5024	5038
32	5051	5065	5079	5092	5105	5119	5132	5145	5159	5172
33	5185	5198	5211	5224	5237	5250	5263	5276	5289	5302
34	5315	5328	5340	5353	5366	5378	5391	5403	5416	5428
35	5441	5453	5465	5478	5490	5502	5514	5527	5539	5551
36	5563	5575	5587	5599	5611	5623	5635	5647	5658	5670
37	5682	5694	5705	5717	5729	5740	5752	5763	5775	5786
38	5798	5809	5821	5832	5843	5855	5866	5877	5888	5899
39	5911	5922	5933	5944	5955	5966	5977	5988	5999	6010
40	6021	6031	6042	6053	6064	6075	6085	6096	6107	6117
41	6128	6138	6149	6160	6170	6180	6191	6201	6212	6222
42	6232	6243	6253	6263	6274	6284	6294	6304	6314	6325
43	6335	6345	6355	6365	6375	6385	6395	6405	6415	6425
44	6435	6444	6454	6464	6474	6484	6493	6503	6513	6522
45	6532	6542	6551	6561	6571	6580	6590	6599	6609	6618
46	6628	6637	6646	6656	6665	6675	6684	6693	6702	6712
47	6721	6730	6739	6749	6758	6767	6776	6785	6794	6803
48	6812	6821	6830	6839	6848	6857	6866	6875	6884	6893
49	6902	6911	6920	6928	6937	6946	6955	6964	6972	6981
50	6990	6998	7007	7016	7024	7033	7042	7050	7059	7067
51	7076	7084	7093	7101	7110	7118	7126	7135	7143	7152
52	7160	7168	7177	7185	7193	7202	7210	7218	7226	7235
53	7243	7251	7259	7267	7275	7284	7292	7300	7308	7316
54	7324	7332	7340	7348	7356	7364	7372	7380	7388	7396

Table II

Values of trigonometric functions for real numbers

θ	sin θ	cos θ	tan θ	cot θ	sec θ	csc θ
0.00	0.0000	1.0000	0.0000	.....	1.000	.....
0.01	0.0100	1.0000	0.0100	99.997	1.000	100.00
0.02	0.0200	0.9998	0.0200	49.993	1.000	50.00
0.03	0.0300	0.9996	0.0300	33.323	1.000	33.34
0.04	0.0400	0.9992	0.0400	24.987	1.001	25.01
0.05	0.0500	0.9988	0.0500	19.983	1.001	20.01
0.06	0.0600	0.9982	0.0601	16.647	1.002	16.68
0.07	0.0699	0.9976	0.0701	14.262	1.002	14.30
0.08	0.0799	0.9968	0.0802	12.473	1.003	12.51
0.09	0.0899	0.9960	0.0902	11.081	1.004	11.13
0.10	0.0998	0.9950	0.1003	9.967	1.005	10.02
0.11	0.1098	0.9940	0.1104	9.054	1.006	9.109
0.12	0.1197	0.9928	0.1206	8.293	1.007	8.353
0.13	0.1296	0.9916	0.1307	7.649	1.009	7.714
0.14	0.1395	0.9902	0.1409	7.096	1.010	7.166
0.15	0.1494	0.9888	0.1511	6.617	1.011	6.692
0.16	0.1593	0.9872	0.1614	6.197	1.013	6.277
0.17	0.1692	0.9856	0.1717	5.826	1.015	5.911
0.18	0.1790	0.9838	0.1820	5.495	1.016	5.586
0.19	0.1889	0.9820	0.1923	5.200	1.018	5.295
0.20	0.1987	0.9801	0.2027	4.933	1.020	5.033
0.21	0.2085	0.9780	0.2131	4.692	1.022	4.797
0.22	0.2182	0.9759	0.2236	4.472	1.025	4.582
0.23	0.2280	0.9737	0.2341	4.271	1.027	4.386
0.24	0.2377	0.9713	0.2447	4.086	1.030	4.207
0.25	0.2474	0.9689	0.2553	3.916	1.032	4.042
0.26	0.2571	0.9664	0.2660	3.759	1.035	3.890
0.27	0.2667	0.9638	0.2768	3.613	1.038	3.749
0.28	0.2764	0.9611	0.2876	3.478	1.041	3.619
0.29	0.2860	0.9582	0.2984	3.351	1.044	3.497
0.30	0.2955	0.9553	0.3093	3.233	1.047	3.384
0.31	0.3051	0.9523	0.3203	3.122	1.050	3.278
0.32	0.3146	0.9492	0.3314	3.018	1.053	3.179
0.33	0.3240	0.9460	0.3425	2.920	1.057	3.086
0.34	0.3335	0.9428	0.3537	2.827	1.061	2.999
0.35	0.3429	0.9394	0.3650	2.740	1.065	2.916
0.36	0.3523	0.9359	0.3764	2.657	1.068	2.839
0.37	0.3616	0.9323	0.3879	2.578	1.073	2.765
0.38	0.3709	0.9287	0.3994	2.504	1.077	2.696
0.39	0.3802	0.9249	0.4111	2.433	1.081	2.630
θ	sin θ	cos θ	tan θ	cot θ	sec θ	csc θ

Table II

Values of trigonometric functions for real numbers

θ	$\sin\theta$	$\cos\theta$	$\tan\theta$	$\cot\theta$	$\sec\theta$	$\csc\theta$
0.40	0.3894	0.9211	0.4228	2.365	1.086	2.568
0.41	0.3986	0.9171	0.4346	2.301	1.090	2.509
0.42	0.4078	0.9131	0.4466	2.239	1.095	2.452
0.43	0.4169	0.9090	0.4586	2.180	1.100	2.399
0.44	0.4259	0.9048	0.4708	2.124	1.105	2.348
0.45	0.4350	0.9004	0.4831	2.070	1.111	2.299
0.46	0.4439	0.8961	0.4954	2.018	1.116	2.253
0.47	0.4529	0.8916	0.5080	1.969	1.122	2.208
0.48	0.4618	0.8870	0.5206	1.921	1.127	2.166
0.49	0.4706	0.8823	0.5334	1.875	1.133	2.125
0.50	0.4794	0.8776	0.5463	1.830	1.139	2.086
0.51	0.4882	0.8727	0.5594	1.788	1.146	2.048
0.52	0.4969	0.8678	0.5726	1.747	1.152	2.013
0.53	0.5055	0.8628	0.5859	1.707	1.159	1.978
0.54	0.5141	0.8577	0.5994	1.668	1.166	1.945
0.55	0.5227	0.8525	0.6131	1.631	1.173	1.913
0.56	0.5312	0.8473	0.6269	1.595	1.180	1.883
0.57	0.5396	0.8419	0.6410	1.560	1.188	1.853
0.58	0.5480	0.8365	0.6552	1.526	1.196	1.825
0.59	0.5564	0.8309	0.6696	1.494	1.203	1.797
0.60	0.5646	0.8253	0.6841	1.462	1.212	1.771
0.61	0.5729	0.8196	0.6989	1.431	1.220	1.746
0.62	0.5810	0.8139	0.7139	1.401	1.229	1.721
0.63	0.5891	0.8080	0.7291	1.372	1.238	1.697
0.64	0.5972	0.8021	0.7445	1.343	1.247	1.674
0.65	0.6052	0.7961	0.7602	1.315	1.256	1.652
0.66	0.6131	0.7900	0.7761	1.288	1.266	1.631
0.67	0.6210	0.7838	0.7923	1.262	1.276	1.610
0.68	0.6288	0.7776	0.8087	1.237	1.286	1.590
0.69	0.6365	0.7712	0.8253	1.212	1.297	1.571
0.70	0.6442	0.7648	0.8423	1.187	1.307	1.552
0.71	0.6518	0.7584	0.8595	1.163	1.319	1.534
0.72	0.6594	0.7518	0.8771	1.140	1.330	1.517
0.73	0.6669	0.7452	0.8949	1.117	1.342	1.500
0.74	0.6743	0.7385	0.9131	1.095	1.354	1.483
0.75	0.6816	0.7317	0.9316	1.073	1.367	1.467
0.76	0.6889	0.7248	0.9505	1.052	1.380	1.452
0.77	0.6961	0.7179	0.9697	1.031	1.393	1.437
0.78	0.7033	0.7109	0.9893	1.011	1.407	1.422
0.79	0.7104	0.7038	1.009	0.9908	1.421	1.408
θ	$\sin\theta$	$\cos\theta$	$\tan\theta$	$\cot\theta$	$\sec\theta$	$\csc\theta$

Table II

Values of trigonometric functions for real numbers

θ	$\sin\theta$	$\cos\theta$	$\tan\theta$	$\cot\theta$	$\sec\theta$	$\csc\theta$
0.80	0.7174	0.6967	1.030	0.9712	1.435	1.394
0.81	0.7243	0.6895	1.050	0.9520	1.450	1.381
0.82	0.7311	0.6822	1.072	0.9331	1.466	1.368
0.83	0.7379	0.6749	1.093	0.9146	1.482	1.355
0.84	0.7446	0.6675	1.116	0.8964	1.498	1.343
0.85	0.7513	0.6600	1.138	0.8785	1.515	1.331
0.86	0.7578	0.6524	1.162	0.8609	1.533	1.320
0.87	0.7643	0.6448	1.185	0.8437	1.551	1.308
0.88	0.7707	0.6372	1.210	0.8267	1.569	1.297
0.89	0.7771	0.6294	1.235	0.8100	1.589	1.287
0.90	0.7833	0.6216	1.260	0.7936	1.609	1.277
0.91	0.7895	0.6137	1.286	0.7774	1.629	1.267
0.92	0.7956	0.6058	1.313	0.7615	1.651	1.257
0.93	0.8016	0.5978	1.341	0.7458	1.673	1.247
0.94	0.8076	0.5898	1.369	0.7303	1.696	1.238
0.95	0.8134	0.5817	1.398	0.7151	1.719	1.229
0.96	0.8192	0.5735	1.428	0.7001	1.744	1.221
0.97	0.8249	0.5653	1.459	0.6853	1.769	1.212
0.98	0.8305	0.5570	1.491	0.6707	1.795	1.204
0.99	0.8360	0.5487	1.524	0.6563	1.823	1.196
1.00	0.8415	0.5403	1.557	0.6421	1.851	1.188
1.01	0.8468	0.5319	1.592	0.6281	1.880	1.181
1.02	0.8521	0.5234	1.628	0.6142	1.911	1.174
1.03	0.8573	0.5148	1.665	0.6005	1.942	1.166
1.04	0.8624	0.5062	1.704	0.5870	1.975	1.160
1.05	0.8674	0.4976	1.743	0.5736	2.010	1.153
1.06	0.8724	0.4889	1.784	0.5604	2.046	1.146
1.07	0.8772	0.4801	1.827	0.5473	2.083	1.140
1.08	0.8820	0.4713	1.871	0.5344	2.122	1.134
1.09	0.8866	0.4625	1.917	0.5216	2.162	1.128
1.10	0.8912	0.4536	1.965	0.5090	2.205	1.122
1.11	0.8957	0.4447	2.014	0.4964	2.249	1.116
1.12	0.9001	0.4357	2.066	0.4840	2.295	1.111
1.13	0.9044	0.4267	2.120	0.4718	2.344	1.106
1.14	0.9086	0.4176	2.176	0.4596	2.395	1.101
1.15	0.9128	0.4085	2.234	0.4475	2.448	1.096
1.16	0.9168	0.3993	2.296	0.4356	2.504	1.091
1.17	0.9208	0.3902	2.360	0.4237	2.563	1.086
1.18	0.9246	0.3809	2.427	0.4120	2.625	1.082
1.19	0.9284	0.3717	2.498	0.4003	2.691	1.077
θ	$\sin\theta$	$\cos\theta$	$\tan\theta$	$\cot\theta$	$\sec\theta$	$\csc\theta$

Table II

Values of trigonometric functions for real numbers

θ	sin θ	cos θ	tan θ	cot θ	sec θ	csc θ
1.20	0.9320	0.3624	2.572	0.3888	2.760	1.073
1.21	0.9356	0.3530	2.650	0.3773	2.833	1.069
1.22	0.9391	0.3436	2.733	0.3659	2.910	1.065
1.23	0.9425	0.3342	2.820	0.3546	2.992	1.061
1.24	0.9458	0.3248	2.912	0.3434	3.079	1.057
1.25	0.9490	0.3153	3.010	0.3323	3.171	1.054
1.26	0.9521	0.3058	3.113	0.3212	3.270	1.050
1.27	0.9551	0.2963	3.224	0.3102	3.375	1.047
1.28	0.9580	0.2867	2.341	0.2993	3.488	1.044
1.29	0.9608	0.2771	3.467	0.2884	3.609	1.041
1.30	0.9636	0.2675	3.602	0.2776	3.738	1.038
1.31	0.9662	0.2579	3.747	0.2669	3.878	1.035
1.32	0.9687	0.2482	3.903	0.2562	4.029	1.032
1.33	0.9711	0.2385	4.072	0.2456	4.193	1.030
1.34	0.9735	0.2288	4.256	0.2350	4.372	1.027
1.35	0.9757	0.2190	4.455	0.2245	4.566	1.025
1.36	0.9779	0.2092	4.673	0.2140	4.779	1.023
1.37	0.9799	0.1994	4.913	0.2035	5.014	1.021
1.38	0.9819	0.1896	5.177	0.1931	5.273	1.018
1.39	0.9837	0.1798	5.471	0.1828	5.561	1.017
1.40	0.9854	0.1700	5.798	0.1725	5.883	1.015
1.41	0.9871	0.1601	6.165	0.1622	6.246	1.013
1.42	0.9887	0.1502	6.581	0.1519	6.657	1.011
1.43	0.9901	0.1403	7.055	0.1417	7.126	1.010
1.44	0.9915	0.1304	7.602	0.1315	7.667	1.009
1.45	0.9927	0.1205	8.238	0.1214	8.299	1.007
1.46	0.9939	0.1106	8.989	0.1113	9.044	1.006
1.47	0.9949	0.1006	9.887	0.1011	9.938	1.005
1.48	0.9959	0.0907	10.983	0.0910	11.029	1.004
1.49	0.9967	0.0807	12.350	0.0810	12.390	1.003
1.50	0.9975	0.0707	14.101	0.0709	14.137	1.003
1.51	0.9982	0.0608	16.428	0.0609	16.458	1.002
1.52	0.9987	0.0508	19.670	0.0508	19.695	1.001
1.53	0.9992	0.0408	24.498	0.0408	24.519	1.001
1.54	0.9995	0.0308	32.461	0.0308	32.476	1.000
1.55	0.9998	0.0208	48.078	0.0208	48.089	1.000
1.56	0.9999	0.0108	92.620	0.0108	92.626	1.000
1.57	1.0000	0.0008	1255.8	0.0008	1255.8	1.000
1.58	1.0000	−0.0092	−108.65	−0.0092	−108.65	1.000
1.59	0.9998	−0.0192	−52.067	−0.0192	−52.08	1.000
1.60	0.9996	−0.0292	−34.233	−0.0292	−34.25	1.000
θ	sin θ	cos θ	tan θ	cot θ	sec θ	csc θ

Table III

Values of trigonometric functions for angles

Measure	Sin	Cos	Tan	Cot	Sec	Csc	
0° 00′	0.0000	1.0000	0.0000	.....	1.000	.....	**90° 00′**
10′	0.0029	1.0000	0.0029	343.8	1.000	343.8	**50′**
20′	0.0058	1.0000	0.0058	171.9	1.000	171.9	**40′**
30′	0.0087	1.0000	0.0087	114.6	1.000	114.6	**30′**
40′	0.0116	0.9999	0.0116	85.94	1.000	85.95	**20′**
50′	0.0145	0.9999	0.0145	68.75	1.000	68.76	**10′**
1° 00′	0.0175	0.9998	0.0175	57.29	1.000	57.30	**89° 00′**
10′	0.0204	0.9998	0.0204	49.10	1.000	49.11	**50′**
20′	0.0233	0.9997	0.0233	42.96	1.000	42.98	**40′**
30′	0.0262	0.9997	0.0262	38.19	1.000	38.20	**30′**
40′	0.0291	0.9996	0.0291	34.37	1.000	34.38	**20′**
50′	0.0320	0.9995	0.0320	31.24	1.001	31.26	**10′**
2° 00′	0.0349	0.9994	0.0349	28.64	1.001	28.65	**88° 00′**
10′	0.0378	0.9993	0.0378	26.43	1.001	26.45	**50′**
20′	0.0407	0.9992	0.0407	24.54	1.001	24.56	**40′**
30′	0.0436	0.9990	0.0437	22.90	1.001	22.93	**30′**
40′	0.0465	0.9989	0.0466	21.47	1.001	21.49	**20′**
50′	0.0494	0.9988	0.0495	20.21	1.001	20.23	**10′**
3° 00′	0.0523	0.9986	0.0524	19.08	1.001	19.11	**87° 00′**
10′	0.0552	0.9985	0.0553	18.07	1.002	18.10	**50′**
20′	0.0581	0.9983	0.0582	17.17	1.002	17.20	**40′**
30′	0.0610	0.9981	0.0612	16.35	1.002	16.38	**30′**
40′	0.0640	0.9980	0.0641	15.60	1.002	15.64	**20′**
50′	0.0669	0.9978	0.0670	14.92	1.002	14.96	**10′**
4° 00′	0.0698	0.9976	0.0699	14.30	1.002	14.34	**86° 00′**
10′	0.0727	0.9974	0.0729	13.73	1.003	13.76	**50′**
20′	0.0756	0.9971	0.0758	13.20	1.003	13.23	**40′**
30′	0.0785	0.9969	0.0787	12.71	1.003	12.75	**30′**
40′	0.0814	0.9967	0.0816	12.25	1.003	12.29	**20′**
50′	0.0843	0.9964	0.0846	11.83	1.004	11.87	**10′**
5° 00′	0.0872	0.9962	0.0875	11.43	1.004	11.47	**85° 00′**
10′	0.0901	0.9959	0.0904	11.06	1.004	11.10	**50′**
20′	0.0929	0.9957	0.0934	10.71	1.004	10.76	**40′**
30′	0.0958	0.9954	0.0963	10.39	1.005	10.43	**30′**
40′	0.0987	0.9951	0.0992	10.08	1.005	10.13	**20′**
50′	0.1016	0.9948	0.1022	9.788	1.005	9.839	**10′**
6° 00′	0.1045	0.9945	0.1051	9.514	1.006	9.567	**84° 00′**
10′	0.1074	0.9942	0.1080	9.255	1.006	9.309	**50′**
20′	0.1103	0.9939	0.1110	9.010	1.006	9.065	**40′**
30′	0.1132	0.9936	0.1139	8.777	1.006	8.834	**30′**
40′	0.1161	0.9932	0.1169	8.556	1.007	8.614	**20′**
50′	0.1190	0.9929	0.1198	8.345	1.007	8.405	**10′**
7° 00′	0.1219	0.9925	0.1228	8.144	1.008	8.206	**83° 00′**
10′	0.1248	0.9922	0.1257	7.953	1.008	8.016	**50′**
20′	0.1276	0.9918	0.1287	7.770	1.008	7.834	**40′**
30′	0.1305	0.9914	0.1317	7.596	1.009	7.661	**30′**
40′	0.1334	0.9911	0.1346	7.429	1.009	7.496	**20′**
50′	0.1363	0.9907	0.1376	7.269	1.009	7.337	**10′**
8° 00′	0.1392	0.9903	0.1405	7.115	1.010	7.185	**82° 00′**
10′	0.1421	0.9899	0.1435	6.968	1.010	7.040	**50′**
20′	0.1449	0.9894	0.1465	6.827	1.011	6.900	**40′**
30′	0.1478	0.9890	0.1495	6.691	1.011	6.765	**30′**
40′	0.1507	0.9886	0.1524	6.561	1.012	6.636	**20′**
50′	0.1536	0.9881	0.1554	6.435	1.012	6.512	**10′**
9° 00′	0.1564	0.9877	0.1584	6.314	1.012	6.392	**81° 00′**
	Cos	**Sin**	**Cot**	**Tan**	**Csc**	**Sec**	**Measure**

Table III

Values of trigonometric functions for angles

Measure	Sin	Cos	Tan	Cot	Sec	Csc	
9° 00′	0.1564	0.9877	0.1584	6.314	1.012	6.392	81° 00′
10′	0.1593	0.9872	0.1614	6.197	1.013	6.277	50′
20′	0.1622	0.9868	0.1644	6.084	1.013	6.166	40′
30′	0.1650	0.9863	0.1673	5.976	1.014	6.059	30′
40′	0.1679	0.9858	0.1703	5.871	1.014	5.955	20′
50′	0.1708	0.9853	0.1733	5.769	1.015	5.855	10′
10° 00′	0.1736	0.9848	0.1763	5.671	1.015	5.759	80° 00′
10′	0.1765	0.9843	0.1793	5.576	1.016	5.665	50′
20′	0.1794	0.9838	0.1823	5.485	1.016	5.575	40′
30′	0.1822	0.9833	0.1853	5.396	1.017	5.487	30′
40′	0.1851	0.9827	0.1883	5.309	1.018	5.403	20′
50′	0.1880	0.9822	0.1914	5.226	1.018	5.320	10′
11° 00′	0.1908	0.9816	0.1944	5.145	1.019	5.241	79° 00′
10′	0.1937	0.9811	0.1974	5.066	1.019	5.164	50′
20′	0.1965	0.9805	0.2004	4.989	1.020	5.089	40′
30′	0.1994	0.9799	0.2035	4.915	1.020	5.016	30′
40′	0.2022	0.9793	0.2065	4.843	1.021	4.945	20′
50′	0.2051	0.9787	0.2095	4.773	1.022	4.876	10′
12° 00′	0.2079	0.9781	0.2126	4.705	1.022	4.810	78° 00′
10′	0.2108	0.9775	0.2156	4.638	1.023	4.745	50′
20′	0.2136	0.9769	0.2186	4.574	1.024	4.682	40′
30′	0.2164	0.9763	0.2217	4.511	1.024	4.620	30′
40′	0.2193	0.9757	0.2247	4.449	1.025	4.560	20′
50′	0.2221	0.9750	0.2278	4.390	1.026	4.502	10′
13° 00′	0.2250	0.9744	0.2309	4.331	1.026	4.445	77° 00′
10′	0.2278	0.9737	0.2339	4.275	1.027	4.390	50′
20′	0.2306	0.9730	0.2370	4.219	1.028	4.336	40′
30′	0.2334	0.9724	0.2401	4.165	1.028	4.284	30′
40′	0.2363	0.9717	0.2432	4.113	1.029	4.232	20′
50′	0.2391	0.9710	0.2462	4.061	1.030	4.182	10′
14° 00′	0.2419	0.9703	0.2493	4.011	1.031	4.134	76° 00′
10′	0.2447	0.9696	0.2524	3.962	1.031	4.086	50′
20′	0.2476	0.9689	0.2555	3.914	1.032	4.039	40′
30′	0.2504	0.9681	0.2586	3.867	1.033	3.994	30′
40′	0.2532	0.9674	0.2617	3.821	1.034	3.950	20′
50′	0.2560	0.9667	0.2648	3.776	1.034	3.906	10′
15° 00′	0.2588	0.9659	0.2679	3.732	1.035	3.864	75° 00′
10′	0.2616	0.9652	0.2711	3.689	1.036	3.822	50′
20′	0.2644	0.9644	0.2742	3.647	1.037	3.782	40′
30′	0.2672	0.9636	0.2773	3.606	1.038	3.742	30′
40′	0.2700	0.9628	0.2805	3.566	1.039	3.703	20′
50′	0.2728	0.9621	0.2836	3.526	1.039	3.665	10′
16° 00′	0.2756	0.9613	0.2867	3.487	1.040	3.628	74° 00′
10′	0.2784	0.9605	0.2899	3.450	1.041	3.592	50′
20′	0.2812	0.9596	0.2931	3.412	1.042	3.556	40′
30′	0.2840	0.9588	0.2962	3.376	1.043	3.521	30′
40′	0.2868	0.9580	0.2994	3.340	1.044	3.487	20′
50′	0.2896	0.9572	0.3026	3.305	1.045	3.453	10′
17° 00′	0.2924	0.9563	0.3057	3.271	1.046	3.420	73° 00′
10′	0.2952	0.9555	0.3089	3.237	1.047	3.388	50′
20′	0.2979	0.9546	0.3121	3.204	1.048	3.356	40′
30′	0.3007	0.9537	0.3153	3.172	1.049	3.326	30′
40′	0.3035	0.9528	0.3185	3.140	1.049	3.295	20′
50′	0.3062	0.9520	0.3217	3.108	1.050	3.265	10′
18° 00′	0.3090	0.9511	0.3249	3.078	1.051	3.236	72° 00′
	Cos	Sin	Cot	Tan	Csc	Sec	Measure

Table III

Values of trigonometric functions for angles

Measure	Sin	Cos	Tan	Cot	Sec	Csc	
18° 00′	0.3090	0.9511	0.3249	3.078	1.051	3.236	**72° 00′**
10′	0.3118	0.9502	0.3281	3.047	1.052	3.207	**50′**
20′	0.3145	0.9492	0.3314	3.018	1.053	3.179	**40′**
30′	0.3173	0.9483	0.3346	2.989	1.054	3.152	**30′**
40′	0.3201	0.9474	0.3378	2.960	1.056	3.124	**20′**
50′	0.3228	0.9465	0.3411	2.932	1.057	3.098	**10′**
19° 00′	0.3256	0.9455	0.3443	2.904	1.058	3.072	**71° 00′**
10′	0.3283	0.9446	0.3476	2.877	1.059	3.046	**50′**
20′	0.3311	0.9436	0.3508	2.850	1.060	3.021	**40′**
30′	0.3338	0.9426	0.3541	2.824	1.061	2.996	**30′**
40′	0.3365	0.9417	0.3574	2.798	1.062	2.971	**20′**
50′	0.3393	0.9407	0.3607	2.773	1.063	2.947	**10′**
20° 00′	0.3420	0.9397	0.3640	2.747	1.064	2.924	**70° 00′**
10′	0.3448	0.9387	0.3673	2.723	1.065	2.901	**50′**
20′	0.3475	0.9377	0.3706	2.699	1.066	2.878	**40′**
30′	0.3502	0.9367	0.3739	2.675	1.068	2.855	**30′**
40′	0.3529	0.9356	0.3772	2.651	1.069	2.833	**20′**
50′	0.3557	0.9346	0.3805	2.628	1.070	2.812	**10′**
21° 00′	0.3584	0.9336	0.3839	2.605	1.071	2.790	**69° 00′**
10′	0.3611	0.9325	0.3872	2.583	1.072	2.769	**50′**
20′	0.3638	0.9315	0.3906	2.560	1.074	2.749	**40′**
30′	0.3665	0.9304	0.3939	2.539	1.075	2.729	**30′**
40′	0.3692	0.9293	0.3973	2.517	1.076	2.709	**20′**
50′	0.3719	0.9283	0.4006	2.496	1.077	2.689	**10′**
22° 00′	0.3746	0.9272	0.4040	2.475	1.079	2.669	**68° 00′**
10′	0.3773	0.9261	0.4074	2.455	1.080	2.650	**50′**
20′	0.3800	0.9250	0.4108	2.434	1.081	2.632	**40′**
30′	0.3827	0.9239	0.4142	2.414	1.082	2.613	**30′**
40′	0.3854	0.9228	0.4176	2.394	1.084	2.595	**20′**
50′	0.3881	0.9216	0.4210	2.375	1.085	2.577	**10′**
23° 00′	0.3907	0.9205	0.4245	2.356	1.086	2.599	**67° 00′**
10′	0.3934	0.9194	0.4279	2.337	1.088	2.542	**50′**
20′	0.3961	0.9182	0.4314	2.318	1.089	2.525	**40′**
30′	0.3987	0.9171	0.4348	2.300	1.090	2.508	**30′**
40′	0.4014	0.9159	0.4383	2.282	1.092	2.491	**20′**
50′	0.4041	0.9147	0.4417	2.264	1.093	2.475	**10′**
24° 00′	0.4067	0.9135	0.4452	2.246	1.095	2.459	**66° 00′**
10′	0.4094	0.9124	0.4487	2.229	1.096	2.443	**50′**
20′	0.4120	0.9112	0.4522	2.211	1.097	2.427	**40′**
30′	0.4147	0.9100	0.4557	2.194	1.099	2.411	**30′**
40′	0.4173	0.9088	0.4592	2.177	1.100	2.396	**20′**
50′	0.4200	0.9075	0.4628	2.161	1.102	2.381	**10′**
25° 00′	0.4226	0.9063	0.4663	2.145	1.103	2.366	**65° 00′**
10′	0.4253	0.9051	0.4699	2.128	1.105	2.352	**50′**
20′	0.4279	0.9038	0.4734	2.112	1.106	2.337	**40′**
30′	0.4305	0.9026	0.4770	2.097	1.108	2.323	**30′**
40′	0.4331	0.9013	0.4806	2.081	1.109	2.309	**20′**
50′	0.4358	0.9001	0.4841	2.066	1.111	2.295	**10′**
26° 00′	0.4384	0.8988	0.4877	2.050	1.113	2.281	**64° 00′**
10′	0.4410	0.8975	0.4913	2.035	1.114	2.268	**50′**
20′	0.4436	0.8962	0.4950	2.020	1.116	2.254	**40′**
30′	0.4462	0.8949	0.4986	2.006	1.117	2.241	**30′**
40′	0.4488	0.8936	0.5022	1.991	1.119	2.228	**20′**
50′	0.4514	0.8923	0.5059	1.977	1.121	2.215	**10′**
27° 00′	0.4540	0.8910	0.5095	1.963	1.122	2.203	**63° 00′**
	Cos	**Sin**	**Cot**	**Tan**	**Csc**	**Sec**	**Measure**

Table III

Values of trigonometric functions for angles

Measure	Sin	Cos	Tan	Cot	Sec	Csc	
27° 00′	0.4540	0.8910	0.5095	1.963	1.122	2.203	**63° 00′**
10′	0.4566	0.8897	0.5132	1.949	1.124	2.190	**50′**
20′	0.4592	0.8884	0.5169	1.935	1.126	2.178	**40′**
30′	0.4617	0.8870	0.5206	1.921	1.127	2.166	**30′**
40′	0.4643	0.8857	0.5243	1.907	1.129	2.154	**20′**
50′	0.4669	0.8843	0.5280	1.894	1.131	2.142	**10′**
28° 00′	0.4695	0.8829	0.5317	1.881	1.133	2.130	**62° 00′**
10′	0.4720	0.8816	0.5354	1.868	1.134	2.118	**50′**
20′	0.4746	0.8802	0.5392	1.855	1.136	2.107	**40′**
30′	0.4772	0.8788	0.5430	1.842	1.138	2.096	**30′**
40′	0.4797	0.8774	0.5467	1.829	1.140	2.085	**20′**
50′	0.4823	0.8760	0.5505	1.816	1.142	2.074	**10′**
29° 00′	0.4848	0.8746	0.5543	1.804	1.143	2.063	**61° 00′**
10′	0.4874	0.8732	0.5581	1.792	1.145	2.052	**50′**
20′	0.4899	0.8718	0.5619	1.780	1.147	2.041	**40′**
30′	0.4924	0.8704	0.5658	1.767	1.149	2.031	**30′**
40′	0.4950	0.8689	0.5696	1.756	1.151	2.020	**20′**
50′	0.4975	0.8675	0.5735	1.744	1.153	2.010	**10′**
30° 00′	0.5000	0.8660	0.5774	1.732	1.155	2.000	**60° 00′**
10′	0.5025	0.8646	0.5812	1.720	1.157	1.990	**50′**
20′	0.5050	0.8631	0.5851	1.709	1.159	1.980	**40′**
30′	0.5075	0.8616	0.5890	1.698	1.161	1.970	**30′**
40′	0.5100	0.8601	0.5930	1.686	1.163	1.961	**20′**
50′	0.5125	0.8587	0.5969	1.675	1.165	1.951	**10′**
31° 00′	0.5150	0.8572	0.6009	1.664	1.167	1.942	**59° 00′**
10′	0.5175	0.8557	0.6048	1.653	1.169	1.932	**50′**
20′	0.5200	0.8542	0.6088	1.643	1.171	1.923	**40′**
30′	0.5225	0.8526	0.6128	1.632	1.173	1.914	**30′**
40′	0.5250	0.8511	0.6168	1.621	1.175	1.905	**20′**
50′	0.5275	0.8496	0.6208	1.611	1.177	1.896	**10′**
32° 00′	0.5299	0.8480	0.6249	1.600	1.179	1.887	**58° 00′**
10′	0.5324	0.8465	0.6289	1.590	1.181	1.878	**50′**
20′	0.5348	0.8450	0.6330	1.580	1.184	1.870	**40′**
30′	0.5373	0.8434	0.6371	1.570	1.186	1.861	**30′**
40′	0.5398	0.8418	0.6412	1.560	1.188	1.853	**20′**
50′	0.5422	0.8403	0.6453	1.550	1.190	1.844	**10′**
33° 00′	0.5446	0.8387	0.6494	1.540	1.192	1.836	**57° 00′**
10′	0.5471	0.8371	0.6536	1.530	1.195	1.828	**50′**
20′	0.5495	0.8355	0.6577	1.520	1.197	1.820	**40′**
30′	0.5519	0.8339	0.6619	1.511	1.199	1.812	**30′**
40′	0.5544	0.8323	0.6661	1.501	1.202	1.804	**20′**
50′	0.5568	0.8307	0.6703	1.492	1.204	1.796	**10′**
34° 00′	0.5592	0.8290	0.6745	1.483	1.206	1.788	**56° 00′**
10′	0.5616	0.8274	0.6787	1.473	1.209	1.781	**50′**
20′	0.5640	0.8258	0.6830	1.464	1.211	1.773	**40′**
30′	0.5664	0.8241	0.6873	1.455	1.213	1.766	**30′**
40′	0.5688	0.8225	0.6916	1.446	1.216	1.758	**20′**
50′	0.5712	0.8208	0.6959	1.437	1.218	1.751	**10′**
35° 00′	0.5736	0.8192	0.7002	1.428	1.221	1.743	**55° 00′**
10′	0.5760	0.8175	0.7046	1.419	1.223	1.736	**50′**
20′	0.5783	0.8158	0.7089	1.411	1.226	1.729	**40′**
30′	0.5807	0.8141	0.7133	1.402	1.228	1.722	**30′**
40′	0.5831	0.8124	0.7177	1.393	1.231	1.715	**20′**
50′	0.5854	0.8107	0.7221	1.385	1.233	1.708	**10′**
36° 00′	0.5878	0.8090	0.7265	1.376	1.236	1.701	**54° 00′**
	Cos	**Sin**	**Cot**	**Tan**	**Csc**	**Sec**	**Measure**

Table III

Values of trigonometric functions for angles

Measure	Sin	Cos	Tan	Cot	Sec	Csc	
36° 00′	0.5878	0.8090	0.7265	1.376	1.236	1.701	**54° 00′**
10′	0.5901	0.8073	0.7310	1.368	1.239	1.695	**50′**
20′	0.5925	0.8056	0.7355	1.360	1.241	1.688	**40′**
30′	0.5948	0.8039	0.7400	1.351	1.244	1.681	**30′**
40′	0.5972	0.8021	0.7445	1.343	1.247	1.675	**20′**
50′	0.5995	0.8004	0.7490	1.335	1.249	1.668	**10′**
37° 00′	0.6018	0.7986	0.7536	1.327	1.252	1.662	**53° 00′**
10′	0.6041	0.7969	0.7581	1.319	1.255	1.655	**50′**
20′	0.6065	0.7951	0.7627	1.311	1.258	1.649	**40′**
30′	0.6088	0.7934	0.7673	1.303	1.260	1.643	**30′**
40′	0.6111	0.7916	0.7720	1.295	1.263	1.636	**20′**
50′	0.6134	0.7898	0.7766	1.288	1.266	1.630	**10′**
38° 00′	0.6157	0.7880	0.7813	1.280	1.269	1.624	**52° 00′**
10′	0.6180	0.7862	0.7860	1.272	1.272	1.618	**50′**
20′	0.6202	0.7844	0.7907	1.265	1.275	1.612	**40′**
30′	0.6225	0.7826	0.7954	1.257	1.278	1.606	**30′**
40′	0.6248	0.7808	0.8002	1.250	1.281	1.601	**20′**
50′	0.6271	0.7790	0.8050	1.242	1.284	1.595	**10′**
39° 00′	0.6293	0.7771	0.8098	1.235	1.287	1.589	**51° 00′**
10′	0.6316	0.7753	0.8146	1.228	1.290	1.583	**50′**
20′	0.6338	0.7735	0.8195	1.220	1.293	1.578	**40′**
30′	0.6361	0.7716	0.8243	1.213	1.296	1.572	**30′**
40′	0.6383	0.7698	0.8292	1.206	1.299	1.567	**20′**
50′	0.6406	0.7679	0.8342	1.199	1.302	1.561	**10′**
40° 00′	0.6428	0.7660	0.8391	1.192	1.305	1.556	**50° 00′**
10′	0.6450	0.7642	0.8441	1.185	1.309	1.550	**50′**
20′	0.6472	0.7623	0.8491	1.178	1.312	1.545	**40′**
30′	0.6494	0.7604	0.8541	1.171	1.315	1.540	**30′**
40′	0.6517	0.7585	0.8591	1.164	1.318	1.535	**20′**
50′	0.6539	0.7566	0.8642	1.157	1.322	1.529	**10′**
41° 00′	0.6561	0.7547	0.8693	1.150	1.325	1.524	**49° 00′**
10′	0.6583	0.7528	0.8744	1.144	1.328	1.519	**50′**
20′	0.6604	0.7509	0.8796	1.137	1.332	1.514	**40′**
30′	0.6626	0.7490	0.8847	1.130	1.335	1.509	**30′**
40′	0.6648	0.7470	0.8899	1.124	1.339	1.504	**20′**
50′	0.6670	0.7451	0.8952	1.117	1.342	1.499	**10′**
42° 00′	0.6691	0.7431	0.9004	1.111	1.346	1.494	**48° 00′**
10′	0.6713	0.7412	0.9057	1.104	1.349	1.490	**50′**
20′	0.6734	0.7392	0.9110	1.098	1.353	1.485	**40′**
30′	0.6756	0.7373	0.9163	1.091	1.356	1.480	**30′**
40′	0.6777	0.7353	0.9217	1.085	1.360	1.476	**20′**
50′	0.6799	0.7333	0.9271	1.079	1.364	1.471	**10′**
43° 00′	0.6820	0.7314	0.9325	1.072	1.367	1.466	**47° 00′**
10′	0.6841	0.7294	0.9380	1.066	1.371	1.462	**50′**
20′	0.6862	0.7274	0.9435	1.060	1.375	1.457	**40′**
30′	0.6884	0.7254	0.9490	1.054	1.379	1.453	**30′**
40′	0.6905	0.7234	0.9545	1.048	1.382	1.448	**20′**
50′	0.6926	0.7214	0.9601	1.042	1.386	1.444	**10′**
44° 00′	0.6947	0.7193	0.9657	1.036	1.390	1.440	**46° 00′**
10′	0.6967	0.7173	0.9713	1.030	1.349	1.435	**50′**
20′	0.6988	0.7153	0.9770	1.024	1.398	1.431	**40′**
30′	0.7009	0.7133	0.9827	1.018	1.402	1.427	**30′**
40′	0.7030	0.7112	0.9884	1.012	1.406	1.423	**20′**
50′	0.7050	0.7092	0.9942	1.006	1.410	1.418	**10′**
45° 00′	0.7071	0.7071	1.0000	1.000	1.414	1.414	**45° 00′**
	Cos	**Sin**	**Cot**	**Tan**	**Csc**	**Sec**	**Measure**

Table IV

Logarithms of values of trigonometric functions (four places)

Measure	L Sin	d	L Tan	cd	L Cot	d	L Cos	
0° 0′	.....		.....		.....	0	10.0000	90° 0′
10′	7.4637	3011	7.4637	3011	12.5363	0	10.0000	50′
20′	7.7648	1760	7.7648	1761	12.2352	0	10.0000	40′
30′	7.9408	1250	7.9409	1249	12.0591	0	10.0000	30′
40′	8.0658	969	8.0658	969	11.9342	0	10.0000	20′
50′	8.1627	792	8.1627	792	11.8373	1	10.0000	10′
1° 0′	8.2419	669	8.2419	670	11.7581	0	9.9999	89° 0′
10′	8.3088	580	8.3089	580	11.6911	0	9.9999	50′
20′	8.3668	511	8.3669	512	11.6331	0	9.9999	40′
30′	8.4179	458	8.4181	457	11.5819	1	9.9999	30′
40′	8.4637	413	8.4638	415	11.5362	0	9.9998	20′
50′	8.5050	378	8.5053	378	11.4947	1	9.9998	10′
2° 0′	8.5428	348	8.5431	348	11.4569	0	9.9997	88° 0′
10′	8.5776	321	8.5779	322	11.4221	1	9.9997	50′
20′	8.6097	300	8.6101	300	11.3899	0	9.9996	40′
30′	8.6397	280	8.6401	281	11.3599	1	9.9996	30′
40′	8.6677	263	8.6682	263	11.3318	0	9.9995	20′
50′	8.6940	248	8.6945	249	11.3055	1	9.9995	10′
3° 0′	8.7188	235	8.7194	235	11.2806	1	9.9994	87° 0′
10′	8.7423	222	8.7429	223	11.2571	0	9.9993	50′
20′	8.7645	212	8.7652	213	11.2348	1	9.9993	40′
30′	8.7857	202	8.7865	202	11.2135	1	9.9992	30′
40′	8.8059	192	8.8067	194	11.1933	1	9.9991	20′
50′	8.8251	185	8.8261	185	11.1739	1	9.9990	10′
4° 0′	8.8436	177	8.8446	178	11.1554	0	9.9989	86° 0′
10′	8.8613	170	8.8624	171	11.1376	1	9.9989	50′
20′	8.8783	163	8.8795	165	11.1205	1	9.9988	40′
30′	8.8946	158	8.8960	158	11.1040	1	9.9987	30′
40′	8.9104	152	8.9118	154	11.0882	1	9.9986	20′
50′	8.9256	147	8.9272	148	11.0728	2	9.9985	10′
5° 0′	8.9403	142	8.9420	143	11.0580	1	9.9983	85° 0′
10′	8.9545	137	8.9563	138	11.0437	1	9.9982	50′
20′	8.9682	134	8.9701	135	11.0299	1	9.9981	40′
30′	8.9816	129	8.9836	130	11.0164	1	9.9980	30′
40′	8.9945	125	8.9966	127	11.0034	2	9.9979	20′
50′	9.0070	122	9.0093	123	10.9907	1	9.9977	10′
6° 0′	9.0192	119	9.0216	120	10.9784	1	9.9976	84° 0′
10′	9.0311	115	9.0336	117	10.9664	2	9.9975	50′
20′	9.0426	113	9.0453	114	10.9547	1	9.9973	40′
30′	9.0539	109	9.0567	111	10.9433	1	9.9972	30′
40′	9.0648	107	9.0678	108	10.9322	2	9.9971	20′
50′	9.0755	104	9.0786	105	10.9214	1	9.9969	10′
7° 0′	9.0859	102	9.0891	104	10.9109	2	9.9968	83° 0′
10′	9.0961	99	9.0995	101	10.9005	2	9.9966	50′
20′	9.1060	97	9.1096	98	10.8904	1	9.9964	40′
30′	9.1157	95	9.1194	97	10.8806	2	9.9963	30′
40′	9.1252	93	9.1291	94	10.8709	2	9.9961	20′
50′	9.1345	91	9.1385	93	10.8615	1	9.9959	10′
8° 0′	9.1436	89	9.1478	91	10.8522	2	9.9958	82° 0′
10′	9.1525	87	9.1569	89	10.8431	2	9.9956	50′
20′	9.1612	85	9.1658	87	10.8342	2	9.9954	40′
30′	9.1697	84	9.1745	86	10.8255	2	9.9952	30′
40′	9.1781	82	9.1831	84	10.8169	2	9.9950	20′
50′	9.1863	80	9.1915	82	10.8085	2	9.9948	10′
9° 0′	9.1943		9.1997		10.8003		9.9946	81° 0′
	L Cos	d	L Cot	cd	L Tan	d	L Sin	Measure

Table IV

Logarithms of values of trigonometric functions (four places)

Measure	L Sin	d	L Tan	cd	L Cot	d	L Cos	
9° 0′	9.1943	79	9.1997	81	10.8003	2	9.9946	**81° 0′**
10′	9.2022	78	9.2078	80	10.7922	2	9.9944	**50′**
20′	9.2100	76	9.2158	78	10.7842	2	9.9942	**40′**
30′	9.2176	75	9.2236	77	10.7764	2	9.9940	**30′**
40′	9.2251	73	9.2313	76	10.7687	2	9.9938	**20′**
50′	9.2324	73	9.2389	74	10.7611	2	9.9936	**10′**
10° 0′	9.2397	71	9.2463	73	10.7537	3	9.9934	**80° 0′**
10′	9.2468	70	9.2536	73	10.7464	2	9.9931	**50′**
20′	9.2538	68	9.2609	71	10.7391	2	9.9929	**40′**
30′	9.2606	68	9.2680	70	10.7320	3	9.9927	**30′**
40′	9.2674	66	9.2750	69	10.7250	2	9.9924	**20′**
50′	9.2740	66	9.2819	68	10.7181	3	9.9922	**10′**
11° 0′	9.2806	64	9.2887	66	10.7113	2	9.9919	**79° 0′**
10′	9.2870	64	9.2953	67	10.7047	3	9.9917	**50′**
20′	9.2934	63	9.3020	65	10.6980	2	9.9914	**40′**
30′	9.2997	61	9.3085	64	10.6915	3	9.9912	**30′**
40′	9.3058	61	9.3149	63	10.6851	2	9.9909	**20′**
50′	9.3119	60	9.3212	63	10.6788	3	9.9907	**10′**
12° 0′	9.3179	59	9.3275	61	10.6725	3	9.9904	**78° 0′**
10′	9.3238	58	9.3336	61	10.6664	2	9.9901	**50′**
20′	9.3296	57	9.3397	61	10.6603	3	9.9899	**40′**
30′	9.3353	57	9.3458	59	10.6542	3	9.9896	**30′**
40′	9.3410	56	9.3517	59	10.6483	3	9.9893	**20′**
50′	9.3466	55	9.3576	58	10.6424	3	9.9890	**10′**
13° 0′	9.3521	54	9.3634	57	10.6366	3	9.9887	**77° 0′**
10′	9.3575	54	9.3691	57	10.6309	3	9.9884	**50′**
20′	9.3629	53	9.3748	56	10.6252	3	9.9881	**40′**
30′	9.3682	52	9.3804	55	10.6196	3	9.9878	**30′**
40′	9.3734	52	9.3859	55	10.6141	3	9.9875	**20′**
50′	9.3786	51	9.3914	54	10.6086	3	9.9872	**10′**
14° 0′	9.3837	50	9.3968	53	10.6032	3	9.9869	**76° 0′**
10′	9.3887	50	9.4021	53	10.5979	3	9.9866	**50′**
20′	9.3937	49	9.4074	53	10.5926	4	9.9863	**40′**
30′	9.3986	49	9.4127	51	10.5873	3	9.9859	**30′**
40′	9.4035	48	9.4178	52	10.5822	3	9.9856	**20′**
50′	9.4083	47	9.4230	51	10.5770	4	9.9853	**10′**
15° 0′	9.4130	47	9.4281	50	10.5719	3	9.9849	**75° 0′**
10′	9.4177	46	9.4331	50	10.5669	3	9.9846	**50′**
20′	9.4223	46	9.4381	49	10.5619	4	9.9843	**40′**
30′	9.4269	45	9.4430	49	10.5570	3	9.9839	**30′**
40′	9.4314	45	9.4479	48	10.5521	4	9.9836	**20′**
50′	9.4359	44	9.4527	48	10.5473	4	9.9832	**10′**
16° 0′	9.4403	44	9.4575	47	10.5425	3	9.9828	**74° 0′**
10′	9.4447	44	9.4622	47	10.5378	4	9.9825	**50′**
20′	9.4491	42	9.4669	47	10.5331	4	9.9821	**40′**
30′	9.4533	43	9.4716	46	10.5284	3	9.9817	**30′**
40′	9.4576	42	9.4762	46	10.5238	4	9.9814	**20′**
50′	9.4618	41	9.4808	45	10.5192	4	9.9810	**10′**
17° 0′	9.4659	41	9.4853	45	10.5147	4	9.9806	**73° 0′**
10′	9.4700	41	9.4898	45	10.5102	4	9.9802	**50′**
20′	9.4741	40	9.4943	44	10.5057	4	9.9798	**40′**
30′	9.4781	40	9.4987	44	10.5013	4	9.9794	**30′**
40′	9.4821	40	9.5031	44	10.4969	4	9.9790	**20′**
50′	9.4861	39	9.5075	43	10.4925	4	9.9786	**10′**
18° 0′	9.4900		9.5118		10.4882		9.9782	**72° 0′**
	L Cos	d	L Cot	cd	L Tan	d	L Sin	Measure

Table IV

Logarithms of values of trigonometric functions (four places)

Measure	L Sin	d	L Tan	cd	L Cot	d	L Cos	
18° 0′	9.4900	39	9.5118	43	10.4882	4	9.9782	72° 0′
10′	9.4939	38	9.5161	42	10.4839	4	9.9778	50′
20′	9.4977	38	9.5203	42	10.4797	4	9.9774	40′
30′	9.5015	37	9.5245	42	10.4755	5	9.9770	30′
40′	9.5052	38	9.5287	42	10.4713	4	9.9765	20′
50′	9.5090	36	9.5329	41	10.4671	4	9.9761	10′
19° 0′	9.5126	37	9.5370	41	10.4630	5	9.9757	71° 0′
10′	9.5163	36	9.5411	40	10.4589	4	9.9752	50′
20′	9.5199	36	9.5451	40	10.4549	5	9.9748	40′
30′	9.5235	35	9.5491	40	10.4509	4	9.9743	30′
40′	9.5270	36	9.5531	40	10.4469	5	9.9739	20′
50′	9.5306	35	9.5571	40	10.4429	4	9.9734	10′
20° 0′	9.5341	34	9.5611	39	10.4389	5	9.9730	70° 0′
10′	9.5375	34	9.5650	39	10.4350	4	9.9725	50′
20′	9.5409	34	9.5689	38	10.4311	5	9.9721	40′
30′	9.5443	34	9.5727	39	10.4273	5	9.9716	30′
40′	9.5477	33	9.5766	38	10.4234	5	9.9711	20′
50′	9.5510	33	9.5804	38	10.4196	4	9.9706	10′
21° 0′	9.5543	33	9.5842	37	10.4158	5	9.9702	69° 0′
10′	9.5576	33	9.5879	38	10.4121	5	9.9697	50′
20′	9.5609	32	9.5917	37	10.4083	5	9.9692	40′
30′	9.5641	32	9.5954	37	10.4046	5	9.9687	30′
40′	9.5673	31	9.5991	37	10.4009	5	9.9682	20′
50′	9.5704	32	9.6028	36	10.3972	5	9.9677	10′
22° 0′	9.5736	31	9.6064	36	10.3936	5	9.9672	68° 0′
10′	9.5767	31	9.6100	36	10.3900	6	9.9667	50′
20′	9.5798	30	9.6136	36	10.3864	5	9.9661	40′
30′	9.5828	31	9.6172	36	10.3828	5	9.9656	30′
40′	9.5859	30	9.6208	35	10.3792	5	9.9651	20′
50′	9.5889	30	9.6243	36	10.3757	6	9.9646	10′
23° 0′	9.5919	29	9.6279	35	10.3721	5	9.9640	67° 0′
10′	9.5948	30	9.6314	34	10.3686	6	9.9635	50′
20′	9.5978	29	9.6348	35	10.3652	5	9.9629	40′
30′	9.6007	29	9.6383	34	10.3617	6	9.9624	30′
40′	9.6036	29	9.6417	35	10.3583	5	9.9618	20′
50′	9.6065	28	9.6452	34	10.3548	6	9.9613	10′
24° 0′	9.6093	28	9.6486	34	10.3514	5	9.9607	66° 0′
10′	9.6121	28	9.6520	33	10.3480	6	9.9602	50′
20′	9.6149	28	9.6553	34	10.3447	6	9.9596	40′
30′	9.6177	28	9.6587	33	10.3413	6	9.9590	30′
40′	9.6205	27	9.6620	34	10.3380	5	9.9584	20′
50′	9.6232	27	9.6654	33	10.3346	6	9.9579	10′
25° 0′	9.6259	27	9.6687	33	10.3313	6	9.9573	65° 0′
10′	9.6286	27	9.6720	32	10.3280	6	9.9567	50′
20′	9.6313	27	9.6752	33	10.3248	6	9.9561	40′
30′	9.6340	26	9.6785	32	10.3215	6	9.9555	30′
40′	9.6366	26	9.6817	33	10.3183	6	9.9549	20′
50′	9.6392	26	9.6850	32	10.3150	6	9.9543	10′
26° 0′	9.6418	26	9.6882	32	10.3118	7	9.9537	64° 0′
10′	9.6444	26	9.6914	32	10.3086	6	9.9530	50′
20′	9.6470	25	9.6946	31	10.3054	6	9.9524	40′
30′	9.6495	26	9.6977	32	10.3023	6	9.9518	30′
40′	9.6521	25	9.7009	31	10.2991	7	9.9512	20′
50′	9.6546	24	9.7040	32	10.2960	6	9.9505	10′
27° 0′	9.6570		9.7072		10.2928		9.9499	63° 0′
	L Cos	d	L Cot	cd	L Tan	d	L Sin	Measure

Table IV

Logarithms of values of trigonometric functions (four places)

Measure	L Sin	d	L Tan	cd	L Cot	d	L Cos	
27° 0′	9.6570	25	9.7072	31	10.2928	7	9.9499	**63° 0′**
10′	9.6595	25	9.7103	31	10.2897	6	9.9492	**50′**
20′	9.6620	24	9.7134	31	10.2866	7	9.9486	**40′**
30′	9.6644	24	9.7165	31	10.2835	6	9.9479	**30′**
40′	9.6668	24	9.7196	30	10.2804	7	9.9473	**20′**
50′	9.6692	24	9.7226	31	10.2774	7	9.9466	**10′**
28° 0′	9.6716	24	9.7257	30	10.2743	6	9.9459	**62° 0′**
10′	9.6740	23	9.7287	30	10.2713	7	9.9453	**50′**
20′	9.6763	24	9.7317	31	10.2683	7	9.9446	**40′**
30′	9.6787	23	9.7348	30	10.2652	7	9.9439	**30′**
40′	9.6810	23	9.7378	30	10.2622	7	9.9432	**20′**
50′	9.6833	23	9.7408	30	10.2592	7	9.9425	**10′**
29° 0′	9.6856	22	9.7438	29	10.2562	7	9.9418	**61° 0′**
10′	9.6878	23	9.7467	30	10.2533	7	9.9411	**50′**
20′	9.6901	22	9.7497	29	10.2503	7	9.9404	**40′**
30′	9.6923	23	9.7526	30	10.2474	7	9.9397	**30′**
40′	9.6946	22	9.7556	29	10.2444	7	9.9390	**20′**
50′	9.6968	22	9.7585	29	10.2415	8	9.9383	**10′**
30° 0′	9.6990	22	9.7614	30	10.2386	7	9.9375	**60° 0′**
10′	9.7012	21	9.7644	29	10.2356	7	9.9368	**50′**
20′	9.7033	22	9.7673	28	10.2327	8	9.9361	**40′**
30′	9.7055	21	9.7701	29	10.2299	7	9.9353	**30′**
40′	9.7076	21	9.7730	29	10.2270	8	9.9346	**20′**
50′	9.7097	21	9.7759	29	10.2241	7	9.9338	**10′**
31° 0′	9.7118	21	9.7788	28	10.2212	8	9.9331	**59° 0′**
10′	9.7139	21	9.7816	29	10.2184	8	9.9323	**50′**
20′	9.7160	21	9.7845	28	10.2155	7	9.9315	**40′**
30′	9.7181	20	9.7873	29	10.2127	8	9.9308	**30′**
40′	9.7201	21	9.7902	28	10.2098	8	9.9300	**20′**
50′	9.7222	20	9.7930	28	10.2070	8	9.9292	**10′**
32° 0′	9.7242	20	9.7958	28	10.2042	8	9.9284	**58° 0′**
10′	9.7262	20	9.7986	28	10.2014	8	9.9276	**50′**
20′	9.7282	20	9.8014	28	10.1986	8	9.9268	**40′**
30′	9.7302	20	9.8042	28	10.1958	8	9.9260	**30′**
40′	9.7322	20	9.8070	27	10.1930	8	9.9252	**20′**
50′	9.7342	19	9.8097	28	10.1903	8	9.9244	**10′**
33° 0′	9.7361	19	9.8125	28	10.1875	8	9.9236	**57° 0′**
10′	9.7380	20	9.8153	27	10.1847	9	9.9228	**50′**
20′	9.7400	19	9.8180	28	10.1820	8	9.9219	**40′**
30′	9.7419	19	9.8208	27	10.1792	8	9.9211	**30′**
40′	9.7438	19	9.8235	28	10.1765	9	9.9203	**20′**
50′	9.7457	19	9.8263	27	10.1737	8	9.9194	**10′**
34° 0′	9.7476	18	9.8290	27	10.1710	9	9.9186	**56° 0′**
10′	9.7494	19	9.8317	27	10.1683	8	9.9177	**50′**
20′	9.7513	18	9.8344	27	10.1656	9	9.9169	**40′**
30′	9.7531	19	9.8371	27	10.1629	9	9.9160	**30′**
40′	9.7550	18	9.8398	27	10.1602	9	9.9151	**20′**
50′	9.7568	18	9.8425	27	10.1575	8	9.9142	**10′**
35° 0′	9.7586	18	9.8452	27	10.1548	9	9.9134	**55° 0′**
10′	9.7604	18	9.8479	27	10.1521	9	9.9125	**50′**
20′	9.7622	18	9.8506	27	10.1494	9	9.9116	**40′**
30′	9.7640	17	9.8533	26	10.1467	9	9.9107	**30′**
40′	9.7657	18	9.8559	27	10.1441	9	9.9098	**20′**
50′	9.7675	17	9.8586	27	10.1414	9	9.9089	**10′**
36° 0′	9.7692		9.8613		10.1387		9.9080	**54° 0′**
	L Cos	**d**	**L Cot**	**cd**	**L Tan**	**d**	**L Sin**	**Measure**

Table IV

Logarithms of values of trigonometric functions (four places)

Measure	L Sin	d	L Tan	cd	L Cot	d	L Cos	
36° 0′	9.7692	18	9.8613	26	10.1387	10	9.9080	54° 0′
10′	9.7710	17	9.8639	27	10.1361	9	9.9070	50′
20′	9.7727	17	9.8666	26	10.1334	9	9.9061	40′
30′	9.7744	17	9.8692	26	10.1308	10	9.9052	30′
40′	9.7761	17	9.8718	27	10.1282	9	9.9042	20′
50′	9.7778	17	9.8745	26	10.1255	10	9.9033	10′
37° 0′	9.7795	16	9.8771	26	10.1229	9	9.9023	53° 0′
10′	9.7811	17	9.8797	27	10.1203	10	9.9014	50′
20′	9.7828	16	9.8824	26	10.1176	9	9.9004	40′
30′	9.7844	17	9.8850	26	10.1150	10	9.8995	30′
40′	9.7861	16	9.8876	26	10.1124	10	9.8985	20′
50′	9.7877	16	9.8902	26	10.1098	10	9.8975	10′
38° 0′	9.7893	17	9.8928	26	10.1072	10	9.8965	52° 0′
10′	9.7910	16	9.8954	26	10.1046	10	9.8955	50′
20′	9.7926	15	9.8980	26	10.1020	10	9.8945	40′
30′	9.7941	16	9.9006	26	10.0994	10	9.8935	30′
40′	9.7957	16	9.9032	26	10.0968	10	9.8925	20′
50′	9.7973	16	9.9058	26	10.0942	10	9.8915	10′
39° 0′	9.7989	15	9.9084	26	10.0916	10	9.8905	51° 0′
10′	9.8004	16	9.9110	25	10.0890	11	9.8895	50′
20′	9.8020	15	9.9135	26	10.0865	10	9.8884	40′
30′	9.8035	15	9.9161	26	10.0839	10	9.8874	30′
40′	9.8050	16	9.9187	25	10.0813	11	9.8864	20′
50′	9.8066	15	9.9212	26	10.0788	10	9.8853	10′
40° 0′	9.8081	15	9.9238	26	10.0762	11	9.8843	50° 0′
10′	9.8096	15	9.9264	25	10.0736	11	9.8832	50′
20′	9.8111	14	9.9289	26	10.0711	11	9.8821	40′
30′	9.8125	15	9.9315	26	10.0685	10	9.8810	30′
40′	9.8140	15	9.9341	25	10.0659	11	9.8800	20′
50′	9.8155	14	9.9366	26	10.0634	11	9.8789	10′
41° 0′	9.8169	15	9.9392	25	10.0608	11	9.8778	49° 0′
10′	9.8184	14	9.9417	26	10.0583	11	9.8767	50′
20′	9.8198	15	9.9443	25	10.0557	11	9.8756	40′
30′	9.8213	14	9.9468	26	10.0532	12	9.8745	30′
40′	9.8227	14	9.9494	25	10.0506	11	9.8733	20′
50′	9.8241	14	9.9519	25	10.0481	11	9.8722	10′
42° 0′	9.8255	14	9.9544	26	10.0456	12	9.8711	48° 0′
10′	9.8269	14	9.9570	25	10.0430	11	9.8699	50′
20′	9.8283	14	9.9595	26	10.0405	12	9.8688	40′
30′	9.8297	14	9.9621	25	10.0379	11	9.8676	30′
40′	9.8311	13	9.9646	25	10.0354	12	9.8665	20′
50′	9.8324	14	9.9671	26	10.0329	12	9.8653	10′
43° 0′	9.8338	13	9.9697	25	10.0303	12	9.8641	47° 0′
10′	9.8351	14	9.9722	25	10.0278	11	9.8629	50′
20′	9.8365	13	9.9747	25	10.0253	12	9.8618	40′
30′	9.8378	13	9.9772	26	10.0228	12	9.8606	30′
40′	9.8391	14	9.9798	25	10.0202	12	9.8594	20′
50′	9.8405	13	9.9823	25	10.0177	13	9.8582	10′
44° 0′	9.8418	13	9.9848	26	10.0152	12	9.8569	46° 0′
10′	9.8431	13	9.9874	25	10.0126	12	9.8557	50′
20′	9.8444	13	9.9899	25	10.0101	13	9.8545	40′
30′	9.8457	12	9.9924	25	10.0076	12	9.8532	30′
40′	9.8469	13	9.9949	26	10.0051	13	9.8520	20′
50′	9.8482	13	9.9975	25	10.0025	12	9.8507	10′
45° 0′	9.8495		10.0000		10.0000		9.8495	45° 0′
	L Cos	d	L Cot	cd	L Tan	d	L Sin	Measure

SELECTED ANSWERS

Section 1-1, page 3

1.a. Q, R **c.** W, I, Q, R **e.** Irrational numbers, R **g.** I, Q, R **i.** Irrational numbers, R **k.** Q, R **3.** Irrational **5.** Irrational **7.** Rational In exercises 8–11 answers may vary.

9. $\begin{matrix} 1, & 2, & 3, & \ldots, & n, & \ldots \\ \updownarrow & \updownarrow & \updownarrow & & \updownarrow & \\ -1, & -2, & -3, & \ldots, & -n, & \ldots \end{matrix}$ **11.** $\begin{matrix} 1, & 2, & 3, & \ldots, & n, & n+1, & \ldots \\ \updownarrow & \updownarrow & \updownarrow & & \updownarrow & \updownarrow & \\ 0, & -1, & 1, & \ldots, & -\frac{n}{2}, & \frac{n}{2}, & \ldots \end{matrix}$ where n is even.

12.a. Commutative property for addition **c.** Inverse property for multiplication **e.** Additive inverse property **g.** Additive identity property **i.** Commutative property for multiplication **13.** $-36, -1195, 9,$ **14.** $-\frac{1}{35}, -\frac{29}{6}, -2.2762$

15. $2\pi, 2\sqrt{2}, \pi, 9, 10$ **17.** 15, 15, 0, 0 **19.** $-\frac{6}{35}, \frac{35}{6}, -0.0640512$

20. 105, 105, 1, 1.

Section 1-2, page 7

1.a. Theorem 2 **c.** Theorem 1 **2.a.** ab^{-1} **c.** $b \cdot b^{-1}$ **3.a.** $\frac{-a}{b}, -\frac{a}{b}, \frac{-a}{-b}, \frac{a}{b}$

5. No. $a = b$ or $a = -b$, where $a, b \neq 0$ **7.a.** $-36, -1195, 9, -57$ **8.a.** π **c.** -7

e. 1 **g.** 3 **i.** $-6 + \sqrt{3}$ **k.** -14 **9.** 12, 12, Yes **11.a.** $\frac{-22}{63}$ **c.** $\sqrt{2}$ **e.** $-\frac{2}{15}$

Section 1-3, page 11

1.a. $1.9 < 1.99, 1.9 \leqslant 1.99, 1.99 > 1.9, 1.99 \geqslant 1.9$ **c.** $1.09 < 1.9, 1.09 \leqslant 1.9, 1.9 > 1.09, 1.9 \geqslant 1.09$ **e.** $-1.9 < 1.09, -1.9 \leqslant 1.09, 1.09 > -1.9, 1.09 \geqslant -1.9$ **3.** No, yes, no, yes. **5.** Answers may vary: $2 < 3$ and $2 \cdot 4 < 3 \cdot 4$, $2 < 3$ and $3(-4) < 2(-4)$, $2 \neq 3$ and $2 \cdot 0 = 3 \cdot 0$ **7.** (i) The points for $a + c$ and $b + c$ on the real number line are at a distance c to the right of the points for a and b, respectively. (ii) The points for $a + c$ and $b + c$ are at a distance $|c|$ to the left of the points for a and b, respectively. **9.a.** 7 **c.** 8 **e.** 3 **g.** 1 **i.** 5 **k.** 1 **10.a.** $\{4, -4\}$ **c.** $\{0\}$ **e.** $\{1, -7\}$ **g.** $\{8, 2\}$ **11.a.** $\{3, 4, 5, 6, 7\}$ **c.** $\{x\colon\ x \in I \text{ and } x < -6\} \cup \{x\colon\ x \in I \text{ and } x > 4\}$ **12.a.** Graph: the ray whose endpoint is 2 and all points on the number line to the right of 2. **c.** Graph: the half-line with endpoint at -1 and all points on the number line to the left of -1. **13.a.** $\{x\colon\ |x| > 4\}$

Chapter 1 Review Exercises, page 12

1.a. $(3 \cdot \sqrt{5}) \cdot \frac{2}{15} = 3\left(\sqrt{5} \cdot \frac{2}{15}\right)$ **b.** $3\left(\sqrt{5} + \frac{2}{15}\right) = 3\sqrt{5} + 3 \cdot \frac{2}{15}$ **2.** None in I, $-\frac{1}{2}$ in Q and R **3.a.** Theorem 5 **b.** Theorem 6 **c.** Theorem 3

4. $\{-1, 0, 1, 2, 3, 4, 5, 6, 7\}$ **5.** Graph: the open segment on the number line whose endpoints are -4 and 2.

Section 2-1, page 16

1.a. $n = 3, a_3 = \frac{1}{2}, a_2 = \pi, a_1 = 3, a_0 = \sqrt{2}$ **c.** $n = 3, a_3 = \frac{4}{9}, a_2 = 0, a_1 = 3, a_0 = -2$

e. $n = 2, a_2 = -1, a_1 = 0, a_0 = 1$ **g.** $n = 4, a_4 = a_2 = a_1 = 0, a_3 = 4, a_0 = -3$ **i.** $n = 0,$

$a_0 = 0$ **3.a.** 3 **b.** 0 **c.** 3 **d.** 1 **e.** 2 **g.** 3 **h.** 100 **j.** 3; **f., i.** no degree
5. Sum: **a.** $9x^3 + 3x + 16$ **c.** 1 **e.** $24x^8 - 1$; Difference: **a.** $x^3 + 6x^2 - 7x + 2$
c. $6x^5 + 4x^2 - 14x + 15$ **e.** $-24x^8 + 1$ **6.** 1st: **a.** 3 **c.** 5 **e.** None; 2nd: **a.** 3
c. 5 **e.** 8; Sum: **a.** 3 **c.** 0 **e.** 8; Difference: **a.** 3 **c.** 5 **e.** 8 **7.a.** $-2x^3 - 8x^2$
c. $5x^4 + 2x^3 - 8x + 13$ **e.** $2x^3 - 6x^2$ **g.** $14x^7 - 6x^4 - 4x^3 + 4x^2 - 3x - 5$ **i.** 0
9. All $a_2x^2 + a_1x + a_0$ where $a_2 \neq 0$; all $a_1x + a_0$ where $a_1 \neq 0$; all a_0 except $a_0 = 0$; 0
11.a. $2x^5 + x^4 + 7x^3 + 4x^2 - x + 2$, $-4x^3 - 2x^2 + 7x + 5$, $7x^3 + 4x^2 + 7x - 6$
12.a. Since $a_n = a_n, a_{n-1} = a_{n-1}, \ldots, a_1 = a_1$, and $a_0 = a_0$, then $p = p$. **c.** If $p = q$ and $q = r$, then $a_n = b_n, a_{n-1} = b_{n-1}, \ldots, a_1 = b_1, a_0 = b_0$ and $b_n = c_n, b_{n-1} = c_{n-1}, \ldots, b_1 = c_1, b_0 = c_0$. Thus $a_n = c_n, a_{n-1} = c_{n-1}, \ldots, a_1 = c_1, a_0 = c_0$ and $p = r$.
e.
$$\begin{aligned}(p + q) + r &= ((a_n + b_n)x^n + \cdots + (a_1 + b_1)x + (a_0 + b_0)) + (c_nx^n + \cdots + c_1x + c_0)\\ &= ((a_n + b_n) + c_n)x^n + \cdots + ((a_1 + b_1) + c_1)x + ((a_0 + b_0) + c_0)\\ &= (a_n + (b_n + c_n))x^n + \cdots + (a_1 + (b_1 + c_1))x + (a_0 + (b_0 + c_0))\\ &= (a_nx^n + \cdots + a_1x + a_0) + ((b_n + c_n)x^n + \cdots + (b_1 + c_1)x + (b_0 + c_0))\\ &= p + (q + r)\end{aligned}$$
13. $(3x^2 - 9x - 1) + (4x^2 + 5x + 4) = 7x^2 - 4x + 3$. Hence $(7x^2 - 4x + 3) - (4x^2 + 5x + 4) = 3x^2 - 9x - 1$.

Section 2-2, page 21

1. $28x^7 + 32x^3$ **3.** $51x^3 - 68x^2 + 34x - 119$ **5.** $-14x^6 + 6x^5 - 4x^4 + 8x^3$
7. $6x^2 + 2x - 28$ **9.** $56x^2 + 123x + 52$ **11.** $6x^2 + 11x - 35$ **13.** $51x^2 + 232x - 28$
15. $x^2 - x - 6$ **17.** $9x^2 - 16$ **19.** $4x^9 + 2x^8 - x^7$ **21.** $x^2 - 25$
23. $2x^4 + 9x^3 + 13x^2 + 4x - 3$ **25.** $30x^6 - 5x^5 = 33x^4 + 45x^3 - 22x^2 + 17x - 12$
27. $x^7 + x^5 - 10x^3 + 8x$ **29.** $x^3 + 6x^2 + 12x + 8$
31. $3x^{10} + x^8 - 19x^6 + 27x^5 + 4x^4 + 9x^3 - 17x^2 - 36x - 12$
33. $x^3 - 3rx^2 + 3r^2x - r^3$ **35.** $x^3 - 2x^2 + 3x - 4$, 5
37. $\frac{1}{3}x^5 - \frac{5}{9}x^3 - x^2 + \frac{25}{27}x + \frac{5}{31}, -\frac{17}{27}x - \frac{22}{3}$ **39.** $x^3 - 5, -2x^2 + 1$
41. $x^4 - 4x^3 + 15x^2 - 56x + 209, -780x - 210$

Section 2-3, page 24

1. 2, 4; 2, 16; $x^3 + 3x + 2$, $x - 2$, $x^2 + 2x + 7$, 16 **3.** −8, 14, −28; 4, 14;
$4x^3 + x^2 + 3$, $x + 2$, $4x^2 - 7x + 14$, −25 **5.** $x^2 + x + 4$, 3 **7.** $x^3 + 2x^2 + 4x + 8$, 13
9. $3x^2 + 11x + 29$, 92 **11.** $x^3 - 4$, −1 **13.a.** $x^2 + x - 6$, 0 **c.** $x^2 + 3x + 2$, 12
e. $x^2 - 2x - 3$, 12 **14.a.** $x^2 - 2x - 3$, −6 **c.** $x^2 - 1$, −6 **e.** $x^2 - 5x + 9$, −21
15.a. $x^2 - \frac{1}{2}x + 1, \frac{1}{2}$ **c.** $x^2 - \frac{5}{4}x + \frac{19}{16}, -\frac{13}{64}$ **e.** $x^2 - 2x + \frac{5}{2}, -\frac{7}{4}$ **17.** a, b, e, h.

Section 2-4, page 28

1. Prime **3.** Composite **5.** Prime **7.** Neither **9.** Composite **11.** Composite
12. (3) $(x - 5)(x + 5)(x^2 + 25)$ (9) $2(x - 3)(x + 3)$ (11) $(5x + 6)(25x^2 - 30x + 36)$
13.a. $(4x - 5)(2x + 3)$ **c.** $(4x + 3)(2x - 5)$ **15.** $\frac{1}{105}(70x^2 + 60x + 63)$
17. a, d, g, h **19.** a, b, c, d, e, f, g, h

Section 2-5, page 30

1.a. 4 **c.** 3 **e.** 4 **g.** 0 **i.** 2 **k.** 2 **3.** $2x^3y^2 + 10x^2y^3 + 3x^2y^2 + 7x + 6y$
5. $3x^2y + 4xy^2 + 7y^2 + 5xy + x - y$ **7.** $x^2y^2z - 2xz$ **9.** $4x^3 - xy^2$
11. $4x^4 - 4x^2y^2 + y^4$ **13.** $2x^2 - 2y^2$ **15.** $x^2 - y^2$ **17.** $49x^2 + 70xy + 25y^2$
19. $4x^3y - 8x^2y^3 + 4xy^4$ **21.** $3x^2 + 15xy - 6x$ **23.** $3x^2 + 11xy - 6x + 8y - 20y^2$
25. $x^3 - y^3$ **27.** $4x^3 + 5x^2y + 2x^2z - 2xy^2 + 4xz^2 + 4x^2yz - 3xy^2z + 2xyz^2 + 4xyz + 2y^2z - 3yz^2 - 3y^3 + 2z^3$ **28.a.** $xy(x - y)$ **c.** $(4x - 3y)(x - y)$ **29.a.** $3y$
c. $8x$ **e.** $7x^2y^2$ **31.** $6xy^2(1 + 3x^2y - y)$ **33.** $8x(x + 1)$ **35.** $2y(3x - 2y)$
37. $2xy^2(3xy - 4 + 2y)$ **39.** $9xy(4x - 2y + 1)$ **41.** $(2x + 1)(x - y)$
43. $(x^2 - 2)(y^2 - 5y + 1)$ **45.** $(4y - x)(x + y)$ **47.** $\frac{1}{6}x^2y(2x^2y^2 - y - 4x)$
49. $(x - 1)(x + 1)(y + 1)(y + 1)$

Section 2-6, page 33

1. $(y - x)(y + x)$ **3.** $(7 - y)(7 + y)$ **5.** $(3x + 3y - 4)(3x + 3y + 4)$
7. $(2x - y)(2x + y)$ **9.** $(2x - 3y)(2x + 3y)$ **11.** $3x(x - y)(x + y)$
13. $(2x - 5y)(4x^2 + 10xy + 25y^2)$ **15.** $(2x + 5y)(4x^2 - 10xy + 25y^2)$
17. $(3x - 5y)(3x + 5y)(y - x)(y + x)$ **19.** $\frac{1}{1000}(2x - 3y)(4x^2 + 6xy + 9y^2)$

21. $(2x - y - 1)(4x^2 - 4xy + y^2 + 2x - y + 1)$ **23.** $(y - x)(y^2 + xy + x^2)$ **25.** $x^3(xy^2 + 6)(x^2y^4 - 6xy^2 + 36)$ **27.** $(x - 7)(x - 1)$ **29.** $(y - 2)^2$ **31.** $2(x - 1)(x - 3)$ **33.** $(y + 2)^2$ **35.** $(4x - 3y)(2x - 5y)$ **37.** $(x - 2y)^2$ **39.** $(x - 5y)^2$ **41.** $(5x + y)^2$ **43.** $9x(5 + 3y)(1 - y)$ **45.** $(8x - y)(x - 9y)$ **47.** $y(5x + 7y)^2$ **49.** $(4x - 7y)^2$ **51.** $(x + y + 4)(x + y - 3)$ **53.** $(9x - 11y)^2$ **55.** $(x + 1 - y)(x + 1 + y)$ **57.** $(x^2 + 1)(x + 2)$ **59.** $(x^2 + 1)(x - 2)$ **61.** $(3 - 2x - 3y)(3 + 2x + 3y)$ **63.** $(x - y)(x^2 + xy + y^2 - x)$ **65.** $(2 + x + y)(x - y)$ **67.** $(y - 2)(x + y)(x - y)$ **69.** $(x + y)(x^2 - xy + y^2 + 1)$ **71.** $(x + 3 + y)(x + 3 - y)$

Section 2-7, page 38

1.a. Yes. $8 \cdot 3 = 12 \cdot 2$ c. Yes. $(x - 3)(2x + 8) = (x - 3)(2)(x + 4)$ e. Yes. $(x - 1)(x + 1) = x^2 - 1$ **2.a.** $\frac{35}{79}$ c. $\frac{5}{(2x - 6)}$ e. $\frac{(x - 3)}{(x - 4)}$ g. $\frac{2x}{3y}$ i. $\frac{(x + 2)}{(x + 1)}$ k. $\frac{(x - 3)}{(x - 4)}$ m. $\frac{(x^2 - 4)}{(x^2 + 4)}$ o. 1 q. $\frac{1}{(x + 2)}$ **3.a.** $\frac{(2x^2 - 18x + 28)}{(2x^2 + 17x + 21)}$ c. $\frac{4}{x^2}$ e. $\frac{(6x^2 - 3x - 6)}{(2x^3 + 5x^2 + 3x)}$ g. $\frac{9x}{22y}$ i. $\frac{(x - 2y)}{x}$ k. $\frac{(xy - 3x^2)}{(y^2 - 2xy)}$ m. 1 o. $\frac{(x + 2)}{(x^3 + 2x^2 + x)}$ q. 1 s. 1 u. $\frac{(x^2 - 4x + 3)}{(3x^2 + 2x - 1)}$ **4.a.** $\frac{31}{36}$ c. $\frac{(24x - 34)}{(6x^2 - 17x + 5)}$ e. $\frac{(7x - 2)}{(6x^2 - 5x - 4)}$ g. $\frac{(3x + 3)}{(x^3 + 4x^2 + x - 6)}$ i. $\frac{(x^2 - xy - y^2)}{(xy)(x - y)}$ k. $\frac{(4x^2 - 7x + 4)}{(x - 1)}$ m. $\frac{(y - x)}{xy}$ o. $\frac{-2y}{(x^2 - y^2)}$ q. $\frac{(3x^2 - 8x + 1)}{(x^3 - x^2 - 14x + 24)}$ s. $\frac{(-x + 12)}{(x^2 - 9)}$ u. $\frac{(2x^2 - 7x + 7)}{(x^2 - 2x)}$ w. $\frac{(2x^2 + 16x - 9)}{(2x^3 + 9x^2 + 4x - 15)}$ y. $\frac{(-3x^2 - 17x + 19)}{(2x^3 - 3x^2 - 8x + 12)}$

Section 2-8, page 43

1. $\frac{1}{7}$ **3.** $\frac{(2x^2 - 13x - 70)}{(9x^2 + 12x + 4)}$ **5.** $\frac{(y + x)}{(y - x)}$ **7.** $\frac{(2x^2 - 5xy + 2y^2)}{(x^2 - y^2)}$ **9.** $\frac{(-x^2 + 7x + 6)}{(-x^3 + 2x^2 + 3x - 6)}$ **11.** $\frac{2}{(1 + x)}$ **13.** $\frac{(xy^2 + y)}{(x^2y - x)}$ **15.** -1 **17.** 1 **19.** $\frac{-1}{(x^2 + xy)}$ **21.** $\frac{(x^2 - 8x + 16)}{(x^2 - 2x + 1)}$ **23.** Since $\frac{p}{q} \cdot \frac{t}{r} \cdot \frac{r}{t} = \frac{p}{q}$, then $\frac{p}{q} \div \frac{r}{t} = \frac{p}{q} \cdot \frac{t}{r}$ by definition of division of rational expressions.

Chapter 2 Review Exercises, page 44

1.a. 2 **b.** 2 **c.** 0 **d.** 1 **e.** 0 **2.a.** $5x^2 + 4x + 18$ **b.** $-5x^2 - 4x - 18$ **3.** $216x^3 - 756x^2 + 882x - 343$ **4.** $q = x^3 + 6x - 5, r = 15x^3 - 43x^2 + 31x - 1$ **5.a.** $x^4 + 2x^3 + 7x^2 + 20x + 40, r = -75$ **b.** $x^4 - 3x^3 + 12x^2 - 30x + 90, r = -275$ **6.a.** $(x - 1)(x + 1)(x - \sqrt{5})(x + \sqrt{5})$ **b.** $(x - 1)(x + 1)(x^2 - 5)$ **7.** $4x^2y - 3xy^2 + 8xy - 5y, 4x^4y^2 - 3x^3y^4 + 7x^3y^3 - 20x^2y^2 + 15xy^4 - 35xy^2$ **8.** $xy(x - y)$ **9.** $(6y + x)(36y^2 - 6xy + x^2)$ **10.** $3y(4x + y)(2x - y)$ **11.** $(x - y)(x + y)(3y - 4)$ **12.** $\frac{(x^2 - y^2)}{(2x^2 - 3y^2)}$ **13.** $\frac{(xy - y^2)}{(x^2 - xy)}$ **14.** $\frac{(x^2 + 2xy - y^2)}{(x^2y - xy^2)}$ **15.** $\frac{3y}{(2x^2 + 7xy + 6y^2)}$ **16.a.** -1 **b.** $\frac{(xy^2 + 2xy + y^2 + 3y + x + 2)}{(x^3y + x^2 + 2xy + x + y)}$

Section 3-1, page 47

1. A segment with endpoints 0 and 4. **3.** A line through the origin, making a 45° angle with the x-axis and passing through quadrants I and III. **5.** The set containing the x- and y-axes. **7.** Graph: the set of points on, and to the right of, the line $x = 2$. **9.** Graph: the intersection set of the points on, and to the right of the line $x = 2$, and the points on, and below, the line $y = -1$. **11.a.** Replacement set: open circles at every integer. Solution set: points at $-6, -5, -4, -3, -2, -1, 4, 5$, and 6, all points for integers less than -6, and all points for integers greater than 6. **b.** Replacement set: real number line. Solution set: two open rays or half-lines; one determined by an open circle at $-\frac{1}{2}$ and all points to the left of $-\frac{1}{2}$ on the number line, the other determined by an open circle at $3\frac{1}{2}$ and all points to the right of $3\frac{1}{2}$ on the number line. **14.c.** Graph: an open segment—all points on the number line between -4 and 4. **d.** Graph: the vertical strip between the lines $x = 4$ and $x = -4$.

Section 3-2, page 50 **1.** $\{2\}$ **3.** $\{3\}$ **5.** $\{5\}$ **7.** $\{1 - \pi\}$ **9.** $\left\{\frac{18}{5}\right\}$ **11.** $\left\{\frac{10}{\pi - 2}\right\}$ **13.** $\left\{\frac{10}{31}\right\}$ **15.** $\left\{\frac{16}{3}\right\}$ **17.** Let r be a member of the solution set of the sentence $7x + 3 = 8$. Using the theorems for the real numbers we have $7r = 5$ and $r = \frac{5}{7}$. Hence, r is a member of the solution set for the sentence $x = \frac{5}{7}$. Conversely, if r is a real number in the solution set for $x = \frac{5}{7}$, then $r = \frac{5}{7}$, $7r = 5$ and $7r + 3 = 8$. r is in the solution set for $7x + 3 = 8$. Since both equations have the same solution set, they are equivalent. **19.** Use the procedure of Exercise 17 to show that if r is a solution of $3x - 7 = 6x + 12$, then r is a solution of $x = -\frac{19}{3}$, and *vice versa*. **21.** $h = \frac{V}{b}$ **23.** $h = \frac{3V}{\pi r^2}$ **25.** $g = \frac{2S}{t^2}$ **27.** $x > \frac{5}{2}$: open ray to the right of $\frac{5}{2}$. **29.** $x > \frac{5}{7}$: open ray to the right of $\frac{5}{7}$. **31.** $x > -\frac{2}{7}$: open ray to the right of $-\frac{2}{7}$. **33.** $x > -\frac{4}{3}$: open ray to the right of $-\frac{4}{3}$. **35.** $x < -\frac{13}{8}$: open ray to the left of $-\frac{13}{8}$. **37.** $x \leqslant 6$: ray to the left, endpoint 6. **39.** True for all $x \in R$; number line. **41.** $x < -4$: ray to the left, endpoint -4. **43.** First solve the linear inequality, then test one other point in the inequality. If the statement is true, the ray contains the point, and if false, it does not contain the point. (1) $2x - 5 = 0$ implies $x = \frac{5}{2}$. Replace x by 0, which gives $-5 > 0$, a false statement. Hence, the open ray extends to the right of $\frac{5}{2}$; (2) $11 - 3x = 7$ implies $x = \frac{4}{3}$. Replace x by 2, which gives $5 \leqslant 7$, a true statement. Hence, the ray contains 2 and extends to the right of $\frac{4}{3}$; 29–63 can be done in a similar fashion. **45.** A real number r is a member of the solution set for $x + b = d$

iff $r + b = d$

iff $r + b + (-b) = d + (-b)$

iff $r = d - b$

iff r is a member of the solution set for $x = d - b$. **47.** A real number r is a member of the solution set for $ax + b = cx + d$

iff $ar + b = cr + d$

iff $ar + (-cr) + d + (-b) = ar + (-cr) + d + (-b)$

iff $r(a - c) = d - b$

iff $r(a - c) \cdot \frac{1}{a - c} = (d - b) \cdot \frac{1}{a - c}$

iff $r = \frac{d - b}{a - c}$

iff r is a member of the solution set for $x = \frac{d - b}{a - c}$.

Section 3-3, page 54

1. A line in the plane determined by points at (0, 1) and (2, 0). **3.** A line in the plane determined by points at (0, −1) and (2, 2). **5.** A line in the plane determined by points at (1, 2) and (3, 1). **7.** A line in the plane determined by points at (0, 2) and (1, 0). **9.** A line in the plane determined by points at (0, 3) and (3, 0). **11.** An element (r_1, r_2) of $R \times R$ is a member of the solution set for $ax + by$

$$\text{iff } ar_1 + br_2 = c$$

$$\text{iff } (-ar_1) + ar_2 + br_2 = (-ar_1) + c$$

$$\text{iff } br_2 = -ar_1 + c$$

$$\text{iff } \frac{1}{b} \cdot br_2 = \frac{1}{b}(-ar_1 + c)$$

$$\text{iff } r_2 = -\frac{a}{b}r_2 + \frac{c}{b}$$

iff (r_1, r_2) is a member of the solution set for $y = -\frac{a}{b}x + \frac{c}{b}$. **15.** The points to the right of, but excluding, the line in Exercise 3. **17.** The points to the left of, but excluding, the line in Exercise 5. **19.** The points to the right of, but excluding, the line in Exercise 7. **21.** The points to the right of, and including, the line in Exercise 9. **23.** Graph: a triangle determined by points at (0, 0), $\left(0, 2\frac{1}{2}\right)$, and (5, 0) and its interior. **25.** Graph: a triangle determined by points at (−2, 3), (1, −2), and (5, 6) and its interior.

Section 3-4, page 57

Points may vary in Exercises 1–8. **1.** (0, 5), $\left(\frac{5}{4}, 0\right)$; −4 **3.** $\left(0, \frac{3}{4}\right)$, (3, 0); $-\frac{1}{4}$ **5.** (0, 0), (4, −1); $\frac{1}{4}$ **7.** $\left(\frac{5}{3}, 0\right)$, $\left(\frac{5}{3}, 1\right)$; has no slope **9.a.** $x = 0$ **c.** $x + y = 0$ **e.** $(-1)x + 2y = 0$ **g.** $y = 0$ **11.** $y = (-2)x + \frac{5}{3}$; −2; $\left(0, \frac{5}{3}\right)$; Graph: a line in the plane determined by points at $\left(0, 1\frac{2}{3}\right)$ and $\left(1, -\frac{1}{3}\right)$. **13.** $y = -2x + \left(-\frac{5}{3}\right)$; −2; $\left(0, -\frac{5}{3}\right)$; Graph: a line in the plane determined by points at $\left(-1, \frac{1}{3}\right)$ and $\left(0, -2\frac{2}{3}\right)$. **14.** $y = \frac{1}{2}x + 0$; $\frac{1}{2}$; (0, 0); Graph: a line in the plane determined by points at (0, 0) and (2, 1). **15.** $y = \frac{1}{2}x + 4$; $\frac{1}{2}$; (0, 4); Graph: a line in the plane determined by points at (−2, 3) and (0, 4). **17.** $y = -\frac{1}{2}x + (-4)$; $-\frac{1}{2}$; (0, −4); Graph: a line in the plane determined by points at (−2, −3) and (0, −4). **19.** $y = \frac{5}{2}x + (-5)$; $\frac{5}{2}$; (0, −5); Graph: a line in the plane determined by points at (0, −5) and (2, 0). **21.** $x = \frac{3}{2}$; the line has no slope or intercept; Graph: a vertical line in the plane determined by the points at $\left(1\frac{1}{2}, -3\right)$ and $\left(1\frac{1}{2}, 0\right)$. **22.a.** $x + (-1)y = 0$ **c.** $x + 0y = 3$ **e.** $x + y = -2$ **g.** $x + 3y = -3$ **23.** L_1 and L_4 are parallel; L_1 and L_3, L_3 and L_4 are pairs of perpendicular lines. **25.** $3x + y = 9$ **27.** $\left(0, \frac{c}{b}\right)$, $\left(\frac{c}{a}, 0\right)$ **29.** $\left(0, \frac{c}{b}\right)$

Section 3-5, page 63

1. $\begin{cases} x = -3 \\ y = 5 \end{cases}$ **3.** $\begin{cases} x = -\frac{12}{5} \\ y = \frac{9}{4} \end{cases}$ **5.** $\begin{cases} x = -6 \\ y = -2 \end{cases}$ **7.** $\begin{cases} x = \frac{7}{3} \\ y = -\frac{14}{5} \end{cases}$

In Exercises 9–11 both equations determine the same line. **9.** Graph: a line in the plane determined by points at $\left(0, -\frac{3}{4}\right)$ and $\left(\frac{3}{2}, 0\right)$. **11.** Graph: a line in the plane determined by points at $\left(0, \frac{3}{7}\right)$ and (3, 0). **13.** $b_2 = 0, c_1a_2 \neq a_1c_2$ **15.** 122, 273 **17.** 45 ft., 90 ft. **19.** 13′4″, 6′8″ **21.** −23, −14 **23.** 135 mph, 15 mph **25.** 450 mph, 75 mph **27.** Yes; $\frac{35}{24}$ miles

Chapter 3 Review Exercises, page 65

1. The strip, and its boundary, bounded by the x-axis and the lines $x = 1$ and $x = -1$. **2.** $x = \frac{23}{12}$ **3.** A ray determined by a point at $3\frac{1}{10}$ and all points on the number line to the right of $3\frac{1}{10}$. **4.** A line in the plane determined by points at $\left(0, -2\frac{2}{3}\right)$ and (4,0). **5.** The set of points to the right of, and including, the line determined by points at $\left(0, \frac{-5}{3}\right)$ and (5, 0). **7.** $7x + 3y = -5$ **8.** $2x - y = 1$ **9.** $\left\{\frac{31}{47}, -\frac{6}{47}\right\}$ **10.** $\{(x, y)\colon\ 2x - 3y = 9\}$ **11.** ϕ **12.** 3, 106

Section 4-1, page 68

1.a. 1 **c.** 1 **e.** $-\frac{1}{27}$ **g.** $\frac{1}{9}$ **i.** 1 **k.** $\frac{8}{125}$ **m.** $\frac{108}{31}$ **o.** 32 **q.** 1, 010, 101.01 **2.a.** $\frac{1}{x}, x \neq 0$ **c.** $u^4, u \neq 0$ **e.** $\frac{a^5}{b^3}, b \neq 0$ **g.** $\frac{1}{s^2}, s \neq 0$ **i.** $\frac{z^3 + z^2}{z^5}, z \neq 0$ **k.** $x, y \neq 0, x \neq 0$ **m.** $\frac{1}{x^5}, x, y \neq 0$ **o.** $\frac{v^6}{u^4w^8}, u, v, w \neq 0$ **q.** $x^3y^5, x, y \neq 0$ **3.a.** $\frac{1}{2}$ **c.** $\frac{121}{2}$ **e.** $\frac{81}{64}$ **g.** $\frac{25}{36}$ **i.** 1 **4.a.** $(2x)^2$ **c.** $\left(\frac{x}{y}\right)^2, y \neq 0$ **e.** $\frac{1}{xy}, x, y \neq 0$ **g.** $\frac{1}{(x^2 - y)^2}, y \neq x^2$ **5.a.** $19^{1/7}$ is the number which, when raised to the seventh power, is 19. **c.** $(-216)^{1/3}$ is the number which, when cubed, is equal to −216. **e.** $32^{1/8}$ is the positive number which, when raised to the eighth power is 32. **6.a.** 2 **b.** −3 **c.** Not defined **e.** 3 **f.** Not defined **7.a.** 5 **c.** 11 **e.** −3 **g.** −5 **i.** Not defined **k.** 0.1 **9.** $(-2)^3 = -8$; thus −2 is a solution and $(x - (-2))$ is a factor. Use synthetic division to find $x^2 - 2x + 4$ as the other factor; that is $(x + 2)(x^2 - 2x + 4) = x^3 + 8 = 0$. By the quadratic formula, when $x^2 - 2x + 4 = 0$, $x = 1, \pm\sqrt{-3}$, which are not real numbers.

Section 4-2, page 72

1.a. $\frac{25}{16}$ **c.** $\frac{1}{8}$ **e.** $\frac{8}{27}$ **g.** 0.02 **i.** −9 **k.** $\frac{1}{4}$ **m.** $\frac{1}{9}$ **o.** $-\frac{2}{3}$ **2.a.** No **c.** Yes **e.** No **g.** No **i.** No **3.a.** $t^{5/6}, t \geqslant 0$ **c.** $y^{1/6}, y > 0$ **e.** $(3x)^{8/3}$ **g.** $x^2, x \geqslant 0$ **i.** $(xytz)^{2/5}$ **k.** $x^{-3/10}, x > 0$ **m.** $x^{1/18}, x > 0$ **o.** $x, x \geqslant 0$ **q.** $(xy)^{3/4}, x, y \geqslant 0$ **s.** $(xy)^{1/2}, x, y \geqslant 0$ **u.** $2x$ **5.** Property ii holds in all cases except when $a < 0$ or $b < 0$ and n is an even integer. **7.** Property iv holds in all cases except when $b < 0$ and n or q is an even integer.

Section 4-3, page 76

1.a. $7^{1/5}$ **c.** $4^{3/2}$ **e.** $3^{2/5}$ **g.** $x^{2/2}$ **i.** $(3x^5y)^{1/2}$ **k.** $x^{1/4}$ **m.** $(-5)^2$ **o.** $(x - 2)^{2/2}$ **2.a.** $2\sqrt{6}$ **c.** $-2\sqrt[5]{5}$ **e.** $8xy^2\sqrt{2x}, x \geqslant 0$ **g.** $\frac{2x}{7|y|}\sqrt{3x}, x \geqslant 0, y \neq 0$ **i.** 2 **k.** $4xy^2\sqrt{x}, x \geqslant 0$ **m.** $\frac{7}{12}$ **o.** $\frac{2x}{y}\sqrt[3]{12}, y \neq 0$ **q.** $-4x^2\sqrt[3]{x}$ **s.** $11xy$ **u.** $|x|$ **w.** $\sqrt{x + 2},\ x > -2$ **3.a.** 3 **c.** $|x + 2|$ **e.** $|x - y|\sqrt{x + y}, x + y \geqslant 0$ **g.** $\frac{|x|}{|3x - 2|}, x \neq \frac{2}{3}$

Section 4-4, page 81

1.a. $3\sqrt{3}$ **b.** $2\sqrt[3]{y}$ **c.** $2\sqrt{3}$ **d.** $xy\sqrt[3]{y}(3 - 2y)$ **e.** $5\sqrt{2}$ **g.** $-31\sqrt[3]{2}$ **i.** $23|y|\sqrt{x} - 3|y|, x \geqslant 0$ **k.** $3\sqrt{2}$ **m.** 5 **2.a.** $4\sqrt{3}$ **c.** $12x, x \geqslant 0$ **e.** $3x - x\sqrt[3]{7}$ **g.** $82 - 12\sqrt{42}$ **i.** −37 **k.** 43 **m.** $-29 - 5\sqrt{15}$ **3.a.** $\left(\frac{\sqrt{2}}{2}\right) + 0.3|x|(34 - \sqrt{2})$ **c.** $0.64z\sqrt[3]{z} - 0.21\sqrt[3]{z} + 0.04z^2\sqrt[3]{z}$ **e.** $-0.32x^2$ **g.** $z^2(9\sqrt[3]{z} - 1)$

i. $\sqrt{x+y}\,(2x+2y-1)$ 4.a. $\sqrt{2}-9$ c. $\dfrac{504\sqrt{3}-464\sqrt{5}}{27}$ e. 8 5.a. $\sqrt{3}$
c. $\dfrac{\sqrt[3]{25}}{5}$ e. $\dfrac{2-\sqrt{2}}{2}$ g. $\dfrac{21-\sqrt[3]{49}}{7}$ i. $\dfrac{5\sqrt{7}-28}{7}$ k. $\dfrac{\sqrt{6}-3\sqrt{3}}{3}$ m. $\dfrac{\sqrt[3]{9}}{3}$ o. $\dfrac{-\sqrt[3]{4}}{2}$
q. $\dfrac{\sqrt{2x}}{2}, x \geqslant 0$ s. $\dfrac{\sqrt[3]{ab^2xz^2}}{bz}, b, z \neq 0$ u. $\dfrac{\sqrt[5]{1296x^4y}}{3x}, x \neq 0$ 6.a. $-2-2\sqrt{3}$
c. $-2\sqrt{5}-2\sqrt{7}$ e. -1 g. $\dfrac{15\sqrt{2}}{2}$ i. $\dfrac{5\sqrt{2}-4\sqrt{3}}{2}$ 7.a. $\sqrt[3]{49}+\sqrt[3]{28}+\sqrt[3]{16}$
c. $\dfrac{-12+6\sqrt[3]{2}-3\sqrt[3]{4}}{10}$ e. $\dfrac{\sqrt{\sqrt{5}+1}}{2}$ 8.a. $\dfrac{2\sqrt[3]{2}+3\sqrt[3]{3}}{97}$ c. $\dfrac{a\sqrt[3]{b}-c\sqrt[3]{d}}{a^3b-c^3d}$

Chapter 4 Review Exercises, page 83

1. 43,196.378 2. $-6, 3$ 3. If $7^{1/5}=x$, then $x^5=7$; if $(-7)^{1/5}=y$, then $y^5=(-7)$. 4. $(-7)^{2/3}$, $13^{3/4}$, $9^{2/5}$ 5. True for all real numbers greater than zero. 6. Yes, 9 7. $5^{7/3}$, $x^{2/6}$ 8.a. $2x\,|y|\sqrt{x}, x \geqslant 0$ b. $\dfrac{3u^2}{v}\sqrt{3u}, v \neq 0$ 9.a. $9+\sqrt[3]{3}$ b. $\dfrac{1}{|y|}(-17\sqrt{x}+5), y \neq 0, x \geqslant 0$ 10.a. $-116-8\sqrt{15}$ b. $x(7-\sqrt[4]{2}), x \geqslant 0$ 11.a. $\dfrac{-1-\sqrt{3}}{3}$ b. $\dfrac{\sqrt[3]{25}+\sqrt[3]{15}+\sqrt[3]{9}}{2}$.

Section 5-1, page 85

1. $\{-6, 14\}$ 3. $\{17, 20\}$ 5. $\{-6, -31\}$ 7. $\{-\frac{1}{3}, \frac{2}{7}\}$ 9. $\{\frac{1}{6}, \frac{13}{5}\}$ 11. $\{-\frac{7}{2}, -\frac{1}{3}\}$ 13. $\{-5, 8\}$ 15. $\{-11, 7\}$ 17. $\{-4, 40\}$ 19. $\{17, -17\}$ 21. $\{9, -\frac{65}{6}\}$ 23. 18, 19 or $-19, -18$ 25. 14″, 24″ 27. At most 2; the graph of ax^2+bx+c will cross the x-axis in at most two places, corresponding to the solutions of $0=ax^2+bx+c$.

Section 5-2, page 90

1. $\{5, -7\}$ 3. $\{1, 9\}$ 5. $\{-9, -4\}$ 7. $\{3, -4\}$ 9. $\{4, -5\}$ 11. $\{-2, -6\}$ 13. $\{-9, -2\}$ 15. $\{-\frac{2}{5}, \frac{1}{3}\}$ 17. $\{\frac{5}{3}, -\frac{2}{3}\}$ 19. $\{-\frac{6}{5}, -\frac{3}{5}\}$ 21. $\{\frac{5}{4}, -\frac{7}{9}\}$ 23. $\{\frac{7}{6}, \frac{3}{4}\}$ 25. 4, $\{1, 3\}$ 27. -8 29. 12, $\{2+\sqrt{3}, 2-\sqrt{3}\}$ 31. -4 33. 4, $\{1, \frac{1}{3}\}$ 35. 24, $\{-1+\sqrt{6}, -1-\sqrt{6}\}$ 37. 0, $\{\frac{7}{11}\}$ 39. 25, $\{-3, -\frac{1}{2}\}$ 41. 0, $\{\frac{3}{2}\}$ 43. 60, $\{\frac{3+\sqrt{15}}{2}, \frac{3-\sqrt{15}}{2}\}$ 45. 37, $\{\frac{-5+\sqrt{37}}{6}, \frac{-5-\sqrt{37}}{6}\}$ 47. 6 ft, 8 ft 49. $r_1, r_2 = \dfrac{(-b)^2-(\sqrt{b^2-4ac})^2}{(2a)^2} = \dfrac{b^2-(b^2-4ac)}{4a^2} = \dfrac{c}{a}$ 51. $3x^2-10x+3=0$

Section 5-3, page 94

1. $\{r\colon -5<r<5\}$; Graph: an open segment, all points between -5 and 5.
3. $\{r\colon -2 \leqslant r \leqslant 2\}$; Graph: the segment determined by -2 and 2.
5. $\{r\colon -7<r<3\}$; Graph: an open segment, all points between -7 and 5.
7. $\{r\colon -5<r<3\}$; Graph: an open segment, all points between -5 and 3.
9. $\{r\colon r \leqslant -4 \text{ or } r \geqslant \frac{3}{2}\}$; Graph: the union of two rays; -4 and all points to the left, $\frac{3}{2}$ and all points to the right. 11. ϕ 13. $\{r\colon -1 \leqslant r \leqslant 4\}$; Graph: the segment determined by -1 and 4. 15. $\{r\colon -\frac{3}{7}<r<\frac{5}{2}\}$; Graph: an open segment, all points

between $-\frac{3}{7}$ and $\frac{5}{2}$. **17.** $\left\{r\colon \frac{2}{5} < r < \frac{7}{3}\right\}$; Graph: an open segment, all points between $\frac{2}{5}$ and $\frac{7}{3}$. **19.** $\left\{r\colon \frac{-5-\sqrt{13}}{2} \leqslant r \leqslant \frac{-5+\sqrt{13}}{2}\right\}$; Graph: a segment. **21.** ϕ **23.** ϕ **27.a.** $\{r\colon -\sqrt{a} < r < \sqrt{a}\}$ **c.** $\{r\colon -\sqrt{a} \leqslant r \leqslant \sqrt{a}\}$ **29.** The following inequalities are equivalent:

$$ax^2 + bx + c < 0$$

$$(-1)(ax^2 + bx + c) > (-1) \cdot 0$$

$$-ax^2 - bx - c > 0$$

If the coefficient of x^2 in an inequality is negative multiply each member of the inequality by -1 (and reverse the sense of the inequality). Then apply the appropriate result of this section.

Section 5-4, page 96 **1.** $|x+4| < 5$ **3.** $|2x - 11| < 7$ **5.** $|x+2| > 9$ **7.** $|x+7| > 4$ **9.** An open segment, all points between 1 and 7. **11.** The segment determined by -11 and -3. **13.** An open segment, all points between 1 and 2. **15.** The union of two open rays: all points to the left of -1 together with all points to the right of $\frac{1}{7}$. **17.** The segment determined by -5 and 5. **19.** The union of two open rays; all points to the left of -1 together with all points to the right of 5. **21.** The union of two open rays; all points to the left of -2 together with all points to the right of $-\frac{2}{3}$. **23.** The union of two rays; $\frac{1}{2}$ and all points to the left, $6\frac{1}{2}$ and all points to the right. **25.** The union of two open rays; all points to the left of $\frac{5}{6}$ together with all points to the right of $1\frac{5}{6}$. **27.** (11) $x^2 + 14x + 33 \leqslant 0$ (12) $x^2 - 9 \leqslant 0$ (18) $x^2 - 3x + 2 < 0$ (22) $3x^2 - 4x + 2 < 0$

Section 5-5, page 99 **1.** $\{25\}$ **3.** $\{14\}$ **5.** $\{3\}$ **7.** $\{-11\}$ **9.** $\{\sqrt{119}, -\sqrt{119}\}$ **11.** $\left\{\frac{85}{36}\right\}$ **13.** $\{\sqrt{5}, -\sqrt{5}\}$ **15.** $\left\{-\frac{1}{3}\right\}$ **17.** $\{2\}$ **19.** $\{1\}$ **21.** $\{1, 3\}$ **23.** ϕ **25.** $\left\{\frac{23}{18}\right\}$ **27.** $\{4\}$ **29.** ϕ

Section 5-6, page 104 **1.a.** Parabola **c.** Line **e.** Parabola **2.a.** Graph: parabola with y-axis for vertical axis of symmetry, "low" point at (0, 0), and containing (–1, 2) and (1, 2). **c.** Graph: line determined by (1, 2) and (0, 0). **e.** Graph: parabola with y-axis for vertical axis of symmetry, "low" point at (0, 0), and containing (–2, 2) and (2, 2). **3.** "High" point: a, e; "low" point; b, c, d, f **5.** If $a < 0$, the parabola opens downward and the vertex is the highest point; if $a > 0$, the parabola opens upward and the vertex is the lowest point. **7.a.** $x = 0$, (0, 2) **b.** $x = 0$, (0, –3) **c.** $x = 0$, (0, 5) **d.** $x = 0$, (0, –7) **9.** $x = 0$ is the axis of symmetry for each parabola. Vertices: **a.** (0, 0) **b.** (0, 1) **c.** (0, 2) **d.** (0, 3) **e.** (0, –1) **f.** (0, –2) **g.** (0, –3) **11.** (0, 1); same parabola as in Exercise 10. **13.** $x = -\frac{1}{2}$, $\left(-\frac{1}{2}, -\frac{25}{4}\right)$ **14.** $x = \frac{7}{2}$, $\left(\frac{7}{2}, \frac{9}{4}\right)$ **15.** $x = \frac{1}{4}$, $\left(\frac{1}{4}, -\frac{49}{8}\right)$ **19.** $x = 1$, (1, 1) **20.** $x = 1$, (1, –1) **21.** $x = 1$, (1, –2) **25.** $r = 0$ because $r^2 \geqslant 0$, and $r^2 = 0$ if and only if $r = 0$; $r = 0$ because $-r^2 \leqslant 0$, and $-r^2 = 0$ if and only if $r = 0$. **27.** Distributive property; additive identity (addition of $0 = 1 - 1$); associative property of addition and the addition of two numbers; distributive property (applied several times); distributive property, $x = 1$, (1, – 3).

Section 5-7, page 109

1. Circle: $r = 5$, center (0, 0). **3.** Circle: $r = 7$, center (0, 0). **5.** $\emptyset$ **7.** Circle: $r = 1$, center (5, 0). **9.** Circle: $r = 2$, center (5, –3). **11.** $\{(-7, 5)\}$ **13.** $\emptyset$ **15.** $\{(2, -3)\}$ **17.** Circle: center (2, –5), radius 4. **19.** Parabola: vertex $\left(-1\frac{1}{2}, \frac{1}{2}\right)$, axis of symmetry $x = -1\frac{1}{2}$, two other points (–3, 5), (0, 5). **21.** Union of two horizontal lines: one containing (0, 3), the other containing (0, –3). **23.** Parabola: vertex $\left(-\frac{5}{3}, \frac{1}{3}\right)$, axis of symmetry $x = -\frac{5}{3}$, two other pts. (0, –8), $\left(-\frac{10}{3}, -8\right)$. **25.** Union of two intersecting lines: one determined by (0, 0) and (5, 5), the other containing (0, 0) and (–5, 5). **27.** Circle: center (0, 0), radius 4. **29.** Interior of the circle with center (1, –3) and radius 2. **31.** Graph: ellipse with (5, 0) and (–5, 0) as endpoints of the major axis and (0, 3) and (0, –3) as endpoints of the minor axis. **33.** Graph: ellipse with (0, 5) and (0, –5) as endpoints of the major axis and (3, 0) and (–3, 0) as endpoints of the minor axis. **35.** Graph: ellipse with (0, 6) and (0, –6) as endpoints of the major axis and $(\sqrt{6}, 0)$, and $(-\sqrt{6}, 0)$ as endpoints of the minor axis. **37.** Graph: ellipse with (0, 10) and (0, –10) as endpoints of the major axis and (1, 0) and (–1, 0) as endpoints of the minor axis. **39.** Graph: ellipse with (4, 0) and (–4, 0) as endpoints of the major axis and (0, 1) and (0, –1) as endpoints of the minor axis. **41.** Graph: ellipse with (11, 0) and (–11, 0) as endpoints of the major axis and (0, 4) and (0, –4) as endpoints of the minor axis. **43.** Graph: ellipse with $\left(0, 2\frac{1}{2}\right)$ and $\left(0, -2\frac{1}{2}\right)$ as endpoints of the major axis and $\left(1\frac{1}{2}, 0\right)$ and $\left(-1\frac{1}{2}, 0\right)$ as endpoints of the minor axis. **45.** Graph: circle with center at the origin and radius of length $1\frac{1}{2}$ which contains points for $\left(0, 1\frac{1}{2}\right)$, $\left(1\frac{1}{2}, 0\right)$, $\left(0, -1\frac{1}{2}\right)$, and $\left(-1\frac{1}{2}, 0\right)$. **47.** Graph: ellipse with (4, 0) and (–4, 0) as endpoints of the major axis and (0, 3) and (0, –3) as endpoints of the minor axis. **49.** Graph: interior of the ellipse with (0, 6) and (0, –6) as endpoints of the major axis and $(\sqrt{6}, 0)$ and $(-\sqrt{6}, 0)$ as endpoints of the minor axis $[\sqrt{6} \approx 2.4]$. **51.** Graph: ellipse with (4, 0) and (–4,0) as endpoints of the major axis and (0, 1) and (0, –1) as endpoints of the minor axis together with its exterior.

Section 5-8, page 112

In Exercises 1–15 the equations represent hyperbolas.
1. Vertices: (2, 0) and (–2, 0); endpoints of the conjugate axis: (0, 3) and (0, –3).
3. Vertices: (0, 3) and (0, –3); endpoints of the conjugate axis: (1, 0) and (–1, 0).
5. Vertices: (2, 0) and (–2, 0); endpoints of the conjugate axis: (0. 7) and (0, –7).
7. Vertices: (1, 0) and (–1, 0); endpoints of the conjugate axis: (0, 10) and (0, –10).
9. Vertices: (0, 2) and (0, –2); endpoints of the conjugate axis: (7, 0) and (–7, 0).
11. Vertices: $\left(\frac{5}{2}, 0\right)$ and $\left(-\frac{5}{2}, 0\right)$; endpoints of the conjugate axis: (0, 2) and (0, –2). **13.** Vertices: $(\sqrt{20}, 0)$ and $(-\sqrt{20}, 0)$; endpoints of the conjugate axis: $(0, \sqrt{10})$ and $(0, -\sqrt{10})$. **15.** Vertices: $(\sqrt{20}, 0)$ and $(-\sqrt{20}, 0)$; endpoints of the conjugate axis: (0, 4) and (0, –4). **17.** Graph: ellipse with $(\sqrt{20}, 0)$ and $(-\sqrt{20}, 0)$ as endpoints of the major axis and (0, 4) and (0, –4) as endpoints of the minor axis.
23. Given $\frac{x^2}{a^2} - \frac{y^2}{b^2} = 1$, we can solve for y^2. The following sentences are equivalent:

$$\frac{x^2}{a^2} - \frac{y^2}{b^2} = 1$$

$$\frac{y^2}{b^2} = \frac{x^2}{a^2} - 1$$

$$y^2 = b^2\left(\frac{x^2}{a^2} - \frac{a^2}{a^2}\right)$$

$$y^2 = \frac{b^2}{a^2}(x^2 - a^2)$$

$$y = \frac{b}{a}\sqrt{x^2 - a^2} \quad \text{or} \quad y = -\frac{b}{a}\sqrt{x^2 - a^2}.$$

As x becomes quite large, $\frac{a^2}{x^2}$ approaches 0 and $1 - \frac{a^2}{x^2}$ approaches 1. Thus, $\frac{b}{a}|x|\sqrt{1 - \frac{a^2}{x^2}}$ approaches $\frac{b}{a}|x|$ in value. Notice that the equations $y = \pm\frac{b}{a}x$ describe the asymptotes for the hyperbola.

25. $\sqrt{[x - (-\sqrt{2})]^2 + (y - \sqrt{2})^2} - \sqrt{(x - \sqrt{2})^2 + (y + \sqrt{2})^2} = 2\sqrt{2}$

$(x + \sqrt{2})^2 + (y - \sqrt{2})^2 = 8 + 4\sqrt{2}\sqrt{(x - \sqrt{2})^2 + (y + \sqrt{2})^2} + (x - \sqrt{2})^2 + (y + \sqrt{2})^2$

$x^2 + 2\sqrt{2}x + 2 + y^2 - 2\sqrt{2}y + 2 = 8 + 4\sqrt{2}\sqrt{x - \sqrt{2})^2 + (y + \sqrt{2})^2} + x^2 - 2\sqrt{2}x + 2 + y^2 + 2\sqrt{2}y + 2$

$4\sqrt{2}x - 4\sqrt{2}y - 8 = 4\sqrt{2}\sqrt{(x - \sqrt{2})^2 + (y + \sqrt{2})^2}$

$x - y - \sqrt{2} = \sqrt{(x - \sqrt{2})^2 + (y + \sqrt{2})^2}$

$x^2 - 2xy - 2\sqrt{2}x + y^2 + 2\sqrt{2}y + 2 = x^2 - 2\sqrt{2}x + 2 + y^2 + 2\sqrt{2}y + 2$

Chapter 5 Review Exercises, page 113

1. $\{\frac{3}{2}, -5\}$ **2.** $\{\frac{1}{2}, -3\}$ **3.** $\{\frac{9}{4}, -\frac{11}{2}\}$ **4.a.** $\{r: -9 < r < 7\}$; Graph: an open segment, all points between −9 and 7. **b.** $\{r: r \leqslant -7 \text{ or } r \geqslant 9\}$; Graph: the union of two rays, −7 and all points to the left together with 9 and all points to the right. **5.a.** $|x - 1| \leqslant 5$ **b.** $|x - 1| > 5$ **6.a.** $\{2\}$ **b.** $\{1, 5\}$ **7.** Graph: parabola with the y-axis for vertical axis of symmetry, "low" point (0, 0), and containing the points (6, 6) and (−6, 6). **8.** Graph: parabola with the line $x = 4$ the vertical axis of symmetry, "low" point $\left(4, -\frac{2}{3}\right)$, and containing (2, 0) and (6, 0). **9.** Graph: circle, center (3, −4), radius 5. **10.** Graph: ellipse with (0, 7) and (0, −7) as endpoints of the major axis and (3, 0) and (−3, 0) as endpoints of the minor axis. **11.** Graph: hyperbola with vertices at (3, 0) and (−3, 0) and endpoints of the conjugate axis at (0, 7) and (0, −7).

Section 6-1, page 118

1.a. 16 **c.** $\{2, 4\}$ **g.** $\{4, 6\}$ **3.a.** $\{2, 4, 6\}$; $\{2, 4, 6\}$; $\{2\}$; $\{2, 4, 6\}$; $\{2, 4\}$; $\{2, 6\}$ **b.** $\{3, 4, 7\}$; $\{5\}$; $\{3, 5, 7\}$; $\{3, 7\}$; $\{3, 5, 7\}$; $\{3, 5\}$ **e.** s_1, s_2, s_4, s_6 **f.** $s_1^{-1}, s_3^{-1}, s_5^{-1}, s_6^{-1}$ **5.** Domain: $\{0, 1, 2, 3, 4\}$, range: $\{-1, 0, 1, 2\}$, inverse: $\{(x, y) \in I \times I: 0 \leqslant y \leqslant 4 \text{ and } -1 \leqslant x \leqslant 2\}$ **7.** Domain: $\{-2, 0, 2\}$, range: $\{-2, 0, 2\}$, inverse:

$$\{(x, y) \in I \times I: y^2 + x^2 = 4\}$$

Graph and inverse: points for (−2, 0), (0, 2), (2, 0), and (0, −2) on a lattice. **9.** Domain: R, range: $\{\pi\}$, inverse:

$$\{(x, y): x = \pi\}$$

Graph: horizontal line containing $(0, \pi)$ where $\pi \approx 3.14$; Inverse: vertical line containing $(\pi, 0)$ **11.** Domain: $\{x: x \geqslant 0\}$, range: R, inverse

$$\{(x, y): |x| = y\}$$

Graph: union of two rays determined by (0, 0) and (1, 1) and by (0, 0) and (1, −1);

Inverse: union of two rays determined by (0, 0) and (−1, 1) and by (0, 0) and (1, 1) **13.** Domain and range: $\{-5, -4, -3, 0, 3, 4, 5\}$, inverse:

$$\{(x, y) \in I \times I: y^2 + x^2 = 25\}$$

Graph and inverse: points for (−5, 0), (−4, 3), (−3, 4), (0, 5), (3, 4), (4, 3), (5, 0), (4, −3), (3, −4), (0, −5), (−3, −4), and (−4, −3) on a lattice. **15.** Domain: R; range: $\{y: y > 1 \text{ or } y < -1\}$, inverse:

$$\{(x, y): -1 \leqslant y \leqslant 1 \text{ and } x > 1, \text{ or } x < -1\}.$$

17. Domain: R; range: $\{y \in R: y \geqslant 0\}$; Graph: union of two rays with common endpoint (1, 0); one determined by (1, 0) and (0, 1), the other determined by (1, 0) and (2, 1). **19.** Domain: $\{x \in R: x \geqslant -1\}$; range: $\{y \in R: y \geqslant 0\}$. **21.** Domain: R; range: $\{y \in R: 0 \leqslant y < 1\}$. **23.** $s_1, s_1^{-1}, s_6, s_6^{-1}, g_4, g_5$; No **24.** Each vertical line in the plane for $R \times R$ contains at most one point of the graph of the relation.

Section 6-2, page 121

1. $\{(x, y): x = y\}$, Graph: line determined by (0, 0) and (2, 2). **2.a.** Graph: line determined by (0, 0) and (1, 2) **c.** Graph: line parallel to x-axis with y-intercept −2. **3.a.** R **c.** −2 **5.a.** Graph: line determined by (0, 5) and (1, 7); $b = 5, m = 2$. **c.** Graph: line determined by (0, −5) and (1, −3); $b = -5, m = 2$ **e.** Graph: line determined by (0, 0) and (2, 1); $b = 0, m = \frac{1}{2}$. **g.** Graph: line determined by (0, 0) and (2, −1); $b = 0, m = -\frac{1}{2}$ **i.** Graph: horizontal line containing (0, −4); $b = -4$, $m = 0$ **k.** Graph: horizontal line containing (0, 5); $b = 5, m = 0$ **l.** Graph: line determined by (0, 2) and (1, 3); $b = 2, m = 1$. **6.a.** Graph: line determined by $\left(0, 1\frac{2}{3}\right)$ and $\left(2\frac{1}{2}, 0\right)$; $-\frac{2}{3}x + \frac{5}{3}$. **c.** Graph: line determined by $\left(0, 1\frac{2}{3}\right)$ and $\left(-2\frac{1}{2}, 0\right)$; $\frac{2}{3}x + \frac{5}{3}$. **e.** Graph: vertical line containing $\left(\frac{2}{3}, 0\right)$; not a function so no corresponding polynomial exists. **7.** $\{mx + r: r \in R\}$ **9.** $\{r(x - c) + d: r \in R\}$ **11.** The vertical lines are not graphs of functions. The graphs of linear functions comprise the set of non-vertical lines. **13.** Graph of f^{-1}: line determined by $\left(0, \frac{1}{2}\right)$ and (1, 0). **15.** 12 and 13

Section 6-3, page 123

1. Graph: parabola with y-axis for vertical axis of symmetry, "low" point at (0, 0), and containing (−1, 5) and (1, 5); (0, 0), $x = 0$. **3.** Graph: parabola with y-axis for vertical axis of symmetry, "high" point at (0, 0), and containing (−5, −5) and (5, −5); (0, 0), $x = 0$. **5.** Graph: parabola with line for $x = 2$ as vertical axis of symmetry, "low" point at (2, −4), and containing (0, 0) and (4, 0); (2, −4), $x = 2$. **7.** Graph: parabola with y-axis for vertical axis of symmetry, "low" point at (0, 4), and containing (−1, 6) and (1, 6); (0, 4), $x = 0$. **9.** Graph: parabola with line for $x = 3\frac{1}{2}$ as vertical axis of symmetry, "low" point at $\left(3\frac{1}{2}, -9\frac{1}{4}\right)$, and containing (0, 3) and (7, 3); $\left(\frac{7}{2}, -\frac{37}{4}\right)$, $x = \frac{7}{2}$. **11.** Graph: parabola with line for $x = 1\frac{1}{4}$ as vertical axis of symmetry, "high" point at $\left(1\frac{1}{4}, -\frac{7}{8}\right)$, and containing (0, −4) and $\left(2\frac{1}{2}, -4\right)$; $\left(\frac{5}{4}, -\frac{7}{8}\right)$, $x = \frac{5}{4}$. **13.** Graph: parabola with line for $x = -1$ as vertical axis of symmetry, "low" point at (−1, −9), and containing (−2, −5) and (0, −5); (−1, −9), $x = -1$. **15.** $r = \frac{4}{3}$ or $r = -\frac{5}{4}$ **17.** $r = 0$ or $r = \frac{1}{\pi^2}$ **19.** $\frac{11}{12}x^2 + \left(\frac{-47}{12}\right)x + 2$

Section 6-4, page 126

1. If a is near -5, then $\left|\frac{1}{a+5}\right|$ is quite large; if $|a|$ is large, then $\frac{1}{a+5}$ is near zero.

3. If a is near $\frac{5}{2}$, then $\left|\frac{1}{2a-5}\right|$ is large; if $|a|$ is large, $f(a)$ is near zero.

5. $\frac{x-4}{x^2+5x-6} = \frac{x-4}{(x+6)(x-1)}$; if a is near -6 or 1, then $\left|\frac{a-4}{(a+6)(a-1)}\right|$ is large.

$\frac{x-4}{x^2+5x+6} = \frac{\frac{1}{x}-\frac{4}{x^2}}{1+\frac{5}{x}+\frac{6}{x^2}}$; if $|a|$ is large, the value of the function is near zero. The zero is 4. **7.** $\frac{-1}{x}$; if a is near zero, then $\left|\frac{-1}{a}\right|$ is large; if $|a|$ is large, then $\frac{-1}{a}$ is near zero. **9.** $\frac{(x-8)(x+3)}{2(x+3)} = \frac{1}{2}x - 4$. The graph is a line with the point for $\left(-3, -\frac{11}{2}\right)$ deleted. Graph: a line determined by points at $(0, -4)$ and $(8, 0)$ with the point $\left(-3, -5\frac{1}{2}\right)$ deleted **11.** $\frac{4x^2+5x-6}{x^2-2x-24} = \frac{(4x-3)(x+2)}{(x-6)(x+4)} = \frac{4+\frac{5}{x}-\frac{6}{x^2}}{1-\frac{2}{x}-\frac{24}{x^2}}$; if a is near 6 or -4, $|f(a)|$ is quite large. As a increases in absolute value, $f(a)$ approaches 4. The zeros are $\frac{3}{4}$ and -2. **13.** $\frac{(x+4)(x+1)(x-3)}{(x+4)(x-3)} = x+1$ if $x \neq -4$ and $x \neq 3$. The graph is a line with points for $(-4, -3)$ and $(3, 4)$ deleted. Graph: a line determined by points at $(-1, 0)$ and $(0, 1)$ with the points for $(-4, -3)$ and $(3, 4)$ deleted
15. $\frac{(x+6)(x-4)}{x+2} = \frac{x^2+2x-24}{x+2} = x + \frac{-24}{x+2}$; if a is near -2, then $|f(a)|$ is large; if $|a|$ is large, then $f(a)$ is near a. The zeros are -6 and 4. **17.** $\frac{(x+3)(x-2)}{(x+8)(x-6)}$; if a is near -8 or 6, then $|f(a)|$ is large; if $|a|$ is large, then $\frac{a^2+a-6}{a^2+2a-48} = \frac{1+\frac{1}{a}+\frac{-6}{a^2}}{1+\frac{2}{a}+\frac{-48}{a^2}}$ is near 1. The zeros are -3 and 2. **19.** $\frac{x+3}{(x+4)(x-1)} = \frac{\frac{1}{x}+\frac{3}{x^2}}{1+\frac{3}{x}-\frac{4}{x^2}}$; if a is near -4 or 1, then $|f(a)|$ is large; if a is large, then $\frac{\frac{1}{a}+\frac{3}{a^2}}{1+\frac{3}{a}-\frac{4}{a^2}}$ is near zero. The zero is -3.

21.a. If $n < m$ then

$$\frac{a_nx^n + a_{n-1}x^{n-1} + \cdots + a_1x + a_0}{b_mx^m + b_{m-1}x^{m-1} + \cdots + b_1x + b_0} = \frac{\frac{a_n}{x^{m-n}} + \frac{a_{n-1}}{x^{m-n+1}} + \cdots + \frac{a_1}{x^{m-1}} + \frac{a_0}{x^m}}{b_m + \frac{b_{m-1}}{x} + \cdots + \frac{b_1}{x^{m-1}} + \frac{b_0}{x^m}}$$

If $|a|$ is large, $f(a)$ is near zero. Also $|f(a)|$ is very large for a near the zeros of the polynomial function associated with the denominator. **c.** If $m = n$ then

$f(x) = \frac{a_n + \frac{a_{n-1}}{x} + \cdots + \frac{a_1}{x^{n-1}} + \frac{a_0}{x^n}}{b_n + \frac{b_{n-1}}{x} + \cdots + \frac{b_1}{x^{n-1}} + \frac{b_0}{x^n}}$. If $|a|$ is large, then $f(a)$ is near $\frac{a_n}{b_n}$. If a is near the zeros for the polynomial function for the denominator, then $|f(a)|$ is large.

Section 6-5, page 130 **1.a.** $\left\{\frac{3}{2}\right\}$ **c.** $\left\{\frac{4}{3}\right\}$ **e.** $\left\{\frac{9}{2}\right\}$ **g.** $\left\{\frac{13}{3}\right\}$ **i.** $\left\{-\frac{5}{4}\right\}$ **j.** $\{12\}$ **k.** $\{-10\}$ **m.** $\{0\}$ **o.** $\{3, -7\}$ **q.** $\left\{\frac{8}{3}, -2\right\}$ **s.** $\left\{\frac{7}{5}, -2\right\}$ **u.** $\left\{-6, \frac{1}{6}\right\}$ **w.** $\{-4\}$ **2.a.** Graph parabola with line for $x = -\frac{1}{2}$ as vertical axis of symmetry, "low" point at $\left\{-\frac{1}{2}, 3\frac{3}{4}\right\}$, and containing $(-2, 6)$ and $(1, 6)$, no point on x-axis. **c.** Graph: parabola with line for $x = 1$ as vertical axis of symmetry, "high" point at $(1, -1)$, and containing $(-1, -5)$ and $(3, -5)$; no point on x-axis. **3.** $\left\{-\frac{b}{m}\right\}$ **4.a.** $\{4\}$ **c.** $\{0, 5\}$ **e.** $\{2, -2\}$ **g.** $\left\{-\frac{3}{2}\right\}$ **i.** $\{0, 5\}$ **k.** $\{2\}$ **m.** $\emptyset$ **o.** $\emptyset$ **5.** A polynomial with degree zero is also an element $r \in R$ such that $r = 0$. The polynomial function with degree zero is a constant function with no zero values; hence the set of zero is $\emptyset$. **7.** If $b^2 - 4ac > 0$, then $\left\{\frac{-b \pm \sqrt{b^2 - 4ac}}{2a}\right\}$ is the set of zeros; if $b^2 - 4ac = 0$, then $\left\{-\frac{b}{2a}\right\}$ is the set of zeros; if $b^2 - 4ac < 0$, then $\emptyset$ is the set of zeros. **9.** 49 and 63 **11.** $\frac{4}{13}$ **13.a.** $\frac{6}{17}$ **c.** $\frac{10}{3}$

Chapter 6 Review Exercises, page 132 **1.** Domain: $\{3, 4, 5\}$, range: $\{1, 2\}$, $w^{-1} = \{(1, 3), (2, 3), (2, 4), (1, 5), (2, 5)\}$ **2.** $g(0) = 2, g(1) = 1, g(-1) = 3, g(2) = 0, g\left(-\frac{1}{2}\right) = \frac{5}{2}$. Domain: R, range: $\{y\colon y \geqslant 0\}$. Graph: union of two rays with common endpoint $(2, 0)$, one determined by $(2, 0)$ and $(1, 1)$, the other by $(2, 0)$ and $(3, 1)$. **3.** $f(0) = 9, f(3) = 36, f(-3) = 54, f\left(\frac{1}{2}\right) = \frac{17}{2}, f\left(-\frac{1}{2}\right) = \frac{23}{2}$. Domain: R. **4.a.** Graph: line determined by $(0, 3)$ and $(1, -1)$; **b.** Graph: line determined by $(0, -3)$ and $(-4, -2)$ **c.** Graph: line determined by $(0, 0)$ and $(4, 1)$ **d.** Graph: horizontal line containing $(0, 4)$ **5.** Graph: parabola with line for $x = -1$ as vertical axis of symmetry, "high" point at $(-1, 9)$ and containing $\left(-2\frac{1}{2}, 0\right)$ and $\left(\frac{1}{2}, 0\right)$. **b.** Graph: a line determined by points at $\left(-2\frac{1}{2}, 0\right)$ and $(0, 5)$ with the point for $(3, 11)$ deleted. **7.a.** $\frac{7}{4}$ **b.** $-2, 1$ **c.** $\emptyset$ **8.** 5

Section 7-1, page 135 **1.a.** $-i$ **c.** i **e.** $-i$ **g.** 1 **i.** -1 **3.** $2 + (-8)i$ **5.** $0 = 0 + 0i$ **7.** $-1 + 15i$ **9.** $1 + i$ **11.** $0 = 0 + 0i$ In Exercises 12–17, substitute the given "numbers" in the equation. Real solutions are given. **13.** none **15.** 1, 3, −1 **17.** $\sqrt{3}, -\sqrt{3}$

Section 7-2, page 137 In Exercises 1-14 the real part is listed first. **1.** 2, 4 **3.** −5, 0 **5.** 0, −3 **7.** 14, 6 **9.** 11, 9 **11.** −3, −3 **13.** 6, 0 **15.a.** $-4 - 3i$ **c.** $5 - 2i$ **e.** $2i$ **17.** If $a + bi = c + di$, then $a = c$ and $b = d$. But $c = a$ and $d = b$, so $c + di = a + bi$. **19.** $(a + 0i) + (0 + bi) = (a + 0) + (0 + b)i = a + bi$

Section 7-3, page 140 In Exercises 1–20 the real part is listed first. **1.** 14, 22 **3.** −8, −25 **5.** 34, −13 **7.** 40, −42 **9.** 15, −3 **11.** 168, 101 **13.** −46, 9 **15.** 1, 0 **17.** 0, 0 **19.** −6, 9 **21.a.** $\frac{1}{3}$ **c.** $-\frac{1}{5}i$ **e.** $\frac{1}{4}i$

Section 7-4, page 143 **1.** $4 - 7i, \frac{4}{65} + \frac{-7}{65}i$ **3.** $-2 - 7i, \frac{-2}{53} + \frac{-7}{53}i$ **5.** $4 + 7i, \frac{4}{65} + \frac{7}{65}i$ In Exercises 7–24 the real part is listed first. **7.** $\frac{2}{15}, -\frac{1}{15}$ **9.** $-\frac{2}{3}, \frac{4}{3}$ **11.** $-\frac{27}{17}, -\frac{28}{17}$ **13.** $0, -\frac{1}{2}$ **15.** $-\frac{3}{2}, -\frac{1}{2}$ **17.** $\frac{6}{5}, \frac{3}{5}$ **19.** $\frac{582}{689}, \frac{401}{689}$ **21.** 0, 3 **23.** $\frac{8}{85}, \frac{36}{85}$ **25.** Let

$z = a + bi$; then $z \cdot \bar{z} = (a + bi)(a - bi) = a^2 + b^2 + [ab + a(-b)]\, i = a^2 + b^2$. Since $a \in R$ and $b \in R$, $a^2 + b^2 \in R$ and $z \cdot \bar{z}$ is a real number.
27. $\bar{\bar{z}} = \overline{\overline{(a + bi)}} = \overline{a + (-b)i} = a - (-b)i = a + bi = z$.

Section 7-5, page 145

5. (1) 5, 5, 5, 0, 3, 5, 5 (3) 3, 4, 6, 4 **7.** $\sqrt{29}, \sqrt{5}, \sqrt{145}, \frac{\sqrt{29}}{29}, \frac{\sqrt{5}}{5}, \frac{\sqrt{145}}{5}$
9. $\sqrt{13}, 2\sqrt{5}, \sqrt{113}$ **15.** Graph: interior of a circle with center at the origin and radius 3. **17.** Graph: exterior of a circle with center at -3 and radius 2.
19. $z \cdot \bar{z} = (a + bi)(a - bi) = a^2 + b^2 = (\sqrt{a^2 + b^2})^2 = (|z|)^2$

Chapter 7 Review Exercises, page 146

1. $5 + 4i$ **2.a.** 12, -3 **b.** 12, 1 **c.** -1, 1 **d.** 0, 0 **3.a.** 1, 31 **b.** -21, 2 **c.** 1, 0 **d.** 9, 46 **4.a.** $-4i, -\frac{1}{4}i$ **b.** $3 + 7i, \frac{3}{58} + \frac{7}{58}i$ **5.a.** $-\frac{2}{41}, -\frac{23}{41}$ **b.** $-\frac{4}{3}, \frac{1}{3}$
7. 7, $\sqrt{89}, \sqrt{85}$ **8.** 10

Section 8-1, page 150

1.a. 2 **3.** 3 **5.** 241 **7.** 2 **9.** 45 **11.** (a) $x - 1$; 2 **13.** 0, 0, 12, 12, 12, 0
15. $\frac{1}{2}, 0, -\frac{13}{64}, -\frac{9}{2}, -\frac{7}{4}, -\frac{63}{64}$ **17.a.** 12 **c.** 0 **e.** 0 **g.** 0 **i.** 64 **k.** 270
19. 1, 2, -2, -3 **21.** If b is a zero of $q(x)$ then $q(b) = 0$. By the factor theorem there exists a polynomial q_1 such that $(x - b)q_1 = q$. But

$$p = q(x - a) = [q_1(x - b)](x - a) = [q_1(x - a)](x - b).$$

Thus, $(x - b)$ is a factor of p. By the factor theorem $p(b) = 0$, and b is a zero of $p(x)$.
23. If $(x - a_1)$ is a factor of p, then $p = q_1(x - a_1)$ since r_1 must equal 0. Similarly, if $(x - a_2)$ is a factor of q_1, then $q_1 = q_2(x - a_2)$. Hence

$$p = [q_2(x - a_2)](x - a_1) = (x - a_2)[q_2(x - a_1)]$$

and $x - a_2$ is a factor of p.

Section 8-2, page 153

In Exercises 1–3 the real part is listed first. **1.** 0, 8; 0, -8; 8, 0; -8, 0
3. -6, 0; 0, 6; -20, 5 **5.a.** $(x + 1)\left[x - \left(\frac{1}{2} + \frac{\sqrt{3}}{2}i\right)\right]\left[x - \left(\frac{1}{2} - \frac{\sqrt{3}}{2}i\right)\right]$
b. $(4x - 3)\left[x - \left(-\frac{3}{8} + \frac{3\sqrt{3}}{8}i\right)\right]\left[x - \left(-\frac{3}{8} - \frac{3\sqrt{3}}{8}i\right)\right]$
c. $(x + 1)(x + 1)[x - (1 + \sqrt{2}\,i)][x - (1 - \sqrt{2}\,i)]$ **7.** $\left\{\frac{3}{2}\right\}, \left\{\frac{3}{2}\right\}$
9. $\emptyset, \{4i, -4i\}$ **11.** $\{\sqrt{5}, -\sqrt{5}\}, \{\sqrt{5}, -\sqrt{5}\}$ **13.** $\emptyset, \{3 + 3i, 3 - 3i\}$
15. $(3x + 4)(2x - 3), (3x + 4)(2x - 3), (3x + 4)(2x - 3)$; $(2x - 3)^2, (2x - 3)^2, (2x - 3)^2$; prime, $[x - (1 + \sqrt{2})][x - (1 - \sqrt{2})]$, $[x - (1 + \sqrt{2})][x - (1 - \sqrt{2})]$.
17. Prime, prime, $(x - \sqrt{5}\,i)(x + \sqrt{5}\,i)$; prime, prime, $[x - (3 + 3i)][x - (3 - 3i)]$; prime, prime, $[2x - (1 + 2\sqrt{5}\,i)][2x - (1 - 2\sqrt{5}\,i)]$ **19.** $3 + 2i, -\frac{1}{2}$ **21.** Let $a + bi = z$. Then $a - bi = \bar{z}$. We are given that z is a solution; hence

$$a_n z^n + a_{n-1} z^{n-1} + \cdots a_1 z + a_0 = 0.$$

Also

$$\begin{aligned} a_n(\bar{z})^n + a_{n-1}(\bar{z})^{n-1} + \cdots + a_1(\bar{z}) + a_0 &= a_n\overline{z^n} + a_{n-1}\overline{z^{n-1}} + \cdots + a_1\bar{z} + a_0 \\ &= \overline{a_n z^n} + \overline{a_{n-1} z^{n-1}} + \cdots + \overline{a_1 z} + \overline{a_0} \\ &= \overline{a_n z^n + a_{n-1} z^{n-1} + \cdots + a_1 z + a_0} \\ &= \bar{0} \quad \text{(since } z \text{ is a zero)} \\ &= 0. \end{aligned}$$

Section 8-3, page 159

1. $\left\{\frac{1}{2}, -\frac{1}{2}, 1, -1\right\}$ **3.** $\left\{\frac{1}{4}, -\frac{1}{4}, \frac{1}{2}, -\frac{1}{2}, \frac{3}{4}, -\frac{3}{4}, \frac{3}{2}, -\frac{3}{2}, 1, -1, 3, -3\right\}$
5. (1) $-\frac{1}{2}$ (2) $\frac{1}{3}$ (3) $\frac{1}{2}$ (4) 1 or -2 **7.** 2, -3 **9.** $-5, -3, -2, 4$ **11.** $\frac{2}{3}, -\frac{3}{5}$
13. 2, -3, $\frac{3}{5}, -\frac{4}{5}$ **15.** -1, 2, $-\frac{1}{2}$ **17.** By Theorem 2 of this section, if $\frac{p}{q}$ is a rational zero, then p is a factor of a_0 and q is a factor of $a_n = 1$. Therefore

$$\frac{p}{q} = \frac{p}{1} = p \text{ or } \frac{p}{q} = \frac{p}{-1} = -p.$$

Thus, $\frac{p}{q}$ is an integer and is a factor of a_0. **19.** $(x-1)^2(x-\sqrt{3})(x+\sqrt{3})$
21. $(x-2i)(x+2i)(x-\sqrt{5}\,i)(x+\sqrt{5}\,i)$
23. $(x+1)(x-4)(x+3)\left[x-\left(\frac{5}{2}+\frac{\sqrt{3}}{2}i\right)\right]\left[x-\left(\frac{5}{2}-\frac{\sqrt{3}}{2}i\right)\right]$ **25.** Pair linear factors of the type $[x-(a+bi)]$ and $[x-(a-bi)]$ and multiply. These products form quadratic factors with real coefficients:

$$[x-(a+bi)]\,[x-(a-bi)] = x^2 - 2ax + (a^2+b^2).$$

Section 8-4, page 162

1. 0.20 **3.** 1.09 **5.** -2.8 and 2.5; -2.78 and 2.49 **7.** 0.65, -0.53, 2.88 **9.** 1.732

Chapter 8 Review Exercises, page 162

1. $-4, -14, 1, -43, -\frac{11}{4}, -\frac{101}{4}$ **2.** 1, 3, -2 **3.a.** $\{2+3i, 2-3i\}$
b. $\{\sqrt{5}, -\sqrt{5}, i, -i\}$ **4.** $\frac{2}{3}, -\frac{1}{3}, -1, 2$ **5.** $\{-1, 3\}$; $\{-1, 3, -2i, 2i\}$ **6.** 0.77

Section 9-1, page 164

1.a. Graph: like Figure 1, going through the point (1, 10). **b.** Graph: like Figure 2 going through the point $(-1, 10)$. **c.** Graph: same as (b). **d.** Graph: same as (a). **3.** $y = 3^{-x}$ and $y = 2^{-x}$ are reflections of the graphs of $y = 3^x$ and $y = 2^x$, respectively, about the y-axis. **7.** The graph of $x = 3^y$ is a reflection of the graph of $y = 3^x$ about the line $y = x$. **8.** Some sample points: **a.** $\left(-1, \frac{1}{5}\right)$, (0, 1), (1, 5)
c. $\left(-1, \frac{4}{3}\right)$, (0, 1), $\left(1, \frac{3}{4}\right)$ **e.** $\left(-1, \frac{1}{\pi}\right)$, (0, 1), $(1, \pi)$

Section 9-2, page 167

1.a. $2^3 = 8$ **c.** $\log_{25} 5 = \frac{1}{2}$ **e.** $4^{-2} = \frac{1}{16}$ **g.** $2^1 = 2$ **i.** $\log_a c = b$ **2.a.** 2 **b.** $\frac{1}{3}$
c. 100 **d.** $\frac{1}{2}$ **e.** $\frac{1}{3}$ **6.a.** Greater than **c.** Less than **e.** Less than
7. a, b, d, e, g, i, j **8.** $b^0 = 1$ and $b^1 = b$ since $b > 0$; hence, $\log_b 1 = 0$ and $\log_b b = 1$ for all $b > 0$, $b \neq 1$.

Section 9-3, page 171

1. $\{x\colon x > 0\}, R$ **3.a.** $2\log_b x - \log_b y$ **d.** $2\log_b x + \log_b y$
h. $\log_b x + \log_b y - \log_b 8 - 2\log_b z$ **l.** $\frac{2}{3}\log_b x + \frac{1}{3}\log_b y - \frac{1}{2}\log_b z$
5.a. 0.9030 **c.** 1.6020 **e.** 1.9084 **g.** -0.3979 **i.** 0.7781 **6.b.** 9 **d.** 8 **f.** 500
h. 16 **j.** 49 **l.** 15 **8.a.** $-6 + 0.3211$ **c.** $-9 + 0.2228$

Section 9-4, page 176

1.a. 7.26 **c.** 12.6 **e.** 54.1 **g.** 0.0903 **i.** 0.000210 **k.** 250 **m.** 0.00598
o. 0.0387 **2.a.** 3.475 (10^1) **c.** 3.21 (10^{-3}) **e.** 5.76 (10^{-5}) **g.** 1.86 (10^5)
i. 1.968 (10^3) **k.** 3.2789 (10^1) **m.** 3.97 (10^5) **o.** 7.9468 (10^3) **3.a.** 0 **d.** 1
g. 1 **j.** -6 **m.** 3 **p.** 7 **4.a.** 0.5740 **d.** 0.7931 **g.** 0.9921 **j.** 0.3979
m. 0.3010 **p.** 0.9832 **5.a.** 0.0899 **d.** 0.1959 + 1 **h.** 0.6561 − 4 **l.** 0.3284 − 2
p. 0.6425 − 2 **t.** Undefined **6.a.** 0.7536 + 2 **c.** 0.5877 + 3 **f.** 0.4669 − 4
h. 0.7101 − 3 **j.** 0.7993 − 5 **l.** 0.8082 − 8

Section 9-5, page 179 **1.a.** 0.2938 + 1 **e.** 0.5129 − 3 **i.** 0.1455 + 6 **m.** 0.4668 + 1 **q.** 0.2963 + 2 **2.a.** 3.962 **e.** 0.002863 **i.** 0.00003485 **m.** 104.0 **q.** 473.4

Section 9-6, page 182 **1.a.** 14520 **d.** $7.405\,(10)^{11}$ **h.** 24.31 **k.** 0.6907 **n.** 42.63 **p.** 11.88 **3.** \$13.19 **5.** 3929 feet **7.** 1196.7 feet **9.** $1.720\,(10^4)$ m.p.h.

Section 9-7, page 185 **1.** 1.383 **3.** 2.206 **5.** 5.412 **7.** 0.4392 **9.** 0.3029 **11.** 18.00

Section 9-8, page 187 **1.a.** $\left\{\frac{1}{3}\right\}$ **c.** $\{2\}$ **e.** $\{1.262\}$ **g.** $\{1.796\}$ **i.** $\{1.036\}$ **k.** $\{2.432\}$ **m.** $\{-1.677\}$ **o.** $\left\{\frac{16}{3}\right\}$ **2.a.** $\{1.584\}$ **c.** $\{1.893\}$ **e.** $\{1.584\}$ **g.** $\{1.167\}$ **i.** $\{0.6643\}$ **3.a.** $\{0.3882\}$ **c.** $\{-2.215\}$ **4.a.** $\{x\colon\ x > 1.223\}$ **c.** $\{x\colon\ x < 0.3693\}$

Section 9-9, page 191 **1.** $2^5 = 32$; $2^7 = 128$; 2^n **3.** 21 **4.** $A = 100\left(\frac{1}{2}\right)^{t/24}$ milligrams; 6.25 milligrams; after 79.73 ≈ 80 days **6.** $A = (100)(10)^{t/3}$; A = 10,000 bacteria; 12 hours

Chapter 9 Review Exercises, page 192 **1.** $8 < \left(\frac{1}{2}\right)^{-\pi} < 10$ **3.** $n = 2, m = 4$ **4.** $\frac{1}{16}$ **5.** $\log_x z = y$ **6.** $r^{-1} = \{(x, y)\colon\ x = \log_3 y\}$ **7.** $\log_4 \frac{a^{1/2}b^2}{c^{2/3}}$ **8.** 10,000 **9.** 3.7654692×10^4 **10.** 8360 **11.** −4 + 0.2428 **12.** −1 + 0.6043 **13.** 0.01122 **14.** 1.63 **15.** 536.91 feet **16.** −4.943 **17.** −5 **18.a.** −7.224 **b.** $-\frac{1}{4}$ **19.** 1.129 **20.** $5\sqrt[4]{2}$ milligrams; 173.7 days

Section 10-1, page 196 **1.** G, 3 × 4; H, 3 × 4; I, 3 × 3; J, 3 × 4; K, 3 × 3; L, 3 × 3. **3.** Answers may vary: [3 0 0], [0 4 0], [0 0 2]. **5.** They are equal. Associative property of matrix addition.

7. $M = \begin{bmatrix} -1 & -1 & -2 \\ 0 & -4 & 0 \\ -3 & -9 & 2 \end{bmatrix}, \begin{bmatrix} -1 & -5 & 2 & -3 \\ -4 & 1 & 0 & -6 \\ -2 & -5 & 3 & -8 \end{bmatrix}, \begin{bmatrix} -4 & 9 & -3 & -2 \\ -7 & -5 & 3 & -8 \\ 0 & -9 & -6 & -3 \end{bmatrix}, \begin{bmatrix} -4 & 1 & -2 \\ -5 & 3 & -8 \\ -7 & -6 & -3 \end{bmatrix}$

Section 10-2, page 199 **1.a.** −39 **c.** 50 **3.a.** −12 **c.** $\begin{bmatrix} 31 & -8 & 12 \\ -7 & -19 & -44 \\ 2 & -4 & 2 \end{bmatrix}$ **d.** $\begin{bmatrix} 16 & 55 & -18 \\ 6 & 0 & 33 \\ 2 & 15 & -2 \end{bmatrix}$ **f.** F

5. The results of *both* are: $\begin{bmatrix} 25 & -29 \\ -21 & -50 \\ 3 & -10 \end{bmatrix}$

Section 10-3, page 204 **1.** $\begin{bmatrix} 1 & -3 & -2 & 9 \\ 3 & 2 & 6 & 20 \\ 4 & -1 & 3 & 25 \end{bmatrix}$, (2, −5, 4) **3.** $\begin{bmatrix} 2 & 3 & 1 & 1 \\ 1 & -2 & -1 & -10 \\ -1 & 4 & 2 & 15 \end{bmatrix}$, (−5, 6, −7)

5. $\begin{bmatrix} 1 & 1 & -1 & -3 \\ 5 & -1 & -1 & 1 \\ 2 & -6 & 3 & 1 \end{bmatrix}$, (14, 26, 43) **7.** $\begin{bmatrix} 5 & 1 & 1 & 4 \\ 1 & 3 & 1 & 3 \\ 1 & 1 & -1 & -1 \end{bmatrix}$, $\left(\frac{2}{5}, \frac{3}{10}, \frac{17}{10}\right)$

9. $\begin{bmatrix} 1 & -1 & 1 & -1 & 0 \\ 3 & 3 & 3 & 3 & -1 \\ -2 & 2 & 1 & 2 & -1 \\ -1 & 1 & -4 & 4 & -1 \end{bmatrix}$, $\left(\frac{1}{6}, \frac{1}{2}, -\frac{1}{3}, -\frac{2}{3}\right)$

Section 10-4, page 207

1. $\begin{bmatrix} \frac{7}{41} & \frac{3}{41} \\ \frac{-2}{41} & \frac{5}{41} \end{bmatrix}$, (3, –2) **3.** $\begin{bmatrix} \frac{14}{5} & -\frac{2}{5} \\ -\frac{6}{5} & \frac{8}{5} \end{bmatrix}$, (–0.4, 0.6)

5. $\begin{bmatrix} \frac{7}{22} & -\frac{3}{22} \\ \frac{1}{11} & -\frac{2}{11} \end{bmatrix}$, (–4, 9) **7.** $\begin{bmatrix} \frac{1}{4} & -\frac{1}{6} \\ \frac{1}{8} & \frac{1}{8} \end{bmatrix}$, $\left(\frac{5}{12}, -\frac{3}{4}\right)$

9. The points on the line $4x_1 + 5x_2 + 2 = 0$.

Section 10-5, page 211

1.a. –8 **c.** –479 **e.** 135 **g.** 0 **i.** –210 **2.a.** 80 **3.a.** (3, –2)
5. Expanded along the 1st row we have:

$$a_{11}|a_{22}| - a_{12}|a_{21}| = a_{11}a_{22} - a_{12}a_{21}.$$

Section 10-6, page 213

1. Graph: a triangle determined by points (–3, 4) (1, –4), and (5, 7) and its interior. **3.** Graph: the triangle with vertices (0, 0), (0, 6), and (4, 0), and its interior. **5.** Graph: the polygon with vertices (0, 0), (0, 5), $\left(5\frac{2}{11}, 3\frac{3}{11}\right)$, and (6, 0), and its interior. **7.** 17, –2; 12.5, 0; 8, 0; 8, –10; $\frac{150}{11}$, 0; 20, –12.
9. 9, –11; 5, –20; 4, –12; 8, –16; 6, –10; 12, –36. **11.** 4 of type 1, 6 of type 2
13. 8 T_2 toys and no T_1 toys **15.** 4 I_1 and $\frac{9}{2} I_2$ items.

Section 10-7, page 216

1.a. $-\frac{4}{3} < x < 4$ **c.** $x > 4$ or $x < -\frac{4}{3}$ **e.** $-2 < x < 6$. **2.a.** Graph: an open segment; all points between $-1\frac{1}{3}$ and 4. **c.** Graph: the union of two open rays; all points to the left of $-1\frac{1}{3}$ together with all points to the right of 4. **e.** Graph: an open segment; all points between –2 and 6. **3.a.** Graph: the strip between, but not including, the lines $x = 4$ and $x = \frac{-4}{3}$. **c.** Graph: the union of the set of points to the right of the line $x = 4$, with the set of points to the left of the line $x = \frac{-4}{3}$.
e. Graph: the strip between, but not including, the lines $x = 6$ and $x = -2$.
5. Graph: the strip between, but not including, the lines $x = 3$ and $x = 1$.
11. Let r be a member of the solution set for the sentence $|x| < a$. Then $|r| < a$. (1) If $r \geqslant 0$, then $|r| = r$ and $r < a$, thus $0 \leqslant r < a$. (2) If $r < 0$, then $|r| = -r$, $-r < a$ and $r > -a$. Hence, $-a < r < 0$. From (1) and (2) we have $-a < r < a$ and r is a member of the solution set for $-a < x < a$. Conversely, if r is a member of the solution set for $-a < x < a$, then $-a < r < a$ and $r < a$ and $-a < r$. We know $|r| = r$ or $|r| = -r$. Since $r < a$, if $|r| = r$, then $|r| < a$. Since $-a < r$, $a > -r$ and if $|r| = -r$, then $a > |r|$. Hence, in either case $|r| < a$, r is a member of the solution set for $|x| < a$ and the sentences are equivalent.

Chapter 10 Review Exercises, page 217

1. $\begin{bmatrix} -4 & 7 & 4 \\ 0 & -6 & 10 \\ 9 & 11 & 1 \end{bmatrix}$, $\begin{bmatrix} -3 & 7 & 4 \\ 0 & -5 & 10 \\ 9 & 11 & 2 \end{bmatrix}$ **2.** $\begin{bmatrix} 15 & -1 & 10 \\ 40 & 43 & -6 \\ -11 & 10 & 12 \end{bmatrix}$, $\begin{bmatrix} -6 & -14 & 8 \\ 31 & 10 & -37 \\ -17 & 21 & 66 \end{bmatrix}$, A

3. $\begin{bmatrix} 1 & 5 & 7 & -6 \\ 3 & -4 & -3 & 15 \\ -5 & 2 & -4 & -20 \end{bmatrix}$, (2, –3, 1) **4.** $\begin{bmatrix} -\frac{1}{10} & \frac{1}{5} \\ \frac{2}{5} & -\frac{7}{15} \end{bmatrix}$, $\left(-\frac{1}{2}, \frac{1}{3}\right)$

5. 104, 171, 1 **6.** Graph: a triangle with vertices $\left(-1, -4\frac{1}{2}\right)$, $\left(-1, 4\frac{1}{2}\right)$ and (8, 0) together with its interior. **7.** Graph (if R is the replacement set): an open ray or half-line, all points to the left of $-\frac{2}{3}$. If $R \times R$ is the replacement set then the graph is the set of all points to the left of the line $x = -\frac{2}{3}$.

Section 11-1, page 220 **3.** Positive: a, c, d; negative: b **4.a.** $\angle APC, \angle APD$ **b.** $\overrightarrow{PA}(I), \overrightarrow{PC}(T); \overrightarrow{PA}(I), \overrightarrow{PD}(T)$ **7.** Six **9.** Eight

Section 11-2, page 224 **1.a.** $\frac{9\pi}{16}$ **c.** $\frac{10\pi}{3}$ **e.** $-\frac{3}{5}$ **2.a.** 0 **b.** π **c.** $\frac{3\pi}{2}$ **3.a.** $\frac{\pi}{6}$ **c.** $-\frac{\pi}{12}$ **e.** $\frac{5\pi}{3}$ **g.** $\frac{5\pi}{2}$ **i.** $\frac{7\pi}{40}$ **4.a.** 540 **c.** $-\frac{337.68}{\pi}$ **e.** 15 **g.** $\frac{360}{\pi}$ **i.** -225 **5.a.** II **c.** II **e.** IV **g.** IV **i.** III **7.** 12 radians **9.** $\frac{15}{2}$ **11.** $\frac{\pi}{3}$ inches; $\frac{5\pi}{4}$ inches **13.** 400π miles

Section 11-3, page 228 **3.b.** $\frac{2\pi}{3}, \frac{5\pi}{3}, \frac{8\pi}{3}, \frac{11\pi}{3}, \frac{14\pi}{3}, -\frac{\pi}{3}, -\frac{4\pi}{3}, -\frac{7\pi}{3}, -\frac{10\pi}{3}, -\frac{13\pi}{3}$ **d.** 1, 3, 5, 7, 9, $-1, -3, -5, -7, -9$ **f.** $2\pi, \frac{9\pi}{2}, 7\pi, \frac{19\pi}{2}, 12\pi, -\frac{\pi}{2}, -3\pi, -\frac{11\pi}{2}, -8\pi, -\frac{21\pi}{2}$ **4.a.** 0 **c.** $-\pi, -\frac{\pi}{2}, 0, \frac{\pi}{2}, \pi, \frac{3\pi}{2}, 2\pi, \frac{5\pi}{2}$ **5.** a, c, d, g, j, 1 **6.a.** $\left\{\frac{\pi}{3} + 2\pi n\right\}$ **d.** $\left\{\frac{\pi}{4} + 2\pi n\right\}$ **j.** $\left\{\frac{3\pi}{2} + 2\pi n\right\}$ **7.a.** (1, 0) **c.** $\left(\frac{1}{2}, \frac{\sqrt{3}}{2}\right)$ **e.** $\left(\frac{\sqrt{3}}{2}, -\frac{1}{2}\right)$ **g.** $\left(-\frac{\sqrt{2}}{2}, \frac{\sqrt{2}}{2}\right)$ **i.** (0, -1) **k.** $\left(\frac{\sqrt{2}}{2}, -\frac{\sqrt{2}}{2}\right)$ **8.b.** (0, 1) **e.** $\left(\frac{\sqrt{2}}{2}, \frac{\sqrt{2}}{2}\right)$ **h.** $\left(-\frac{\sqrt{2}}{2}, \frac{\sqrt{2}}{2}\right)$

Section 11-4, page 232 **1.**

θ	$\sin\theta$	$\cos\theta$	$\tan\theta$	$\cot\theta$	$\sec\theta$	$\csc\theta$
0						
$\frac{\pi}{6}$			$\frac{\sqrt{3}}{3}$	$\sqrt{3}$	$\frac{2\sqrt{3}}{3}$	2
$\frac{\pi}{4}$		$\frac{\sqrt{2}}{2}$		1	$\sqrt{2}$	
$\frac{\pi}{3}$	$\frac{\sqrt{3}}{2}$		$\sqrt{3}$		2	
$\frac{\pi}{2}$	1	0		0		1
$\frac{3\pi}{4}$				-1	$-\sqrt{2}$	$\sqrt{2}$
π	0		0	undefined	-1	
$\frac{3\pi}{2}$	-1	0	undefined		undefined	-1

3.a. 0 **b.** $\frac{3 + 4\sqrt{3}}{6}$ **c.** 0 **5.** $\tan\theta = \sqrt{3}$; $\cot\theta = \frac{\sqrt{3}}{3}$; $\sec\theta = 2$; $\csc\theta = \frac{2\sqrt{3}}{3}$. Any element of the set $\left\{\frac{\pi}{3} + 2\pi n\right\}$. **7.** $\left(\frac{\sqrt{2}}{2}\right)^2 + \left(\frac{\sqrt{2}}{2}\right)^2 = \frac{2}{4} + \frac{2}{4} = \frac{4}{4} = 1$ so $\left(\frac{\sqrt{2}}{2}, \frac{\sqrt{2}}{2}\right)$ is on the unit circle. **8.a.** $\frac{\pi}{6}$ **c.** $\frac{\pi}{4}$ **e.** $\frac{\pi}{4}$

10.a. $\dfrac{\sin\frac{\pi}{3}\cdot\sec\frac{\pi}{3}}{\cos\frac{\pi}{4}} = \dfrac{\frac{\sqrt{3}}{2}\cdot 2}{\frac{\sqrt{2}}{2}} = \dfrac{\sqrt{3}\cdot 2}{\sqrt{2}} = \sqrt{6}$ **11.a.** E **b.** T **c.** F **d.** T

12.a. I, III **b.** II **c.** I **d.** I **13.a.** $\frac{\pi}{3}$ **c.** $\frac{3\pi}{2}$ **e.** $\frac{3\pi}{2}$ **g.** $\frac{\pi}{6}$ **i.** 6 **k.** π

14.a. -1 **b.** Undefined **c.** $2\sqrt{2}$

Section 11-5, page 239

1.a. $\frac{5\pi}{3}$ **c.** $\frac{5\pi}{4}$ **e.** $\frac{\pi}{6}$ **g.** $\frac{5\pi}{4}$ **i.** $\frac{5\pi}{3}$ **2.b.** $\frac{\pi}{3}$ **d.** $\pi + \frac{\pi}{6}$ **f.** $\pi - (\pi - 3)$ **h.** $-\frac{2\pi}{7}$

3.a. $-\sqrt{3}$ **e.** $\frac{2\sqrt{3}}{3}$ **i.** $\frac{1}{2}$ **m.** $-\frac{\sqrt{3}}{3}$ **q.** $-\frac{\sqrt{3}}{3}$ **u.** $\frac{\sqrt{2}}{2}$ **4.a.** $-\cot\frac{\pi}{3}$ **c.** $\csc\frac{\pi}{2}$

e. $\sec\frac{\pi}{2}$ **g.** $-\cot\frac{3\pi}{8}$ **i.** $-\cot\frac{\pi}{3}$ **5.b.** $-\sin 0$ **d.** $\sin\frac{\pi}{4}$ **f.** $-\cos\frac{5\pi}{12}$ **h.** $-\sin(6\pi - 18)$

6.a. Undefined **c.** $-\frac{3}{2}$ **e.** $-\frac{\sqrt{3}}{2}$

Chapter 11 Review Exercises, page 240

1. Geometry: a set of points, the union of two rays with a common endpoint; trigonometry: the figure obtained by rotating a ray about its endpoint.
2. Answers may vary. **a.** $\angle GOA$ (or $\angle GOC$); $\angle BOA$ (or $\angle BOC$, $\angle IOA$, $\angle IOC$); $\angle AOC$ (or $\angle AOA$, $\angle COC$, $\angle COA$, $\angle FOA$ etc.); $\angle KOA$ (or $\angle KOC$); $\angle DOA$ (or $\angle DOC$) **b.** $\overrightarrow{OA}$ (or $\overrightarrow{OC}$) and $\overrightarrow{OG}$; $\overrightarrow{OA}$ and $\overrightarrow{OB}$; $\overrightarrow{OA}$ and $\overrightarrow{OC}$; $\overrightarrow{OA}$ and $\overrightarrow{OF}$; $\overrightarrow{OA}$ and $\overrightarrow{OK}$; $\overrightarrow{OA}$ and $\overrightarrow{OD}$. **3.a.** Positive **b.** Negative **c.** Positive **4.** $\frac{9\pi}{2}$ feet **5.** $\frac{1}{2}$ **6.** 36

7. $\frac{\pi}{2}, -\frac{3\pi}{2}$ **8.a.** $(-1, 0)$ **b.** $\left(\frac{\sqrt{3}}{2}, \frac{1}{2}\right)$ **c.** $(0, 1)$ **d.** $\left(\frac{\sqrt{3}}{2}, \frac{1}{2}\right)$ **9.** $x^2 + y^2 = 1$

10.a. $\frac{2\sqrt{6}}{5}$ **b.** $\frac{2\sqrt{6}}{25}$ **c.** $\frac{1}{5}$ **d.** 1 **e.** $2\sqrt{6}$ **f.** 1 **11.** $\frac{\sqrt{3}}{3}$ **12.** $\frac{\pi}{3}$

13.	Either	True	False
a.	I, III	II	IV
b.	All	none	none
c.	1, III	IV	II

14.a. II **b.** IV **c.** III **15.** $3\frac{1}{2}$ **16.a.** $\frac{\pi}{4}$ **b.** $\frac{\pi}{6}$ **c.** 3 **17.a.** $\frac{\pi}{4}$ **b.** $\frac{\pi}{6}$ **c.** 0

18.a. $\frac{\sqrt{2}}{2}$ **b.** $\frac{\sqrt{3}}{3}$ **c.** -1

Section 12-1, page 245

1. First use similar triangles OUV and OCW to establish $c = \cot\theta$. We have $\frac{c}{x} = \frac{1}{y}$, $c = \frac{x}{y} = \frac{\cos\theta}{\sin\theta} = \cot\theta$. Observe that as θ increases from 0 to π, c takes on all real number values but is undefined for $\theta \in \{\pi n\}$. Thus the domain of the cotangent function is $\{\theta\colon\ \theta \neq \pi n\}$ and the range of the function is R. **4.a.** sin, cos **c.** tan, cot, sec, csc **e.** sin, cos **5.** Answers may vary. **a.** tan **c.** $y = x^2 + 1$ **e.** $y = -x^2$ **g.** $y = \dfrac{1}{\left(x - \frac{1}{2}\right)(x + 1)}$ **i.** cot **6.** Figure 9: yes, 2; Figure 10: no; Figure 11: yes, 3; Figure 12: yes, 2π; Figure 13: no; Figure 14: yes, 2; Figure 15: no; Figure 16: yes, 3; Figure 17: yes, π **8.** None **9.a.** 2π **c.** 2π **e.** π

Section 12-2, page 250

3. and 4.

θ	0	$\frac{\pi}{6}$	$\frac{\pi}{4}$	$\frac{\pi}{3}$	$\frac{\pi}{2}$	$\frac{2\pi}{3}$	$\frac{3\pi}{4}$	$\frac{5\pi}{6}$	π
$\cos\theta$	1	$\frac{\sqrt{3}}{2}$	$\frac{\sqrt{2}}{2}$	$\frac{1}{2}$	0	$-\frac{1}{2}$	$-\frac{\sqrt{2}}{2}$	$-\frac{\sqrt{3}}{2}$	-1

6. Figure 9: C and E or B and D; Figure 10: A and B; Figure 11: C and D or D and F; Figure 12: C and F; Figure 13: A and C

Section 12-3, page 255

1.a. (*i*) Domain: set of all automobiles, range: set of all manufacturers; (*ii*) yes, only one manufacturer per automobile; (*iii*) to each manufacturer corresponds the automobiles he manufactured; (*iv*) domain: set of all manufacturers, range: automobiles manufactured; (*v*) no, more than one automobile for each manufacturer.
c. (*i*) Domain: set of all children, range: set of all fathers; (*ii*) yes, only one father per child; (*iii*) each father corresponds to his children; (*iv*) domain: set of all fathers, range: set of all children; (*v*) no, there can be more than one child per father.
e. (*i*) Domain: $\{a, b\}$, range: $\{b, c\}$; (*ii*) no, b and c both correspond to a; (*iii*) $r^{-1} = \{(b, a), (c, b), (c, a)\}$; (*iv*) domain: $\{b, c\}$, range: $\{a, b\}$; (*v*) no, a and b both correspond to c. **g.** (*i*) Domain: R, range: $\{y\colon\ y \geq 0\}$; (*ii*) yes, only one member of the range for each member of domain; (*iii*) $r^{-1} = \{(x, y)\colon\ x = y^2\}$; (*iv*) domain: $\{x\colon\ x \geq 0\}$, range: R; (*v*) no, two members of the range correspond to most elements of the domain. **2.a.** $\left\{\frac{\pi}{6} + 2\pi n, \frac{5\pi}{6} + 2\pi n\right\}$ **d.** $\{1\}$ **g.** $\left\{\frac{\pi}{6} + 2\pi n, \frac{11\pi}{6} + 2\pi n\right\}$ **j.** $\left\{\frac{\pi}{4} + 2\pi n, \frac{5\pi}{4} + 2\pi n\right\}$ **m.** $\{\pi + 2\pi n\}$ **p.** $\{-1\}$
3.a. (*i*) Domain: $\{x\colon\ x \neq 0\}$, range: $\{y\colon\ y \neq 0\}$; (*ii*) yes, one-to-one correspondence between members of the domain and range; (*iii*) $\left\{(x, y)\colon\ x = \frac{1}{y}\right\}$; (*iv*) domain: $\{x\colon\ x \neq 0\}$, range: $\{y\colon\ y \neq 0\}$; (*v*) yes, one-to-one correspondence between members of the domain and range; (*vi*) each element of the range of r corresponds to only one element of the domain. **c.** (*i*) Domain: R, range: $\{y\colon\ -1 \leq y \leq 1\}$; (*ii*) yes, to each element in the domain there corresponds only one element of the range; (*iii*) $\{(x, y)\colon y = \sin^{-1}x\}$; (*iv*) domain: $\{x\colon\ -1 \leq x \leq 1\}$, range: R; (*v*) no, to each domain element corresponds more than one range element. **5.** Answers may vary: $\{(x, y)\colon\ y = x^2\}$ **7.** Answers may vary: $\{(x, y)\colon\ x^2 + y^2 = 1\}$ **9.** Figures 6 and 11. **13.a.** Associate only one voter to each congressman. **b.** Answers may vary: $\{(4, -5), (\pi, \sqrt{2}), (4, \log_{10} 9), (33, 3^{5.6})\}$ **c.** Answers may vary: $\{(x, y)\colon\ y = \sqrt{-x - 1}\}$

Section 12-4, page 260

1.a. $\left\{\frac{5\pi}{4} + 2\pi n, \frac{7\pi}{4} + 2\pi n\right\}$ **d.** $\left\{\frac{3\pi}{4} + 2\pi n, \frac{5\pi}{4} + 2\pi n\right\}$ **f.** $\left\{\frac{\pi}{3} + 2\pi n, \frac{5\pi}{3} + 2\pi n\right\}$ **g.** $\left\{\frac{\pi}{3} + 2\pi n, \frac{4\pi}{3} + 2\pi n\right\}$ **j.** $\left\{\frac{\pi}{6} + 2\pi n, \frac{5\pi}{6} + 2\pi n\right\}$ **l.** $\left\{\frac{\pi}{6} + 2\pi n, \frac{7\pi}{6} + 2\pi n\right\}$
2.a. $\frac{3\pi}{4}$ **c.** $\frac{\pi}{6}$ **e.** $-\frac{\pi}{4}$ **g.** $\frac{\pi}{3}$ **i.** 0 **k.** $-\frac{\pi}{3}$ **m.** $\frac{5\pi}{6}$ **o.** $-\frac{\pi}{6}$ **q.** $\frac{\pi}{2}$ **3.a.** 0 **c.** $\frac{1}{2} - \frac{\pi}{6}$ **e.** $\frac{\pi}{6}$ **g.** Not possible **i.** $\frac{\pi}{6}$ **k.** $-\frac{\pi}{3}$ **4.a.** 1 **c.** $\frac{\pi}{4}$ **g.** Undefined **h.** Undefined
5.a. Domain: R, range: $\{y\colon\ y \neq \pi n\}$; domain: R, range: $\{y\colon\ 0 < y < \pi\}$
7.a. Domain: $\{x\colon\ x \geq 1 \text{ or } x \leq -1\}$, range: $y\colon\ y \neq \frac{\pi}{2} + n\pi$; domain: $\{x\colon\ x \geq 1 \text{ or } x \leq -1\}$, range: $y\colon\ 0 \leq y \leq \pi, y \neq \frac{\pi}{2}$ **8.a.** Let $y = \cot^{-1}a$. Then $a = \cot y$. Also $\tan^{-1}\frac{1}{a} = \tan^{-1}\left(\frac{1}{\cot y}\right) = \tan^{-1}(\tan y) = y = \cot^{-1}a$.

Section 12-5, page 265

1.a. π **d.** $\frac{2\pi}{3}$ **g.** 4π **j.** π **m.** 2 **2.a.** Graph: part of sine curve containing $(0, 0)$, $\left(\frac{\pi}{2}, 4\right)$, $(\pi, 0)$, $\left(\frac{3\pi}{2}, -4\right)$, and $(2\pi, 0)$. **d.** Graph: part of sine curve in part (c) reflected about x-axis and containing $(0, 0)$, $\left(\frac{\pi}{2}, -\frac{1}{4}\right)$, $(\pi, 0)$, $\left(\frac{3\pi}{2}, \frac{1}{4}\right)$, $(2\pi, 0)$. **g.** Graph: part of cotangent curve in part (e) reflected about x-axis, containing $\left(\frac{\pi}{4}, -2\right)$, $\left(\frac{\pi}{2}, 0\right)$, $\left(\frac{3\pi}{4}, 2\right)$, and asymptotic to vertical lines for $x = 0$ and $x = \pi$. **3.a.** Graph: part of cosine curve containing $(0, 4)$, $(\pi, 0)$, $(2\pi, -4)$, $(3\pi, 0)$, and $(4\pi, 4)$. **c.** Graph: part of cosine curve containing $\left(0, \frac{1}{4}\right)$, $(\pi, 0)$, $\left(2\pi, -\frac{1}{4}\right)$, $(3\pi, 0)$, and $\left(4\pi, \frac{1}{4}\right)$. **4.** a, c, d, e, f, g, h, i **5.a.** 2π **b.** 2 **c.** $\frac{4\pi}{3}$ **d.** 2 **e.** 4π

Section 12-6, page 268

1.a. 2π **d.** $\frac{1}{3}, 2\pi$ **g.** $-\frac{1}{2}$ **j.** $\frac{4}{3}, \frac{\pi}{3}$ **2.a.** $5, 4\pi, 2$ **d.** $3, \pi, -\frac{1}{2}$ **g.** $3, 2, -\frac{2}{\pi}$ **k.** $\frac{1}{2}, 2, \frac{2}{\pi}$ **m.** Undefined, $\frac{\pi}{2}, -2$ **3.** Graphs are given over one primitive period. **a.** Graph: part of sine curve containing $(0, 0)$, $\left(\frac{\pi}{6}, 2\right)$, $\left(\frac{\pi}{3}, 0\right)$, $(\pi, -2)$, and $\left(\frac{2\pi}{3}, 0\right)$. **c.** Graph: part of sine curve shifted $\frac{\pi}{8}$ units to the left and containing $\left(-\frac{\pi}{8}, 0\right)$, $\left(\frac{\pi}{8}, 3\right)$, $\left(\frac{3\pi}{8}, 0\right)$, $\left(\frac{5\pi}{8}, -3\right)$, $\left(\frac{7\pi}{8}, 0\right)$.

Chapter 12 Review Exercises, page 269

1.a. Domain: $\{x: x \neq n\pi\}$; range: R **b.** Domain: R; range: $\{y: y \geqslant 0\}$ **2.a.** A function f is said to be a periodic function if and only if $f(x) = f(x + k)$ for some nonzero real number k and for all x in the domain of f. **b.** The smallest positive period is called the primitive period of f. **5.** $y = \tan^{-1}x$. When $x = 0$, $y = \pi n$, so more than one element of the range corresponds to one element in the domain; therefore, $\tan^{-1}$ is not a function.
6.a. $\left\{x: x \neq \frac{\pi}{2} + n\pi\right\}$ **b.** $\{y: y \geqslant 1 \text{ or } y \leqslant -1\}$ **c.** $y = \sec^{-1}x$ **d.** $\{x: x \geqslant 1 \text{ or } x \leqslant -1\}$ **e.** $\left\{y: y \neq \frac{\pi}{2} + n\pi\right\}$ **f.** Yes **g.** No **7.** $\left\{\frac{\pi}{3} + 2\pi n, \frac{5\pi}{3} + 2\pi n\right\}$ **8.** Answers may vary: (1, 2), (2, 3), (3, 4) **9.a.** $\frac{\pi^2}{9}$ **b.** $\sqrt{2}$ **c.** 1 **10.** Domain: R; range: $\left\{y: -\frac{\pi}{2} < y < \frac{\pi}{2}\right\}$; Graph: Figure 6 (p. 261). **11.a.** π **b.** $\frac{\pi}{3}$ **c.** 2π **d.** 2 **12.a.** 2π **b.** 2π **13.** 1, undefined **14.** Graph over one primitive period: part of tangent curve containing $\left(-\frac{1}{4}, 3\right)$, $(0, 0)$, $\left(\frac{1}{4}, 3\right)$ and asymptotic to vertical lines $x = -\frac{1}{2}$ and $x = \frac{1}{2}$. **15.** $2, \pi, -\frac{\pi}{4}$

Section 13-1, page 274

2.a. $R, \{-1\}$ **c.** R, R **e.** $R, \left\{-\frac{1}{2} \pm \frac{\sqrt{3}}{2}i\right\}$ **g.** $R, \left\{x: x = \frac{\pi}{2} + 2\pi n\right\}$ **3.b.** $\{\alpha: \alpha \neq \pi + 2\pi n\}$ **d.** R **4.a.** $R, \left\{x: x = \frac{3\pi}{2} + 2\pi n\right\}$ **c.** $R, \{11, -1\}$ **e.** $R, \left\{\frac{2}{3} \pm \frac{\sqrt{11}}{3}i\right\}$ **6.** Answers may vary; let **a.** $x = 6$ **c.** $x = 1$ **e.** $x = 12$

g. $x = -1$ **7.** a, e, g, i **9.a.** $\sin^2\theta + \cos^2\theta = 1$; $\sin^2\theta + \cos^2\theta - \cos^2\theta = 1 - \cos^2\theta$; $\sin^2\theta = 1 - \cos^2\theta$

Section 13-2, page 279

1.a. $\dfrac{1}{\cos^2\theta\,\sin^2\theta} = \sec^2\theta\,\csc^2\theta$ **c.** $\cos x$ **f.** $\dfrac{1}{\cos^2 x\,\sin^2 x} = \sec^2 x\,\csc^2 x$ **2.a.** $\sec^2\theta$ **d.** $\sin^2\varphi$ **g.** $\dfrac{\sqrt{1-\sin^2 x} + \sin^3 x}{(\sin x)(1-\sin^2 x)}$ **i.** $|\sec\varphi|$ **3.a.** $\sin^2\theta + \cos^2\theta = 1$, so adding $-\sin^2\theta$ to both members gives $\cos^2\theta = 1 - \sin^2\theta$. **d.** $\tan y\sin y + \cos y = \dfrac{\sin y}{\cos y}\cdot\dfrac{\sin y}{1} + \dfrac{\cos^2 y}{\cos y} = \dfrac{\sin^2 y + \cos^2 y}{\cos y} = \dfrac{1}{\cos y} = \sec y$ **4.a.** $\sin^2\theta + \cos^2\theta = 1$, so $\dfrac{\sin^2\theta}{\cos^2\theta} + \dfrac{\cos^2\theta}{\cos^2\theta} = \dfrac{1}{\cos^2\theta}$ or $\tan^2\theta + 1 = \sec^2\theta$ **d.** $-2\sin^3 x = 1 - 2(1-\cos^2 x) = 1 - 2 + 2\cos^2 x = 2\cos^2 x - 1$ **g.** $(\tan x + \cot x)^2 = \tan^2 x + 2\tan x\cot x + \cot^2 x = \tan^2 x + 2 + \cot^2 x = (\tan^2 x + 1) + (1 + \cot^2 x) = \sec^2 x + \csc^2 x$

j. $\dfrac{1-\sin\theta}{1+\sin\theta} = \dfrac{1-\sin\theta}{1+\sin\theta}\cdot\dfrac{1-\sin\theta}{1-\sin\theta} = \dfrac{1 - 2\sin\theta + \sin^2\theta}{1-\sin^2\theta} = \dfrac{1}{\cos^2\theta} - \dfrac{2\sin\theta}{\cos^2\theta} + \dfrac{\sin^2\theta}{\cos^2\theta} = \sec^2\theta - 2\tan\theta\sec\theta + \tan^2\theta$ **5.a.** $\dfrac{1-\cos x}{1+\cos x} = \dfrac{1-\cos x}{1+\cos x}\cdot\dfrac{1-\cos x}{1-\cos x} = \dfrac{1 - 2\cos x + \cos^2 x}{1-\cos^2 x} = \dfrac{1}{\sin^2 x} - \dfrac{2\cos x}{\sin^2 x} + \dfrac{\cos^2 x}{\sin^2 x} = \csc^2 x - 2\csc x\cot x + \cot^2 x = (\csc x - \cot x)^2 = (\cot x - \csc x)^2$ **c.** $\dfrac{1}{\sec y - \tan y}\cdot\dfrac{\sec y + \tan y}{\sec y + \tan y} = \dfrac{\sec y + \tan y}{\sec^2 y - \tan^2 y} = \sec y + \tan y$ **e.** $\dfrac{\sin\omega + \tan\omega}{1+\cos\omega}\cdot\dfrac{1-\cos\omega}{1-\cos\omega} = \dfrac{\sin\omega + \tan\omega - \sin\omega\cos\omega - \cos\omega\tan\omega}{\sin^2\omega} = \dfrac{\tan\omega - \sin\omega\cos\omega}{\sin^2\omega} = \dfrac{\sin\omega - \sin\omega\cos^2\omega}{\sin^2\omega\cos\omega} = \dfrac{1-\cos^2\omega}{\sin\omega\cos\omega} = \dfrac{\sin^2\omega}{\sin\omega\cos\omega} = \dfrac{\sin\omega}{\cos\omega} = \tan\omega$

g. $\sin^2 t\sec^2 t + \sin^2 t\csc^2 t = \dfrac{\sin^2 t}{\cos^2 t} + 1 = \tan^2 t + 1 = \sec^2 t$

i. $(1-\cos^2\theta)(1+\cos^2\theta) = \sin^2\theta(1+\cos^2\theta) = \sin^2\theta + \sin^2\theta(1-\sin^2\theta) = \sin^2\theta + \sin^2\theta - \sin^4\theta = 2\sin^2\theta - \sin^4\theta$

6.a. $\dfrac{\sin\theta}{1+\cos(-\theta)} - \cot(-\theta) = \dfrac{\sin\theta}{1+\cos\theta} + \cot\theta = \dfrac{\sin\theta}{1+\cos\theta} + \dfrac{\cos\theta}{\sin\theta} = \dfrac{\sin^2\theta + \cos\theta + \cos^2\theta}{(1+\cos\theta)\sin\theta} = \dfrac{1+\cos\theta}{(1+\cos\theta)\sin\theta} = \csc\theta = \csc(\pi-\theta)$ **c.** $\dfrac{1-\cos^6(-\varphi)}{\sin^2(\pi-\varphi)} = \dfrac{1-\cos^6\varphi}{\sin^2\varphi} = \dfrac{1-(1-\sin^2\varphi)^3}{\sin^2\varphi} = \dfrac{1-(1-3\sin^2\varphi + 3\sin^4\varphi - \sin^6\varphi)}{\sin^2\varphi} = \dfrac{3\sin^2\varphi - 3\sin^4\varphi + \sin^6\varphi}{\sin^2\varphi} = 3 - 3\sin^2\varphi + \sin^4\varphi = 3 - 3\sin^2(\pi-\varphi) + \sin^4(\pi-\varphi)$

7.a. $\dfrac{1}{\sec(-\theta) - \tan(\pi+\theta)} = \dfrac{1}{\sec\theta - \tan\theta} = \dfrac{\sec\theta + \tan\theta}{\sec^2\theta - \tan^2\theta} = \dfrac{\sec\theta + \tan\theta}{1} = \sec\theta + \tan\theta$ **c.** $\dfrac{\sec(-n)}{1-\cos(-n)} = \dfrac{\sec n}{1-\cos n}\cdot\dfrac{1+\cos n}{1+\cos n} = \dfrac{\sec n + \sec n\cos n}{1-\cos^2 n} = \dfrac{\sec n + 1}{\sin^2 n} = \dfrac{\sec n + 1}{[-\sin(-n)]^2} = \dfrac{\sec n + 1}{\sin^2(-n)}$

Section 13-3, page 283

1.a. $\dfrac{63}{65}$ **d.** $\dfrac{56}{65}$ **e.** $\dfrac{56}{33}$ **2.a.** 0.5019 **d.** 0.9987 **e.** 47.45 **3.a.** 0 **d.** $-\dfrac{1}{2}$ **e.** $\dfrac{\sqrt{3}}{3}$ **4.a.** $\dfrac{\sqrt{2}-\sqrt{6}}{4}$ **d.** $\dfrac{\sqrt{6}-\sqrt{2}}{4}$ **e.** $-2+\sqrt{3}$ **5.a.** $2-\sqrt{3}$ **d.** $\dfrac{\sqrt{6}-\sqrt{2}}{4}$ **g.** $\dfrac{\sqrt{3}}{2}$ **j.** $\dfrac{\sqrt{6}-\sqrt{2}}{4}$ **m.** $-\dfrac{\sqrt{6}+\sqrt{2}}{4}$ **7.** $\tan(\pi+u) = \dfrac{\tan\pi + \tan u}{1 - \tan\pi\tan u} =$

$\dfrac{0+\tan u}{1-0\cdot\tan u}=\tan u$ **9.a.** $\cos\theta$ **b.** $\sin\alpha$ **10.a.** $\cos\left(\dfrac{\pi}{2}-y\right)=$ $\cos\dfrac{\pi}{2}\cos y+\sin\dfrac{\pi}{2}\sin y=0\cdot\cos y+1\cdot\sin y=\sin y$

Section 13-4, page 287 **1.** $\sin 2\alpha=\dfrac{10}{13}$, $\tan 2\alpha=\dfrac{18720}{16511}$ **3.** $\sin\dfrac{\alpha}{2}=\dfrac{\sqrt{2+\sqrt{3}}}{2}$; $\cos\dfrac{\alpha}{2}=\dfrac{\sqrt{2-\sqrt{3}}}{2}$; $\tan\dfrac{\alpha}{2}=2+\sqrt{3}$ **5.** $\dfrac{\sqrt{2+\sqrt{3}}}{2}$, $\dfrac{\sqrt{2-\sqrt{2+\sqrt{3}}}}{2}$

7. $\left|\tan\dfrac{\theta}{2}\right|=\sqrt{\dfrac{1-\cos\theta}{1+\cos\theta}\cdot\dfrac{1-\cos\theta}{1-\cos\theta}}=\sqrt{\dfrac{1-2\cos\theta+\cos^2\theta}{1-\cos^2\theta}}=$ $\sqrt{\dfrac{1-2\cos\theta+\cos^2\theta}{\sin^2\theta}}=\sqrt{\csc^2\theta-2\cot\theta\csc\theta+\cot^2\theta}=\sqrt{(\csc\theta-\cot\theta)^2}=$ $|\csc\theta-\cot\theta|$ **9.** $\cot 2\theta=\dfrac{\cos 2\theta}{\sin 2\theta}=\dfrac{\cos^2\theta-\sin^2\theta}{2\sin\theta\cos\theta}=\dfrac{\cot\theta}{2}-\dfrac{\tan\theta}{2}=\dfrac{\cot\theta-\tan\theta}{2}=$ $\dfrac{\cot^2\theta-1}{2\cot\theta}$ **11.a.** $\cos^4\alpha-\sin^4\alpha=(\cos^2\alpha-\sin^2\alpha)(\cos^2\alpha+\sin^2\alpha)=(\cos 2\alpha)(1)=$ $\cos 2\alpha$ **b.** $\dfrac{1-\tan^2\frac{x}{2}}{1+\tan^2\frac{x}{2}}=\dfrac{1-\frac{1-\cos x}{1+\cos x}}{1+\frac{1-\cos x}{1+\cos x}}=\dfrac{1+\cos x-(1-\cos x)}{1+\cos x+1-\cos x}=\dfrac{2\cos x}{2}=\cos x$

Section 13-5, page 291 **1.a.** $\left\{\dfrac{\pi}{6}+2\pi n,\dfrac{5\pi}{6}+2\pi n,\dfrac{\pi}{2}+2\pi n\right\}$ **d.** $\emptyset$ **g.** $\left\{\dfrac{\pi}{4}+\pi n\right\}$ **j.** $\left\{\dfrac{\pi}{4}+\pi n\right\}$ **2.a.** $\left\{\dfrac{\pi}{2}+2\pi n\right\}$ **d.** $\left\{\tan^{-1}\left(\dfrac{2\sqrt{3}}{3}\right)\right\}$ **g.** $\left\{\dfrac{4\pi}{3}+4\pi n\right\}$ **j.** $\left\{\pi n,\dfrac{2\pi}{3}+2\pi n,\dfrac{4\pi}{3}+2\pi n\right\}$ **3.a.** $\left\{\dfrac{\pi}{6}+\pi n,\dfrac{5\pi}{6}+\pi n\right\}$ **c.** $\left\{\dfrac{3\pi}{4}+\pi n\right\}$ **e.** $\left\{\cos^{-1}\left(\dfrac{(1+\sqrt{5})}{2}\right)\right\}$ **g.** $\left\{\theta\colon\ \theta=\cos\dfrac{-13}{5}\text{ and }2\pi n<\theta<\dfrac{\pi}{2}+2\pi n\right\}$

Section 13-6, page 295 **1.a.** 0.6816 **c.** 19.670 **e.** 0.0200 **g.** 1.0065 **i.** 0.2857 **k.** 1.189 **m.** 0.8321 **o.** 0.4843 **2.a.** 1.851 **d.** −1.084 **g.** −0.5875 **j.** $\{0.804+2\pi n, 2.338+2\pi n\}$ **m.** No possible answer **p.** No possible answer **3.a.** $\left\{\pi n,\dfrac{\pi}{6}+2\pi n,\dfrac{5\pi}{6}+2\pi n\right\}$ **c.** $\left\{\dfrac{3\pi}{2}+2\pi n\right\}$ **e.** True for all x **g.** $\{\pm 2.48+2\pi n\}$ **5.a.** $\left\{\dfrac{\pi}{2}+2\pi n,\dfrac{3\pi}{2}+2\pi n\right\}$ **6.a.** $\{1.946+2\pi n, 4.338+2\pi n\}$ **c.** $\{\pi n\}$ **e.** Same as part d **7.** $\{2.035+\pi n\}$

Chapter 13 Review Exercises, page 295 **1.** $\{x\colon\ -1\leqslant x\leqslant 1\}$ **2.a.** $\dfrac{10}{(2x-5)(2x+5)}-\dfrac{1(2x+5)}{(2x-5)(2x+5)}=\dfrac{10-2x-5}{(2x-5)(2x+5)}=$ $\dfrac{-1(2x-5)}{(2x-5)(2x+5)}=\dfrac{-1}{(2x+5)}$ **b.** If $x=1$, the equation is not true.
3. $(\cos\theta,\sin\theta)$ is on the unit circle, $x^2+y^2=1$, so $\cos^2\theta+\sin^2\theta=1$.
4.a. $(\sin x+\cos x)^2=\sin^2 x+2\sin x\cos x+\cos^2 x=1+2\sin x\cos x=1+\sin 2x$
b. $\dfrac{\csc\theta+\cot\theta}{\frac{1}{1-\cos\theta}}=\left(\dfrac{1}{\sin\theta}+\dfrac{\cos\theta}{\sin\theta}\right)(1-\cos\theta)=\dfrac{1-\cos^2\theta}{\sin\theta}=\dfrac{\sin^2\theta}{\sin\theta}=\sin\theta$

5.a. $\frac{\sqrt{2}+\sqrt{6}}{4}$ **b.** $-\sqrt{3}$ **6.** 0.884 **7.** $\frac{3}{5}, \frac{3\sqrt{10}}{10}$ **8.a.** $\left\{\frac{\pi}{2} + \pi n:\ n \in I\right\}$
b. $\tan^{-1}\frac{1}{2}$ **c.** $\left\{\frac{3\pi}{2} + 2\pi n:\ n \in I\right\}$ **9.a.** 1.193 **b.** 0.816

Section 14-1, page 300 **1.a.** $\frac{\sqrt{3}}{2}$ **c.** 0.1419 **e.** $\sqrt{2}$ **g.** 0 **i.** $\frac{\sqrt{3}}{3}$ **k.** -1 **2.a.** 0.8411 **d.** 1.000 **e.** 0.1611 **h.** 0.4030 **l.** 0.9507 **p.** 1.204 **3.a.** 86° 20′ **c.** 45° 34′ **e.** 49° **g.** 14° 46′ **i.** 28° 2′ **4.a.** 1.507 **c.** 0.795 **e.** 0.855 **g.** 0.258 **i.** 0.489
6. Using known trigonometric identities, it is possible to write a function of any angle in terms of some function of an angle with measure between 0° and 45°. For example, $\cos\left(\frac{\pi}{2} - a\right) = \sin\alpha$, so if $\frac{\pi}{2} - \alpha$ (or $90° - \alpha$) is greater than $\frac{\pi}{4}$ (45°), then α is less than $\frac{\pi}{4}$ (45°). Similar identities deal with angles of larger measure.

Section 14-2, page 305 **1.a.** 78° **c.** 84° **e.** 341° **g.** 188° **i.** 37° **k.** 158° **2.a.** $\frac{\sqrt{2}}{2}$ **c.** $\sqrt{2}$ **e.** -0.2126 **g.** 1.423 **i.** 1.091 **k.** $\frac{\sqrt{2}}{2}$ **m.** -0.3420 **o.** -6.123 **3.a.** $\frac{\sqrt{2}+\sqrt{6}}{4}$ **d.** $-\frac{\sqrt{3}}{2}$ **g.** $-\frac{\sqrt{2}}{2}$ **j.** $\frac{\sqrt{2}}{2}$ **4.a.** -4.720 **d.** -0.1404 **5.a.** 0.8480 **c.** $\frac{\sqrt{2}}{2}$ **e.** 0 **g.** 3.487

Section 14-3, page 308

1.	x	y	r	$\sin\theta$	$\cos\theta$	$\tan\theta$	$\cot\theta$	$\sec\theta$	$\csc\theta$
a.	2	2	$2\sqrt{2}$	$\frac{\sqrt{2}}{2}$	$\frac{\sqrt{2}}{2}$	1	1	$\sqrt{2}$	$\sqrt{2}$
f.	-7	-5	$\sqrt{74}$	$-\frac{5\sqrt{74}}{74}$	$-\frac{7\sqrt{74}}{74}$	$\frac{5}{7}$	$\frac{7}{5}$	$-\frac{\sqrt{74}}{7}$	$-\frac{\sqrt{74}}{5}$
l.	5	-13	$\sqrt{194}$	$-\frac{13\sqrt{194}}{194}$	$\frac{5\sqrt{194}}{194}$	$-\frac{13}{5}$	$-\frac{5}{13}$	$\frac{\sqrt{194}}{5}$	$-\frac{\sqrt{194}}{13}$

2.a. Length of side opposite $\angle A$, c. **c.** Length of side adjacent to $\angle B$, a. **e.** Length of side opposite $\angle A$, a.

3.	$\sin\angle A$	$\cos\angle A$	$\tan\angle A$	$\cot\angle A$	$\sec\angle A$	$\csc\angle A$	$\sin\angle B$	$\cos\angle B$	$\tan\angle B$	$\cot\angle B$	$\sec\angle B$	$\csc\angle B$
a.	$\frac{3}{5}$	$\frac{4}{5}$	$\frac{3}{4}$	$\frac{4}{3}$	$\frac{5}{4}$	$\frac{5}{3}$	$\frac{4}{5}$	$\frac{3}{5}$	$\frac{4}{3}$	$\frac{3}{4}$	$\frac{5}{3}$	$\frac{5}{4}$
g.	$\frac{\sqrt{2}}{2}$	$\frac{\sqrt{2}}{2}$	1	1	$\sqrt{2}$	$\sqrt{2}$	$\frac{\sqrt{2}}{2}$	$\frac{\sqrt{2}}{2}$	1	1	$\sqrt{2}$	$\sqrt{2}$

5. $-\frac{7\sqrt{74}}{74}$ **7.** $\frac{5}{12}$ **9.** $-\frac{\sqrt{3}}{2}$ **11.** $-\frac{7}{6}$ **13.** $\frac{3}{5}$ **15.** $\frac{12}{13}$ or $-\frac{12}{13}$ **16.** Answers are listed $\sin\theta$, $\cos\theta$, and $\tan\theta$: **a.** $\frac{2\sqrt{13}}{13}, \frac{3\sqrt{13}}{13}, \frac{2}{3}$ **d.** $\frac{8\sqrt{113}}{113}, \frac{7\sqrt{113}}{113}, \frac{8}{7}$ **g.** $\frac{\sqrt{10}}{10}, \frac{3\sqrt{10}}{10}, \frac{1}{3}$ **j.** $-\frac{2\sqrt{13}}{13}, -\frac{3\sqrt{13}}{13}, \frac{2}{3}$ **17.** $\sin\angle A = \frac{\text{length of the side opposite } \angle A}{\text{length of the hypotenuse}} = \frac{\text{length of the side adjacent } \angle B}{\text{length of the hypotenuse}} = \cos\angle B$ because the side opposite $\angle A$ is the side adjacent to $\angle B$.

Section 14-4, page 313 **1.a.** $m_r \angle B = \frac{\pi}{3}, a = 3, b = 3\sqrt{2}$ **c.** $m \angle B = 30°, b = 2, c = 4$ **e.** $m \angle A = 45°$, $b = 2, m \angle B = 45°$ **g.** $m \angle A = 30°, a = 2, m \angle B = 60°$ **i.** $m_r \angle A = \frac{\pi}{4}, a = 3$, $c = 3\sqrt{2}$ **2.a.** $c = \sqrt{130.09}$ **d.** $b = 94.1$ **g.** $m \angle B = 2°21'$ **j.** $c = 16.181$ **m.** $m_r \angle A = 1.107$ **p.** $m_r \angle B = 0.481$ **s.** $b = 1.6209$ **3.** 76.170 feet **5.** 1300 miles **7.** He may use the 20-foot-long ladder. **9.** $\sqrt{2}$ hours; 45° **11.** 1145.8 feet

Section 14-5, page 317 **1.a.** 7.045 **c.** 0.590 **e.** 15.85 **g.** 0.6942 **i.** 1.026 **k.** 15.50 **2.a.** 62° 14′ **c.** 30° 59′ **e.** 63° 22′ **g.** 70° 58′ **i.** 77° 40′ **k.** 59° 32′ **3.** 38° 11′ **5.a.** 23° 40′ **7.** $\log(\sec 63°14') = \log(\cos 63°14')^{-1} = -\log(\cos 63°14')$ **9.** $\log(\csc \angle A) = \log(\sin \angle A)^{-1} = -\log(\sin \angle A)$

Section 14-6, page 321 **1.a.** 4.074 **d.** 2.95 **g.** 75° 7′ **j.** 8.12 **m.** 39.39 **o.** Two solutions: 20.94 or 13.72 **2.a.** 180; 270 **b.** North **3.** 598 miles **5.** 3.9 hours **7.** $m < A = 22°11'$; $m < B = 100°18'$; $m < C = 57°31'$ **9.** 2.21 miles **11.** 178.9 light years **13.** 151 miles **17.** 1341 miles **20.** Use a transit to measure $\angle A$ and $\angle C$, and to measure $\overline{AC}$. Since $m(\angle B) = 180° - [m(\angle A) + m(\angle C)]$ and $\frac{\sin \angle C}{m(\overline{AB})} = \frac{\sin \angle B}{m(\overline{AC})}$ then $m(\overline{AB}) = \frac{m(\overline{AC}) \sin \angle C}{\sin \angle B}$.

Section 14-7, page 327 **1.** Parts b, c, and e are vector quantities. **9.** 81° 28′ **11.** 0°, 180°

Chapter 14 Review Exercises, page 327 **1.** The cosine of an angle is the x coordinate of the point of intersection of the terminal side of the angle with the unit circle where the angle is in standard position and the unit circle centered at the origin. **2.a.** 0.8134 **b.** 0.9993 **c.** 0.9781 **d.** $-\frac{\sqrt{2}}{2}$ **3.** 1.252 **4.a.** 0.7353 **b.** 0.3590 **c.** 20.47 **5.** 51° 33′ **6.** 4.531 **7.** 0.9877 **8.** $\sin \angle A = -\frac{2\sqrt{13}}{13}, \cos \angle A = \frac{3\sqrt{13}}{13}, \tan \angle A = -\frac{2}{3}, \cot \angle A = -\frac{3}{2}$, $\sec \angle A = \frac{\sqrt{13}}{3}, \csc \angle A = -\frac{\sqrt{13}}{2}$ **9.** $-\sqrt{10}$ **10.a.** Length of the side adjacent to $\angle X, x$ **b.** Length of the side adjacent to $\angle Z, y$ **11.** $a^2 + c^2 = 2 + 10 = 12 = b^2$; $\sin \angle A = \frac{\sqrt{6}}{6}, \cos \angle A = \frac{\sqrt{30}}{6}, \tan \angle A = \frac{\sqrt{5}}{5}, \cot \angle A = \sqrt{5}, \sec \angle A = \frac{\sqrt{30}}{5}$, $\csc \angle A = \sqrt{6}; \sin \angle C = \frac{\sqrt{30}}{6}, \cos \angle C = \frac{\sqrt{6}}{6}, \tan \angle C = \sqrt{5}, \cot \angle C = \frac{\sqrt{5}}{5}$, $\sec \angle C = \sqrt{6}, \csc \angle C = \frac{\sqrt{30}}{5}$

12.

	$\sin\theta$	$\cos\theta$	$\tan\theta$	$\cot\theta$	$\sec\theta$	$\csc\theta$
30°	$\frac{1}{2}$	$\frac{\sqrt{3}}{2}$	$\frac{\sqrt{3}}{3}$	$\sqrt{3}$	$\frac{2\sqrt{3}}{3}$	2
$\frac{\pi}{3}$	$\frac{\sqrt{3}}{2}$	$\frac{1}{2}$	$\sqrt{3}$	$\frac{\sqrt{3}}{3}$	2	$\frac{2\sqrt{3}}{3}$
45°	$\frac{\sqrt{2}}{2}$	$\frac{\sqrt{2}}{2}$	1	1	$\sqrt{2}$	$\sqrt{2}$

13. 10.44 **14.** 152 feet **15.** 47 **16.** 8.40 **17.** 80° 57′ **18.** 7.2 **19.** Not a triangle **20.** Direction: $\frac{\pi}{2} - 0.1975$; speed: 153 miles per hour **21.** Magnitude: 14.81 pounds; direction: 0.602 radians from F_0.

Section 15-1, page 332 **1.a.** II **c.** I **e.** I **g.** IV **l.** II **2.** Any two members of the set in each case where $k \in I$: **a.** $\{(5, 343° + k \cdot 360°)\}$ **c.** $\left\{\left(\sqrt{5}, \frac{\pi}{6} + 2\pi k\right)\right\}$ **e.** $\{1, 265° + k \cdot 360°)\}$ **g.** $\{(10, 2\pi k)\}$ **i.** $\{(3, -30° + n \cdot 360°)$: **3.a.** $(2\sqrt{2}, 45°)$ **c.** $(\sqrt{13}, 236° 19')$ **e.** $(3\sqrt{5}, 26° 34')$ **g.** $(2\sqrt{10}, 71° 34')$ **4.a.** $\left(\frac{\sqrt{2}}{2}, \frac{\sqrt{2}}{2}\right)$ **c.** $(-1.9890, -0.2090)$ **e.** $\left(-\frac{\sqrt{2}}{2}, -\frac{\sqrt{2}}{2}\right)$ **g.** $\left(\frac{3}{2}, \frac{3\sqrt{3}}{2}\right)$ **5.** $(0, x°), x \in R$ **7.a.** $r = 2$ **b.** $2r \cos\theta + r\sin\theta = 4$ **8.a.** $x^2 + y^2 = 4x$ **b.** $y = x$ **c.** $x^2 + y^2 = 49$

Section 15-2, page 335 **1.** Graph: points for the following in the rectangular coordinate plane: **a.** $(3, -5)$ **d.** $(-9, -6)$ **g.** $(-1, 0)$ **j.** $(0, 0)$ **m.** Approximately $(1.9, 2.3)$ **2.a.** $\sqrt{5}$ cis $296° 34'$ **d.** $6\sqrt{2}$ cis $225°$ **g.** $\sqrt{74}$ cis $234° 28'$ **j.** $\sqrt{313}$ cis $42° 43'$ **3.a.** $\frac{3}{2} + \frac{3\sqrt{3}}{2}i$ **d.** $\frac{5\sqrt{2}}{2} + \frac{5\sqrt{2}}{2}i$ **g.** $-0.9925 - 0.1219i$ **j.** $-4.856 + 1.1915i$ **4.** $0 + i = \text{cis}\,\frac{\pi}{2}$, $1 + \sqrt{3}\,i = 2\,\text{cis}\,\frac{\pi}{3}$, $(0 + i)(1 + \sqrt{3}\,i) = -\sqrt{3} + i = 2\,\text{cis}\,\frac{5\pi}{6}$; the absolute value of the product is the product of the absolute values and the argument of the product is the sum of the arguments. **5.a.** $\sqrt{5}$ cis $63° 26'$ **d.** $6\sqrt{2}\,\text{cis}\,\frac{3\pi}{4}$ **g.** $\sqrt{74}$ cis $125° 32'$ **j.** $\sqrt{313}$ cis $317° 17'$ **6.** $r_1 \text{ cis } \theta_1 = r_2 \text{ cis } \theta_2$ if and only if $r_1 = r_2$ and $\theta_1 = \theta_2 + 2\pi n$. *Proof:* If $r_1 = r_2$ and $\theta_1 = \theta_2 + 2\pi n$, then

$$r_1 \text{ cis } \theta_1 = r_2 \text{ cis } (\theta_2 + 2\pi n) = r_2 \text{ cis } \theta_2.$$

If $r_1 \text{ cis } \theta_1 = r_2 \text{ cis } \theta_2$, let $a + bi = r_1 \text{ cis } \theta_1$ and $c + di = r_2 \text{ cis } \theta_2$ so that $a + bi = c + di$ or $a = c$ and $b = d$. Thus

$$r_1 = \sqrt{a^2 + b^2} = \sqrt{c^2 + d^2} = r_2 \quad \text{and} \quad \theta_1 = \tan^{-1}\frac{b}{a} = \tan^{-1}\frac{d}{c} = \theta_2 + 2\pi n.$$

Section 15-3, page 339 **1.a.** $6\,\text{cis}\,\frac{7\pi}{6}, -3\sqrt{3} - 3i$ **c.** 50 cis $155°$, $-45.32 + 21.31i$ **e.** 2 cis $195°$, $-1.9318 - 0.5176i$ **g.** $\sqrt{7957}$ cis $127° 15'$, $-54 + 71i$ **i.** $5\sqrt{149}$ cis $88° 7'$, $2 + 61i$ **2.a.** $\frac{3}{2}$ cis $345°$ **c.** $\frac{4}{5}\,\text{cis}\,\frac{\pi}{6}$ **e.** 2 cis $50°$ **g.** $\frac{5}{13} + \frac{1}{13}i$ **i.** $-\frac{14}{17} + \frac{5}{17}i$ **3.a.** $4\,\text{cis}\,\frac{\pi}{2} = 4i$ **c.** $\frac{1}{16}\,\text{cis}\,\frac{2\pi}{3}$ **e.** $2\sqrt{2}\,\text{cis}\,\frac{3\pi}{4}$ **g.** 125 cis $20° 36'$ **h.** $\text{cis}\,\frac{7\pi}{4}$ **5.a.** $-\frac{3}{2}i$ **c.** $\frac{\sqrt{3}}{10} + \frac{1}{10}i$ **d.** $\frac{5}{2}i$

Section 15-4, page 342 **1.a.** $1, -\frac{1}{2} + \frac{\sqrt{3}}{2}i, -\frac{1}{2} - \frac{\sqrt{3}}{2}i$ **c.** $\text{cis}\,\frac{\pi}{8}, \text{cis}\,\frac{5\pi}{8}, \text{cis}\,\frac{9\pi}{8}, \text{cis}\,\frac{13\pi}{8}$ **e.** $\text{cis}\,\frac{\pi}{2} = i, \text{cis}\,\frac{3\pi}{2} = -i$ **g.** $2, 2i, -2, -2i$ **i.** $1, \text{cis}\,\frac{\pi}{3}, \text{cis}\,\frac{2\pi}{3}, -1, \text{cis}\,\frac{4\pi}{3}, \text{cis}\,\frac{5\pi}{3}$ **2.a.** $\sqrt[4]{2}\,\text{cis}\,\frac{\pi}{12}, \sqrt[4]{2}\,\text{cis}\,\frac{7\pi}{12}, \sqrt[4]{2}\,\text{cis}\,\frac{13\pi}{12}, \sqrt[4]{2}\,\text{cis}\,\frac{19\pi}{12}$ **3.a.** $\sqrt[3]{2}\,\text{cis}\,\frac{\pi}{9}, \sqrt[3]{2}\,\text{cis}\,\frac{7\pi}{9}, \sqrt[3]{2}\,\text{cis}\,\frac{13\pi}{9}$ **c.** $\text{cis}\,\frac{\pi}{4}, \text{cis}\,\frac{5\pi}{4}$

Chapter 15 Review Exercises, page 343 **1.** Answers may vary: $(2\sqrt{2}, 315°), (2\sqrt{2}, 675°)$ **2.** $(1, -\sqrt{3}), \left(-\frac{3\sqrt{3}}{2}, -\frac{3}{2}\right)$ **3.** $10\,\text{cis}\,\frac{2\pi}{3}$ **4.** $1.2252 + 3.8080i, -2.089 + 4.547i$ **5.** $2\sqrt{2}\,\text{cis}\,\frac{\pi}{4}, 2 + 2i$ **6.** $-\frac{\sqrt{2}}{3} - \frac{1}{3}i$ **7.** $a^b \text{ cis } b\alpha$ **8.** $216\,\text{cis}\,\frac{3\pi}{2} = -216i, 4096 \text{ cis } \pi = -4096$ **9.** $4\,\text{cis}\,\frac{\pi}{6}$ **10.** $\text{cis}\,\frac{3\pi}{8}, \text{cis}\,\frac{7\pi}{8}, \text{cis}\,\frac{11\pi}{8}, \text{cis}\,\frac{15\pi}{8}$; Graph: if O corresponds to the origin and X

corresponds to (1, 0), then the roots can be pictured by four points (A, B, C, and D) on a unit circle with center at the origin such that $m_r(\angle AOX) = \frac{3\pi}{8}$, $m_r(BOX) = \frac{7\pi}{8}$, $m_r(\angle COX) = \frac{11\pi}{8}$, and $m_r(DOX) = \frac{15\pi}{8}$ **11.a.** $\text{cis}\,\frac{3\pi}{8}$, $\text{cis}\,\frac{7\pi}{8}$, $\text{cis}\,\frac{11\pi}{8}$, $\text{cis}\,\frac{15\pi}{8}$
b. $\left\{3\,\text{cis}\,\frac{\pi}{6}, 3\,\text{cis}\,\frac{5\pi}{6}, 3\,\text{cis}\,\frac{3\pi}{2} = -3i\right\}$

Section 16-1, page 348 **1.** f_1: $\{1, 2, 3, 4, 5\}$, f_2: $\{1, 2, 3, 4, 5, 6, 7\}$, f_3: $\{1, 2, 3, 4, 5, 6\}$, f_4: $\{1, 2, 3, 4, 5, 6, 7, 8, 9, 10\}$ **3.** f_1, 0; f_3, 23 **4.** f_2, $\frac{1}{3}$; f_4, $\frac{1}{2}$ **5.** 52 **7.** 0, $\frac{3}{4}$ **9.** $27\sqrt{2}$ **11.** $\frac{1}{4\pi}$, $\frac{8}{\pi}$ **15.a.** 220 **c.** -1116 **e.** $\frac{15}{4}$ **g.** $4\sqrt{2}(1-\sqrt{3})$ **i.** 2046 **k.** $\frac{255}{4\pi}$ **l.** $\frac{255}{8}$ **16.a.** $a_{16} = 79$, $S = 664$ **c.** $a_1 = \frac{30}{13}$ **e.** $n = 7$, $a_7 = -8$ **17.a.** $n = 4$, $a_4 = -162$ **c.** $a_6 = \frac{243}{2}$, $S = 182$ **e.** $a_1 = 32$, $a_5 = 2$ **18.a.** 78 **c.** 156 **e.** 24 **19.a** Since $a_1 + d = a_2$ and $a_1 + 2d = a_3$

$$\frac{a_1 + a_3}{2} = \frac{a_1 + a_1 + 2d}{2} = \frac{2a_1 + 2d}{2} = a_1 + d = a_2$$

20.a. 1, 3, 5, 7, 9 **21.a.** Since $a_2 = a_1 r$ and $a_3 = a_1 r^2$, $\sqrt{a_1 a_2} = \sqrt{a_1 \cdot a_1 r^2} = \sqrt{a_1^2 \cdot r^2} = a_1 r = a_2$ **23.** 5000 $(1.08)^{10}$ **25.** \$200 $(1.01)^{16}$ = \$234.51.

Section 16-2, page 353 **1.**

	Terms 5th	8th	10th	jth	$(j-1)$st	$(j+1)$st
a.	15	24	30	$3j$	$3(j-1)$	$3(j+1)$
c.	120	40,320	3,628,800	$j!$	$(j-1)!$	$(j+1)!$

2. Part b

3.

	Terms 5th	7th	10th	kth	$(k-1)$st	$(k+1)$st
a.	$\frac{5}{2}$	$\frac{7}{2}$	5	$\frac{k}{2}$	$\frac{k-1}{2}$	$\frac{k+1}{2}$
c.	25	49	100	k^2	$(k-1)^2$	$(k+1)^2$
e.	2	2	0	$1+(-1)^{k+1}$	$1+(-1)^k$	$1+(-1)^{k+2}$
g.	1	-1	0	$\sin\frac{k\pi}{2}$	$\sin\frac{(k-1)\pi}{2}$	$\sin\frac{(k+1)\pi}{2}$
i.	$\frac{\pi}{\sqrt{5}}$	$\frac{\pi}{\sqrt{7}}$	$\frac{\pi}{\sqrt{10}}$	$\frac{\pi}{\sqrt{k}}$	$\frac{\pi}{\sqrt{k-1}}$	$\frac{\pi}{\sqrt{k+1}}$
k.						

5. $\frac{m}{2} > 100$ iff $m > 200$; $\frac{m}{2} > 1000$ iff $m > 2000$; $\frac{m}{2} > 1{,}000{,}000$ iff $m > 2{,}000{,}000$; $\frac{m}{2} > M$ iff $m > 2M$ **7.** $m^2 > 100$ iff $m > 10$; $m^2 > M$ iff $m > \sqrt{M}$ **9.** Parts **d, f, h, i, j**

Section 16-3, page 357 In Exercises 1–11 each sequence satisfies the definition. **1.** $a = 0.4$, $r = \frac{1}{10}$; $\frac{4}{9}$ **3.** $a = 1$, $r = 3$; No limit **5.** $a = 1$, $r = 0.1$; $\frac{a}{(1-r)} = \frac{10}{9}$ **7.** $a = \frac{1}{4}$, $r = 2$; No limit.

9. $a = 3.1, r = \frac{1}{100}; \frac{310}{99}$ **11.** $a = 1, r = -\frac{1}{3}; \frac{3}{4}$ **13.a.** $0.222 \cdots = 2 \cdot \frac{1}{10} + 2 \cdot \frac{1}{100} + \cdots = 2\left(\frac{1}{10} + \frac{1}{100} + \cdots\right) = 2 \cdot \frac{1}{9} = \frac{2}{9}$ **c.** $0.444 \cdots = 4 \cdot \frac{1}{10} + 4 \cdot \frac{1}{100} + \cdots = 4 \cdot \frac{1}{9} = \frac{4}{9}$

Section 16-4, page 362 **1.** $x^7 + 7x^6y + 21x^5y^2 + 35x^4y^2 + 35x^3y^4 + 21x^2y^5 + 7xy^5 + y^7$ **3.** $32x^5 - 80x^4y + 80x^3y^2 - 40x^2y^3 + 10xy^4 - y^5$ **5.** $81x^4 - 216x^3y + 216x^2y^2 - 96xy^3 + 16y^4$ **7.** $-x^7 - 14x^6y - 84x^5y^2 - 280x^4y^3 - 560x^3y^4 - 672x^2y^5 - 448xy^6 - 128y^7$ **9.** $\frac{1}{243}x^5 - \frac{5}{162}x^4y + \frac{5}{54}x^3y^2 - \frac{5}{36}x^2y^3 + \frac{5}{48}xy^4 - \frac{y^5}{32}$ **11.** $\frac{32}{x^5} - \frac{80}{x^4y} + \frac{80}{x^3y^2} - \frac{40}{x^2y^3} + \frac{10}{xy^4} - \frac{1}{y^5}$ **13.** $\frac{1}{16x^4} + \frac{1}{2x^3y} + \frac{3}{2x^2y^2} + \frac{2}{xy^3} + \frac{1}{y^4}$ **15.** $\frac{32}{243x^5} + \frac{80}{27x^4y} + \frac{80}{3x^3y^2} + \frac{120}{x^2y^3} + \frac{270}{xy^4} + \frac{243}{y^5}$ **17.** $x^{13}, 13x^{12}y, 78x^{11}y^2, 286x^{10}y^3, 1716x^6y^7$ **19.** $1, 13x, 78x^2, 286x^3; 715x^9$ **21.** $x^{10}, -20x^9y, 180x^8y^2, -960x^7y^3$

Section 16-5, page 365 **1.a.** $-\frac{1}{10}, \frac{1}{100}, -\frac{1}{1000}$ **c.** $1, -\frac{2}{5}, \frac{3}{25}, -\frac{4}{125}$ **e.** $1, -\frac{1}{100}, \frac{1}{10{,}000}, -\frac{1}{1{,}000{,}000}$ **2.a.** 1, 0.9, 0.91, 0.909 **c.** $1, \frac{3}{5}, \frac{18}{25}, \frac{86}{125}$ **e.** 1, 0.99, 0.9901, 0.990099 **3.a.** $\frac{1}{11{,}000}$ **c.** $\frac{29}{4500}$ **e.** $\frac{1}{101{,}000{,}000}$ **4.a.** $1, \frac{1}{30}, -\frac{1}{900}, \frac{5}{81{,}000}; 1, \frac{31}{30}, \frac{929}{900}, \frac{16{,}723}{16{,}200}$ **c.** $1, \frac{1}{8}, -\frac{1}{128}, \frac{1}{1024}; 1, \frac{9}{8}, \frac{143}{128}, \frac{1145}{1024}$ **e.** $1, \frac{1}{6}, -\frac{1}{72}, \frac{1}{432}; 1, \frac{7}{6}, \frac{83}{72}, \frac{499}{432}$ **f.** $1, -\frac{1}{6}, -\frac{1}{72}, -\frac{1}{432}; 1, \frac{5}{6}, \frac{59}{72}, \frac{353}{432}$ **5.a.** 1.00995 **c.** 1.03228 **e.** 5.4772

Section 16-6, page 367 **1.a.** Part (2): 10 + 1 is not in the set when 10 is. **2.a.** 3^2; $1 + 3 + 5 + \cdots + (2n - 1) = n^2, n \in N$ **3.** P_n is $1 + 3 + \cdots + (2n - 1) = n^2$ **5.** Let $S = \{n: n \in N$ and P_n is true$\}$. (*i*) $1 = 1^2$. Therefore, P_1 is true and $1 \in S$. (*ii*) Assume P_n is true (or $n \in S$):

$$1 + 3 + \cdots + (2n - 1) = n^2.$$

Then

$$[1 + 3 + \cdots + (2n - 1)] + (2n + 1) = n^2 + (2n + 1) = (n + 1)^2$$

which is P_{n+1}. Therefore, P_{n+1} is true and $(n + 1) \in S$. By the property of mathematical induction, $S = N$ and P_n is true for every $n \in N$. **7.** Let $S = \{n: n \in N$ and P_n is true$\}$. (*i*) $2 = 1(1 + 1)$. Therefore, P_1 is true and $1 \in S$. (*ii*) Assume $n \in S$ or P_n is true:

$$2 + 4 + \cdots + 2n = n(n + 1).$$

Then

$$[(2 + 4 + \cdots + 2n)] + 2(n + 1) = n(n + 1) + 2(n + 1)$$
$$= (n + 1)(n + 2).$$

Thus, P_{n+1} is true, $(n + 1) \in S$, and, by the property of mathematical induction, $S = N$ and P_n is true for every $n \in N$.

Chapter 16 Review Exercises, page 370

1. $\frac{1}{2}, \frac{5}{4}, 2, \frac{11}{4}$ **2.** $a_1 = 7, a_7 = \frac{448}{729}$ **3.** 159 **4.** $\frac{1{,}000{,}019{,}683}{130{,}000{,}000}$ **5.** 5th: 0.2121212121, 8th: 0.2121212121212121, 11th: 0.2121212121212121212121, kth: $\sum_{i=1}^{k} \frac{21}{100}\left(\frac{1}{100}\right)^{i-1}$, $(k-1)$st: $\sum_{i=1}^{k-1} \frac{21}{100}\left(\frac{1}{100}\right)^{i-1}$, $(k+1)$st: $\sum_{i=1}^{k+1} \frac{21}{100}\left(\frac{1}{100}\right)^{i-1}$ **6.** 0.21, 0.2121, 0.212121, 0.21212121; $\frac{7}{33}$ **7.** $x^9, -18x^8y, 144x^7y^2, -672x^6y^3; 5376x^3y^6$ **8.** $1, -\frac{2}{3}a, \frac{5}{9}a^2, -\frac{40}{81}a^3$ **9.** If S is a subset of N such that (1) $1 \in S$ and $n + 1 \in S$ whenever $n \in S$, then $S = N$.

INDEX

Boldface page references indicate formal or informal definitions.